DEFECTS IN TWO-DIMENSIONAL MATERIALS

Materials Today
DEFECTS IN TWO-DIMENSIONAL MATERIALS

Edited by

Rafik Addou

Luigi Colombo

Elsevier
Radarweg 29, PO Box 211, 1000 AE Amsterdam, Netherlands
The Boulevard, Langford Lane, Kidlington, Oxford OX5 1GB, United Kingdom
50 Hampshire Street, 5th Floor, Cambridge, MA 02139, United States

Copyright © 2022 Elsevier Inc. All rights reserved.

No part of this publication may be reproduced or transmitted in any form or by any means, electronic or mechanical, including photocopying, recording, or any information storage and retrieval system, without permission in writing from the publisher. Details on how to seek permission, further information about the Publisher's permissions policies and our arrangements with organizations such as the Copyright Clearance Center and the Copyright Licensing Agency, can be found at our website: www.elsevier.com/permissions.

This book and the individual contributions contained in it are protected under copyright by the Publisher (other than as may be noted herein).

Notices

Knowledge and best practice in this field are constantly changing. As new research and experience broaden our understanding, changes in research methods, professional practices, or medical treatment may become necessary.

Practitioners and researchers must always rely on their own experience and knowledge in evaluating and using any information, methods, compounds, or experiments described herein. In using such information or methods they should be mindful of their own safety and the safety of others, including parties for whom they have a professional responsibility.

To the fullest extent of the law, neither the Publisher nor the authors, contributors, or editors, assume any liability for any injury and/or damage to persons or property as a matter of products liability, negligence or otherwise, or from any use or operation of any methods, products, instructions, or ideas contained in the material herein.

Library of Congress Cataloging-in-Publication Data
A catalog record for this book is available from the Library of Congress

British Library Cataloguing-in-Publication Data
A catalogue record for this book is available from the British Library

ISBN: 978-0-12-820292-0

For information on all Elsevier publications
visit our website at https://www.elsevier.com/books-and-journals

---

Publisher: Matthew Deans
Acquisitions Editor: Kayla Dos Santos
Editorial Project Manager: Rachel Pomery
Production Project Manager: Debasish Ghosh
Designer: Christian J. Bilbow

Typeset by VTeX

Contents

List of contributors ix
About the editors xi
Preface xiii

1. Introduction

RAFIK ADDOU AND LUIGI COLOMBO

2. Physics and theory of defects in 2D materials: the role of reduced dimensionality

HANNU-PEKKA KOMSA AND
ARKADY V. KRASHENINNIKOV

2.1 Introduction 7
2.2 Classification of defects 8
2.3 Insights into the atomic structures of defects from scanning tunneling and transmission electron microscopy experiments 10
2.4 Production of defects in two-dimensional materials under electron and ion irradiation 11
2.5 Examples of defects in two-dimensional materials 12
2.6 Theoretical aspects of the physics of defects in bulk crystalline solids and two-dimensional materials 18
2.7 Calculations of defect formation energies and electronic structure using the supercell approach 21
2.8 Electronic structure of 2D materials with defects 26
2.9 Point defects and vibrational properties of 2D materials from atomistic simulations 32
2.10 Conclusions and outlook 34
References 37

3. Defects in two-dimensional elemental materials beyond graphene

PAOLA DE PADOVA, BRUNO OLIVIERI, CARLO OTTAVIANI,
CLAUDIO QUARESIMA, YI DU,
MIECZYSŁAW JAŁOCHOWSKI, AND MARIUSZ KRAWIEC

3.1 Introduction 43
3.2 Borophene 45
3.3 Silicene 49
3.4 Germanene 52
3.5 Stanene 55
3.6 Plumbene 57
3.7 Phosphorene 59
3.8 Arsenene (h-As) and Antimonene (h-Sb) 64
3.9 Bismuthene 70
3.10 Selenene and tellurene 72
3.11 Gallenene 75
3.12 Hafnene 77
3.13 Conclusions and outlook 78
References 79

4. Defects in transition metal dichalcogenides

STEPHEN MCDONNELL AND PETRA REINKE

4.1 Introduction 89
4.2 Point defects 89
4.3 Impurities 94
4.4 Line defects 104
4.5 Control of defects and their applications 107
4.6 Summary 109
References 110

5. Realization of electronic grade graphene and h-BN

VITALIY BABENKO AND STEPHAN HOFMANN

5.1 Challenges overview: growth, transfer, and integration 119
5.2 Apparatus and methodology overview 121
5.3 Scalable growth by chemical vapor deposition 124

5.4 Material optimization 136
5.5 Conclusions and outlook 149
References 149

6. Realization of electronic-grade two-dimensional transition metal dichalcogenides by thin-film deposition techniques

YU-CHUAN LIN, RICCARDO TORSI, NICHOLAS A. SIMONSON, AZIMKHAN KOZHAKHMETOV, AND JOSHUA A. ROBINSON

6.1 Current challenges in transition metal dichalcogenide synthesis 159
6.2 Current synthesis techniques 161
6.3 Controlling nucleation and crystal growth 170
6.4 Materials engineering 175
6.5 Summary 182
References 183

7. Materials engineering – defect healing & passivation

YU LI HUANG, REBEKAH CHUA, AND ANDREW THYE SHEN WEE

7.1 Introduction 195
7.2 Defect formation and healing in 2D TMDs 196
7.3 Defect engineering by chemical treatment and applications 201
7.4 Defect control by external sources 207
7.5 Future perspectives 213
References 214

8. Nonequilibrium synthesis and processing approaches to tailor heterogeneity in 2D materials

DAVID B. GEOHEGAN, KAI XIAO, ALEX A. PURETZKY, YU-CHUAN LIN, YILING YU, AND CHENZE LIU

8.1 Introduction 221
8.2 Non-equilibrium synthesis – effects of chemical potential on the heterogeneity of 2D materials 224
8.3 Strain induced phenomena in 2D materials 234

8.4 Heterogeneity introduced by the self-assembly of nanoscale 'building blocks' 241
8.5 The effects of kinetic energy on defects and doping: hyperthermal implantation for the formation of Janus monolayers 247
8.6 Summary and outlook 250
References 252

9. Two-dimensional materials under ion irradiation: from defect production to structure and property engineering

MAHDI GHORBANI-ASL, SILVAN KRETSCHMER, AND ARKADY V. KRASHENINNIKOV

9.1 Introduction 259
9.2 Response of two-dimensional materials to ion irradiation: theoretical aspects 261
9.3 Experiments on ion irradiation of two-dimensional materials 273
9.4 Applications 285
9.5 Summary, challenges, and outlook 289
References 291

10. Tailoring defects in 2D materials for electrocatalysis

LEPING YANG, YUCHI WAN, AND RUITAO LV

10.1 Introduction 303
10.2 Defect-tailored 2D electrocatalysts for hydrogen evolution reaction (HER) 304
10.3 Defect-tailored 2D electrocatalysts for oxygen evolution reaction (OER) 311
10.4 Defect-tailored 2D electrocatalysts for nitrogen reduction reaction (NRR) 317
10.5 Defect-tailored 2D electrocatalysts for carbon dioxide reduction reaction (CO$_2$RR) 325
10.6 Challenges and perspectives of defect engineering for 2D electrocatalysts 330
References 332

11. Devices and defects in two-dimensional materials: outlook and perspectives

AMRITESH RAI, ANUPAM ROY, AMITHRAJ VALSARAJ, SAYEMA CHOWDHURY, DEEPYANTI TANEJA, YAGUO WANG, LEONARD FRANK REGISTER, AND SANJAY K. BANERJEE

11.1 Introduction 339

11.2 Defect characterization in 2D TMDs using ultrafast pump-probe spectroscopy 341

11.3 Devices fabricated on 2D CVD-grown TMDs 348

11.4 Devices fabricated on MBE-grown TMDs 355

11.5 2D van der Waals (vdW) heterostructures 359

11.6 Enhancing 2D device performance using defect engineering 364

11.7 Theoretical investigation of defects in 2D TMDs 378

References 387

12. Concluding remarks

RAFIK ADDOU AND LUIGI COLOMBO

Index 411

List of contributors

Rafik Addou University of Texas at Dallas, Richardson, TX, United States

Vitaliy Babenko Department of Engineering, University of Cambridge, Cambridge, United Kingdom

Sanjay K. Banerjee Microelectronics Research Center and Department of Electrical and Computer Engineering, The University of Texas at Austin, Austin, TX, United States

Sayema Chowdhury Microelectronics Research Center and Department of Electrical and Computer Engineering, The University of Texas at Austin, Austin, TX, United States

Rebekah Chua Department of Physics, National University of Singapore, Singapore, Singapore

Luigi Colombo University of Texas at Dallas, Richardson, TX, United States

Paola De Padova Istituto di Struttura della Materia-CNR (ISM-CNR), Rome, Italy
LNF-INFN, Frascati, Rome, Italy

Yi Du ISEM-AIIM and Innovation Campus, University of Wollongong, Wollongong, NSW, Australia

David B. Geohegan Functional Hybrid Nanomaterials Group, Center for Nanophase Materials Sciences, Oak Ridge National Laboratory, Oak Ridge, TN, United States

Mahdi Ghorbani-Asl Helmholtz-Zentrum Dresden-Rossendorf, Institute of Ion Beam Physics and Materials Research, Dresden, Germany

Stephan Hofmann Department of Engineering, University of Cambridge, Cambridge, United Kingdom

Yu Li Huang Department of Physics, National University of Singapore, Singapore, Singapore

Mieczysław Jałochowski Institute of Physics, Maria Curie-Sklodowska University, Lublin, Poland

Hannu-Pekka Komsa Department of Applied Physics, Aalto University, Aalto, Finland

Azimkhan Kozhakhmetov The Pennsylvania State University, University Park, PA, United States

Arkady V. Krasheninnikov Helmholtz-Zentrum Dresden-Rossendorf, Institute of Ion Beam Physics and Materials Research, Dresden, Germany
Department of Applied Physics, Aalto University, Aalto, Finland

Mariusz Krawiec Institute of Physics, Maria Curie-Sklodowska University, Lublin, Poland

Silvan Kretschmer Helmholtz-Zentrum Dresden-Rossendorf, Institute of Ion Beam Physics and Materials Research, Dresden, Germany

Yu-Chuan Lin The Pennsylvania State University, University Park, PA, United States
Functional Hybrid Nanomaterials Group, Center for Nanophase Materials Sciences, Oak Ridge National Laboratory, Oak Ridge, TN, United States

Chenze Liu Functional Hybrid Nanomaterials Group, Center for Nanophase Materials Sciences, Oak Ridge National Laboratory, Oak Ridge, TN, United States

Ruitao Lv State Key Laboratory of New Ceramics and Fine Processing, School of Materials Science and Engineering, Tsinghua University, Beijing, China

Stephen McDonnell Department of Materials Science and Engineering, University of Virginia, Charlottesville, VA, United States

Bruno Olivieri CNR-ISAC, Rome, Italy

Carlo Ottaviani Istituto di Struttura della Materia-CNR (ISM-CNR), Rome, Italy

Alex A. Puretzky Functional Hybrid Nanomaterials Group, Center for Nanophase Materials Sciences, Oak Ridge National Laboratory, Oak Ridge, TN, United States

Claudio Quaresima Istituto di Struttura della Materia-CNR (ISM-CNR), Rome, Italy

Amritesh Rai Microelectronics Research Center and Department of Electrical and Computer Engineering, The University of Texas at Austin, Austin, TX, United States

Leonard Frank Register Microelectronics Research Center and Department of Electrical and Computer Engineering, The University of Texas at Austin, Austin, TX, United States

Petra Reinke Department of Materials Science and Engineering, University of Virginia, Charlottesville, VA, United States

Joshua A. Robinson The Pennsylvania State University, University Park, PA, United States

Anupam Roy Microelectronics Research Center and Department of Electrical and Computer Engineering, The University of Texas at Austin, Austin, TX, United States

Nicholas A. Simonson The Pennsylvania State University, University Park, PA, United States

Deepyanti Taneja Microelectronics Research Center and Department of Electrical and Computer Engineering, The University of Texas at Austin, Austin, TX, United States

Andrew Thye Shen Wee Department of Physics, National University of Singapore, Singapore, Singapore

Riccardo Torsi The Pennsylvania State University, University Park, PA, United States

Amithraj Valsaraj Microelectronics Research Center and Department of Electrical and Computer Engineering, The University of Texas at Austin, Austin, TX, United States

Yuchi Wan State Key Laboratory of New Ceramics and Fine Processing, School of Materials Science and Engineering, Tsinghua University, Beijing, China

Yaguo Wang Department of Mechanical Engineering and Texas Materials Institute, The University of Texas at Austin, Austin, TX, United States

Kai Xiao Functional Hybrid Nanomaterials Group, Center for Nanophase Materials Sciences, Oak Ridge National Laboratory, Oak Ridge, TN, United States

Leping Yang State Key Laboratory of New Ceramics and Fine Processing, School of Materials Science and Engineering, Tsinghua University, Beijing, China

Yiling Yu Functional Hybrid Nanomaterials Group, Center for Nanophase Materials Sciences, Oak Ridge National Laboratory, Oak Ridge, TN, United States

About the editors

Rafik Addou is a research scientist at the University of Texas at Dallas, USA, where he leads efforts on understanding the interface and surface science of graphene, transition metal dichalcogenides, and other emerging 2D materials for nano- and optoelectronics. He earned a BSc in Physics from Mohamed Premier University, Oujda, Morocco, and MSc in Materials Physics from Aix-Marseille University, France. In 2010, he received his PhD degree in Materials Science from Ecole des Mines (Nancy, France) in association with Empa Materials Science and Technology Laboratory (Thun, Switzerland). Before joining UTD, Dr. Addou was at the University of South Florida (Tampa FL, USA) as a postdoctoral research fellow in Physics where he studied the surface physics of graphene.

Luigi Colombo is the Director of Strategic Programs and Adjunct Professor in the Department of Materials Science & Engineering at the University of Texas at Dallas, USA. For almost 40 years, he worked on a variety of materials research and development programs and device integration at Texas Instruments in Dallas, TX, USA. From 2008 to 2013, in collaboration with the Ruoff group at the University of Texas at Austin, USA, he discovered and developed a large area graphene film growth using a catalytic CVD process on Cu substrates.

Preface

This book, *Defects in Two-Dimensional Materials*, provides a *review of* the fundamental physics and chemistry of defects in 2D materials and their effects on the chemical, electronic, opto-electronic, and mechanical and other physical properties relevant to applications. The primary objective is to review defects present in a variety of 2D materials, such as transition metal dichalcogenides (TMD), graphene, hexagonal boron nitride (h-BN), and elemental 2D materials beyond graphene (e.g. silicene, phosphorene, bismuthene, tellurene, hafnene, etc.). Techniques to characterize defects in 2DM will be introduced throughout the book. The discussions also highlight how understanding and controlling these defects can provide a pathway to novel applications requiring defect engineering, enabling scientists to tune the properties of 2D materials to realize specialized applications.

Large number of review papers and articles are being published every year on conventional and newly discovered/synthetized 2D materials from theory to experiment and applications. This book makes it easy for experts and non-experts to keep up with the latest reports, discoveries, and developments that involve defects in 2D materials.

The book is designed to cover growth, theory of defects, identification and characterization of defects, defect healing and passivation, defect creation by irradiation and annealing, defect engineering for catalysis and the role of defects in electronic devices. This book is suitable for both academia and industry working in the disciplines of materials science and engineering and may be of interest to physicists, chemists, and electrical engineers as well as beginners to the field.

The editors would like to thank all the contributors of the book from all over the world. Especially the main principal investigators for accepting to participate to this effort: Arkady V. Krasheninnikov (*Helmholtz-Zentrum Dresden-Rossendorf, Germany*), Paola de Padova (ISM-CNR, Italy), Stephen McDonnell and Petra Reinke (University of Virginia, USA), *Vitaliy Babenko* and Stephan Hofmann (*University of Cambridge*, UK), Andrew T.S. Wee (National University of Singapore), Joshua Robinson (Penn State University, USA), David Geohegan (Oak Ridge National Lab, USA), Ruitao Lv (Tsinghua, China) and Sanjay Banerjee (University of Texas at Austin, USA). The editors also thank Silvan Kretschmer (*Helmholtz-Zentrum Dresden-Rossendorf, Germany*) for the book cover.

Finally, Rafik Addou thanks his wife, Sara, daughter, Yasmin, parents, Joudia and Mimoun, and sisters, Hanan, Wissam and Sanaa for patience and unconditional support. Luigi Colombo would like to thank his patient wife, Sumico, during the editing of this book.

Rafik Addou
Research Scientist
Luigi Colombo
Adjunct Professor & Director of Strategic Projects
August 2021

CHAPTER

1

Introduction

Rafik Addou and Luigi Colombo

University of Texas at Dallas, Richardson, TX, United States

Two-dimensional (2D) materials have been extensively studied over several decades but especially since the isolation of graphene now over 15 years ago by Nobel Laureates Konstantin Novoselov and Andre K. Geim [1]. This discovery has led to a significant level of effort across the world not only to understand the fundamental properties of graphene but also reintroduce and expand the study to other two-dimensional materials such as hexagonal boron nitride (h-BN) and transition metal dichalcogenides (TMDs) [2]. The search for new and "better" 2D materials has also led to discovery and growth of new elemental two-dimensional materials. The expansion to other two-dimensional materials was driven principally by the need to cover the electromagnetic spectrum with materials from the far infrared (IR) to the ultra-violet (UV) to meet stringent current and future applications requirements.

Graphene with its zero band gap can be useful for many applications but high performance electronic devices typically need a band gap to achieve low device off currents. Elemental 2D materials [3] with narrow band gaps could be used for some electronic applications requiring narrow bandgaps, < 1 eV, but it is the TMDs with their tunable bandgaps and potential for heterostructure fabrication [4] that are attractive for many electronic and optoelectronic applications including transistors to help scale complementary metal oxide semiconductor (CMOS) devices. Finally, hexagonal boron nitride(h-BN), sometimes referred to as white graphite, an insulator which is the 2D material with the highest bandgap, about 6 eV, is a highly sought material because of its exceptional low interface trap density and ability to screen charges from substrates [5].

From a materials science perspective, the success of the electronics and optoelectronics industry is largely due to the availability of large low defect density single crystals of Si, III–V, and II–VI compounds. Further, the electronic properties of these materials have been controlled by intrinsic and extrinsic doping at levels below 10^{14} cm^{-3}. The ability to control impurities, doping and alloying of these materials has enabled the electronics industry over the last seven decades to harness electronic transport and electro-optical effects with unparalleled success. While defect chemistry and control are very mature in these cubic systems, the scientific community is now faced with a similar challenge for 2D materials. This book reviews the status of defects and defect chemistry in 2D materials as well as challenges and opportunities for the fabrication of new devices and applications.

Defects in Two-Dimensional Materials
https://doi.org/10.1016/B978-0-12-820292-0.00007-0

During the early days of 2D materials development the focus was on using small very high-quality exfoliated graphene from graphitic carbon sources to develop basic understanding of the fundamental properties. However, as manufacturing growth processes were developed to address the lack of large area films of any of the 2D materials and evaluate them for various applications, another challenge came to light, defects. As will be described in detail in this book, even though 2D materials are atomically thin, there is a richness of defects as in the case of 3D crystals.

The success of the electronics industry is based on the simultaneous creation of the highest crystalline quality materials in conjunction with precise control of defects. Control of defects in semiconductors has enabled the industry to fabricate many different types of electronic devices from simple transistors to CMOS devices to high-electron mobility transistors (HEMT) using III–V compound heterostructures, and high-performance infrared detectors using II–VI compounds. The fabrication of these devices was enabled by the growth of the highest quality bulk crystals and/or thin films and/or their heterostructures and control of interfaces and point defects to manipulate transport of electrons and holes, and band structure. The basic understanding of defects and defect chemistry in the bulk, and interfaces of semiconducting heterostructures and semiconductors with dielectrics has led to the fabrication of not only high-performance devices but also reliable devices supporting a few trillion-dollar electronics industry.

The success of 2D materials for electronic applications will also largely depend on our ability to not only grow the highest quality materials but also control defects and surfaces to a similar degree as in traditional electronic materials in production today. Under ideal conditions, defects in 2D materials are like those in 3D semiconductor materials with one major distinction, 2D materials have two surfaces and "no bulk". In other words, surfaces play a critical role in the defect formation, types, characteristics, and methods to control them. The hope for many applications is to take advantage of the low surface state density due to the sp2 bonding nature of the 2D materials, and the atomically thin nature for electrostatic control of scaled transistors.

The book starts with a chapter by *H.-P. Komsa and A. V. Krasheninnikov* on an overview of recent advances and current understanding of the physics of defects in 2DMs. In this chapter the authors discuss the changes in the theoretical description of native and extrinsic defects, and the reduced dimensionality of some defects: for example, a grain boundary is a line defect in 2D, and edge dislocation is a point defect. Interstitial atoms do not exist in the majority of 2D materials (e.g. graphene, h-BN, MoS_2), as it is energetically more favorable for the atom to take an adatom position rather than be embedded into the atomic network. The authors also discuss the plausibility that various species can be adsorbed at the reactive dangling bonds of the atoms next to vacancies (e.g., hydrogen or nitrogen atoms) in 2D MoS_2 and other TMDs. These "defects" can give rise to shallow occupied states close to conduction band minimum (CBM) or empty states next to valence band minimum (VBM). The authors further discuss how the geometry of 2D materials requires making changes in the theoretical description of charged point and line defects, as the screening/electrostatics is strongly anisotropic and inhomogeneous, and the results may depend on the shape of the simulation cell and electrostatic correction scheme used. The authors further state that the environment should also be carefully accounted for, both in modifying the screening and by choosing the values of atom chemical potentials matching the experimental situation.

The third chapter by *Paola de Padova et al.* on "Defects in two-dimensional elemental materials beyond graphene" presents an extensive review of elemental 2D materials beyond graphene such as Borophene, Silicene, Germanene, Stanene, Plumbene, Phosphorene, Arsenene, Antimonene Bismuthene, Selenene and Tellurene. Here, the authors review different types of structural point defects, including Stone-Wales, single vacancy, and bi-vacancies, as well as grain boundaries (lines) and adatoms as described based on theoretical and experimental published reports. The discussion extends to highlighting that some of these emerging 2D materials show non-trivial topological behavior due to the presence of defects that could be of great interest for their potential use in several electronic applications.

The fourth chapter is by *Stephen McDonnell and Petra Reinke* on "Defects in transition metal dichalcogenides" provides the reader with an overview of the role of defects on the properties of TMD materials. The authors list the various types of defects unique to 2D materials and provide examples from the experimental database and point out that the community is still a long way from synthesizing truly high-purity films comparable to silicon. In the long term, achieving high purity material will significantly simplify property engineering of TMDs, but in the near term, we must continue to thoroughly characterize the materials and continue to consider both random/uncontrolled and engineered defects when interpreting their functional properties.

To complement the chapters on defects, two chapters have been dedicated to crystal growth of graphene, h-BN and TMDs. While Chapter 5 written by *Vitaliy Babenko and Stephan Hofmann* reviews the growth of electronic grade graphene and h-BN, Chapter 6 written by *Yu-Chuan Lin et al.* reviews the growth of electronic grade TMD thin films. Over the last decade or so a large effort has been dedicated to the development of large-scale crystal growth of 2D materials. Bottom-up thin film approaches are being developed for 2D materials to meet industrial requirements and extensive efforts are also being dedicated to the understanding and control of defects in synthetic materials as opposed to naturally occurring 2D materials like graphite and some MoS_2. Hexagonal boron nitride and many other TMDs have not been found in nature. These processes are developed to replace the exfoliated process for graphene, h-BN and TMDs extensively used to generate samples for fundamental studies. Today, wafer scale processes such as catalytic chemical vapor deposition of graphene and h-BN have been developed and numerous vapor deposition processes are being developed for TMDs. The authors of these two chapters review and describe in depth various processes used to grow large area 2D materials films with controlled defect densities to meet the many device performance requirements.

The previous chapters on growth are followed by Chapter 7 by *David B. Geohegan et al.* on "Nonequilibrium synthesis and processing approaches to tailor heterogeneity in 2D materials" where defects are studied with respect to nucleation and growth where stochastic variations in chemical potential, temperature, flux of different species push the synthesis environment out of equilibrium. Cooperative effects, such as strain accumulation due to coalescence with other crystalline domains during growth, can also induce both localized and long-range heterogeneities. In the case of 2D TMD materials such effects are manifested as changes in optoelectronic properties. The authors go on to describe a synergistic approach to reveal the synthetic origins of heterogeneity in 2D TMD materials that involves a combination of: (1) temporally- and spatially-resolved in situ diagnostics of growth environment, using primarily optical spectroscopic and electron microscopy techniques, (2) a correlation between spectroscopic maps of optoelectronic properties and atomistic characterization of heterogeneity, using primarily Z-contrast scanning transmission electron microscopy, and (3) correlated theory of electronic and vibrational properties, leading to computational modeling simulations of synthesis dynamics. The authors then discuss recent progress on the role of kinetic energy on defect generation in atomically thin 2D TMD crystals for the formation of Janus TMD monolayers.

In Chapter 8 "Materials engineering – defect healing & passivation" the authors *Yu Li Huang et al.* review strategies on defect healing and passivation as applied to TMD materials. They discuss the formation of intrinsic defects and self-healing with small molecules (e.g., O_2) as well as via chemical treatments, e.g., decoration with organic layers, which result in improved device. Thermal annealing, electron beam irradiation, plasma treatment, and encapsulation, are also discussed. They conclude the chapter with an outlook on the challenges and opportunities for defect minimization in 2D materials, with a focus on TMDs.

Mahdi Ghorbani-Asl et al. in Chapter 9 present a review on *"Two-dimensional materials under ion irradiation: from defect production to structure and property engineering"* where they discuss the effects of ion irradiation of 2D materials and the role of reduced dimensionality. They go on to compare the impact of ions on 3D and 2D materials. The primary objective of this chapter is to review and discuss the effect of ion beams on 2D materials and how they can be used to engineer their structure and properties. It should be noted that the work reviewed in this chapter is mostly on graphene and MoS_2 given the limited data on other 2D materials.

Chapter 10 by *Leping Yang et al.* is dedicated to "Tailoring defects in 2D materials for electrocatalysis" where the authors summarize the state-of-the-art advances in defect engineering of 2D materials for electrocatalytic applications involving hydrogen evolution reaction (HER), oxygen evolution reaction (OER), carbon dioxide reduction reaction (CO_2RR), and nitrogen reduction reaction (NRR). The challenges that the community faces in this very fertile field are tuning surface electronic states to optimize the reaction energy barriers, the incorporation of sufficient active sites to improve the surface electrochemical reactivity, and enhancing the electrical conductivity to accelerate electron transport among the reaction interfaces.

The last chapter of the book before the summary by *A. Rai et al.* on "Devices and defects in two-dimensional materials: outlook and perspectives" presents a comprehensive review on 2D materials-based devices and effects of defects, interfaces, interfacial defects, and dielectrics on electronic device characteristics and transport. They report on devices fabricated on CVD and MBE grown 2D films and using heterostructures formed using a variety of 2D materials. They present a plethora of data on the effect of many dielectrics on the device properties of TMDs based devices. They also discuss the fabrication of Van der Waals heterostructures where any atomically thin 2D material can be stacked, rotated (twisted) to achieve the desired band structure thus leading to unprecedented advancement in electronics. The chapter concludes with a review on the theoretical investigation of defects in TMDs.

References

[1] Scientific Background on the Nobel Prize in Physics 2010, Class for Physics of the Royal Swedish Academy of Sciences. https://www.nobelprize.org/uploads/2018/06/advanced-physicsprize2010.pdf, 2010.

[2] K.S. Novoselov, D. Jiang, F. Schedin, T.J. Booth, V.V. Khotkevich, S.V. Morozov, A.K. Geim, Two-dimensional atomic crystals, Proceedings of the National Academy of Sciences of the United States of America 102 (30) (2005) 10451–10453.

[3] N.R. Glavin, R. Rao, V. Varshney, E. Bianco, A. Apte, A. Roy, E. Ringe, P.M. Ajayan, Emerging applications of elemental 2D materials, Advanced Materials 32 (7) (2020).

[4] F.N. Xia, H. Wang, D. Xiao, M. Dubey, A. Ramasubramaniam, Two-dimensional material nanophotonics, Nature Photonics 8 (12) (2014) 899–907.

[5] C.R. Dean, A.F. Young, I. Meric, C. Lee, L. Wang, S. Sorgenfrei, K. Watanabe, T. Taniguchi, P. Kim, K.L. Shepard, J. Hone, Boron nitride substrates for high-quality graphene electronics, Nature Nanotechnology 5 (10) (2010) 722–726.

CHAPTER
2

Physics and theory of defects in 2D materials: the role of reduced dimensionality

Hannu-Pekka Komsa[a] *and Arkady V. Krasheninnikov*[b,a]

[a]Department of Applied Physics, Aalto University, Aalto, Finland [b]Helmholtz-Zentrum Dresden-Rossendorf, Institute of Ion Beam Physics and Materials Research, Dresden, Germany

2.1 Introduction

Defects in crystalline solids are unavoidable. Indeed, the second law of thermodynamics states that a certain amount of disorder should be present in any material at finite temperatures, provided that the system is in thermodynamic equilibrium. Moreover, the concentration of defects in synthetic materials can be well above the equilibrium value depending upon the preparation process. Typically materials grown under near equilibrium conditions at temperatures close to or at the melting point tend to have lower defect densities than material grown under non-equilibrium conditions. Defects can also appear due to external factors, such as pressure or irradiation with energetic particles. Defects strongly influence the behavior of crystalline solids, so much so that both scientists and engineers have devoted their careers to not only reduce them, but control them with great accuracy [1]. Specifically, defects can completely govern the electronic, optical, thermal, and mechanical properties of the solid.

Although the word "defect" has negative connotation, imperfections in crystals do not always have detrimental effects on materials properties [2], with the most prominent example being the doping of semiconductors by controllable introduction of impurities using ion implantation or color centers in wide-gap semiconductors [3]. Other examples include the pinning of magnetic vortices on defects in type-II superconductors [4], or the possibility to control mechanical strength and ductility of metals by introducing dislocations, grain boundaries and impurities [5]. To emphasize the versatile and important role of defects, it is tempting to quote from the Ashcroft-Mermin's textbook [6]: "Like human defects, those of crystals come in a seemingly endless variety, many dreary and depressing, and a few fascinating."

Defects in Two-Dimensional Materials
https://doi.org/10.1016/B978-0-12-820292-0.00008-2

As for irradiation-induced defects, they can easily be produced in irradiation-hostile environments, such as open space or fission/fusion reactors. Moreover, the treatments of solids with beams of energetic ions and electrons have been shown to be very useful for tailoring properties of materials after synthesis. These aspects have stimulated huge interest not only in the physics of defects, but also in their production mechanisms under irradiation, including knock-on or ballistic damage, electronic excitations and, in low-dimensional materials, irradiation-induced chemical etching.

Nano-structured materials also have defects and impurities. The reduced dimensionality of these systems, however, makes the physics of defects quite different from that in bulk systems. This is particularly relevant to two-dimensional materials (2DMs), which since the isolation [7] of a single sheet of graphene by A. Geim and K. Novoselov, have been at the forefront of research due to a unique combination of their electronic, optical, and mechanical properties and potential applications in nanoelectronics, photonics, catalysis, as well as energy storage and conversion, and more, see Refs. [8–11] for an overview. The family of the experimentally synthesized 2DMs includes now about a hundred members, with the most prominent ones among them (in addition to graphene) being hexagonal boron nitride (h-BN) [12] and transition metal dichalcogenides (TMDs) [13].

In 2DMs, even when the formation energy of defects is high as in the case of graphene, so that the equilibrium concentration of defects is negligible over a wide range of temperatures, imperfections can appear due to the interaction with the environment [14–16], as 2DMs have a very high surface-to-volume ratio, Fig. 2.1(a,b). Moreover, any species on the surface of 2DMs can have strong effect on the material properties, while the processes on the surfaces of bulk systems are normally completely ignored in the context of point (and frequently line) defects.

In this chapter, we first give a brief overview of the types of defects in bulk systems and 2DMs with the main focus on the differences originating from the reduced dimensionality of the latter. We briefly discuss how the geometry of 2DMs also makes it possible to deliberately introduce defects with nearly atomic spatial resolution using focused ion and electron beams, which also helps in imaging the defects using scanning probe and transmission electron microscopy. We also discuss the changes which should be made in the theoretical description of native and extrinsic defects. Finally we bring forth the challenges in this field and the issues which still lack basic understanding. We stress that our main goal is not to give a complete overview of all possible defects in all 2DMs, but to analyze general trends and illustrate them by a few examples.

2.2 Classification of defects

Prior to discussing defects in 2DMs, it is instructive to give a general definition of defects in crystalline materials and their classification. The simplest notion of a defect in a crystalline solid is the idea of a structural imperfection (e.g., missing atom or presence of an impurity), that is a deviation from the perfect order (periodicity). A more rigorous definition can be formulated as follows: "A structural defect is a configuration in which an atom (or group of atoms) does not satisfy the structure rules pertaining to the ideal reference system or material".

2.2. Classification of defects

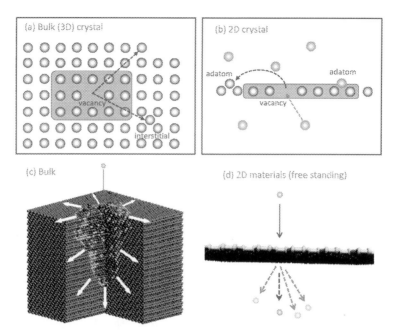

FIGURE 2.1 (a) Schematic illustration of a vacancy formation in a bulk crystalline solid. The effects of the environment are normally neglected. The grey area represents the supercell used in defect simulations. (b) Formation of a vacancy in a 2DM. Unless vacuum is perfect, the reactive dangling bonds at the vacancy can pick up atoms from the environment, making the equilibrium vacancy concentration dependent on the chemical environment. Note that in ideal 2DMs, like graphene or h-BN, adatoms play the role of the interstitials as in bulk crystals. (c) Impact of an energetic ion onto a bulk target. The ion kinetic energy is transferred ballistically to the target atoms, which results in atom displacement, along with temperature and pressure rise, and ultimately to defect formation. The atoms are colored according to their kinetic energy from blue (zero energy) to high (red) energies. (d) Impact of an ion onto a 2DM. While all the ion energy is deposited into the bulk systems, only a fraction of ion energy is transferred to the 2D target. Besides, a considerable amount of the deposited energy is taken away by sputtered atoms.

Defects can be classified according to their **dimensionality**. Correspondingly, in **bulk** systems, one can differentiate between:

- Point defects or 0D defects, e.g., vacancies;
- Linear defects or 1D defects, e.g., dislocations;
- Planar defects or 2D defects, e.g., grain boundaries;
- Volume defects or 3D defects, e.g., voids.

We note here that edge dislocations could also be considered 2D defects due to the added or removed half-plane or a 3D defect due to finite thickness of the half-plane. The given examples from bulk solids may not be appropriate in the case of 2DMs. The edge dislocations in 2DMs are 0D defects, since the dislocation line cannot exist perpendicular to the layer, and the same is true for disclinations. Screw dislocations cannot be present in the case of monolayers either, but can connect layers in multilayer structures [17–19]. Similarly, grain boundaries in 2DMs become 1D defects. Voids in 2D systems are holes formed by, e.g., agglomeration of vacancies.

Point defects can be further characterized according to the **chemical content**:

- Intrinsic defects, e.g., vacancies, self-interstitials;
- Extrinsic defects, e.g., impurity atoms;
- Antisite defects in compound solids;
- Isotopes.

Note that impurity atoms can be in both substitutional and interstitial positions, that is, they can occupy the positions of host atoms, or be between them. As for isotopes, all the atoms in this case are chemically equivalent, but different masses of the isotope atoms affect vibrational and indirectly electronic properties, giving rise, e.g., to the isotope effect in superconductors.

Point defects can also be classified according to the local number of atoms in a certain volume (area in 2DMs): When locally there are fewer atoms than in the pristine lattice, one can talk about vacancy-type defects, and when the opposite holds, about interstitials. At the same time, the number of atoms can be the same as in the perfect lattice, but they can be arranged in a different way, due to, e.g. bond rotations (note that such configurations have a higher energy). Examples of such defects, also referred to as Wigner defects [20], are bound vacancy-interstitial pairs in graphite [20] and silicon [21], or Stone-Wales defects [22] in the sp^2-hybridized carbon systems. The latter can relatively easily be annealed, as, for example, contrary to vacancies and interstitials, they do not require migration of the defects, and they may store a considerable amount of energy.

One can also classify defects according to their **origin**: for instance native defects (vacancies and interstitials) may naturally exist in the specimen at finite temperatures (it is assumed that the system is in equilibrium) or they can be induced by high-temperature annealing in a specific atmosphere or irradiation–the so-called "irradiation-induced defects". This wording is frequently used to emphasize that the concentration of defects is well above the equilibrium value, while it is fundamentally impossible to say if a particular vacancy (or any other defect) was produced due to a thermal fluctuation or the impact of an energetic particle. In the case of high-temperature annealing, such as vacancies in multi-component II–VI and III–V systems annealed under different pressures of one of the components, the defect concentration may be close to equilibrium at the elevated temperature, but well above the equilibrium concentration at room temperature.

2.3 Insights into the atomic structures of defects from scanning tunneling and transmission electron microscopy experiments

Various experimental techniques can be used to probe defects in 2DMs, see Chapter 3 for details. Defects and impurities can be detected using optical spectroscopy [23] or by X-ray absorption spectroscopy and related methods (XAFS, XANES) based on probing the defect-induced electronic states by exciting core electrons in the atoms next to the defect [24]. Also Raman spectroscopy can be used to identify the presence of defects in the sample and to assess defect concentration [25,26].

High-resolution transmission electron microscopy (HR-TEM) and scanning transmission electron microscopy (STEM) have proven to be very efficient tools to get insights into the atomic structure of defects. In fact, the quick progress in the investigation of defects in 2DMs can be partly credited to the recent impressive developments [27–29] in aberration-corrected HR-TEM (as evident from the 2020 Kavli Prize in Nanoscience awarded exactly for these developments) and the very nature of any 2D system. The TEM studies of bulk materials assume fabrication of a thin (yet preserving the morphology of the bulk system) sample, followed by the reconstruction of its atomic structure from the TEM image, where one spot can correspond to a column of atoms. TEM characterization of 2DMs does not require this, as the structure is already atomically thin, and every atom can directly be "seen". Moreover, by focusing electron beam on specific areas (essentially on single atom), and using electron energy loss spectroscopy (EELS), it is possible now to identify impurity atoms in substitutional positions [30] and get the information on the local electronic structure/bonding configurations [29,31], magnetic states [32], and even phonons [33].

The bombardment of the sample by the energetic electrons (in the 30-200 keV range) may lead to sample damage. On one hand, fast or excessive damage may hinder the experiments, but on the other hand, the ability to create defects on demand is ideal when one wants to study defects.

Nowadays, even the full three-dimensional structure, yielding information about rippling and bending around defects can be obtained using STEM with tilting (tomography) and computer-aided reconstruction techniques taking advantage of (S)TEM image simulations [34,35]. Recently, it has also become possible to extract charge density and electric field maps around defects using electron holography and electron ptychography [36,37].

The unique geometry of 2DMs (surface only), enables one to directly "see" the defects using scanning probe microscopy (scanning tunneling microscopy, STM, and atomic force microscopy, AFM, which can both be referred to as scanning probe microscopy, SPM) techniques. By scanning the STM tip bias, denoted scanning tunneling spectroscopy (STS), it is also possible to obtain energy-resolved information on the electronic states close to the tip and, for instance, find the energies of defect states within the band gap. The obtained spectra often closely resemble the local density of states, which can be readily compared to those obtained from electronic structure calculations. Since STM/STS yields spatially-resolved and energy-resolved information about the electron density, it thus complements the more direct structural information obtained from (S)TEM. STM is also largely non-destructive method, and thus enables in depth studies of electron dynamics in/around the defects, e.g., defect charging or Tomonaga-Luttinger liquid [38,39]. On the other hand, the interpretation of STM images is often more complicated, as it probes tunneling current to local electronic states in a certain energy range, not the atoms, and may require comparison with the calculated images.

2.4 Production of defects in two-dimensional materials under electron and ion irradiation

The reduced dimensionality of 2DMs also affect the behavior of these systems under impacts of energetic particles, such as electrons and ions. Here we briefly touch upon the main aspects of the irradiation response of 2DMs, while the detailed analysis of the experimental data and theoretical results is presented in Chapter 11. There are also several recent comprehensive review articles [40–43] on the subject.

Although the effects of irradiation on bulk materials are well understood, for an overview see, e.g., Refs. [1,44,45], the response of 2DMs can be quite different, as schematically illustrated in Fig. 2.1(c,d). Due to the planar geometry of the target, the sputtering yield and the average energy taken away from the system by the sputtered atoms and thus energy deposition is different, and the development of collisional cascades is suppressed. As a result, one can expect that although the number of the produced defects in 2DMs consisting of a single or a few layers of atoms will first naturally increase with ion energy, it will drop at higher energies due to a smaller probability for the ion to displace an atom [46]. For bulk materials, this implies that the ion will penetrate deeper into the material, but overall more damage in the target will be created at higher ion energies. The situation is different for 2D systems, especially for the free-standing sheets (deposited on a grid or suspended over a trench in the substrate): at higher energies ions should go through the 2D system without producing much damage through ballistic collisions, unless defects at higher ion energies start appearing again due to the enhanced energy deposition into electronic excitations followed by the conversion of the excitation energy into defects. The geometry of 2DMs also makes the displacement of atoms through ballistic collisions much easier than in bulk solids also under electron irradiation, specifically in the transmission electron microscopy experiments [47,48].

The situation is different for the supported 2DMs [49–53]. At low ion energies the substrate should decrease the amount of damage created by the energetic ions by simply stopping the displaced atoms. On the other hand, for light ions with medium energies, the defect production in 2D systems can be governed by the backscattered ions and atoms sputtered from the substrate rather than by the direct ion impacts [54]. The evolution of defects in the 2DMs, e.g., vacancies, may also be dominated by the interaction of defects with the substrate [55–57]. The environment will strongly affect the behavior of the irradiation-induced defects – additional defects can appear (e.g., due to the interaction with the reactive species like oxygen molecules) or the other way around, disappear – an example is the self-healing of vacancies in graphene due to the dissociation of hydrocarbon molecules [58].

2.5 Examples of defects in two-dimensional materials

2.5.1 Point defects

SPM and TEM studies of 2D materials, especially when combined with atomistic simulations, have provided a plethora of unique data on the behavior of point and line defects. For example, the migration of vacancies, their coalescence, and transformations to agglomerations of non-hexagonal rings have been observed in graphene in real time with atomic resolution [48,65,66]. Likewise, the atomic structures of point intrinsic [62,67,68] and extrinsic [69] defects in 2D TMDs and other inorganic 2D materials such as h-BN [70] or silica bilayers [71] have been obtained, along with deep insights into their behavior.

Examples of point defects in 2DMs are shown in Fig. 2.2. In some cases, removal of an atom or atoms gives rise to a pronounced reconstruction of the crystal lattice near the defect,

2.5. Examples of defects in two-dimensional materials

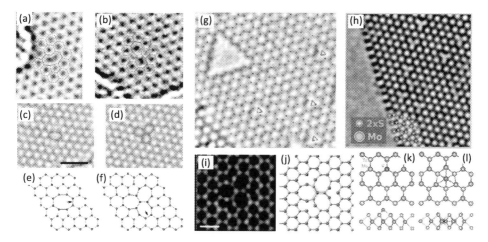

FIGURE 2.2 Examples of point defects in 2DMs. (a) HR-TEM image showing a single vacancy in graphene (from Ref. [59]). (b) HR-TEM image of a Stone-Wales defect in graphene, formed by rotating a carbon-carbon bond by 90°. Reprinted with permission from Ref. [59]. TEM images (c,d) and the corresponding atomic structures (e,f) of the reconstructed di-vacancies in graphene. Reprinted with permission from Ref. [60]. (g) Boron single vacancies (red triangles) and larger defects in h-BN as revealed by HR-TEM. Reprinted with permission from Ref. [61]. (h) Single vacancies in MoS_2. Reprinted with permission from Ref. [62]. (i) TEM image of a rotational defect and its atomic structure (j) in WSe_2. Reprinted with permission from Ref. [63]. Note that as compared to the pristine structure, 6 Se atoms are missing. (k,l) Schematic illustration if a V adatom and interstitial in $MoTe_2$. Reprinted with permission from Ref. [64].

such as in the case of graphene as shown in Fig. 2.2(a,c,d). When h-BN and MoS_2 are irradiated by electron beams, boron vacancy is the predominant type found in h-BN, Fig. 2.2(g), whereas S vacancies are found in MoS_2, Fig. 2.2(h). In both cases, the structural reconstruction is relatively minor. As a distinct example of structural reconstruction, rotational defects have been observed in various materials. Well-known examples of such defects are the Stone-Wales defects in graphene (graphite) [22], obtained by a 90° rotation of carbon-carbon bonds, Fig. 2.2(b). Another example is the so-called "flower" defects [72,73] obtained by six 90° rotations of carbon-carbon bonds. Bond rotations can also give rise to reconstructions of vacancies in graphene, Fig. 2.2(c–f), and in TMDs, Fig. 2.2(i,j). A difference between 2DMs and bulk materials is that interstitials do not exist in many 2DMs, as it is energetically favorable for the extra atom to be on the surface, Fig. 2.2(k), rather than take the interstitial position, e.g., in the middle of the hexagon in graphene lattice. However, one example to the contrary is $MoTe_2$, which has large lattice constant and thus interstitial space large enough to accommodate atoms of, e.g. 3d transition metals [64], Fig. 2.2(l). Substitutional impurities can obviously be present in any 2DMs.

The STM technique has also been useful for finding the signatures of point defects in 2DMs. As mentioned, the interpretation of STM images is often more complicated and requires a comparison with the calculated images. Few examples of defects whose identity is fairly well established are presented in Fig. 2.3. Fig. 2.3(a–b) shows the STS spectra recorded at pristine regions of WS_2 and at point defects, clearly showing the band gap and band edges, as well as the emergence of two defect states within the gap. Fig. 2.3(c) shows the STM images recorded at the energy range enclosing the defect states. Comparison of the defect state energies to STS spectra and the simulated STM images with the experimental ones allows identification of the defect as an S vacancy. The electronic structure of S vacancy in the top (facing the tip) and bottom (facing the substrate) sides of the WS$_2$ sheet are very similar and thus they are difficult to distinguish via STM/STS, but possible using non-contact AFM with CO tip [38].

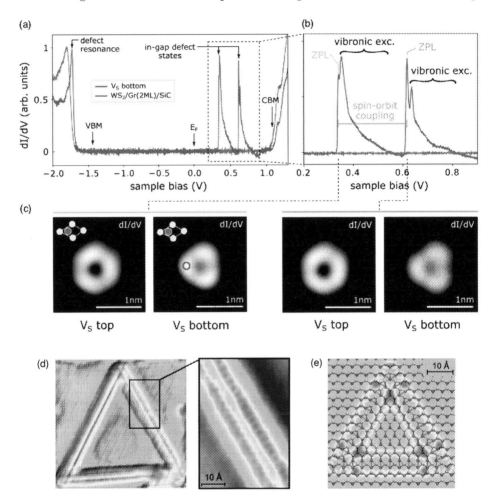

FIGURE 2.3 STM images and STS spectra of point and line defects in 2DMs. (a,b) Experimental STS of WS$_2$ at pristine parts and over a defect site, showing emergence of two defect states. (c) The corresponding STM images for energy range enclosing the defect states. Based on the comparison to the simulations, the defect can be identified as an S vacancy. Reprinted with permission from Ref. [38]. (d,e) Experimental and simulated STM images of a 4|4P mirror-twin boundary in MoTe$_2$. Reprinted with permission from Ref. [74].

Contrary to bulk systems, disclinations and edge dislocations in 2DMs are not line, but point defects. Disclination corresponds to addition or removal of a wedge of material, Fig. 2.4(a), and leads to a positive or negative Gaussian curvature of the 2D sheet and consequently strong buckling [65,75,76]. An edge dislocation corresponds to addition or removal of one unit-cell wide strip of material, Fig. 2.4(a). Edge dislocation can also be constructed by placing two disclinations next to each other and consequently leads to strain dipole and possibly local buckling at the two disclinations, Fig. 2.4(c).

Many other examples of point defects will be given for a broad range of free-standing and supported 2DMs in the following chapters.

2.5.2 Line defects

TEM and STM experiments have also made it possible to identify the atomic structure and properties of line defects in 2D systems, such as linear agglomerations of vacancies and grain boundaries, see Refs. [76,77,82] for an overview. As for grain boundaries, they are particularly important for understanding the mechanical properties of polycrystalline 2DMs grown by chemical vapor deposition and related techniques, as well as for electronic transport in these systems and even optical characteristics. Grain boundaries normally appear when islands of the growing material with different orientations of the crystal lattice coalesce.

Depending on the misorientation angle and growth conditions, the morphology of the boundaries can be quite different [76,77,82,83]. For instance, the grain boundaries in graphene consisting mostly from non-hexagonal rings, including pentagons, heptagons and octagons, can be straight or serpentine, Fig. 2.4(a) and Fig. 2.5(b), depending on the misorientation angle between the grains.

Examples of lines formed from aligned vacancies and grain boundaries are presented in Fig. 2.5. An STM image of an extended line defect (vacancy line) in graphene is shown in Fig. 2.5(a) [84]. A grain boundary in graphene representing an aperiodic line of defects which stitch the two grains together as revealed by HR-TEM, is presented in Fig. 2.5(b) [83]. The pentagons (blue), heptagons (red) and distorted hexagons (green) in the grain boundary are outlined. A curious case with characteristics of point and line defects is that of missing row of atoms (or unit cells) of finite length, which in the case of 2DMs can equally well be considered as a pair of dislocations, classified as point defects above. As an example of such a case in graphene, Fig. 2.4(d) shows HR-TEM image of a dislocation pair. Fig. 2.4(e) presents the filtered image aimed to measure the projected area of the hexagon (i.e., orientation of the sheet) and comparison with simulated models confirms the buckling, Fig. 2.4(f–i). It is clear that the "flat" theoretical structure does not match the experimental observations.

Experimental STM images and the corresponding atomic models of line defects in the epitaxial borophene [79] are shown in Fig. 2.5(e,f). The formation energy of such line defects in borophene is relatively low, so that under suitable growth conditions, their concentration can be high. These line defects energetically favor spatially periodic self-assembly that gives rise to the formation of new borophene phases, blurring the distinction between borophene crystals and defects.

The structure of grain boundaries [67,85–87] in TMDs can also vary. Low-angle grain boundaries consist of a row of dislocations with the structure similar to those in graphene, i.e., consisting of 5- and 7-membered rings (in the projected view) [86]. Contrary to elemental 2DMs, the odd-numbered rings in TMDs will necessarily contain either metal-metal or chalcogen-chalcogen bonds. Large-angle grain boundaries in MoS_2 were reported to consist of 8- and 4-membered rings [87]. At 60° the grain boundaries correspond to mirror twin

16   2. Physics and theory of defects in 2D materials: the role of reduced dimensionality

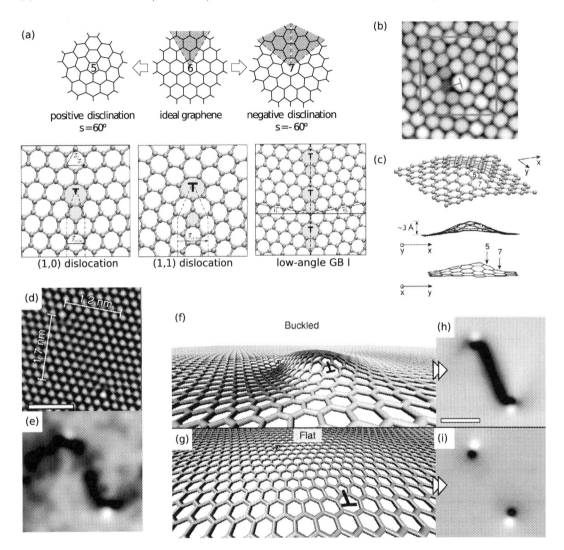

FIGURE 2.4 Disclinations and dislocations. (a) Structural models for disclination, edge dislocations, and grain boundaries in graphene. (b) Filtered HR-TEM image of an isolated dislocation in graphene and (c) the structural model highlighting the buckling of the sheet. (d) HR-TEM image of dislocation pair in graphene and (e) filtered version aimed to highlight the orientation of the sheet (i.e., smaller projected area per hexagon). (f,h) Structural model and simulated image when accounting for the buckling. (g,i) Structural model and simulated image when structure is constrained flat. Panels (a,c) are reprinted with permission from Ref. [75]. Panels (b,d–i) are reprinted with permission from Ref. [65].

boundaries (MTBs). TEM images of three observed mirror twin boundaries (MTBs) with different atomic structures are shown in panels Fig. 2.5(g–i). These are: MTB in $MoSe_2$ [bright field TEM image, [81]], MTB in MoS_2 [annular dark field STEM image, [67]], and MTB in WSe_2 [annular dark field STEM image, [63]]. Corresponding atomic structures are shown be-

2.5. Examples of defects in two-dimensional materials

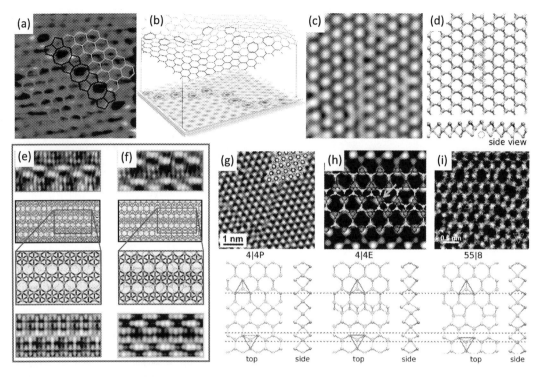

FIGURE 2.5 Examples of line defects and dislocations in 2DMs. (a) STM image of an extended line defect (vacancy line) in graphene. Courtesy of M. Batzill and J. Lahiri. (b) A serpentine-type grain boundary in graphene representing an aperiodic line of defects which stitch the two grains together as revealed by HR-TEM. The sketch of the atomic structure is also presented. The pentagons (blue), heptagons (red) and distorted hexagons (green) in the grain boundary are outlined. Reprinted with permission from Ref. [77]. (c) HR-TEM image of a vacancy line in MoS_2 and the corresponding atomic model (d). Reprinted with permission from Ref. [78]. (e,f) Experimental STM images and the corresponding atomic models of line defects in the epitaxial borophene [79]. Reprinted with permission from Ref. [80]. (g) Mirror twin boundary in $MoSe_2$ (bright field TEM image). Reprinted with permission from Ref. [81]. (h) Mirror twin boundary in MoS_2 (annular dark field STEM image). Reprinted with permission from Ref. [67]. (i) Mirror twin boundary in WSe_2 (annular dark field STEM image). Reprinted with permission from Ref. [63]. Corresponding atomic structures are shown below each panel. The triangles highlight the registry of the lattice at the two sides of the boundaries. The dashed horizontal lines highlight strain in the perpendicular direction, when the lattices are aligned at the top.

low each panel. The most common MTBs in TMDs are 4|4P and 4|4E, with the 55|8 one observed in W chalcogenides under certain conditions [88]. The large-angle grain boundary consisting of 4- and 8-membered rings can also be considered as short segment of MTBs joined by 8-membered rings. Mirror twin boundaries have also been reported to appear in $MoSe_2$ and WSe_2 as a corollary of chalcogen atom deficiency during the growth [89]. In semiconducting TMD, grain boundaries and other one-dimensional defects can be conducting channels [87,89–94], while electronic transport in the direction perpendicular to the boundaries is normally suppressed. First-principles calculations indicate that some lines defects in TMDs can exhibit magnetism [89,92,95], thus adding new functionalities to the system.

HR-TEM image of a vacancy line in MoS_2 and the corresponding atomic model are shown in Fig. 2.5(c,d) [78]. Such defect appeared due to the migration and agglomeration of isolated vacancies [78,96].

Moreover, a number of in-situ TEM experiments [63,78,89,97] showed that agglomeration of vacancies in TMDs gives rise to the development of grains embedded into a larger grain with a different orientation of the crystal lattice of the material. An interesting and somewhat unexpected observation is that depending on the chemical content of TMDs, line defects with different morphologies are formed. Sputtering of chalcogen atoms in WSe_2 and WS_2 results in the development of rotational defects and eventually mirror grains with the boundaries consisting of 5- and 8-member rings [63]. At the same time, under similar conditions formation of different line structures in $MoSe_2$ was reported [89,97]. Alternatively, deposition of extra Mo can give rise to MTBs in $MoTe_2$ and $MoSe_2$ [74]. An example of triangular inversion domain surrounded by MTB loop is shown in Fig. 2.3(d,e).

2.6 Theoretical aspects of the physics of defects in bulk crystalline solids and two-dimensional materials

2.6.1 Defect formation energy

Formation of a defect in the crystalline solid is always energetically unfavorable. Otherwise defects would spontaneously appear in the solid even at zero temperature, so that its lowest energy atomic structure would be different. That is, the energy required to create a defect in the pristine material, referred to as defect formation energy E_f, is always positive.

The simplest example is a vacancy in an elemental solid:

$$E_f = E_{tot}[vac] - E_{tot}[0] + \mu_{host}. \tag{2.1}$$

Here $E_{tot}[vac]$ and $E_{tot}[0]$ are the total energy of the system (at $T = 0$) with and without a vacancy, respectively, and μ_{host} is the chemical potential of the atom in the pristine system. It is presumed here that there is always a sink for the removed atom, so that $\mu_{host} = E_{tot}[0]/N$, where N is the total number of atoms in the system. This is the so-called **Schottky** defect: the atom has diffused to the surface of the crystal, where it has found a sink, Fig. 2.1, or alternatively a vacancy nucleated at the surface and diffused into the bulk of the crystal. Note that for the sake of simplicity, it is assumed here that the defect is neutral, that is, there is no extra charge localized on the defect. This is normally the case in the metallic systems, but not necessarily in semiconductors and insulators.

For interstitials (both self-interstitials and foreign atoms)

$$E_f = E_{tot}[I] - E_{tot}[0] - \mu_I. \tag{2.2}$$

Here $E_{tot}[I]$ and $E_{tot}[0]$ are the total energy of the system (at $T = 0$) with and without the interstitial atom, respectively, and μ_I is the chemical potential of the interstitial atom in the reference material. One can also consider a **Frenkel** defect [5]: in this case the atom takes the position of the interstitial, and the formation energy is, to the first order, a sum of vacancy and interstitial formation energies.

2.6.2 Gibbs free energy of defect formation

In a more accurate approach, the Gibbs free energy must be evaluated. For example, in the case of the interstitial, Gibbs free energy of formation ΔG_f is defined as

$$\Delta G_f(P, T) = F_I(\Omega_I, T) - F_0(\Omega_0, T) + P V_f - \mu_I(T). \tag{2.3}$$

Here Ω_I and Ω_0 are the volumes of the system with and without interstitial, P is external pressure, $V_f = \Omega_I - \Omega_0$, $F_I(\Omega_I, T)$ and $F_0(\Omega_0, T)$ are the Helmholtz free energies of the system with and without the interstitial, and $\mu_I(T)$ is the chemical potential (Gibbs free energy per atom) in the reference system. In the case of 2D materials, V_f and Ω are areas and P is force per edge length.

For the sake of direct comparison of the electronic and vibrational contributions to the Gibbs energy upon defect formation at a given temperature, Eq. (2.3) can also be written as

$$\Delta G_f(P, T) = E_f - \Delta F_{el}(T) - \Delta F_{ph}(T), \tag{2.4}$$

where ΔF_{el} and ΔF_{ph} are the corresponding changes in free energy associated with the electronic structure and phonons, and where it is assumed that the supercell is sufficiently large so that the pressure term $P V_f$ is small.

ΔF_{ph} can be evaluated directly by carrying out DFT molecular dynamics simulations and thermodynamic integration, which come at a high computational cost [98]. Without account for the anharmonic effects, i.e., at low (e.g., room) temperatures and when the material is dynamically stable, free energy F can be expressed as

$$\Delta F_{ph}(\Omega, T) = \sum_i \left\{ \frac{\hbar \omega_i}{2} + k_B T \ln \left[1 - \exp \left(-\frac{\hbar \omega_i}{k_B T} \right) \right] \right\} \tag{2.5}$$

where the mode frequencies ω_i are obtained by finding the eigenvalues of the dynamical matrix [99,100]. We note that in order to obtain accurate ΔF_{ph} in the case of 2DMs the quadratic dispersion of the flexural acoustic phonon mode needs to be captured correctly.

At low temperatures and with continuous density of states, the electronic contribution to free energy can be shown [99] to behave as

$$F_{el}(\Omega, T) = \frac{1}{2} T k_B \sum_i [(1 - f_i) \ln(1 - f_i) + f_i \ln f_i] \tag{2.6}$$

where the sum runs over all electronic states with energies ϵ_i and $f_i(\epsilon_i, T)$ is the Fermi–Dirac occupation numbers of energy levels ϵ_i.

2.6.3 Equilibrium concentration of defects

In the dilute limit, the equilibrium concentration n of defects, e.g., vacancies or interstitials in a solid at temperature T can be evaluated [5] through

$$n(T) = A_g \exp(-G_f/k_B T), \tag{2.7}$$

where A_g is the geometry factor, k_B is the Boltzmann constant and G_f is the Gibbs free energy of defect formation defined above. For example, for octahedral and tetrahedral interstitials in FCC and HCP metals $A_g = 1$ and 2, respectively. It is assumed that the system (the host material for the interstitial) is in the thermodynamic equilibrium with the reference system. If the entropic phonon and electronic contributions are neglected, e.g., when they are much smaller than E_f,

$$n(T) \approx A_g \exp(-E_f/k_B T). \tag{2.8}$$

Formation energies of defects in 2D materials vary in a wide range. For example E_f of a single vacancy of graphene is about 7 eV [60], which indicates that equilibrium concentration of vacancies is negligible even at elevated (up to 1000 K) temperatures. On the other hand, E_f in borophene is relatively low, less than 1 eV (with respect to the lowest energy phase) [101], so that vacancies can easily appear, agglomerate and form even new phases [79], which can be referred to as alloys of boron atoms with vacancies [102].

Eq. (2.8), where E_f is assumed to be independent of T, reflects the balance of the configurational entropy, which lowers the free energy of the system, and the cost of defect creation. When the formation energy of the defects (vacancy or interstitial) is high and temperature is low, Eq. (2.8) normally gives the qualitatively and even quantitatively correct concentration as a function of T, and makes it possible to determine E_f from measuring the concentration of defects at several temperatures. It should be pointed out that this approach may give rise to errors, especially at high temperatures [98], when anharmonic effects become substantial.

When calculating the formation energies of native defects in bulk solids, the corresponding chemical potentials need to be defined via equilibrium with the crystal. As illustrated in Fig. 2.1, this is equivalent to considering that the elements are exchanged with the material surface and, when the surface contributions are neglected or assuming an infinite solid, this essentially corresponds to adding or removing one atom to the bulk reservoir. In more practical considerations of the most likely defect candidates, with bulk systems one can often ignore what is outside the material surface, as the reactions or adsorption related to the environment only affects a thin layer of the material close to its surface. The time required to reach equilibrium with the sample interior may be excessively long. Under equilibrium and with given chemical potentials, there is no fundamental difference in how to treat the defects in the case of 2DMs. However, the immediate proximity of the environment changes the situation in practice. The defects are immediately accessible to any species present in the environment.

We note that even if the sample is assumed to be in vacuum, it is not possible to calculate formation energies corresponding to perfect vacuum conditions. The pressure dependence of a chemical potential is $\mu(p) = \mu(p_0) + k_B T \ln(p/p_0)$ and thus chemical potential of any element in a theoretical perfect vacuum is $-\infty$, which consequently makes the formation energies of all defects $\pm\infty$. In practice this means that, while atoms occasionally desorb from the surface, additional atoms are never adsorbed to the surface, eventually leading to the breakdown of the material.

2.7 Calculations of defect formation energies and electronic structure using the supercell approach

2.7.1 Assessment of defect formation energies

In order to assess the formation energy of defects and get insights into the electronic structure of the system with defects, the vast majority of studies are nowadays carried out using density-functional theory (DFT) calculations, which are also referred to as "first-principles" or "ab initio" approaches, and supercell geometry. In principle, defect formation energies can also be evaluated using the analytical potential calculations, but normally the accuracy of this approach is rather low [103]. The DFT calculations give access to total energy, single-particle orbitals (wave functions and energies), charge density, magnetic moments, mechanical properties, etc. Since DFT yields accurate total energies, it is also expected to provide accurate geometries and forces. As a result, the vibrational modes related to defects can also be accessed and consequently the corresponding contributions to the free energy can be simulated.

When calculating the structure and properties of periodic solids, the pristine host is modeled using unit cell subject to the periodic boundary conditions. Defects are commonly modeled within the supercell approach, where a single defect is introduced to a box of a sufficiently large piece of pristine material. The periodic construction unavoidably leads to a periodic array of defects with the density of defects being commonly much higher than what is found in the experiments. However, having exactly the same defect concentration in simulations as in the experiments is often not necessary. When the defect property of interest is defect-specific (e.g., formation energy, formation volume), it is only required that the property is converged with respect to increasing supercell size. Moreover, usually any property that depends on defect concentration can be mapped to such defect-specific property, e.g., magnetization from magnetic moments, strain from formation volume, and carrier concentration from ionization energies. In practice this means repeating the calculations for increasing supercell size until the relevant properties converge. A great advantage of 2DMs is that the supercell only needs to be increased in two directions instead of three in bulk systems.

Convergence of most properties is relatively fast for neutral defects. However, due to the long range of Coulomb interactions, the convergence becomes poor when the defects are charged. The formation energy of the defect X in charge state q is defined as

$$E_{\mathrm{f}}[X^q] = E_{\mathrm{tot}}[X^q] - E_{\mathrm{tot}}[\text{pristine}] - \sum_i n_i \mu_i + q[\epsilon_F + \epsilon_v] \tag{2.9}$$

where E_{tot} is the total energy of defective or pristine system, and μ_i are the atomic chemical potentials. ϵ_v is the valence band maximum (VBM) of the pristine system and ϵ_F is the Fermi-level position with respect to the VBM.

The supercell must be charge-neutral in order to obtain well-defined energy, since summation of interactions from charged cells would lead to an infinite total energy. To achieve in the case of charged defects, compensating/background charge is added to the cell. Moreover, in the case of 2DMs, the screening is highly anisotropic and inhomogeneous. The Coulomb interaction between defect images in the in-plane and out-of-plane directions is different, all of which makes the electrostatics behave in a highly non-trivial fashion, Fig. 2.6(a). The problem can be circumvented by extrapolating the results to the dilute limit, adding energy corrections accounting for the spurious electrostatic interactions, or adopting computational cell that is non-periodic in all three dimensions [104–107]. One should also be aware that if a homogeneous background is assumed, in the case of negatively charged defect and positive background charge, it leads to attractive potentials for electrons inside the vacuum region. For a sufficiently large vacuum region, some electron density can become confined in the center of vacuum and lead to incorrect results [108]. This problem can be avoided by confining the background charge to within the 2D sheet [109,110].

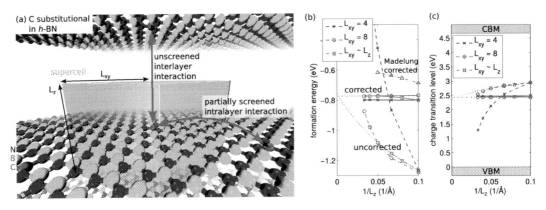

FIGURE 2.6 (a) Illustration of the computational setup for charged defect in 2DM. (b) Sensitivity of the formation energies to the supercell/vacuum size and the effect of the corrections. (c) The dependence of defect levels on supercell/vacuum size. Reprinted with permission from Ref. [106].

The formation energy in Eq. (2.9) is a function of the atomic chemical potentials and the Fermi-level position. As for the atomic chemical potentials, the number of free parameters is usually one less than the number of species in the material, since equilibrium dictates that, e.g., in MoS_2, $\mu_{Mo} + 2\mu_S = \mu_{MoS2}$, where μ_{MoS2} is the energy of the host, and thus allows elimination of one unknown. The atomic chemical potentials, corresponding to $\partial G(P,T)/\partial N$, can be linked to the experimental conditions either by explicitly calculating the temperature and pressure dependencies or by exploiting experimental data available in the thermochemical tables.

As an example, the formation energies of native defects as a function of atomic chemical potentials and charge state fixed to neutral are shown in Fig. 2.7(b), and as a function of Fermi-level and fixed atomic chemical potentials in Fig. 2.7(e,f). The atomic structures for vacancies and adatoms are shown in Fig. 2.7(a).

The formation energies and the stable charge states are governed by the position of the Fermi-level, but the Fermi-level is defined by the defect concentrations and their charge states. Thus, one may finally determine self-consistently the Fermi-level position and the dominant defect types and concentrations. As evident from Fig. 2.7(c,d), such calculation for MoS_2 shows that only S vacancies are expected to be present under thermal equilibrium, whether the growth conditions are closer to S- or Mo-rich conditions and also rather insensitive to the S partial pressure.

2.7. Calculations of defect formation energies and electronic structure using the supercell approach

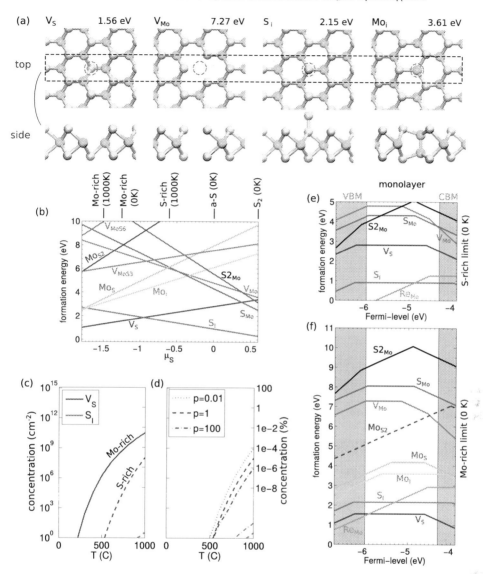

FIGURE 2.7 (a) Atomic structures for the S and Mo vacancies and adatoms. (b) The formation energies of native defects as a function of S chemical potential. The experimental conditions yielding a particular value are shown above the plot. (c,d) Defect concentrations as a function of temperature (c) in the S- and Mo-rich limits and (d) for a few different S_2 partial pressures. (e,f) Formation energies of native defects in the S- and Mo-rich limits as a function of Fermi-level. Only the stable charge states are shown. Reprinted with permission from Ref. [111].

2.7.2 First-principles approaches for calculating defect states

DFT is a ground-state theory, and thus in principle incapable of describing excitations. The DFT single-particle, or Kohn-Sham levels, are only a mathematical construct formally without strict physical meaning, a mapping between real interacting particles and fictitious non-interacting particles in effective potential. In practice, however, these single-particle states are a reasonably good approximation for the physical orbital structure and quasi-particle energies.

The energies for the initial and final states of an "excitation" involving two ground states with a localized charge added to or removed from the system can be calculated safely. Such approach gives rather reliable ionization energies in the case of atoms and molecules. In principle adding/removing electron to/from the defect state should also lead to reliable total energy, provided that the total energies are corrected for the finite-size supercell errors. On the other hand, when calculating the energy difference between neutral and charged systems, one still needs to assign energy (chemical potential ϵ_F) for the electron in the reference state, cf. Eq. (2.9). Unfortunately, the same electron addition/removal is not reliable when the charge is added to delocalized conduction or valence band states of the host. Thus, if the energies of the delocalized states cannot be estimated this way, then one cannot consider electron exchange between defect and valence or conduction band, and in principle one can only take the electron from another localized state of defect, atom, etc. Alternatively, one has to combine information from different levels of theory, such as e.g. taking defect levels from DFT calculations and band edge positions calculated at higher level of theory. In addition, in the case of 2D materials it is possible to use the vacuum level as a reference to align these different bits of information [111–114].

In Fig. 2.7(e,f), the kinks where the stable charge state changes to another also correspond to the thermodynamic charge transition levels [99]. One can readily extract ionization energy from the energy difference between the charge transition level and the corresponding band edges. Note, that in this case the structural geometry is relaxed to the energy minimum of a given charge state. When the structure is fixed, one gains access to the vertical transitions, or optical charge transition levels (discussed further in Sect. 2.9.2) [99]. The defect states can also be observed in the single-particle band structure, as shown in Fig. 2.8(a) in the case of S vacancy in MoS_2 [115,116]. Comparison of the energy of the single-particle state within the gap in Fig. 2.8 and the position of the charge transition level in Fig. 2.7 shows fairly close correspondence. Indeed, in order to localize additional charge on the defect, there needs to be a defect state within the band gap. However, the energy of the KS level does not correspond exactly to the charge transition level (and they shift upon changing occupation) and they can thus only provide an initial guess of what charge states might be stable. Since the charge transition levels are obtained from the total energy differences, they are also affected by the finite-size supercell errors. The single-particle levels are affected in a similar way. The effect of finite supercell size errors on the formation energies and charge transition levels is shown in Fig. 2.6(b). Nevertheless, some defect properties may be rather insensitive to these effects. For example, the atomic structure of defects or their magnetic moments are likely independent of the shift of the defect states, as long as the defect states remain within the gap.

In order to more rigorously model the electron addition and removal, i.e. quasi-particle (QP) excitations, many-body Green's function approach, such as the so-called "GW" method [118], are required [119]. Due to their extremely high computational costs, the GW calculations are often limited to relatively small systems, but recently even defective systems have been calculated due to both advances in the methodology as well as in the computational power.

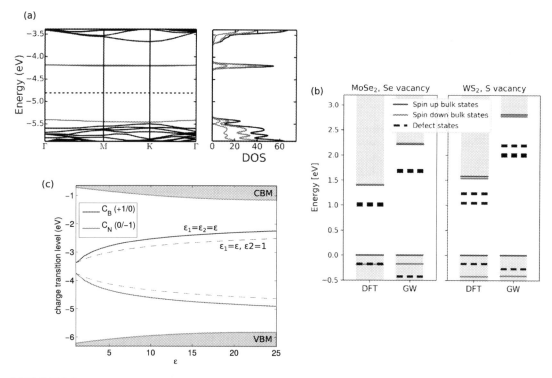

FIGURE 2.8 (a) DFT band structure and density of states of S vacancy in MoS$_2$. Reprinted with permission from Ref. [116]. (b) Comparison of the defect levels and band edges calculated at the DFT and GW levels of theory, and in the case of MoSe$_2$ and WS$_2$. Reprinted with permission from Ref. [117]. (c) Schematic drawing illustrating the change of defect level positions and QP band edges in the case of C$_B$ and C$_N$ defects in h-BN with changing dielectric constant of the environment, either on only one side ($\varepsilon_1 = \varepsilon$, dashed line) or on both sides ($\varepsilon_1 = \varepsilon_2 = \varepsilon$, solid line) of the sheet.

The smaller supercells in the case of 2D materials also helps, and at least h-BN and MoS$_2$ have already been considered [113,116,117,120].

The fundamental gap, corresponding to the energy difference in electron addition to conduction band and removal from valence band, can be obtained from the QP band structure calculations or experimentally via scanning tunneling spectroscopy. The optical gap corresponds to the lowest energy optical (bright) transition, which is essentially the fundamental gap minus exciton binding energy, and can be obtained via optical (absorption) spectroscopies. With bulk materials, due to the small exciton binding energy, the two gaps are very similar, but in the case of 2DMs one needs to be more careful not to mix them. Curiously and dangerously, the DFT band gap of many TMDs is close to the experimental optical gap due to cancellation of errors arising from DFT band gap underestimation and missing excitonic effects [121]. Naturally, the defect levels calculated without excitonic effects should only be compared to the fundamental band edges without excitonic effects and there is no way to properly position them within an optical gap.

Comparison of the defect level positions within the band gap as calculated within DFT and GW are shown in Fig. 2.8(b). In this case, it appears that the defect levels follow the conduction band edge when using GW formalism, and thus correct results could be obtained already via comparison of the defect level to the DFT band edge. However, it may not be clear a priori how well the levels follow band edge, or which edge, thus leading to ambiguity in the interpretation of pure DFT results.

Since the modification of the environment affects screening, it in turn changes the QP energies of the defects states and band edges, as well as the formation energies of charged defects. The scaling of the fundamental band gap and exciton binding energy with dielectric constant of the environment has been studied both experimentally and computationally [121–124]. In Fig. 2.10(c) we show schematically how the charge transition levels shift in the case of $(+1/0)$ level of C_B defect and in the case of $(0/-1)$ level of C_N defect in h-BN, and also the shift of the band edges.

Charge neutral excitations can be obtained by solving the Bethe-Salpeter equation (BSE), which describes the interaction between the electron and hole quasiparticles (QP) and is thus applied on top of the GW QP electronic structure. Within the BSE approach, only electron-hole pairs are considered, and thus the obtained spectra are missing trions, biexcitons etc. Extension of the BSE for calculation of three-particle spectra has been developed and applied to TMDs [125], but not yet applied to defective systems. Alternatively applying GW+BSE on top of the electronic structure of a charged defect should yield the appropriate emission spectra, but have not been done yet to the best of our knowledge.

Charge neutral excitations have also been explored via total energy differences. Consider for example a defect with two defect states in the gap, one occupied and one empty. Within DFT, the initially occupied state can be forced empty and the initially empty state can be forced to be occupied. While there is no reason for KS-DFT to describe such excited system correctly, it has been found to yield rather accurate intra-defect excitation energies at least in bulk materials [126]. It seems to be less clear whether this approach works well in 2DMs with strong excitonic effects.

The change in screening due to environment has much smaller effect on the neutral excitations. Even though the fundamental gap changes strongly, the exciton binding energy changes equally in the opposite direction, and finally the optical gap is nearly unchanged [121,127].

2.8 Electronic structure of 2D materials with defects

2.8.1 Defect-induced modifications of electronic states

In addition to structural modifications, defects also lead to modifications of the electronic structure around them. In metals or semimetals, such as graphene, the defect-induced perturbation of the lattice tends to lead to resonant or anti-resonant states within host-like states. Similar to bulk, scattering by the defects may significantly modify electron transport [6,128] and e.g. provide catalytically active sites [129]. The modifications of the optical properties are ofter minor unless the concentration of defects is very high.

In semiconductors and insulators, localized defect states can emerge within the forbidden gap, which may significantly modify the optical properties in addition to the electrical ones. Some 2D materials are more susceptible to appearance of deep defect states than others. Naturally, if the band gap is very large, like in the case h-BN, it is very likely that any defect state energy will fall into the forbidden gap. However, the number of 2D materials with large band gap of, say, more than 3 eV is relatively small. Majority of the non-metallic ones have a small or moderate band gap, i.e., they are semiconductors. Among them, e.g. phosphorene was found to be fairly resilient to deep states from native defects [130], since grain boundaries and Stone-Wales defect showed no states and P vacancy and adatom yielded states near VBM. The trends among chalcogen vacancies in TMDs were studied by Pandey et al. [131]. It was found that, if the valence and conduction bands originate from bonding and antibonding states between the metal and chalcogen elements, then there will likely be in-gap defect states (since dangling bond states will fall inside the gap), see Fig. 2.9. Group-6 (H-phase MoX_2, WX_2) and group-10 TMDs (T-phase NiX_2, PdX_2, PtX_2) behave this way. On the other hand, if valence and conduction bands originate from different atomic orbitals, deep states were unlikely. This is the case with group-4 TMDs (1T-phases of TiX_2, ZrX_2, HfX_2). Similar considerations should also work in the case of other 2DMs with more complex structure/composition.

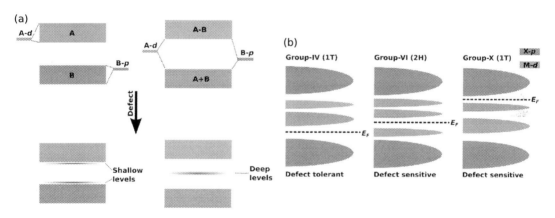

FIGURE 2.9 (a) Construction of band edges from the atomic orbitals and the expected defect level (or dangling bond) positions. (b) Classification of different groups of TMDs to the defect-tolerant and defect-sensitive ones. Reprinted with permission from Ref. [131].

The 2D nature of the host leads to at least two important aspects that can affect electronic structure of defects: confinement of electronic states and inhomogeneous screening. We note here that the defect-related properties in graphene are distinct from those in other 2D materials [128]. However, most of these differences are due to peculiar electronic structure of graphene such as the presence of Dirac cones at the Fermi level, not 2D per se (and thus cannot be generalized to other 2DM).

The wave functions in layered bulk materials are highly anisotropic and often show very small band dispersions in the out-of-plane direction. This suggests that the states are localized within the layers with small interlayer hopping parameter (in the tight-binding language). As a consequence, also the defect wave functions in bulk layered materials tend to be largely localized to layer in which the defect resides. Thus, to a degree, the confinement effect is already present in the bulk of layered phases, although further accentuated by the vacuum potential in the case of pure 2D sheets.

Layered bulk phases also exhibit highly anisotropic, yet homogeneous, dielectric tensor. In the case of 2D materials the screening becomes inhomogeneous, which not only complicates the theoretical description of the screening, but also contributes to the exciting physics of 2D materials. In 2D, the electric potential from a localized charge cannot be described by a simpled screened Coulomb potential $1/\varepsilon r$ even within the plane (and distance r from the charge), but rather the dielectric constant needs to be made distance dependent $\varepsilon(r)$. $\varepsilon(r)$ is similar to the bulk value at small distances but at large distance approaches one, i.e., the interactions become unscreened due to majority of the field lines propagating through the surrounding vacuum. The strong Coulomb interactions at large distances lead to strong excitonic effects, but also have a pronounced effect on some defect properties in 2D, such as on the shallow/deep classification, stability of defect-bound excitons, and on the sensitivity of the defect level positions on the environment. It can also be seen in the strong band bending around charged defects in STS [132,133].

2.8.2 Deep vs. shallow electronic states in 2D materials

In bulk materials, the text book description of shallow defects starts from the effective mass approach and a hydrogenic model for the defect, where e.g. a P atom in place of Si effectively corresponds to one additional proton and one additional electron [1,5,6,134]. The additional proton modifies the CB states, "pulling" one state inside the gap. The energy of this defect state can be to a good approximation solved from that of a hydrogen atom, but the attraction between the proton and electron is screened by the dielectric constant of the host ϵ and the mass replaced by the CB effective mass μ.

$$E_n^{3D} = -\frac{\mu}{n^2\epsilon^2}E_0 \tag{2.10}$$

where E_0 is the Rydberg unit of energy 13.6 eV, and μ and ϵ are given in the units of electron mass and vacuum permittivity, respectively. Eq. (2.10) generally leads to defect ionization energy of few to tens of meVs in traditional semiconductors such as Si or GaAs. The hydrogenic model has also been used to calculate exciton binding energies by simply changing the effective mass in Eq. (2.10) to reduced mass. For instance, in the case of GaAs measured donor ionization energies are 5.8–5.9 meV and exciton binding energy is 4.8 meV, which can be compared to the values from the hydrogenic model of 5.7 meV and 6–7 meV [134], showing especially good agreement for the donor ionization energies. The exciton binding energy calculation is complicated by the degeneracies and warping of the VBM, but a more accurate account improves the theoretical estimate to 4.4 meV, again in good agreement with the experimental value. Therefore, one can often get an estimate for the ionization energy of a shallow donor or acceptor from the exciton binding energy, and vice versa.

When such model is applied to layered materials, the wave function is often confined to a single layer and consequently 2D version of the hydrogenic model may be more suitable:

$$E_n^{2D} = -\frac{\mu}{(n-\frac{1}{2})^2\epsilon^2}E_0. \tag{2.11}$$

If the same material parameters are adopted, one immediately notices that $E_1^{2D} = 4E_1^{3D}$, i.e., the binding energies are expected to be much larger and thus the defects are less shallow. Nevertheless, whether using 3D hydrogenic model or 2D hydrogenic model, binding energy scales linearly with the reduced mass.

As a practical example, let us now apply this model to MoS_2. Calculated effective masses of electrons and holes in monolayer MoS_2 around K-valley are $0.38m_0$ and $0.46m_0$, respectively [135]. Reduced mass for exciton is $0.21m_0$, while for shallow donor/acceptors the reduced mass is directly that of electron/hole. Using Eq. (2.11), we obtain exciton binding energy of 0.050 eV and shallow donor binding energy of 0.091 eV. These can be compared to bulk exciton binding estimated at 0.087 eV [136] and to Re_{Mo} ionization energy at 70–90 meV [137,138]. Moreover the DFT calculated Re_{Mo} ionization energy is 0.06–0.08 eV [111,139]. The agreement is less spectacular than in the case of GaAs, but still reasonably good (also keeping in mind the experimental uncertainties). The smaller lateral extent of the wave functions (due to higher ionization energy) likely makes the effective mass description less reliable.

As mentioned, in the case of free-standing 2D layer, the screening parameter needs to be modified to reflect the reduced screening, which is consequently expected to increase the binding energies. Due to the strongly distance dependent screening, a single parameter cannot be expected to work in all cases, as also demonstrated in the case of excited states of exciton [140]. If the exciton binding energy for free-standing monolayer is 0.63 eV (from GW+BSE calculations [141]), we should use $\epsilon = 4.2$ in Eq. (2.11). When the same value is used for the shallow donor, binding energy of 1.14 eV is obtained! First-principles calculations have resulted in binding energies of around 0.53 eV [139]. The discrepancy again likely reflects the smaller extent of wave function (thus larger screening), and possibly breakdown of the effective-mass description. The lowest energy exciton radius was reported to be about 1 nm [142] and similar radius was found for the Re_{Mo} defect [111], also shown in Fig. 2.10(a). Experimental results are consistent with the deep nature of these defects and further suggesting that the conductivity is due to hopping between defect states rather than thermally activated free carriers [143,144]. Overall, with these defects being as close to hydrogenic as they get in TMDs, it is clear that the hydrogenic "shallow" defects in free-standing 2DM cannot have binding energies of the order of $k_B T$!

When properly accounting for the 2D screening, similar conclusions are still reached. In the case of h-BN encapsulated $MoSe_2$, exciton binding energy is 194 meV and the shallow donor 260 meV (compared to about 1.5 eV band gap) [145]. In the case of graphane (hydrogenated graphene), 1.77 eV for exciton and 2.12 eV for impurity (compared to about 5.4 eV band gap) [146]. In contrast, "deep defects" with highly localized wave function can have any energy for the defect state and thus they can be shallow as long as only the binding energy is concerned.

In the above discussion, the large difference in the binding energies in layered bulk and monolayer materials was assigned to screening. This suggests a novel way to tune the ionization energies just by modifying the dielectric constant of the environment. In the extreme limit, if MoS_2 were surrounded by a material with even higher dielectric constant, the ionization energy can be engineered to be even lower than in bulk MoS_2. It was shown by first principles calculations, that the ionization energy of Re_{Mo} in MoS_2 can be tuned by the substrate/environment from the 0.5 eV in free-standing to about 0.1 eV when surrounded by HfO_2 (a high-κ dielectric) [139], Fig. 2.10(d). On the other hand, the defect level position of a sulfur vacancy was found to be largely independent of the dielectric environment [116] de-

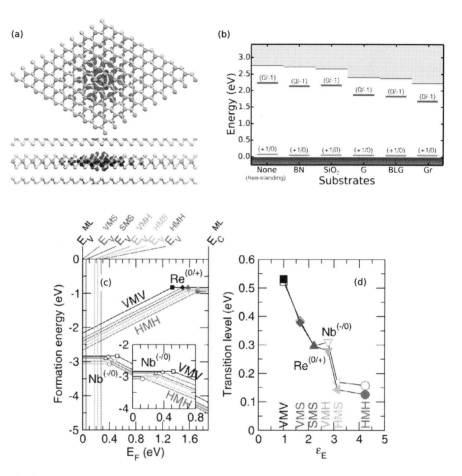

FIGURE 2.10 (a) Shallow defect wave function in bulk MoS$_2$, showing the localization to single layer. (b-d) Modification of the defect level positions and band edges with environment in the case of (b) S vacancy and (c,d) Re$_{Mo}$ and Nb$_{Mo}$ "shallow" defects in MoS$_2$. Reprinted with permission from Refs. [111,116,139].

spite the strong renormalization of the fundamental band gap, which is (i) due to the very localized nature of the defect wave function, and thus depending only on the screening by the host rather than by the environment, and (ii) due to orbital character similar to the band edges, Fig. 2.10(b).

2.8.3 Defect-bound excitons

In traditional semiconductors, the optical transitions can be classified to three categories [134]: (i) "band-to-band" transitions involving transitions between valence and conduction bands, (ii) "free-to-bound" transitions involving transitions between a defect state and valence/conduction band, and (iii) "donor-acceptor pairs" involving transition from defect state of one defect to a defect state of another defect. In addition, there can be transitions between different defect states of the same defect, or transitions to excited electronic or vibronic states. These tend to be more relevant at larger band gap materials and currently of high interest in the context of quantum information technologies [147]. In bulk materials the exciton binding energies are very low (few meV), but at sufficiently low temperatures, the luminescence will be dominated by features arising from excitons: (i) free excitons related to band-to-band transitions and (ii) bound excitons related to free-to-bound transitions. In the latter case, excitons can be bound to either neutral $[D^0X]$ or charged defect $[D^+X]$. For a donor defect, $[D^+X]$ involves two mobile carriers (e+h) and an immobile ion. $[D^0X]$ involves three carriers (2e+h) in addition to the ion, but, especially in the case of deep defects, the electron in the defect state is more localized and more strongly bound.

In 2D materials, as discussed above, the excitonic effects are strongly enhanced by the reduced screening, and thus the absorption and emission spectra are generally dominated by peaks from excitons even in high temperatures. In addition to the high binding energies for excitons, also higher order complexes such as trions and biexcitons become stable [148–150]. Since biexcitons involving four carriers and even larger complexes are stable in 2D materials, it suggests that higher order complexes can also be bound to defects, i.e., trions or biexcitons bound to neutral defect may also be stable at not too low temperature. Similar to the connection of shallow defects and excitons above, the binding energy of defect-bound exciton is expected to be similar in magnitude to trions or biexcitons (or larger due to higher reduced mass). Here, the situation is a bit more complex however due to different types of carriers e.g., immobile ion vs hole on the valence band, or localized defect state vs mobile electron on the conduction band.

As an example, we consider the defect-bound exciton peak in MoS_2, which is often observed at about 0.1–0.2 eV below the lowest energy free exciton (A exciton). The intensity of this peak can be controlled by introducing defects by irradiation and reduced by annealing in sulfur atmosphere, and thus in many papers assigned to excitons bounds to S vacancies [117,151–155].

The temperature-dependence of the bound exciton feature has shown thermal activation energy (15–35 meV) much smaller than the binding energy inferred from the difference of the defect bound and free exciton PL peaks (0.1–0.2 eV) [154,155]. According to Ganchev et al. [156], in the case of e.g. an acceptor-bound exciton the hole is more strongly bound to the acceptor ion A^- than to the electron, explaining why the defect-bound exciton can be dissociated to $[A^0e]$ (neutral defect and electron) at relatively low temperatures. Similarly, according to [154], there are two bound exciton lines arising from recombination of hole with electron in the localized acceptor state (observed at 200 meV below A) or with the electron in the conduction band (observed at around 35 meV below A). It is also worth mentioning, that the defect states can inherit the spin-valley locking present in many TMDs [157]. Consequently, since the valley index of excitons can be controlled by selecting the circular polarization direction of the light [158,159], this could also be used to selectively set or address the spin(s) of the defect state or the defect-bound exciton.

On the other hand, the substrate can also play an important role. In heterostructures of layered materials, the binding energy of interlayer excitons where the electron and hole are localized in different layers is also sizeable [160]. As a consequence, a charged defect in a substrate or in another layer of a heterostructure can bind with charge carriers, exciton, or trions. This, however, depends on the alignment of bands in the heterostructure and location of the impurities, thereby governing where the electrons, holes, and bound charge can reside. For instance, it was found that in the case $MoSe_2/WSe_2$ heterostructure, hole in WSe_2 cannot bind to a neutral donor in MoS_2 layer, but an interlayer exciton can bind to a neutral donor [145].

2.9 Point defects and vibrational properties of 2D materials from atomistic simulations

Defects also affect the vibrational characteristics, such as phonon spectra, thermal conductance and Raman spectra. In the theoretical analysis of the influence of defects on vibrational properties, DFT is normally used, but also analytical potential calculations can be carried out [161]. Although the calculations of these quantities for 2DMs are essentially the same as for the bulk systems, we illustrate here such calculations by the example of how point defects and impurities in MoS_2 affect the Raman spectra of the system and how they affect the line shape of defect-related photoluminescence peaks.

2.9.1 Signatures of defects in Raman spectra

The Raman spectrum is usually calculated via a two-step process: First, the force-constant matrix (or dynamical matrix) of the system is determined, from which the vibrational mode frequencies (eigenvalues) and eigenvectors can be obtained, as discussed previously in Section 2.6.2. A brute-force approach to construct the force-constant matrix involves displacing each atom in the supercell in the three cartesian directions and thus the computational cost grows quickly with increasing supercell size. Second, the change in electronic susceptibility (or dielectric constant) with displacement according to each vibrational mode is evaluated [162]. We note that for defective systems shown in Fig. 2.11(d–f) the Raman spectra was simulated via "Raman-tensor weighted Γ-point DOS" (RGDOS) approach, that largely avoids the second step mentioned above [163].

As an example of the changes observed in the Raman with introduction of defects, the evolution of Raman spectra upon He^+ irradiation of graphene and MoS_2 is shown in Fig. 2.11(a,b). G- and 2D-band peaks are present in pristine graphene and D (and D' mode just above G) arise upon introduction of defects. The intensity ratio D/G is a good estimator for the material quality in the case of graphene [25,26]. MoS_2 shows broadening and shift of the E' and A'_1 modes and some additional broad features, but does not show any distinct new peaks. Qualitatively similar results were obtained in electron-irradiated samples [164]: the shifts of the MoS_2 peaks upon electron irradiation are shown in Fig. 2.11(c), where the DFT simulations of the first-order Raman modes of the system and comparison with the experimental data indicated that these new modes are associated with S vacancies, Fig. 2.11 (c). Note that the material was found to be stable at vacancy concentrations of up to 6%.

Since the E' and A'_1 modes are already present in pristine MoS_2, it is clear that they are just modified upon introduction defects and do not correspond to local vibrational modes. Analytical potential calculations [161] showed that indeed S vacancies lead to modification of the host modes in the way that they become anti-resonant around the defect sites, Fig. 2.11(d). This confinement of the host modes leads to shift of the mode and activation

2.9. Point defects and vibrational properties of 2D materials from atomistic simulations

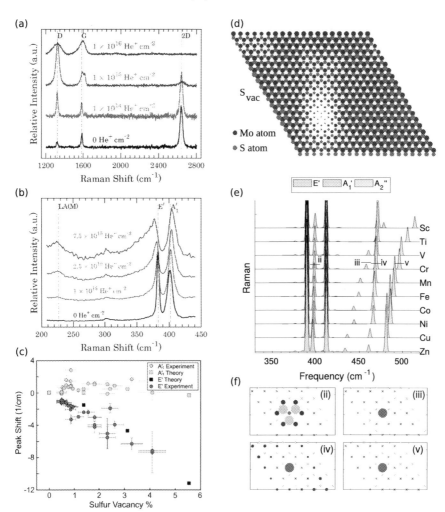

FIGURE 2.11 (a–b) Evolution of the Raman spectra of He$^+$ irradiated graphene (a) and MoS$_2$ (b). Reprinted with permission from Ref. [165]. (c) Raman peak shifts as functions of S vacancy concentration in 2D MoS$_2$: experiment and theory. Reprinted with permission from Ref. [164]. (d) E'-like mode in MoS$_2$ modified by the presence of S vacancy. Size of the circle denotes the mode amplitude at each atom. Reprinted with permission from Ref. [161]. (e) Simulated Raman spectra (via mass-approximation and using Raman intensity-weighted density of states) for 3d transition metal impurities substituting Mo site in MoS$_2$. (f) Illustration of the modes denoted in panel (e). Yellow, blue, and red circles denote Mo, S, and impurity atom. Reprinted with permission from Ref. [163].

of vibrational modes around the Γ-point, that yields the broadening. Some impurities can induce Raman-active local vibration modes. For instance, it was estimated in Ref. [163] via mass-approximation, that 3d metal impurities substitutional to Mo site in MoS$_2$ show two Raman-active peaks and two IR-active (A''$_2$-derived modes) that shift with the impurity atom mass and can thus be used to identify the impurity, Fig. 2.11(e,f).

2.9.2 Phonon contributions to defect-related photo-luminescence spectra in 2D materials

The excitations and de-excitations can be accompanied by a simultaneous creation or annihilation of phonons. This was already seen in the STS spectra presented in Fig. 2.3(a), where the asymmetric broadening of the defect-related states is due to vibronic excitations, i.e., tunneling to defect states assisted by creation of phonons, whereas the phonon annihilation process is unlikely at low temperatures. In a similar fashion, PL peaks may contain a pronounced phonon sideband, as shown in Fig. 2.12(a) in the case of defect-related PL feature in (bulk) SiC [166]. The phonon sideband arises from the emission from the vibronic ground state of the excited electronic state to the vibronic excited states of the electronic ground state. This is illustrated in Fig. 2.12(b) in the form a configuration coordinate (CC) diagram. The lower and upper parabola correspond to the ground and excited electronic states with the vibrational states shown as horizontal lines, and the configuration coordinate is a generalized coordinate describing the structural changes between the ground and excited state. The sideband features a maximum at a some characteristic vibronic state, or, the average number of phonons involved in the photon emission, called the Huang-Rhys (HR) factor. HR depends on how different the atomic structures are in the ground and excited electronic states, i.e., how far apart they are in the configuration coordinate diagram. The sharp peak corresponds to emission from vibronic ground state to vibronic ground state, i.e., involves no phonons, and thereby denoted zero-phonon line.

The PL lineshape with the phonon sideband can nowadays be calculated from first-principles, but remains computationally demanding [168], as it involves determination of the vibrational modes of the defective supercell, which may furthermore converge slowly with the increasing supercell size. Accurate description of excited states is also computationally demanding, as discussed above, but fortunately the two are fairly detached and can be possibly treated separately. In order to simulate the lineshape it is sufficient to have a proper description of the vibrational modes in the ground and excited state, but not the excitation energy or transition matrix element. In order to simulate the PL peak position and intensity it is sufficient to have a proper description of the electronic excitation, but the vibrations can be neglected.

Among 2DMs, the single-photon emitters are observed at least in h-BN and in WSe_2 [147]. h-BN in particular appears to host few SPEs of clearly distinct origin, but also those of similar origin can exhibit different emission energy due to e.g. strain variations in the sample, see an example in Fig. 2.12(c). Note also, that in this case the phonon sideband appears more complicated due to coupling to the acoustic and optical modes of pristine h-BN instead of just a single local vibrational mode [169]. As a result of all this, the identity of the SPEs in h-BN is still intensely debated [167,170–177]. In WSe_2, they likely arise from excitons localized due to strain variations, being quantum-dot like [147,178].

2.10 Conclusions and outlook

In this chapter, we gave a brief overview of the recent advances and the current understanding of the physics of defects in 2DMs with the main stress on the effects of the reduced

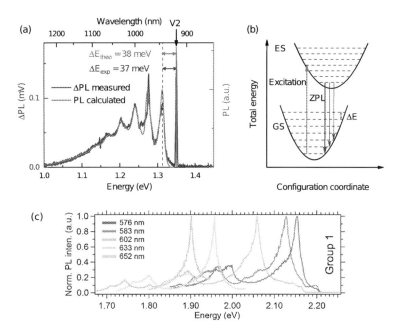

FIGURE 2.12 (a) An example of defect-related PL lineshape, experimental and calculated, from bulk SiC showing sharp zero-phonon line and the broad phonon sideband. (b) Configuration coordinate diagram. Reprinted with permission from Ref. [166]. (c) Selection of PL spectra from different defects in h-BN. Reprinted with permission from Ref. [167].

dimensionality on their behavior. We also discussed the changes which should be made in the theoretical description of native and extrinsic defects.

With regard to the defect types, the geometry of 2DMs excludes the existence of 2D or 3D defects, and also "decreases" the dimensionality of some defects: for example, grain boundary is a line defect in 2D, and edge dislocation is a point defect. Interstitial atoms do not exist in the majority of 2DMs (e.g. graphene, h-BN, MoS_2), as it is energetically more favorable for the atom to take an adatom position than be embedded into the atomic network. At the same time, defects in 2DMs, especially disclinations, can give rise to a large out-of-plane bending of the material.

The geometry of 2DMs means that the environment will strongly affect the behavior of the defects – they can appear, e.g., due to the interaction with the reactive species like oxygen molecules, or disappear, e.g., via self-healing of vacancies in graphene due to the dissociation of hydrocarbon molecules [58]. On the other hand, this makes it possible to easily create defects by chemical treatment or irradiation with energetic particles, ions and electrons, and extremely high concentrations of defects in the material are possible (e.g., concentration of vacancies in TMDs can be up to 20% [179]), being many orders of magnitude higher than the equilibrium concentration at the experimentally accessible temperatures. It is interesting that the crystal lattice imaged under electron microscope is still preserved in this case, contrary to bulk materials: it is of course possible to create a large number of defects in bulk solids by ion irradiation (e.g., in diamond), but vacancies will be accompanied by interstitials, which would make the structure amorphous. In 2DMs, as displaced atoms leave the system, the sample still presents a crystal with a high vacancy concentration, unless holes start appearing.

The environment gives rise to another very interesting aspect of the behavior of defects in 2D systems: the concentration of defects can be much more efficiently changed in 2DMs than in bulk systems. For example, while annealing of irradiated bulk samples normally gives rise to a decrease in vacancy concentration, it may drop or increase in 2DMs depending on the reactive species present. Annealing of 2D MoS_2 in S-rich atmosphere decreases defect concentration [180]. At the same time, although annealing in vacuum at moderate temperatures also deceases the number of defects, annealing at high temperatures gives rise to a higher concentration of vacancies due to thermal evaporation of S atoms [181]. Formation of new features upon annealing, e.g., inversion domains and mirror-twin boundaries in $MoTe_2$ presumably due to the loss of material is also possible [182]. Likewise, annealing of the defective 2D system in a particular environment opens effective routes to post-synthesis introduction of the impurities into the 2DMs [183] and thus facilitates the control over the doping level and electronic structure.

In this context, it should be pointed out that one of the defect-related issues which is not fully understood at the moment is the origin of the n-type doping frequently observed in the semiconducting 2D TMDs. It may originate from impurities [184] (e.g., Re in MoS_2 [69]), often present in TMD samples exfoliated from the bulk natural materials, or metallic contacts [185]. Chalcogen vacancies as a source of doping have also been widely discussed [115,186,187], but it appears that these defects alone cannot be the reason for the observed n-type doping. Indeed, as evident from Fig. 2.8(a,b) and other numerous DFT calculations [111,188,189], vacancies give rise to deep unoccupied states in the gap, i.e., they are acceptors and they cannot donate electrons. However, it is very plausible that various species can be adsorbed at the reactive dangling bonds at the atoms next to vacancies (e.g., hydrogen or nitrogen atoms) in 2D MoS_2 and other TMDs, and they can give rise to shallow occupied states close to CBM or empty states next to VBM [62,188,190], so that bare vacancies are most likely facilitators, but not the source of the doping [190].

The geometry of 2DMs also requires making changes in the theoretical description of charged point and line defects, as the screening/electrostatics is strongly anisotropic and inhomogeneous, and the results may heavily depend on the shape of the simulation cell and electrostatic correction scheme used. The effects of the environment should also be carefully accounted for, both in modifying the screening and also by choosing the values of atom chemical potentials matching the experimental situation.

The unique geometry of 2DMs (surface only) makes it possible to directly "see" the defects and get insight into their atomic structures using various microscopy techniques, such as STM or TEM. Together with other techniques, such as Raman spectroscopy and optical techniques, the identities of the relevant defects can be revealed, but is far from trivial. There is still work to be done to understand, e.g., the origins of the defect-related PL features observed in these materials, especially in those cases where large number of lines is seen such as in WSe_2 [178] and h-BN [170], which can arise, in addition to different defects, from strain variations or phonon-assisted emission processes. The situation is not helped by the fact that simulating these from first principles with predictive accuracy is still computationally very demanding. In addition, 2DMs are known to host also larger carrier complexes than exciton, such as trions and biexcitons. The interaction of these with defects is still largely unexplored territory, both computationally as well as experimentally.

Having direct access to the all the atomic sites of the 2DM from the environment provides unprecedented control over defect behavior, and makes it possible to tailor the characteristics of the defective material for particular applications – catalysis [191,192], optoelectronics [147, 193], and add new functionalities, e.g., memristive behavior [194] and magnetism [195–197], just to mention a few. Numerous examples of defect-mediated structure-property engineering are given in other chapters of this book.

Acknowledgment

The Authors would like to thank for many years of successful collaboration our coworkers, whose contribution to the findings presented in this chapter cannot be overestimated: Ute Kaiser, Kazu Suenaga, Yung-Chang Lin, Ossi Lehtinen, Torbjörn Björkman, Tibor Lehnert, Oleg V. Yazyev, Thomas Michely, Carsten Busse, Matthias Batzill, Jani Kotakoski, Marika Schleberger, Jannik Meyer, Stefan Facsko, Silvan Kretschmer, Mahdi Ghorbani-Asl, Arsalan Hashemi, Xiaohui Hu, Peter and Eli Sutter, Jeyakumar Karthikeyan, and Irina Grigor'eva.

References

[1] R. Tilley, Defects in Solids, Wiley, Canada, 2008.
[2] A. Walsh, A. Zunger, Nature Materials 16 (2017) 964.
[3] M. Stavola (Ed.), Identification of Defects in Semiconductors (Semiconductors and Semimetals, Vol. 51B), Academic Press, San Diego, California, USA, 1999.
[4] P. De Gennes, Superconductivity of Metals and Alloys, W.A. Benjamin, Inc., USA, 1966.
[5] C. Kittel, Introduction to Solid State Physics, John Wiley and Sons, Inc, USA, 2005.
[6] N.W. Ashcroft, N.D. Mermin, Solid State Physics, Harcourt College Publishers, Orlando, Florida, USA, 1976.
[7] K.S. Novoselov, D. Jiang, F. Schedin, T.J. Booth, V. Khotkevich, S. Morozov, A.K. Geim, Proceedings of the National Academy of Sciences of the United States of America 102 (2005) 10451.
[8] Q.H. Wang, K. Kalantar-Zadeh, A. Kis, J.N. Coleman, M.S. Strano, Nature Nanotechnology 7 (2012) 699.
[9] V. Nicolosi, M. Chhowalla, M.G. Kanatzidis, M.S. Strano, J.N. Coleman, Science 340 (2013) 1420.
[10] M. Chhowalla, H.S. Shin, G. Eda, L.-J. Li, K.P. Loh, H. Zhang, Nature Chemistry 5 (2013) 263.
[11] A.C. Ferrari, et al., Nanoscale 7 (2015) 4598.
[12] A. Pakdel, Y. Bando, D. Golberg, Chemical Society Reviews 43 (2014) 934.
[13] A. Kolobov, J. Tominaga, Two-Dimensional Transition Metal Dichalcogenides, Springer, Switzerland, 2016.
[14] X. Zou, B.I. Yakobson, Accounts of Chemical Research 48 (2015) 73.
[15] Z. Lin, B.R. Carvalho, E. Kahn, R. Lv, R. Rao, H. Terrones, M.A. Pimenta, M. Terrones, 2D Materials 3 (2016) 022002.
[16] Z. Hu, Z. Wu, C. Han, J. He, Z. Ni, W. Chen, Chemical Society Reviews 47 (2018) 3100.
[17] L. Chen, B. Liu, A.N. Abbas, Y. Ma, X. Fang, Y. Liu, C. Zhou, ACS Nano 8 (2014) 11543.
[18] P. Sutter, S. Wimer, E. Sutter, Nature 570 (2019) 354.
[19] Y. Liu, et al., Nature 570 (2019) 358.
[20] R. Telling, C. Ewels, A. El-Barbary, M. Heggie, Nature Materials 2 (2003) 333.
[21] M. Tang, L. Colombo, J. Zhu, T. Diaz de la Rubia, Physical Review B 55 (1997) 14279.
[22] A.J. Stone, D.J. Wales, Chemical Physics Letters 128 (1986) 501.
[23] J. Comtet, E. Glushkov, V. Navikas, J. Feng, V. Babenko, S. Hofmann, K. Watanabe, T. Taniguchi, A. Radenovic, Nano Letters 19 (2019) 2516.
[24] C.P. Ewels, M. Glerup, Journal of Nanoscience and Nanotechnology 5 (2005) 1345.
[25] A. Eckmann, A. Felten, A. Mishchenko, L. Britnell, R. Krupke, K.S. Novoselov, C. Casiraghi, Nano Letters 12 (2012) 3925.
[26] M. Lucchese, F. Stavale, E.M. Ferreira, C. Vilani, M. Moutinho, R.B. Capaz, C. Achete, A. Jorio, Carbon 48 (2010) 1592.

[27] D.A. Muller, Nature Materials 8 (2009) 263.

[28] H. Sawada, T. Sasaki, F. Hosokawa, K. Suenaga, Physical Review Letters 114 (2015) 166102.

[29] O.L. Krivanek, et al., Nature 464 (2010) 571.

[30] D. Kepaptsoglou, et al., ACS Nano 9 (2015) 11398.

[31] K. Suenaga, M. Koshino, Nature 468 (2010) 1088.

[32] Y.-C. Lin, P.-Y. Teng, P.-W. Chiu, K. Suenaga, Physical Review Letters 115 (2015) 206803.

[33] O.L. Krivanek, et al., Nature 514 (2014) 209.

[34] C. Hofer, C. Kramberger, M.R.A. Monazam, C. Mangler, A. Mittelberger, G. Argentero, J. Kotakoski, J.C. Meyer, 2D Materials 5 (2018) 045029.

[35] X. Tian, et al., Nature Materials 19 (2020) 867.

[36] S. Fang, Y. Wen, C.S. Allen, C. Ophus, G.G.D. Han, A.I. Kirkland, E. Kaxiras, J.H. Warner, Nature Communications 10 (2019) 1127.

[37] V. Boureau, B. Sklenard, R. McLeod, D. Ovchinnikov, D. Dumcenco, A. Kis, D. Cooper, ACS Nano 14 (2020) 524.

[38] B. Schuler, et al., Physical Review Letters 123 (2019) 076801.

[39] W. Jolie, et al., Physical Review X 9 (2019) 011055.

[40] Z. Li, F. Chen, Applied Physics Reviews 4 (2017) 011103.

[41] M. Schleberger, J. Kotakoski, Materials 11 (2018) 1885.

[42] G.-Y. Zhao, H. Deng, N. Tyree, M. Guy, A. Lisfi, Q. Peng, J.-A. Yan, C. Wang, Y. Lan, Applied Sciences 9 (2019) 678.

[43] R.C. Walker, T. Shi, E.C. Silva, I. Jovanovic, J.A. Robinson, Physica Status Solidi (A) Applications and Materials Science 213 (2016) 3065.

[44] M. Nastasi, J. Mayer, J. Hirvonen, Ion-Solid Interactions – Fundamentals and Applications, Cambridge University Press, Cambridge, Great Britain, 1996.

[45] R. Smith (Ed.), Atomic & Ion Collisions in Solids and at Surfaces: Theory, Simulation and Applications, Cambridge University Press, Cambridge, UK, 1997.

[46] A.V. Krasheninnikov, K. Nordlund, Journal of Applied Physics 107 (2010) 071301.

[47] J.C. Meyer, et al., Physical Review Letters 108 (2012) 196102.

[48] J. Kotakoski, A.V. Krasheninnikov, U. Kaiser, J.C. Meyer, Physical Review Letters 106 (2011) 105505.

[49] W. Li, X. Wang, X. Zhang, S. Zhao, H. Duan, J. Xue, Scientific Reports 5 (2015) 9935.

[50] M. Kalbac, O. Lehtinen, A.V. Krasheninnikov, J. Keinonen, Advanced Materials 25 (2013) 1004.

[51] G. Buchowicz, P.R. Stone, J.T. Robinson, C.D. Cress, J.W. Beeman, O.D. Dubon, Applied Physics Letters 98 (2011) 032102.

[52] D.S. Fox, et al., Nano Letters 15 (2015) 5307.

[53] Q. Ma, et al., Journal of Physics. Condensed Matter 25 (2013) 252201.

[54] S. Kretschmer, M. Maslov, S. Ghaderzadeh, M. Ghorbani-Asl, G. Hlawacek, A.V. Krasheninnikov, ACS Applied Materials & Interfaces 10 (2018) 30827.

[55] S. Standop, O. Lehtinen, C. Herbig, G. Lewes-Malandrakis, F. Craes, J. Kotakoski, T. Michely, A.V. Krasheninnikov, C. Busse, Nano Letters 13 (2013) 1948.

[56] C.L. Wu, H.T. Lin, H.A. Chen, S.Y. Lin, M.H. Shih, C.W. Pao, Materials Today Communications 17 (2018) 60.

[57] H. Cun, M. Iannuzzi, A. Hemmi, J. Osterwalder, T. Greber, Surface Science 634 (2015) 95.

[58] R. Zan, Q. Ramasse, U. Bangert, K.S. Novoselov, Nano Letters 12 (2012) 3936.

[59] J.C. Meyer, C. Kisielowski, R. Erni, M.D. Rossell, M.F. Crommie, A. Zettl, Nano Letters 8 (2008) 3582.

[60] F. Banhart, J. Kotakoski, A.V. Krasheninnikov, ACS Nano 5 (2011) 26.

[61] J.C. Meyer, A. Chuvilin, G. Algara-Siller, J. Biskupek, U. Kaiser, Nano Letters 9 (2009) 2683.

[62] H.-P. Komsa, J. Kotakoski, S. Kurasch, O. Lehtinen, U. Kaiser, A.V. Krasheninnikov, Physical Review Letters 109 (2012) 035503.

[63] Y.-C. Lin, et al., Nature Communications 6 (2015) 6736.

[64] J. Karthikeyan, H.-P. Komsa, M. Batzill, A.V. Krasheninnikov, Nano Letters 19 (2019) 4581.

[65] O. Lehtinen, S. Kurasch, A.V. Krasheninnikov, U. Kaiser, Nature Communications 4 (2013) 2098.

[66] A.W. Robertson, C.S. Allen, Y.a. Wu, K. He, J. Olivier, J. Neethling, A.I. Kirkland, J.H. Warner, Nature Communications 3 (2012) 1144.

[67] W. Zhou, X. Zou, S. Najmaei, Z. Liu, Y. Shi, J. Kong, J. Lou, P.M. Ajayan, B.I. Yakobson, J.-C. Idrobo, Nano Letters 13 (2013) 2615.

[68] J. Hong, et al., Nature Communications 6 (2015) 6293.

[69] Y.-C. Lin, D.O. Dumcenco, H.-P. Komsa, Y. Niimi, A.V. Krasheninnikov, Y.-S. Huang, K. Suenaga, Advanced Materials 26 (2014) 2857.

[70] A.L. Gibb, N. Alem, J.-h. Chen, K.J. Erickson, J. Ciston, A. Gautam, M. Linck, a. Zettl, Journal of the American Chemical Society 135 (2013) 6758.

[71] T. Björkman, S. Kurasch, O. Lehtinen, J. Kotakoski, O.V. Yazyev, A. Srivastava, V. Skakalova, J.H. Smet, U. Kaiser, A.V. Krasheninnikov, Scientific Reports 3 (2013) 3482.

[72] S. Kurasch, J. Kotakoski, O. Lehtinen, V. Skákalová, J. Smet, C.E. Krill, A.V. Krasheninnikov, U. Kaiser, Nano Letters 12 (2012) 3168.

[73] C. Gong, K. He, Q. Chen, A.W. Robertson, J.H. Warner, ACS Nano 10 (2016) 9165.

[74] P.M. Coelho, H.-P. Komsa, H. Coy Diaz, Y. Ma, A.V. Krasheninnikov, M. Batzill, ACS Nano 12 (2018) 3975.

[75] O.V. Yazyev, S.G. Louie, Physical Review B 81 (2010) 195420.

[76] O.V. Yazyev, Y.P. Chen, Nature Nanotechnology 9 (2014) 755.

[77] B.I. Yakobson, F. Ding, ACS Nano 5 (2011) 1569.

[78] H.-p. Komsa, S. Kurasch, O. Lehtinen, U. Kaiser, A.V. Krasheninnikov, Physical Review B 88 (2013) 035301.

[79] X. Liu, Z. Zhang, L. Wang, B.I. Yakobson, M.C. Hersam, Nature Materials 17 (2018) 783.

[80] A.V. Krasheninnikov, Nature Materials 17 (2018) 757.

[81] O. Lehtinen, et al., ACS Nano 9 (2015) 3274.

[82] L. Biró, P. Lambin, New Journal of Physics 15 (2013) 035024.

[83] P.Y. Huang, et al., Nature 469 (2011) 389.

[84] J. Lahiri, Y. Lin, P. Bozkurt, I.I. Oleynik, M. Batzill, Nature Nanotechnology 5 (2010) 326.

[85] A. Azizi, X. Zou, P. Ercius, Z. Zhang, A.L. Elías, N. Perea-López, G. Stone, M. Terrones, B.I. Yakobson, N. Alem, Nature Communications 5 (2014) 4867.

[86] S. Najmaei, Z. Liu, W. Zhou, X. Zou, G. Shi, S. Lei, B. Yakobson, J.-C. Idrobo, P.M. Ajayan, J. Lou, Nature Materials 12 (2013) 754.

[87] A.M. van der Zande, P.Y. Huang, D.A. Chenet, T.C. Berkelbach, Y. You, G.-h. Lee, T.F. Heinz, D.R. Reichman, D.A. Muller, J. Hone, Nature Materials 12 (2013) 554.

[88] H.-P. Komsa, A.V. Krasheninnikov, Advanced Electronic Materials 3 (2017) 1600468.

[89] O. Lehtinen, et al., ACS Nano 9 (2015) 3274.

[90] K.S. Yong, D.M. Otalvaro, I. Duchemin, M. Saeys, C. Joachim, Physical Review B 77 (2008) 205429.

[91] M. Gibertini, N. Marzari, Nano Letters 15 (2015) 6229.

[92] D. Le, T.S. Rahman, Journal of Physics. Condensed Matter 25 (2013) 312201.

[93] T.H. Ly, D.J. Perello, J. Zhao, Q. Deng, H. Kim, G.H. Han, S.H. Chae, H.Y. Jeong, Y.H. Lee, Nature Communications 7 (2016) 10426.

[94] S. Wang, G.-D. Lee, S. Lee, E. Yoon, J.H. Warner, ACS Nano 10 (2016), acsnano. 6b01673.

[95] Z. Zhang, X. Zou, V.H. Crespi, B. Yakobson, ACS nano 7 (2013) 10475.

[96] Y. Han, T. Hu, R. Li, J. Zhou, J. Dong, Physical Chemistry Chemical Physics 17 (2015) 3813.

[97] J. Lin, S.T. Pantelides, W. Zhou, ACS Nano 9 (2015) 5189.

[98] A. Glensk, B. Grabowski, T. Hickel, J. Neugebauer, Physical Review X 4 (2014) 011018.

[99] C. Freysoldt, B. Grabowski, T. Hickel, J. Neugebauer, G. Kresse, A. Janotti, C.G. Van de Walle, Reviews of Modern Physics 86 (2014) 253.

[100] Y. Mishin, M.R. Sorensen, A.F. Voter, Philosophical Magazine A 81 (2001) 2591.

[101] E.S. Penev, S. Bhowmick, A. Sadrzadeh, B.I. Yakobson, Nano Letters 12 (2012) 2441.

[102] Z. Zhang, Y. Yang, G. Gao, B.I. Yakobson, Angewandte Chemie – International Edition 54 (2015) 13022.

[103] M. Ghorbani-Asl, S. Kretschmer, D.E. Spearot, A.V. Krasheninnikov, 2D Materials 4 (2017) 25078.

[104] T.-L. Chan, M.L. Tiago, E. Kaxiras, J.R. Chelikowsky, Nano Letters 8 (2008) 596.

[105] H.-P. Komsa, T.T. Rantala, A. Pasquarello, Physical Review B 86 (2012) 045112.

[106] H.-P. Komsa, N. Berseneva, A.V. Krasheninnikov, R.M. Nieminen, Physical Review X 4 (2014) 031044.

[107] C. Freysoldt, J. Neugebauer, Physical Review B 97 (2018) 205425.

[108] M. Topsakal, S. Ciraci, Physical Review B 85 (2012) 045121.

[109] J. Repp, G. Meyer, S. Paavilainen, F.E. Olsson, M. Persson, Physical Review Letters 95 (2005) 225503.

[110] J. Xiao, K. Yang, D. Guo, T. Shen, H.-X. Deng, S.-S. Li, J.-W. Luo, S.-H. Wei, Physical Review B 101 (2020) 165306.

[111] H.-P. Komsa, A.V. Krasheninnikov, Physical Review B 91 (2015) 125304.

[112] A. Alkauskas, P. Broqvist, A. Pasquarello, Physical Review Letters 101 (2008) 046405.

[113] T.J. Smart, F. Wu, M. Govoni, Y. Ping, Physical Review Materials 2 (2018) 124002.
[114] W. Chen, A. Pasquarello, Journal of Physics: Condensed Matter 27 (2015) 133202.
[115] H. Qiu, et al., Nature Communications 4 (2013) 2642.
[116] M.H. Naik, M. Jain, Physical Review Materials 2 (2018) 084002.
[117] S. Refaely-Abramson, D.Y. Qiu, S.G. Louie, J.B. Neaton, Physical Review Letters 121 (2018) 167402.
[118] L. Hedin, Physical Review 139 (1965) A796.
[119] D. Golze, M. Dvorak, P. Rinke, Frontiers in Chemistry 7 (2019) 377.
[120] C. Attaccalite, M. Bockstedte, A. Marini, A. Rubio, L. Wirtz, Physical Review B 83 (2011) 144115.
[121] H.-P. Komsa, A.V. Krasheninnikov, Physical Review B 86 (2012) 241201.
[122] Y. Lin, X. Ling, L. Yu, S. Huang, A.L. Hsu, Y.-H. Lee, J. Kong, M.S. Dresselhaus, T. Palacios, Nano Letters 14 (2014) 5569.
[123] J. Ryou, Y.-S. Kim, S. KC, K. Cho, Scientific Reports 6 (2016) 29184.
[124] A. Raja, et al., Nature Communications 8 (2017) 15251.
[125] M. Drüppel, T. Deilmann, P. Krüger, M. Rohlfing, Nature Communications 8 (2017) 2117.
[126] P. Deák, B. Aradi, T. Frauenheim, E. Janzén, A. Gali, Physical Review B 81 (2010) 153203.
[127] M. Rohlfing, Physical Review Letters 108 (2012) 087402.
[128] M.I. Katsnelson, Graphene: Carbon in Two Dimensions, Cambridge University Press, Cambridge, 2012.
[129] J. Hong, C. Jin, J. Yuan, Z. Zhang, Advanced Materials 29 (2017) 1606434.
[130] Y. Liu, F. Xu, Z. Zhang, E.S. Penev, B.I. Yakobson, Nano Letters 14 (2014) 6782.
[131] M. Pandey, F.A. Rasmussen, K. Kuhar, T. Olsen, K.W. Jacobsen, K.S. Thygesen, Nano Letters 16 (2016) 2234.
[132] P. Mallet, F. Chiapello, H. Okuno, H. Boukari, M. Jamet, J.-Y. Veuillen, Physical Review Letters 125 (2020) 036802.
[133] C. Murray, C. van Efferen, W. Jolie, J.A. Fischer, J. Hall, A. Rosch, A.V. Krasheninnikov, H.-P. Komsa, T. Michely, ACS Nano 14 (2020) 9176.
[134] P. Yu, M. Cardona, Fundamentals of Semiconductors, Springer, Heidelberg, 2010.
[135] H. Peelaers, C.G. Van de Walle, Physical Review B 86 (2012) 241401.
[136] E. Fortin, F. Raga, Physical Review B 11 (1975) 905.
[137] Q.C. Sun, L. Yadgarov, R. Rosentsveig, G. Seifert, R. Tenne, J.L. Musfeldt, ACS Nano 7 (2013) 3506.
[138] F.D. Brandão, G.M. Ribeiro, P.H. Vaz, J.C. González, K. Krambrock, Journal of Applied Physics 119 (2016) 235701.
[139] J.-Y. Noh, H. Kim, M. Park, Y.-S. Kim, Physical Review B 92 (2015) 115431.
[140] K.S. Thygesen, 2D Materials 4 (2017) 022004.
[141] D.Y. Qiu, F.H. da Jornada, S.G. Louie, Physical Review B 93 (2016) 235435.
[142] M. Goryca, et al., Nature Communications 10 (2019) 4172.
[143] A. Kozhakhmetov, et al., Advanced Materials (2020) 2005159.
[144] H. Gao, et al., Nano Letters 20 (2020) 4095.
[145] M. Danovich, D.A. Ruiz-Tijerina, R.J. Hunt, M. Szyniszewski, N.D. Drummond, V.I. Fal'ko, Physical Review B 97 (2018) 195452.
[146] P. Cudazzo, I.V. Tokatly, A. Rubio, Physical Review B 84 (2011) 085406.
[147] M. Toth, I. Aharonovich, Annual Review of Physical Chemistry 70 (2019) 123.
[148] K.F. Mak, K. He, C. Lee, G.H. Lee, J. Hone, T.F. Heinz, J. Shan, Nature Materials 12 (2013) 207.
[149] C. Mai, A. Barrette, Y. Yu, Y.G. Semenov, K.W. Kim, L. Cao, K. Gundogdu, Nano Letters 14 (2014) 202.
[150] I. Kylänpää, H.-P. Komsa, Physical Review B 92 (2015) 205418.
[151] S. Tongay, et al., Scientific Reports 3 (2013) 2657.
[152] P.K. Chow, R.B. Jacobs-Gedrim, J. Gao, T.-m. Lu, B. Yu, H. Terrones, N. Koratkar, ACS Nano 9 (2015) 1520.
[153] N. Saigal, S. Ghosh, Applied Physics Letters 109 (2016) 122105.
[154] V. Carozo, et al., Science Advances 3 (2017) e1602813.
[155] K. Greben, S. Arora, M.G. Harats, K.I. Bolotin, Nano Letters 20 (2020) 2544.
[156] B. Ganchev, N. Drummond, I. Aleiner, V. Fal'ko, Physical Review Letters 114 (2015) 107401.
[157] Y. Wang, et al., Nano Letters 20 (2020) 2129.
[158] D. Xiao, G.-B. Liu, W. Feng, X. Xu, W. Yao, Physical Review Letters 108 (2012) 196802.
[159] K.F. Mak, K. He, J. Shan, T.F. Heinz, Nature Nanotechnology 7 (2012) 494.
[160] H. Fang, et al., Proceedings of the National Academy of Sciences 111 (2014) 6198.
[161] Z. Kou, A. Hashemi, M.J. Puska, A.V. Krasheninnikov, H.-P. Komsa, npj Computational Materials 6 (2020) 59.

[162] D. Porezag, M.R. Pederson, Physical Review B 54 (1996) 7830.

[163] A. Hashemi, A.V. Krasheninnikov, M. Puska, H.-P. Komsa, Physical Review Materials 3 (2019) 023806.

[164] W.M. Parkin, A. Balan, L. Liang, P.M. Das, M. Lamparski, C.H. Naylor, J.A. Rodríguez-Manzo, A.T. Johnson, V. Meunier, M. Drndić, ACS Nano 10 (2016) 4134.

[165] P. Maguire, et al., Physical Review B 98 (2018) 134109.

[166] Z. Shang, A. Hashemi, Y. Berencén, H.-P. Komsa, P. Erhart, S. Zhou, M. Helm, A.V. Krasheninnikov, G.V. Astakhov, Physical Review B 101 (2020) 144109.

[167] T.T. Tran, C. Elbadawi, D. Totonjian, C.J. Lobo, G. Grosso, H. Moon, D.R. Englund, M.J. Ford, I. Aharonovich, M. Toth, ACS Nano 10 (2016) 7331.

[168] A. Alkauskas, J.L. Lyons, D. Steiauf, C.G. Van de Walle, Physical Review Letters 109 (2012) 267401.

[169] D. Wigger, et al., 2D Materials 6 (2019) 035006.

[170] T.T. Tran, K. Bray, M.J. Ford, M. Toth, I. Aharonovich, Nature Nanotechnology 11 (2016) 37.

[171] S.A. Tawfik, S. Ali, M. Fronzi, M. Kianinia, T.T. Tran, C. Stampfl, I. Aharonovich, M. Toth, M.J. Ford, Nanoscale 9 (2017) 13575.

[172] M.E. Turiansky, A. Alkauskas, L.C. Bassett, C.G. Van de Walle, Physical Review Letters 123 (2019) 127401.

[173] A. Sajid, M.J. Ford, J.R. Reimers, Reports on Progress in Physics 83 (2020) 044501.

[174] A. Gottscholl, et al., Nature Materials 19 (2020) 540.

[175] F. Hayee, et al., Nature Materials 19 (2020) 534.

[176] M. Koperski, D. Vaclavkova, K. Watanabe, T. Taniguchi, K.S. Novoselov, M. Potemski, Proceedings of the National Academy of Sciences 117 (2020) 13214.

[177] Q. Wang, et al., Nano Letters 18 (2018) 6898.

[178] C. Chakraborty, L. Kinnischtzke, K.M. Goodfellow, R. Beams, A.N. Vamivakas, Nature Nanotechnology 10 (2015) 507.

[179] M. Mahjouri-Samani, et al., Nano Letters 16 (2016) 5213.

[180] M.Z. Xie, J.Y. Zhou, H. Ji, Y. Ye, X. Wang, K. Jiang, L.Y. Shang, Z.G. Hu, J.H. Chu, Applied Physics Letters 115 (2019) 121901.

[181] M. Liu, J. Shi, Y. Li, X. Zhou, D. Ma, Y. Qi, Y. Zhang, Z. Liu, Small 13 (2017) 1602967.

[182] H. Zhu, Q. Wang, L. Cheng, R. Addou, J. Kim, M.J. Kim, R.M. Wallace, ACS Nano 11 (2017) 11005.

[183] Q. Ma, et al., ACS Nano 8 (2014) 4672.

[184] R. Addou, et al., ACS Nano 9 (2015) 9124.

[185] C. Gong, L. Colombo, R.M. Wallace, K. Cho, Nano Letters 14 (2014) 1714.

[186] J. Suh, et al., Nano Letters 14 (2014) 6976.

[187] B.W. Baugher, H.O. Churchill, Y. Yang, P. Jarillo-Herrero, Nano Letters 13 (2013) 4212.

[188] A. Singh, A.K. Singh, Physical Review B 99 (2019) 121201.

[189] J.-Y. Noh, H. Kim, Y.-S. Kim, Physical Review B 89 (2014) 205417.

[190] R. Rao, V. Carozo, Y. Wang, A.E. Islam, N. Perea-Lopez, K. Fujisawa, V.H. Crespi, M. Terrones, B. Maruyama, 2D Materials 6 (2019) ab33ab.

[191] H. Li, et al., Nature Materials 15 (2016) 48.

[192] J. Zhang, Y. Zhao, X. Guo, C. Chen, C.-L. Dong, R.-S. Liu, C.-P. Han, Y. Li, Y. Gogotsi, G. Wang, Nature Catalysis 1 (2018) 985.

[193] L. Mennel, J. Symonowicz, S. Wachter, D.K. Polyushkin, A.J. Molina-Mendoza, T. Mueller, Nature 579 (2020) 62.

[194] V.K. Sangwan, D. Jariwala, I.S. Kim, K.-S. Chen, T.J. Marks, L.J. Lauhon, M.C. Hersam, Nature Nanotechnology 10 (2015) 403.

[195] P.M. Coelho, H.-P. Komsa, K. Lasek, V. Kalappattil, J. Karthikeyan, Y. Phan Manh-Huong, A.V. Krasheninnikov, M. Batzill, Advanced Electronic Materials 5 (2019) 1900044.

[196] R. Nair, M. Sepioni, I.-L. Tsai, O. Lehtinen, J. Keinonen, A.V. Krasheninnikov, T. Thomson, a.K. Geim, I.V. Grigorieva, Nature Physics 8 (2012) 199.

[197] R. Nair, I.-L. Tsai, M. Sepioni, O. Lehtinen, J. Keinonen, A.V. Krasheninnikov, A.H. Castro Neto, M.I. Katsnelson, A.K. Geim, I.V. Grigorieva, Nature Communications 4 (2013) 2010.

CHAPTER

3

Defects in two-dimensional elemental materials beyond graphene

Paola De Padova[a,b], Bruno Olivieri[c], Carlo Ottaviani[a], Claudio Quaresima[a], Yi Du[d], Mieczysław Jałochowski[e], and Mariusz Krawiec[e]

[a]Istituto di Struttura della Materia-CNR (ISM-CNR), Rome, Italy [b]LNF-INFN, Frascati, Rome, Italy [c]CNR-ISAC, Rome, Italy [d]ISEM-AIIM and Innovation Campus, University of Wollongong, Wollongong, NSW, Australia [e]Institute of Physics, Maria Curie-Sklodowska University, Lublin, Poland

3.1 Introduction

The awarding of the Nobel Prize for Physics 2010: *"for groundbreaking experiments regarding the two-dimensional material graphene"* [1] to the peculiar experimental discovery of graphene [2–4], one single layer of carbon atoms, contrary to the theories developed hitherto on the impossibility of the existence of two-dimensional (2D) atomic layers, unsupported free-crystals, for reasons related to the divergent contribution of thermal fluctuations in low-dimensional crystal lattices at any finite temperature [5–8], has mobilized scientific research worldwide toward the discovery of new elemental as well as multicomponent 2D materials.

This *per se* discovery, is further enhanced by the consideration that the graphene layer, interpreting the prototype of 2D materials, above all represents the *"antesignanus"* of all extraordinary related physical properties. In graphene, for example, there is the presence of the anomalous quantum Hall effect (QHE), occurring at half-integer filling factors [3,4], as well as ultra-fast electrons, having a $v_{Fermi} \approx 106 \text{ m s}^{-1}$, close to the speed of light. Charge carriers, mimicking relativistic particles with zero rest mass [3], exhibit the linear energy bands π and π^* near the Dirac points, representing the Dirac cones, which meet at the Fermi level, at the two inequivalent K and K′ points (the two A and B sublattices), of the graphene Brillouin zone [9,10].

Defects in Two-Dimensional Materials
https://doi.org/10.1016/B978-0-12-820292-0.00009-4

Silicon finds itself in Group-IVA immediately after carbon, and in its prototype form of 2D material, silicene, single atomic layer of silicon with a honeycomb buckled structure [11]. It was synthesized shortly thereafter, for the first time in 2012, in its prototypal form $3 \times 3/4 \times 4$ on a single crystal Ag (111), moving the gap below the Fermi level of 0.3 eV of its measured Dirac cone [11].

Currently, in addition to graphene, the elemental 2D materials family includes groups IVB, IIIA, IVA, VA and VIA, and in particular hafnene, borophene, gallenene, germanene, stanene, plumbene, phosphorene, arsenene, antimonene, bismuthene, selenene and tellurene. Each one of these new elemental 2D materials possesses intriguing emergent properties, both in fundamental and applied physics of extraordinary importance. More surprising, however, is the fact that none of these elemental 2D materials beyond graphene possesses the structural planarity found in graphene, due to their varying degrees of out-of-plane buckling, which can confer new electronic properties, as for instance, the gap opening with respect to their respective bulk crystals. Unlike carbon, which is stable in the form of sp^3, sp^2 and sp linear bonding, all of the new elemental 2D materials have very limited ability in changing their bonding nature, making their synthesis and stability a major challenge.

Defect control in bulk semiconductor crystals as well as homo- and hetero-epitaxy grown films has been extremely challenging, but the industry has mastered their control, thus enabling electronics engineering. Synthesis of 2D materials has met and is meeting the same challenges in controlling defects. The new 2D materials, today, are designed not only for basic physics, but also for their potential applications in electronics, spintronics, walleytronics and straintronics, where defects can be used to control their mechanical, electronic, and magnetic properties. Fundamental studies of this new class of 2D materials have been carried out to understand the effect of structural defects and their stability on their potential use in nanotechnological applications. Therefore, as will be exhaustively shown in this chapter, extensive research is being performed on the classification and study of defects in elemental 2D materials that could be exploited either to control the electronic properties or discover new properties or effects.

The motivation of this chapter is to provide a comprehensive overview of the most recent artificially synthesized and/or exfoliated 2D elemental materials beyond graphene, such as borophene, silicene, germanene, stanene, plumbene, phosphorene, arsenene, antimonene bismuthene, selenene and tellurene, in addition to the occurrence of their most popular point defects, such as Stone-Wales (SW) [12] single vacancy (SV), divacancies (DVs) and adatoms. Special emphasis will be placed on their state of the art, synthesis and structural defect stability and formation. By analyzing intentionally introduced point defects or "native point defects" in the two-dimensional lattices of such extraordinary materials theoretically and experimentally, one can learn how to modify or engineer their physical properties. The synthesis of gallerene and hafnene is also given.

At first it is important and necessary to establish a vocabulary of defects found in 2D materials. The first defects are the Stone-Wales [12] defects generated by a pure reconstruction of a graphene-like lattice (switching between pentagons, hexagons, and heptagons), no atoms are added or removed in defect reconstruction. When an atom or two atoms are removed from its lattice position, the lattice may relax into a lower energy state by changing the bonding geometry and creating a *single vacancy* and/or *divacancies*. The *adatom defect* involves a new rearrangement of the host lattice, as well as the number of bonds, eventually producing dangling bonds. Lastly, it is worth remembering that point defects can be created artificially, altering the relative properties of pristine 2D materials, under irradiation or heat treatment, where defects can also diffuse and precipitate at external surfaces or within the bulk.

This chapter reviews each 2D elemental material beyond graphene, from group-IIIA, -IVA, -VA and -VIA, presenting the state of the art of the synthesis, and the structural point defects. Furthermore, for each elemental 2D material we describe the structure and the history of its first synthesis and/or exfoliation.

An attempt will be made to report and comment on as much of the published literature as possible, reporting the most significant results. The reader is encouraged to consult the extensive literature cited in the chapter to gain a deeper understanding of the subject matter. In the case of the two elementary materials of group-VIA, selenene and tellurene, no significant literature is cited, having been synthesized only recently. This is also evident in the case of gallenene, group-IIIA, and hafnene, group-IVB, reported at the end of this chapter, where we introduce their synthesis and atomic structure, since, to our knowledge, there is no exhaustive literature concerning the investigation of the possible systematic structural defects formation. Given that 2D elemental materials are relatively new and extremely difficult to synthesize we report the experimental results when available, otherwise we report on the theoretical analysis of their defects.

3.2 Borophene

3.2.1 Synthesis and atomic structure

Boron is an element in the column just before carbon in the Periodic Table and is at the boundary between metals and insulators. It has three valence electrons, which would support its metallicity, nevertheless they are localized enough in a way that insulating states arise. All known structures of boron contain interlinked icosahedral B_{12} clusters, and up to now at least 16 bulk polymorphs have been classified [13]. The bulk rhombohedral form of β-B_{106} of elemental boron was first determined in 1957 by Sands et al., [14] from amorphous boron crystallized in a helium atmosphere after melting it in a "Heliarc" furnace. One year later, McCarty et al. [15] produced red microcrystalline B by pyrolytic decomposition at 1270 K of BI_3, a new crystalline modification of boron measured by x-ray diffraction with a simpler rhombohedral structure, space group $R - 3m$, a = 0.506 nm and $\alpha = 58°4'$. Above a temperature of 1770 K this new modification transforms into the rhombohedral form of boron previously reported by Sands [14]. A recent work revisited the history of boron, showing its extremely complicated chemistry and phase behavior [16,17], where the γ-B_{28} was predicted and observed. Oganov et al., [16] also observed the high-pressure ionic form of elemental boron whose most stable α-rhombohedral and β-rhombohedral phases under ambient conditions were also finally identified [18–20].

A structural transition from planar to tubular in boron clusters was observed experimentally, and theoretically predicted [21]. The boron clusters favor planar 2D structures up to 18 atoms, whereas preferred 3D structures begin with 20 atoms, which has been considered the embryo of the thinnest single-walled boron nanotubes, with a diameter of 0.52 nm [21]. Syn-

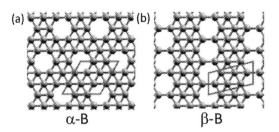

FIGURE 3.1 Draft of the models of sheets α-B and β-B composed by mixtures of hexagons and triangles. The two different unit cells are marked by solid red lines. Adapted from Fig. 1 of Ref. [23]. "Reprinted figure with permission from [H. Tang, et al., Phys. Rev. Lett., 2007, 99, 115501] Copyright (2007) by the American Physical Society."

thesis of single wall pure boron nanotubes with a larger diameter of approximately 3.6 nm has already been reported [22].

In 2007 Tang et al., [23] reported two new boron sheets, α-B and β-B, where hexagons form lines and the most stable phase, α-B, is composed of mixtures of triangular and hexagonal motifs. These structures in terms of a competition of two- and three-centers bonding, can explain the extreme stability of B_{80} fullerenes [24], composed of triangular motifs with pentagonal holes, due to the balance of these two and three center bonds. The α phase sheet can be considered the precursor of B_{80} [23], just as graphene is the precursor of carbon fullerenes [25]. Later, several authors computed the possibility of forming stable boron sheets and boron nanotubes and nanostructures [26–36].

Boron sheets can be obtained by carving different patterns of hexagonal holes within the triangular sheet [23,29]. Fig. 3.1 shows these new α-B and β-B sheets made up of mixtures of hexagons and triangles [23]. They are metallic, flat, and more stable than the buckled triangular B sheet [23]. α-B and β-B sheets were obtained by removing boron atoms from a flat triangular sheet [23].

The density of the hexagonal holes area was described by a global density parameter, η, defined as the ratio of the number of hexagonal holes to the number of atomic sites on the pristine triangular sheet within a unit cell of the decorated boron sheet [23,29]. They showed that the α sheet with a value of $\eta = 1/9$ was the best candidate for the stable boron monolayer sheet [23,29].

So far, no exfoliation process has been developed for B atoms. One could envision an exfoliation process from layered materials with a weak interlayer binding; however, no such type of boron material has been found to exist in nature, which rules out the exfoliation approach. Liu et al., [37] in 2013 investigated a possible fabrication method of 2D boron sheets. They explored the formation of B sheets on metal substrates (Cu, Ag, Au) and metal boride substrates (MgB_2, TiB_2) substrates using first-principles calculations, suggesting that B sheets can be grown on Ag(111) or Au(111) surfaces by B deposition [37]. A few years later, in 2015, the synthesis of borophene layers as an anisotropic 2D boron polymorph was obtained under ultrahigh vacuum (UHV) conditions on a clean Ag(111) surface by evaporating B from a solid source [38]. This paper demonstrated the first experimental evidence of borophene synthesis under UHV conditions on an Ag(111) substrate. During B growth, the silver substrate was maintained between 720 and 970 K under a boron flux between ~ 0.01 to ~ 0.1 ML/min. At a substrate temperature of 820 K and low coverage of B, the STM images revealed two distinct boron phases: a homogeneous phase and a more corrugated "striped" phase. The relative concentration of these phases strongly depends on the deposition rate at which the striped phase is favored (striped-phase B nanoribbons), while at higher deposition rates, homogeneous borophene islands are formed. They also found that Ag(111) was almost completely covered with $\sim$ 1ML of B but they also found B clusters. Furthermore, higher growth temperatures favored the striped phase, suggesting that the homogeneous phase is metastable. High-resolution STM images ($V_{sample} = + 0.1$ V, $I = 500$ pA) revealed a homogeneous phase, which appeared as atomic chains with a 0.30 nm periodicity, periodic vertical buckling, and short- and long-range Moiré rhombohedral patterns [38]. This atomically resolved homogeneous phase demonstrated a chain morphology and Moiré structure in addition to long line defects [38, and Supplementary Materials]. These 1D line defects, which are frequently observed in grown borophene, can appear as antiphase boundaries of interchain lateral distortions, resulting in a possible strain relief and/or phase transition from homogeneous to striped phase [38].

In addition the 2D boron sheet was found to be inert to oxidation, interacting only weakly with the Ag(111) substrate [39]. This was the first experimental evidence to link flat structure models containing hexagonal hollow superstructures with observations of borophene synthesis on Ag(111) substrates [39]. A comprehensive review of the 2D boron structures, properties and applications was reported by Zhang et al. [32]. Unlike the other synthesized 2D elemental materials, beyond graphene, where point defects have been extensively studied, for the newborn borophene sheet, this topic remains unexplored. However, several authors have started to study the role of defects in strain in 2D borophene sheets [40]; simulating vacancy mediated polymorphism [36]; the effect of temperature and strain-rate on mechanical properties [41]; the effect of lattice defects on mechanical anisotropy [42]; and finally, the role of domain boundary and vacancy defects on nanoscale undulations induced by borophene substrate of a single layer [43].

3.2.2 Defects in borophene

Line defects in borophene have been investigated at the atomic scale in ultra-high vacuum (UHV) by scanning tunneling microscopy/spectroscopy (STM/STS) and density functional theory (DFT) [44,45] on borophene synthesized on single crystal Ag(111) [46]. Upon appropriate growth conditions, the borophene phases are found to intermix [46] to accommodate line defects that overlap each other forming structures that match the building block of the other phase. These line defects energetically favor spatially periodic self-assembly that give rise to new phases of borophene, which, in the end, confuse the separation between the borophene sheet and defects [46]. This phenomenon can be considered unique to borophene due to its high in-plane anisotropy and potential to form several energetically and structurally similar polymorphs [46].

At substrate growth temperatures between 710 and 740 K, two distinct phases of borophene are observed, and correspond to structures $\nu_{1/6}$ and $\nu_{1/5}$ with concentrations of HH (hollow hexagons) $\nu = 1/6$ and $1/5$, respectively, where $\nu = n/N$, and n is the number of HHs in an otherwise triangular lattice with N lattice sites) [46].

Fig. 3.2(a) shows the STM image of a self-assembled borophene sheet grown on Ag(111) that contains domains with different periodic structures of $\nu_{1/6}$ and $\nu_{1/5}$ rows, which include

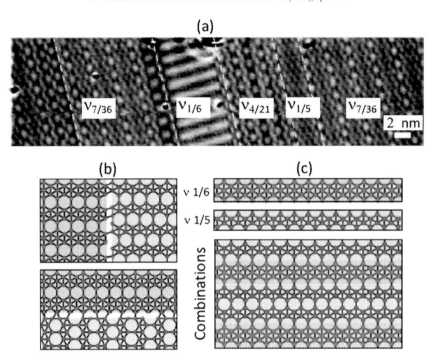

FIGURE 3.2 STM image of a self-assembly borophene sheet grown on Ag(111) containing domains with different periodic structures of $v_{1/6}$ and $v_{1/5}$ rows, which include two new phases of borophene, the $v_{7/36}$ and $v_{4/21}$ sheets (a). A graphic representation of interfacing $v_{1/6}$, shaded red, and $v_{1/5}$ sheets, shaded blue, with mismatched and perpendicular boron rows is shown in (b). The large lattice mismatch causes high interfacial energies. Structures of $v_{1/6}$, top, and $v_{1/5}$, middle, rows, and an example of a new boron phase formed by assembling $v_{1/6}$ and $v_{1/5}$ rows bottom (c). The dashed white lines in (a) represent the boundary of the defect lines of the different borophene phases, marked by the relative labels for the new phase of the borophenes sheet [46]. (See Fig. 8 of Supplementary Information for the structural models and simulated STM images of the $v_{4/21}$ and $v_{7/36}$ sheets). Adapted from Fig. 4 of Ref. [46].

two new phases of borophene, the $v_{7/36}$ and $v_{4/21}$ sheets [46]. In Fig. 3.2(b) it is possible to observe the models of a perfect lattice match along the HH rows at the phase boundaries with a low interface energy, while in (c) the lattice registry along these hollow hexagon rows indicates that the two $v_{1/6}$ and $v_{1/5}$ rows could act as building blocks to reassemble into new 2D borophene sheets. This can occur in part due to the established noncovalent nature of the interactions between borophene and the Ag(111) [46]. In the high-resolution STM image of Fig. 3.2(a), the borophene domains of the different regions defined by line defects with distinct periodic lengths, separated by white dashed lines, are clearly visible. The HH concentration of a periodic assembly of m $v_{1/6}$ and n $v_{1/5}$ units in a supercell can be written as $v = (m + n)/(6m + 5n)$. In this way, two new borophene polymorphs with supercells composed of one $v_{1/6}$ unit assembled with three and $v_{1/6}$ are formed [46]. The prototype case of two-dimensional borophene proposes the interesting issue that line defects can self-assemble into new crystalline phases, blurring the distinctions between perfect and defective crystals, noting that sometimes the defects should not be considered as such [47].

3.3 Silicene

3.3.1 Synthesis and atomic structure

Silicene, a sheet of one Si atom arranged in a buckled honeycomb structure, was the first elemental 2D material of the group-IVA elements to be synthesized [11] after graphene (a sheet of one C atom thick in a flat honeycomb lattice [2]). The aromatic stage of the 2D Si layer was predicted in 1994 by Takeda et al., [48] as a possibility of stage corrugation in Si and Ge analogous to graphite, and in 2007, Guzmán-Verri et al. [49] reported on the electronic properties of Si graphene-like sheets, coined here silicene. Subsequently, in 2009, almost in parallel, Lebègue et al., [50] and Changirov et al., [51] investigated the stability and electronic properties of the two- and one-dimensional honeycomb structures of Si and Ge. Note that planar forms of silicene and germanene, exact analog of graphene, although unstable due to the pseudo-Jahn-Teller (PJT) distortion, have also been predicted by Jose and Datta [52]. However, planar 2D structures of Si and Ge could possibly be synthesized epitaxially on an appropriate substrate to suppress the PJT effect [52].

The compelling experimental evidence for graphene-like 2D Si was obtained by depositing Si on a single Ag(111) crystal in an UHV at low temperature, 490 K, hosting the π Dirac electrons with linear behavior at the K point of the silicene honeycomb lattice [11]. Later, several authors also reported silicene synthesis on supported substrates [53–61].

The model of atomic structure for silicene is reported in Fig. 3.3(c) together with the STM filled-states from clean Ag(111) (1 × 1) surface, (a), and flower-like 3 × 3/4 × 4 silicene, (b). Fig. 3.3(d) on the other hand shows the model determined by using ab initio DFT calculations. A high-resolution STM image and its simulation considering the relaxed geometry and the calculated local electron density of states, according to the Tersoff-Hamann approach [62], are displayed in Fig. 3.3(e).

3.3.2 Defects in silicene

The local structures of six typical point defects in silicene are shown in Fig. 3.4 [63]. The atomic structures (upper panels) and their simulated STM images at $\pm$ 0.5 V_{bias} (lower panels), by *ab initio* calculations [44,45,64], are reported for various local point defects: (1) Stone-Wales (SW) (a); (2) single vacancy (SV)-1 (b) by (55|66) rings; (3) SV-2 (c) with three dangling atoms; divacancy (DV)-1(5|8|5) (d); (4) reconstructed DV-2(555|777) (e); and (5) with Si adatom (f); the numbers 5, 6, 7 and 8 stand for pentagon, hexagon, heptagon and octagon [63,65]. The local SW defect is formed by a 90° rotation of the Si-Si bond, like that of graphene [66]. For silicene, the STM images of simulated filled- and empty-states are different with asymmetric location of the bright spots (lower panels in a), increasing the difficulty in identifying this type of defect, because of the buckling structure of silicene. On the other hand, other point defects such as SV-1 (Fig. 3.4b), DV-1 (Fig. 3.4d), and DV-2 (Fig. 3.4e) give rise to similar phenomena, as observed in the simulated STM images [63]. The SVs in silicene are of two types, i.e. the (SV)-1 (Fig. 3.4b) by (55|66) rings with a hybridization of the central atom sp^3, and the SV-2 (Fig. 3.4c) that produces three dangling atoms, with a distance between A and B atoms of about 0.327 nm. The two types of DVs in silicene, DV-1(5|8|5) and DV-2 (555|777), have energy formation of 3.70 and 2.84 eV respectively, and which are 2.32 eV and 3.18 eV

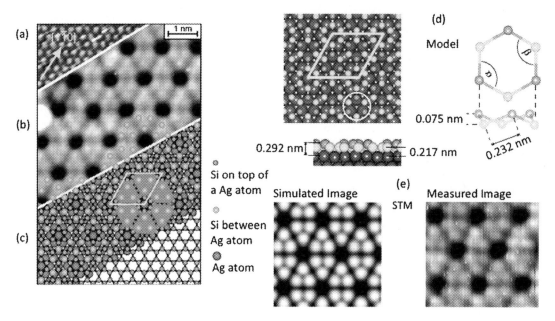

FIGURE 3.3 Filled-states STM images of the initial Ag(111) (1 × 1) ($U_{bias} = -0.2$ V, I = 1.93 nA) (a); (4 × 4) silicene sheet ($U_{bias} = -1.4$ V, I = 0.29 nA) (b). Model of silicene on Ag(111) (c). The Si larger orange balls lie on top of the Ag atoms. The ball-and-stick model for the freestanding silicene layer is visible in the bottom right corner of (c). Top and side view of fully relaxed atomic geometries DFT calculations of silicene (4 × 4) on Ag(111) (d); simulated STM image (left) (e) of the structure shown in (d). High-resolution STM image (right) (e), showing the same hexagonal arrangement of the triangular structure around dark centers as the simulated STM image. The yellow rhombus denotes silicene (4 × 4), with respect Ag(111) (1 × 1). Adapted from Figs. 4 and 5 of Ref. [11]. "Reprinted Figs. 4 and 5 with permission from [P. Vogt, P. De Padova, et al., Phys. Rev. Lett. 108, 155501, 2012] Copyright (2012) by the American Physical Society."

lower than those of two isolated SVs (3.01 eV) respectively, considering that the coalescence of SVs in DVs is highly probable. DV-1(5|8|5) can transform into the energetically favored defect DV-2(555|777) (difference in energy formation 0.86 eV). The localization of DV-1(5|8|5) is clearly shown by the four bright zones in the STM image (Fig. 3.4d), implying a strongly localized electron density around this defect [63]. The image of the DV-2(555|777) is easier to distinguish (with negative bias) due to the distinctive three bright large rings present in the STM image (Fig. 3.4e) [63]. In panel (f) of Fig. 3.4, the Si adatom in silicene is exothermic with a negative formation energy (−0.03 eV). The Si adatom prefers the top silicene site, forming a Si2 dimer, pushing down the original lattice Si embedded perpendicular to the silicene sheet. In this way the Si adatoms form three Si-Si bonds with a sp^3-like tetrahedron configuration, leaving an unpaired electron on top [63]. This Si adatom can be easily identified on both STM images, which show a very bright spot. It is important to note that the formation energies of the point defects in silicene are systemically lower than those of graphene [66].

Interestingly, it has been found that SW defects can be effectively recovered by thermal annealing and that SVs have much higher mobility than DVs, and that two SVs are more likely to coalesce into one DV to lower their energy. Defects such as SW and DVs could modify

3.3. Silicene

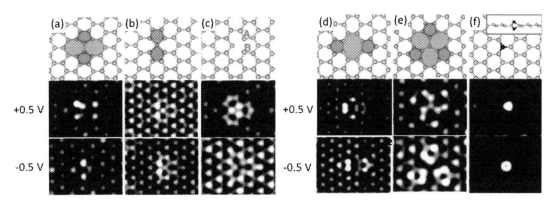

FIGURE 3.4 Silicene atomic structures (top) and their simulated STM images (bottom), including various local point defects with simulated ± 0.5 V bias: SW (a); SV-1 (b) by (55 | 66) rings; SV-2 (c) with three dangling atoms; DV-1(5 | 8 | 5) (d); reconstructed DV-2(555 | 777) (e); and (f) with Si adatom. Adapted from Fig. 1 of Ref. [63]. "Reproduced from Fig. 1 of Ref. [63] with permission from the Royal Society of Chemistry".

the electronic structure, inducing small gaps in the silicene sheet, while the SV defect leads to metallic silicene. Magnetism cannot be induced in silicene, although the magnetic ground state can be achieved at low concentration of point vacancies [63,67]. On the contrary adatoms can play a dramatic role in the electronic structure of silicene due to their stability and ability to act as self-dopant, inducing a long-range spin polarization with a large bandgap making a magnetic silicon semiconductor [63].

Since silicene has been widely investigated, several additional theoretical calculations have predicted the stability and influence of defects (vacancy clusters, extended line defects, double line defects and di-adatoms) on the electronic, mechanical and transport properties of silicene [68–72]. Other types of defects, namely deformed hexagons, have been found in planar silicene on Au(111) thin films [61]. These defects are generated by the frustration of Au adatoms supporting the silicene. Au adatoms on the bare Au(111) surface are located in fcc-hollow sites, and since the silicene lattice is incommensurate with the Au(111) lattice, the Au adatoms linking both subsystems tend to remain in accordance with each other, resulting in deformed hexagons. In addition, other types of common defects, such as SV-1, are also present that develop in pentagon formation (yellow color). Examples of defects are marked in Fig. 3.5.

It is important to add that, in 2016, a systematic investigation on the morphologies of point defects, such as SW, SV, DV, and adatom, has been reported, by means of first-principles calculations joined with STM observations in epitaxial silicene on Ag(111) crystal [73].

The formation energies and possible diffusion behavior of these defects in the three common silicene superstructures on Ag(111), i.e. (4×4), $(\sqrt{13} \times \sqrt{13})$, $(2\sqrt{3} \times 2\sqrt{3})$ were evaluated [73]. It was found that the first silicene reconstruction [11] exhibited the highest defect energy formation for all types of defects, while, on the contrary, the last two, were very highly defective, perfectly in line with the experimental observations [73,74].

52   3. Defects in two-dimensional elemental materials beyond graphene

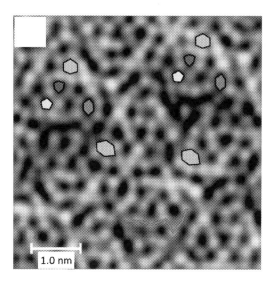

FIGURE 3.5 High-resolution STM image ($V_{bias} = -50$ mV, $I = 2.5$ nA) of silicene in a thin film of Au(111), showing a deformed and flat honeycomb lattice. Different colors mark various kinds of deformations and defects, mainly single vacancies, resulting in the formation of pentagons. Adapted from Fig. 2 of Ref. [61].

3.4 Germanene

3.4.1 Synthesis and atomic structure

Germanene was the first 2D material of group-IVA synthesized after silicene on several substrates, such as: Pt(111) [75], Au(111) [76], Ge$_2$Pt [77], Al(111), [78], MoS$_2$ [79], Cu(111) [80] and Ag(111) [81]. Previously, millimeter-scale crystals of a multilayered hydrogen-terminated germanium graphene analogue (germanane, GeH) from the topochemical deintercalation of CaGe$_2$ were obtained and exfoliated to form a single of a few-layers [82]. Subsequently methyl-terminated (GeCH$_3$) germanene was obtained by replacing the -H termination in GeH with -CH$_3$, with improved structural stability [83]. Ge had already been theoretically predicted to form a buckled honeycomb structure, in conjugation with elemental Si, in 1994, by Shiraishi et al., [48], and in 2009, by Cahangirov et al., [51] and Lebègue et al. [50]. Subsequently, in parallel, experimental synthesis of germanene on supported templates, such as Pt(111), Au(111), Ge$_2$Pt and Al(111) was reported [75–78]. A one atom-thin two-dimensional multiphase Ge film was grown in situ by molecular beam epitaxy on an Au(111) substrate kept at a temperature of ∼490 K and exposed to approximately 1ML of Ge. One of the three phases observed, the $\sqrt{3} \times \sqrt{3}$ reconstructed germanene layer on top of a $\sqrt{7} \times \sqrt{7}$ Au(111) surface, formed nearly flat large domains with a honeycomb structure, as shown in the STM images of Fig. 3.6(a), with in addition a LEED pattern (b) and its one sixth schematic illustration shown in (c) [76].

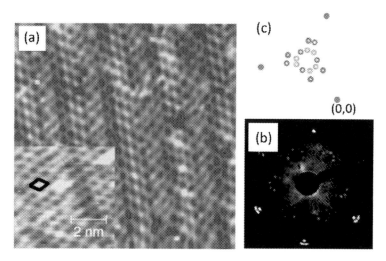

FIGURE 3.6 STM image (16.2 × 16.2) nm² from germanene; in the bottom left corner STM at higher resolution (sample bias: −1.12 V, 1.58 nA); the $\sqrt{7} \times \sqrt{7}$ unit cell is drawn in black) (a); LEED pattern taken on germanene at 59 eV (b); schematic illustration of one sixth of the LEED pattern: filled dots: (0,0) and 1 × 1 substrate integer order spots; the open circles are the spots corresponding to the $\sqrt{7} \times \sqrt{7}$ superstructure (in red), $\sqrt{19} \times \sqrt{19}$ (in green) and 5 × 5 (in blue) (c). Adapted from Fig. 1 of Ref. [76]. "Reprinted Fig. 1 with permission from [M. E. Dávila, et al., New J. Phys., 2014, 16, 095002, https://doi.org/10.1088/1367-2630/16/9/095002] Copyright (2014) licensed by Creative Commons Attribution 3.0 Unported Licence https://creativecommons.org/licenses/by/3.0/."

3.4.2 Defects in germanene

Ab-initio electronic structure and transport calculations of 2D hexagonal germanium have been performed [84]. Four possible structural defects such as SW, SV(5-9) and two (DV) (5-8-5 and 555-777), considered in graphene the most stable defects [66], have been calculated [84]. Fig. 3.7 shows the ball-and-stick model of germanene (a), and its four calculated types of structural defects are modeled in (b), (c), (d) and (e) [84]. The calculated lattice parameter for germanene was found to be 0.408 nm with a Ge-Ge bond of 2.460 Å, while the buckling height, defined as the difference between the z coordinates of sublattices A and B, was $\Delta = 0.07$ nm [84]. To perform calculations, defects are created by removing atoms from A and B of germanene sub-lattices; one atom from the A sublattice for the creation of a single vacancy and one atom from each of the A and B sublattices for the divacancy. In germanene the formation energy for the four defects SW, SV(5-9), DV(5-8-5) and DV(555-777) were 1.81, 2.31, 2.86 and 2.71 eV respectively [84], where the SW has the lowest formation energy, similar to the case of graphene. On the other hand, other defects are substantially different from those found in graphene. In germanene, when an atom is removed, as in Fig. 3.7(c), the SV has the lowest energy formation. In graphene the energy formation of $SV_{graphene}$ is 2.15 eV and is higher than that of the DVs defects [85]. This behavior is explained by the increased relaxation degree of freedom of the SV defect thanks to the buckled germanene structure. None of these point defects in germanene produces a net magnetic moment [85], unlike in graphene where a SV has a local magnetic moment due to the presence of dangling bonds in the reconstructed structure [86]. A SW defect is obtained by rotating of a pair of Ge atoms by 90°

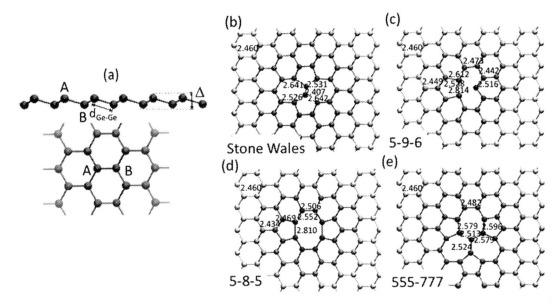

FIGURE 3.7 Ball-and-stick view of the optimized germanene structure (a), where the labels A and B individuate the atoms in the two different sublattices, d $_{Ge-Ge}$ is the distance between two germanium atoms and Δ is associated with the buckling of the atoms; Stone–Wales (55-77) (b); single vacancy (5-9-6) (c); divacancy (5-8-5) (d); divacancy (555-777) (e). The numbers indicated the lengths of the Ge–Ge bond, in Å. The more colorful region is associated with the portion of the crystal lattice that is most affected by the structural defects. Adapted from Figs. 1 and 2 of Ref. [84].

around the center of the Ge-Ge bond, still maintaining the buckled structure, strengthening the structure, and producing two pentagons and two heptagons instead of the four hexagons of germanene [84].

In Fig. 3.7(c) a pentagon–nonagon–hexagon (5-9-6) reconstruction is obtained by creating a SV, by removing two neighboring Ge atoms, DVs on the other hand are formed as pentagon–octagon–pentagon (5-8-5) reconstruction (d). A rotation of one of the bonds in the octagon of (5-8-5) transforms the defect into a reconstruction of three pentagons and three heptagons (555-777) [84].

Interestingly, these predicted defects in germanene show a local reconstruction that can be clearly identified by STM images [84] and should be easily observed experimentally. Among the defects investigated, the SW exhibited the lowest formation energy. In the presence of structural defects, germanene maintains its Dirac cone only when a single vacancy is present, when divacancies or SW defects are present the linear dispersion relation of the electrons, near the Fermi level, is destroyed. Furthermore, these defects create scattering centers, leading to a decrease in current of approximately 42% for the SW and SV, 55% for divacancies 5-8-5 and 68% for divacancies 555-777 [84]. Other authors reported how these defects can affect the structural, electronic, mechanical and transport properties of germanene, evidencing the importance of their investigation [87–89].

3.5 Stanene

3.5.1 Synthesis and atomic structure

After graphene, silicene, and germanene the next element to be investigated in the group IV column was Sn given the potential for new exotic electronic properties. Based on first-principles calculations, it was predicted that 2D Sn films can be a quantum spin Hall insulator with a large bulk gap of 0.3 eV, available for possible applications at room temperature (RT) and showing that QSH states can be tailored by chemical functionalization and external strains [90]. These results have already been observed by using the effective low-energy Hamiltonian involving spin-orbit coupling (SOC) in low-buckled, silicene and 2D Ge and Sn, showing that the effective SOC can open a gap at the Dirac points and establish the QSH effect [91]. Furthermore, the quantum anomalous Hall effect, giant magnetoresistance and perfect spin filter, topological superconductivity, as well as quantum thermal transport for 2D stanene [92–95] were predicted. The successful synthesis of 2D stanene was demonstrated in 2015 by Zhu et al., [96], on a Bi_2Te_3(111) substrate kept at RT by molecular beam epitaxy (MBE). Stanene films have also been grown on different substrates such as Sb(111) [97], InSb(111) [98], PbTe(111) [99], Ag(111) [100] after theoretically establishing that the hexagonal structure is stable [101], and finally on Sb_2Te_3 [102]. Recently, superconductivity in few-layer stanene was also measured [103].

The model of the atomic crystal structure of 2D stanene on Bi_2Te_3(111) is reported in Fig. 3.8 [96]. Stanene, herein, is composed of a biatomic layer of α-Sn(111), in which two triangular sublattices stack together, forming a buckled honeycomb lattice [96]. In this ultrathin Sn film no reconstruction was observed, while STM revealed the Sn(2 × 2) surface reconstruction of thick α-Sn(111) thick films grown heteroepitaxially on InSb(111)A [104]. The grey spheres in Figs. 3.8(a) and (b) are the surface Te atoms of the substrate, while the green and orang spheres represent the lower and upper Sn atoms in the biatomic-layer structure, respectively. In Fig. 3.8(a) the Sn atoms at the bottom atomic layer, called Sn-B (green spheres) are located at the face-centered cubic site of the Te layers, whereas the Sn atoms sit on the top atomic layer, called Sn-A (orange spheres) are located at the hollow position. Fig. 3.8(a) is the side view of the stanene crystal structure model, showing the Sn atoms in Sn-A and Sn-B stanene layers [96].

3.5.2 Defects in stanene

Although, in general, there are not many theoretical studies on the formation of defects in the honeycomb lattice of 2D Sn, a comprehensive classification and study of effects on the electronic and mechanical properties have recently been reported [105–107].

Ab initio calculations have been used to systematically investigate the structures, diffusion behavior and electronic properties of several point defects (SW, SV and DV) in stanene [107]. Fig. 3.9 reports on the geometric structure of a perfect low-buckled stanene structure (a) in the presence of different point defects such as: SW (b); SV-1(55|66) (c); SV-2(3|555) (d); DV-1(5|8|5) (e); and DV-2(555|777) (f). Rotation of a Sn-Sn bond by 90° produces a local SWl defect drawn in Fig. 3.9(b). The SV-1(55|66), which includes a hybridized central tin atom sp^3 has a formation energy of 3 meV lower than SV-2(3|555); by removing one atom from

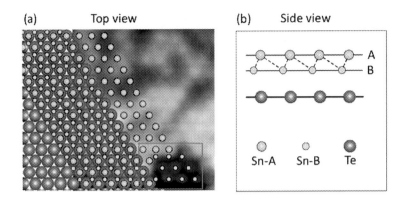

FIGURE 3.8 Top view of the stanene atomic structure model on Bi_2Te_3(111) superimposed on the substrate atoms (gray balls) and on the 2D stanene STM image; the red rectangle includes the bottom Sn atoms (a); side view of the stanene crystal structure model (b). Adapted from Fig. 2 of Ref. [96].

the perfect stanene, the optimization shows that the most stable single vacancy is the SV-2(3|555) defect, which has three bonding atoms. The two types of DVs in stanene, i.e., DV-1(5|8|5) and DV-2(555|777) can be obtained from the coalescence of two SVs, resulting in a considerable reduction of their formation energy [107].

The formation energies, E_f (1.21–1.8 eV), of all these defects in stanene, SW (1.21 eV), SV-1(1.58 eV), SV-2 (1.58 eV), DV-1 (1.75 eV) and DV-2 (1.80 eV) are found to be lower than those in graphene (4.5–8.7 eV) and silicene (2.09–3.77 eV). This is because the binding energy of stanene is smaller, 3.4 eV, compared to 7.90 and 3.96 eV of graphene [66] and silicene [61], respectively, concluding that these point defects are more easily formed in stanene than in graphene and silicene. In contrast, the case of SW in stanene was found to be similar to that of graphene and silicene and can be easily recovered by annealing due to its low reverse barrier [106]. It is worth noting that these point defects in 2D stanene exhibit a non-trivial influence on the electronic properties. The SW defect opens a direct gap in the energy band without reducing carrier mobility, the SV-1(55|66) induces the metallic state of stanene, and the SV-2(3|555), DV-1(5|8|5) and DV-2(555|777) change stanene to an indirect or direct band gap semiconductor [106].

Fig. 3.10 shows point and line defects atomically resolved by STM in ultra-flat stanene on Cu(111) [108]. Stanene was synthesized by depositing Sn atoms on Cu(111) in ultrahigh vacuum keeping the Cu(111) substrate at a low temperature of 200 K in order to obtain high-quality highly strained ultra-flat stanene growth. The ultra-flat honeycomb structure of stanene grown on Cu(111) allowed direct observation of defect structures at single-atomic level, thus opening the door to investigation of the importance of defect engineering in 2D stanene. Two typical point defects, monovacancy and divacancy, are shown in the high resolution STM images Fig. 3.10(a) and (b). The line defect of a stanene domain boundary is also evident in the HR-resolution STM images in Fig. 3.10(c) and (d), while the atomic stacking model is schematically depicted in Fig. 3.10(d), showing, both the lattice mismatch between domains and adsorption sites of Sn atoms. Each unit, red and blue spheres, at the boundary has an Sn atom with four-fold coordination [108].

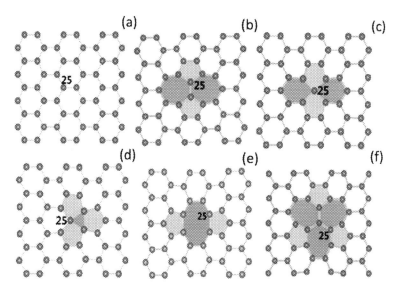

FIGURE 3.9 Geometric structures of perfect and defective stanene: perfect (a); SW (b); SV-1(55|66) (c); SV-2(3|555) (d); DV-1(5|8|5) (e); and DV-2(555|777) (f). (Sn atom number 25 is marked for easy viewing). Adapted from Fig. 1 of Ref. [106]. "Reprinted Fig. 1 with permission from [L. Shen, et al., RSC Advances, 7, 9840, 2017, https://doi.org/10.1039/C6RA28155A] Copyright (2017) licensed by Creative Commons Attribution 3.0 Unported License https://creativecommons.org/licenses/by/3.0/."

3.6 Plumbene

3.6.1 Synthesis and atomic structure

Pb is the heaviest element (Z=82) of group-IVA, and isoelectronic in valence shell with C, Si, Ge and Sn. As a graphene analogue with a honeycomb structure [2], plumbene, is the last of the supported 2D elemental materials, after silicene [11,53–61], germanene [75–81] and stanene [96,100]. The bilayer plumbene had already been theoretically predicted by using first-principle calculations in 2014 by Huang et al., [109] and later several other authors, predicted a class of new QSH insulators in X-decorated plumbene monolayers (PbX; X = H, F, Cl, Br, and I), with an extraordinarily non trivial giant gap in the range of 1.03 to 1.34 eV [110]. The promotion of plumbene to a topological insulator with a large bulk gap (~200 meV) through electron doping [111], its stability and extraordinary mechanical properties [112], as well as the discovery of a new quantum spin Hall phase in bilayer plumbene [113] were theoretically evaluated.

Recently plumbene was successfully experimentally synthesized on different substrates [114,115]. Yuhara et al., [114] reported large-area epitaxial growth of plumbene formed by a segregation process from $Pd_{1-x}Pb_x(111)$ alloys formed by depositing Pb thin films on Pd(111) at RT followed by heating and cooling back to RT. Fig. 3.11(a) reports high-resolution STM images of plumbene on a Pd(111) substrate prepared with 2.2 MLs of Pb deposited at RT on Pd(111) and annealed at 820 K. Fig. 3.11(b) shows the line profile along the white line traced

58 3. Defects in two-dimensional elemental materials beyond graphene

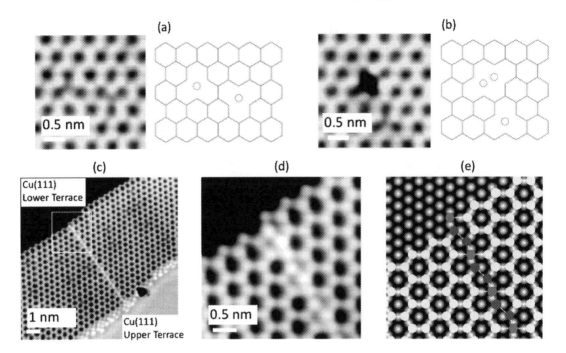

FIGURE 3.10 Atomically resolved STM images of two typical point defects: monovacancy (a) and divacancy (b); STM image (Vs = 0.5 V, I = 0.6 nA) of a boundary of the stanene domain (line defect) close to a step Cu(111) (c); the zoom-in STM image of c gives the detailed boundary structure (d). Atomic stacking model (e) of (d). Adapted from Fig. 4 of Ref. [108].

in the STM image (a). The measured unit cell is ≈ 0.48 nm, consistent with the observed LEED pattern (c), which clearly exhibits a 30° rotated ($\sqrt{3} \times \sqrt{3}$) superstructure, (1/3, 1/3) extra spots, red circles, in addition to (1 × 1) integer ones, yellow circles. A typical structural model of a plumbene overlayer on Pd(111) based on a ($\sqrt{3} \times \sqrt{3}$)R30° unit cell is shown in Fig. 3.11(d) [114].

There is experimental and theoretical evidence of plumbene synthesis on Fe layer on Ir(111) interface as reported by Bihlmayer et al., [115], where the selective hybridization of the Pb p_z states is observed and the Fe minority d states are found to stabilize the flat honeycomb structure of plumbene.

3.6.2 Defects in plumbene

Fig. 3.12 shows STM constant-current images from Pb on Fe/Ir(111) at bias voltages such that defects in plumbene are clearly visible (a–d). The plumbene structure model is drawn in (e) where the defects are located on one of the sublattice Pb atom sites as Pb vacancies. The primitive structural p(2 × 2) primitive unit cell (see white dashed diamond) contains two adsorbed Pb atoms at the fcc and a hcp sites (e), while blue dashed circles mark the position of the Pb vacancy. Note that, the Fe monolayer grows pseudomorphically in fcc stacking on

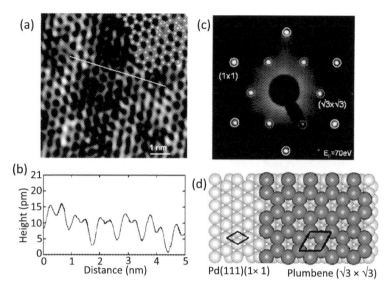

FIGURE 3.11 High-resolution STM image (Us = +0.6 V, I = 200 pA) of plumbene on Pd(111) substrate (a) and line profile (b) along the white line in (a); LEED pattern at Ep=70 eV (c); and top view of the structural model of ($\sqrt{3} \times \sqrt{3}$)R30° plumbene on Pd(111) (d). The red and gray balls are Pb and Pd atoms, respectively. The ($\sqrt{3} \times \sqrt{3}$)R30° of plumbene and the (1 × 1) of Pd(111) are marked. Adapted from Fig. 1 of Ref. [114].

the Ir(111) substrate and the Pb superstructures are commensurate with the substrate, and Pb atoms reside at specific adsorption sites of the Fe/Ir surface [115]. Although, it is difficult to determine whether the fcc or hcp hollow site Pb atoms show the vacancies, the unique direction of triangular features in Fig. 3.12(a–d) show that only one of the hollow sites can allocate the vacancy defect. It is worth noting that a comparable number of defects were observed in all sample preparations, suggesting that they are generated to accommodate the strain relief within the 2D plumbene. Gray scale insets in HR-STM images (c) and (d) show the vacuum DOS obtained from DFT calculations of a p(2 × 2) plumbene honeycomb lattice with a single vacancy [115]. A high density defect density can naturally influence the DOS by modifying the electronic structure of the plumbene, as calculated for a p(2 × 2) supercell of freestanding plumbene with the same structure as obtained for Pb/Fe/Ir(111) in the presene of a vacancy (Supplemental Material of Ref. [115]).

3.7 Phosphorene

3.7.1 Synthesis and atomic structure

Phosphorus (P) is the first pnictogen of the N group, VA, to be solid under ambient conditions. It has several allotropes such as white, red, violet and black where the color corresponds to the fundamental gap [116–119]. At ambient conditions, black phosphorus (BP) is the most thermodynamically stable allotrope with a layered orthorhombic structure, belonging to the

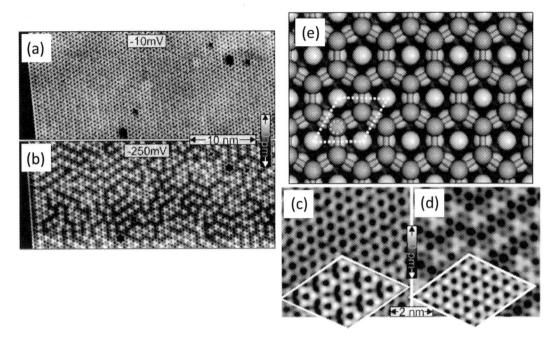

FIGURE 3.12 STM constant-current images (I = 1 nA, T = 8.2 K) of plumbene in Fe/Ir(111) measured at bias voltages of −10 mV (a) and −250 mV (b). In (b) a large density of triangular shaped defects of the honeycomb plumbene lattice is displayed. (c) and (d) magnified views of (a) and (b) STM images, showing in (c) dark triangle defects pointing to the left and in (d) brighter triangles pointing to the right. The insets are the gray-scale simulated density of states STM constant current images obtained from DFT calculations of a p(2 × 2) plumbene unit cell having one missing Pb atom, (e) shows the plumbene structure model with the white dashed p(2 × 2) Pb unit cell. Adapted from Fig. S1 of the Supplemental Material and Fig. 2 of Ref. [115]. "Reprinted Fig. S1 and Fig. 2 with permission from [G. Bihlmayer et al., Phys. Rev. Lett. 124, 126401, 2020] Copyright (2020) by the American Physical Society."

Cmca space group determined back in 1935 [117]. Black phosphorus consists of double layers; the cell is side-centered orthorhombic with a = 0.331 nm, b = 0.438 nm and c = 0.105 nm, with eight atoms in the unit cell [117]. Atoms are arranged in parallel puckered double-floor atomic layers named α-phase. In the individual atomic layer, each P atom has 5 valence electrons from the $3s$ and $3p$ orbitals, joined through covalent bonds to the three neighbors at the same distance as the interatomic distance observed in the P_4 molecule (two covalent bonds lie parallel to the atomic plane and the third one nearly perpendicular makes the connection of phosphorus atoms between the upper and lower layers of the puckered structure [117,119]. Layers are held together by van der Waals forces [120,121]. Bulk black phosphorus exhibits high carrier mobility and a small band gap of 0.33 eV as measured by Keyes [121] on samples prepared from white phosphorus using the method described by Bridgman [122]. The samples were prepared by an irreversible transformation of white P, by applying a hydrostatic pressure of about 12,000 kg/cm^2 at a temperature of 470 K for several minutes. Later, Sun et al., [123] obtained BP both from white and red phosphorus powders, using the high pressure and higher temperature method, i.e. under pressures of 2–5 GPa and temperatures of 470–1070 K for 15 min [123].

In 1963 Jamieson [124] demonstrated that under about 5 GPa BP transforms into a semimetallic layered rhombohedral structure, the β-phase, with space group $R - 3m$. Black phosphorus field-effect transistor was realized by Li et al., [125] in 2014 using few layers black phosphorus crystals with a thickness of a few nanometers. These samples were prepared by mechanical exfoliation from bulk black phosphorus crystals grown under high pressure (about 10 Kbar) and high temperature (1270 K).

The blue phosphorus, observed as the A7 phase of P [124] shares its layered structure and high stability with the black phosphorus allotrope [126]. The hexagonal structure in the plane and the stacking of the bulk layer of the structure are related to graphite, but unlike graphite and black phosphorus, blue phosphorus displays a wide fundamental band gap (around 2 eV) [126]. Thereafter, a new class of monolayer phosphorus allotropes with a non-honeycomb structure and nine new phosphorene polymorphs were predicted [127–129]. Among all these P polymorphs, the puckered (α-phosphorene), deriving from the atomic bilayers of BP, and the buckled (β-phosphorene) monolayer from the blue phosphorus crystals, are the most common allotropes.

There are many synthesis methods to prepare mono and multilayer phosphorene, such as mechanical, liquid and electrochemical exfoliation [125,130–135], plasma assisted and pulsed laser deposition [136,137], and finally chemical vapor deposition [138] with the mechanical exfoliation process the most promising method for devices fabrication [125,131–133].

Epitaxial growth of single layer blue phosphorus (blue phosphorene) was recently shown on Au(111) crystal [139], as well as on Te monolayer functionalized on Au(111) [140]. Fig. 3.13(a) and (b) show the very ordered single layer of phosphorus on Au(111) [139], where the yellow rhombus marks its unit cell with a = b = 14.7 Å; (c) is the line profile traced in (a) and (d) is the optimized structural model for blue phosphorus, with the black rhombus cell unit b1 = b2 = 0.328 nm [139]. The structure of Fig. 3.13 (a) is very similar to the flower-like pattern (see Fig. 3.3) of single layer silicene on Ag(111), the buckled hexagonal arrangement of silicon atoms [11].

3.7.2 Defects in phosphorene

The stability and electronic structures of point defects in phosphorene were systematically investigated by Hu et al., [141] by using density functional theory (DFT) and ab initio molecular dynamics calculations. They found that phosphorene has a wide variety of defects. The different geometric structures of these point defects in phosphorene are depicted in Fig. 3.14 like SW-1 (b), SW-2 (c), SV-(5|9) (d), SV-(55|66) (e), DV-(5|8|5)-1 (f), DV-(5|8|5)-2 (g), DV-(4|10|4) (h), DV-(555|777)-1 (i), DV-(555|777)-2 (j), and DV-(555|777)-3 (k). Fig. 3.14(a) shows the geometric structure of the 5×7 supercell of perfect phosphorene, while the red and blue balls indicate the phosphorus atoms in the two different upper and lower layers of phosphorene. Phosphorus atoms are marked by progressive numbers from top to bottom along with the zig-zag P chains [141].

Point defects in phosphorene are easier to form than in graphene and silicene, due to its low-symmetry atomic lattice [141]; perfect phosphorene (lattice without defects) resulting in less stable with a lower cohesive energy of 3.48 eV/atom [141] compared to graphene

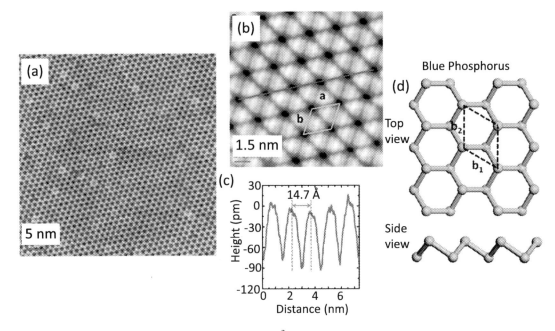

FIGURE 3.13 STM image (Vtip = 0.9 V, 100 × 100 nm^2) (a), and (b) high-resolution STM image (Vtip = 1.0 V, 8 × 8 nm^2) of single layer phosphorus on Au(111). The yellow rhombus marks the unit cell of phosphorus with a = b = 14.7 Å; (c) is the profile of the line along the red line in panel (a), exhibiting the 1.47 nm distance between the dark holes, while (d) is the atomic model of blue phosphorus. The black rhombus is the unit cell, b1 and b2 are the zigzag edges of blue phosphorus, with b1 = b2 = 0.328 nm. Adapted from Figs. 1 and 2 of Ref. [139]. "Reprinted Fig. 1 and Fig. 2 with permission from ACS [J. L. Zhang et al., Nano Lett. 16, 4903, 2016] Copyright (2016) by the American Chemical Society."

(7.90 eV/atom) [68] and silicene (3.96 eV/atom) [65]. Moreover, these defects are easy to distinguish and correlate with the defective atomic structure of phosphorene through simulated STM images [141].

Depending on the types of defects, the electronic and magnetic properties of defective phosphorene can be dramatically modified. SW, DV-(5|8|5)-1, DV-(555|777), and DV-(4|10|4) type defects produced few changes in phosphorene electronic structures, showing a similar band gap values of about 1 eV as found in perfect phosphorene (E_g = 0.905 eV for perfect h-P). On the other hand, the SV-(5|9) and DV-(5|8|5)-2 defects introduce unoccupied localized states into the fundamental band gap opening the gap by up to 3.041 eV; the SV-(5|9) and SV-(55|66) defects induce hole doping and finally, the SV-(5|9) defect can induce a local magnetic moment in phosphorene different from all other defects [141].

It is worth noting that, recently, Kiraly et al., [142] reported the depth profile of several black phosphorus monolayers, to locate the single atomic vacancies by STM and STS investigations combined with calculation of tight-binding electronic structures. They assigned the vacancies to a specific sublattice within the unit cell and showed that single vacancies exhibited strong anisotropic and highly delocalized charge density, with lateral extension up to 20 atomic unit cells [142]. As a consequence of the presence of these vacancies, STS revealed

3.7. Phosphorene 63

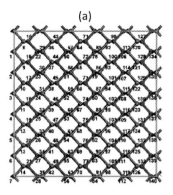

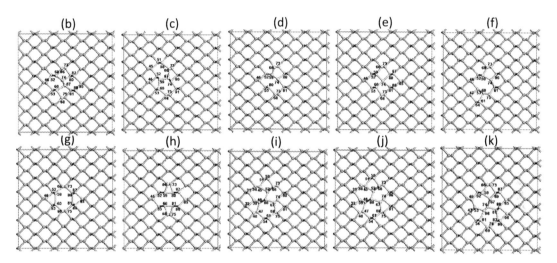

FIGURE 3.14 Geometric structures of 5 × 7 supercell of perfect phosphorene (a); defective phosphorene in: SW-1 (b), SW-2 (c), SV-(5|9) (d), SV-(55|66) (e), DV-(5|8|5)-1 (f), DV-(5|8|5)-2 (g), DV-(4|10|4) (h), DV-(555|777)-1 (i), DV-(555|777)-2 (j), and DV-(555|777)-3 (k) defects. In panel a, the red and blue balls indicate the phosphorus atoms in the two different upper and lower layers of phosphorene. Phosphorus atoms are labeled in increasing numbers. In panels b–h the violet balls are the unaffected phosphorus atoms and the yellow balls are the affected phosphorus atoms in the presence of defects. Adapted from Fig. 1 of Ref. [141]. "Reprinted Fig. 1 with permission from [W. Hu and J. Yang, J. Phys. Chem. C, 119, 20474, 2015] Copyright (2015) by the American Chemical Society."

in-gap resonance states near the valence band edge and strong doping of the bulk BP crystal [142]. Recently, point defects in blue phosphorene have also been investigated by using first-principles calculations [143].

Interesting, the patterning in phosphorene of antidot lattices, which are periodic arrays of holes or nanopores etched in the material, has been used to tailor the electronic properties, by tuning the band gap [144]. Native defects in phosphorus vacancies have been shown to lead to a p-doping in cleaved BP crystals [145]. Annealing of exfoliated BP up to 820 K can also produce characteristic well-aligned craters with their long axis aligned along the zigzag

64 3. Defects in two-dimensional elemental materials beyond graphene

direction [146]. Several other authors also theoretically predicted the formation of defects in phosphorene and their effects on the mechanical properties of phosphorene [147–149].

3.8 Arsenene (h-As) and Antimonene (h-Sb)

3.8.1 Synthesis and atomic structure

As, Sb and Bi are the third, fourth and fifth elements of the VA group of the periodic table. These heavy pnictogens (As, Sb, and Bi) crystallize in a rhombohedral layered structure (space group $R - 3m$) β-phase, this being the most stable allotrope, although several other metastable structures for As and Sb have been reported [150]. Natural bulk As and Sb have three allotropes, which are metallic gray, yellow and black arsenic; gray antimony, black antimony and the explosive antimony phase [150]. The rhombohedral structure (β-phase) of gray arsenic and gray antimony are similar and are the most stable phase. The yellow (nonmetallic) forms of As and Sb are metastable and are prepared by condensing their vapor at very low temperatures [150,151]. Arsenic also forms a polymorph isostructural with black phosphorus [119,132], when prepared by heating amorphous As at 370–445 K in the presence of Hg vapor [151–153], resulting in a layered orthorhombic α-phase (space group $Cmca$). These very old methods used a Hg vapor catalyst for glassy arsenic crystallization [152,153]. In fact, black antimony has the same structure as red phosphorus and explosive antimony often transforms violently into gray antimony under mechanical stress or heating [150].

In 2012 Zhang et al. [154] demonstrated, by first-principles calculations, that the topological and electronic (called topoelectronic) properties of Sb(111) nanofilms can be affected by an interplay between quantum confinement and surface effects, leading to several topoelectronic transitions, as the film thickness is reduced to a few layers. At 7.8 nm (22 bilayer) the phase transition from topological semimetal to a topological insulator was expected, as well as the transformation to a quantum spin Hall phase at 2.7 nm (8 bilayer), and to a normal (topologically trivial) semiconductor at 1.0 nm (3 bilayer) was predicted [154]. On the other hand, electronic structures and band topology, by further thickness reduction to a Sb(111) bilayer in the buckled honeycomb configuration showed a nontrivial topological insulating phase induced by tensile strain, thus opening the way for possible QSH state for Sb thin films using first-principle calculation [155].

After the success of phosphorene synthesis, several authors [129,156–159] have theoretically predicted that these other pnictogens, As, Sb and Bi, could give stable single-layered materials derived from bulk gray arsenic, gray antimony, and bismuth and thus named arsenene (h-As), antimonene (h-Sb) [119,156–158] and bismuthene (h-Bi) [129]. The equilibrium geometry and electronic properties of layered gray As thin films have been studied by Zhu et al., [156] using *ab initio* DFT. They found that, at variance with bulk gray As, layered As is semi-metallic [160], and thin films of As show a large band gap ($E_g \approx 1.71$ eV, probably underestimated by $> \sim 0.4$ eV due to DFT band gap calculation extraction difficulties) that can be influenced by several parameters such as the number of layers, in-layer compressive and/or tensile strains, in layer stacking as well as interlayer distance [156]. Fig. 3.15(a) shows the side view of the (ABC) stack of both the most stable calculated layered structure of bulk rhombohedral gray arsenic and the top view (b) of the buckled honeycomb structure for a

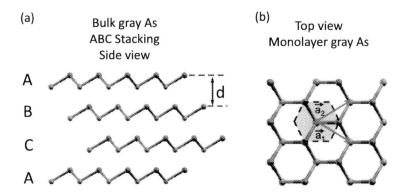

FIGURE 3.15 (a) Side view of the (ABC) stacked layered structure of bulk rhombohedral gray arsenic; (b) buckled honeycomb structure for 1 monolayer of a gray arsenic. Different colors are attributed to atoms at the top and bottom of the nonplanar layers; the shaded region marks the Wigner-Seitz cell. Adapted from Fig. 1 of Ref. [156].

gray arsenic monolayer [156]. According to experimental x-ray diffraction data [161] the calculated interlayer separation in the ABC-stacked bulk system results d = 0.358 nm, while the isolated As monolayer (ML) structure gives an in-plane lattice vectors a = $|a_1|$ = $|a_2|$ = 0.364 nm, very close to that of the bulk structure, a $_{bulk}$ = 0.385 nm, attributing to the equilibrium honeycomb geometry a small buckling [156], approaching the honeycomb lattice of graphene with two atoms per unit cell. The interatomic interactions within the 1 ML of As, on the other hand, are covalent, resulting in a nearest-neighbor distance of 0.253 nm.

Kamal et al., [157] found similar results by means of first-principles DFT calculations showing that arsenene has two stable types of honeycomb structures, buckled- and puckered-type. Both arsenene structures possess indirect gaps tuned by applying a strain. Furthermore, Zhang et al., [162] while identifying these two novel 2D wideband gap semiconductors namely 1 ML As, arsenene, and 1 ML Sb, antimonene concluded that these honeycomb structures are very stable, with bulk-to-monolayer transitions accompanied by large changes in electronic properties, i.e. from semi-metallic to semiconducting with an indirect band-gap of 2.49 and 2.28 eV, respectively. The monolayer β-phase of antimonene consists of buckled hexagonal rings composed of Sb atoms connected through sp^3 bonding, while the bulk antimony can be regarded as formed by ABC stacking of monolayer antimonene [158,162]. The orthorhombic structure (α-phase) is another common allotrope of the bulk Sb similar to that of black phosphorus.

First-principles theoretical calculations reported by Wang et al. [158] predicted several possible allotropes of monolayer antimonene, establishing that both free-standing α-Sb, puckered structure with two atomic sublayers, and β-Sb, buckled hexagonal lattice, and allotropes of antimonene are found stable and semiconducting [158]. The β-Sb phase with a hexagonal lattice and a buckled surface similar to what was predicted for β-P, was found to have bond lengths between neighboring Sb atoms of 0.284 nm and a bond angle of 89.9° with the lattice constant a = 0.412 nm [159], in agreement with the theoretical calculations of Ref. [162].

To date, from an experimental point of view, both arsenene and antimonene have been obtained by applying a series of different preparation techniques, as succinctly reported [163–166]. The first experimental evidence for the synthesis of monolayer arsenene was reported by Shah et al. [163] where arsenene was grown uniformly in Ag(111) single crystal. The films exhibited a (4 × 4) periodicity with respect to the ordered (1 × 1) As layer, a lattice constant of 0.36 nm and electron band dispersions in close agreement with theory for free-standing buckled arsenene, and confirmed by LEED, STM, ARPES and DFT calculations [163]. Previously, non-continuous multilayer arsenene, better considered as a stack of multilayer nanoribbons, (NRs) was successfully synthesized on InAs (001) surfaces using plasma-assisted process [164], while an aqueous shear exfoliation method yielded few-layer exfoliated nanosheets of As [165]. The same process was used for Sb and Bi, the two neighboring layered materials [165], and the ultrasonication of gray arsenic in appropriate solvents produced 2D arsenene nanosheets and nanodots [166].

The first experimental observation of a single bilayer ultrathin film of Sb(111) (antimonene) was made by Lei et al., [167] on Bi_2Te_3 and Sb_2Te_3 substrates by molecular beam epitaxy (MBE). They investigated the films using LEED, XPS, and ARPES [167], finding sharp ARPES spectra, where the interfacial strain effects modulated the band dispersion along the momentum region, and interfacial charge transfer leads to spin splitting [167]. A few months later, a uniformly distributed multilayer antimonene was synthesized on the InSb (001) substrate by the plasma-assisted process. Multilayer antimonene was found to be a non-continuous film as was a pile of multilayer nanoribbons (NRs), emitting orange light at about 610 nm wavelength (about 2.03 eV bandgap) during photoluminescence (PL) measurements at room temperature (RT) [168]. The bandgap opening was caused by the quantum confinement effect of the Sb NRs structure and the turbostratic stacking of the antimonene layers [168]. Furthermore, few antimonene flakes, highly stable under atmospheric conditions and in water over periods of months, were made by Ares et al., [169] by mechanical exfoliation with a minimum layer thickness of ~1.8 nm, consistent with a bilayer of antimony [170]. Liquid-phase [171] and electrochemical exfoliation [172] from bulk Sb also produced a high-quality few-layer of arsenene nanosheets.

In addition, high-quality few-layer antimonene single crystal polygons with a lateral size of 5-10 μm and monolayer thickness of 50 nm were also synthesized on the (001) surface of fluorophlogopite mica ($KMg_3(AlSi_3O_{10})F_2$) through a van der Waals epitaxy growth process [173]. The films were grown in a two-zone tube furnace with separate temperature controls, with antimony powder in one zone at a temperature of about 930 K to provide antimony vapor, and mica substrates were held at about 650 K as substrates to promote the epitaxial growth of antimonene [173].

Recently, the growth and air stability of the antimonene monolayer on a cleaved $PdTe_2$ surface were studied in UHV at 400 K [174]. The atomically smooth Sb film on $PdTe_2$ was probed by STM, as shown in Fig. 3.16, where (a) is the STM image of a large antimonene island on $PdTe_2$ and (b) is an atomic resolution image showing the graphene-like honeycomb lattice. The top view (top) and side view (lower) of the buckled antimonene structure and the line profile along with the blue trace of (b) are shown in (c) and (d) [174]. Antimonene was synthesized on both Ge(111) substrates using solid-source MBE [175] and Ag(111) substrates [176], and on Pb quantum wells [177].

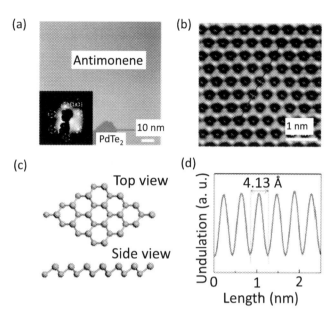

FIGURE 3.16 STM image, −2.0 V, 10 pA, of large antimonene island on PdTe$_2$ (a) and that at atomic resolution, −1.5 V, 200 pA, (b); the inset in (a) shows the 1 × 1 sharp LEED pattern of 1 ML Sb/PdTe$_2$; (c) is the top view (upper) and side view (lower) of the buckled antimonene structure; and (d) is the line profile along with the blue trace of (b). Adapted from Fig. 1 of Ref. [174].

3.8.2 Defects in arsenene and antimonene

As already mentioned, defects at the atomic scale are inevitable. They can be uniformly distributed in materials, interfering with the local lattice arrangement. Defects have been extensively used to control properties of materials such as electrical, mechanical, optical, and chemical properties. To date, defects in emerging 2D materials have been studied theoretically using first-principles calculations. Their type and their effect on geometrical symmetry, ranging from the modification of energy bands, including the energy gap, to the change of electronic features and the onset of magnetic properties have been investigated. We are now at the stage where we could consider taking advantage of their presence to tailor many physical properties [178–183]. To give some examples of the onset of magnetic properties in a 2D material, let us consider the work recently reported by Bafekry et al., [179]; Hu et al., [180] and Liang et al., [181], where upon the influence of the topological defects, including single and double vacancies, Stone–Wales and adatoms, the electronic and magnetic properties of antimonene and arsenene were investigated. Single localized vacancy induced filled spin-up states and empty spin-down states in the fundamental band gap with a magnetic momentum of 0.3 μB [179]. Defect-strain synergy effects, which are very often interconnected, could transform antimonene into a semiconductor, half-metal, or topological insulator [180]. In arsenene, defects can remarkably affect the electronic and magnetic properties, introducing localized states near the Fermi level, a strong local magnetic moment due to dangling bonds, and the generation of unpaired electrons. [181]. Bearing in mind that the onset of new

68   3. Defects in two-dimensional elemental materials beyond graphene

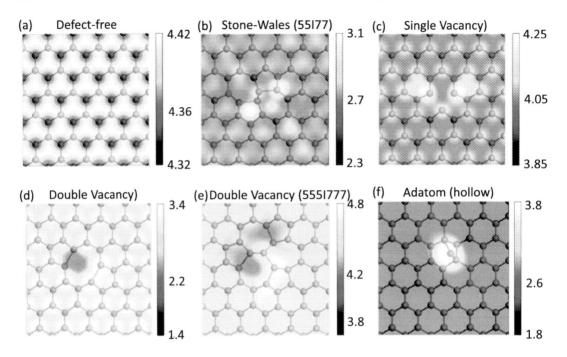

FIGURE 3.17 Atomic structures and the simulated STM images from buckled defect-free and defected h-As: (a) perfect honeycomb structure, (b) Stone–Wales (55|77, SW), (c) single vacancy (SV), (d) double vacancy (5-8-5, DV1), (e) double vacancy (555|777, DV2), and (f) types of adatom (hollow site) defects. STM images were simulated at +2.0 V (voltage between tip-sample), while the color bars are the relative height. Adapted from Fig. 1 of Ref. [184].

physical behavior in 2D elemental materials is due to the rearrangement of the electric charge caused by the presence of defects, which by modifying the number of atoms in the hexagonal rings, could distort the bond angles and lengths and produce dangling bonds and unpaired electrons.

The point defects in both h-As and h-Sb monolayer (ML), have been considered by Sun et al. [184] using the spin-polarized DFT. Fig. 3.17 illustrates the atomic structure and related simulated STM images from 1 ML h-As (h-Sb), considering several configurations without and in the presence of five types of point defects [184]. Fig. 3.17(a) shows the perfect honeycomb buckled structure, while the defective honeycomb structures, from panels b-f, are represented by (b) Stone-Wales (55|77, SW), (c) single vacancy, (d) double vacancy (5-8-5, DV1), (e) double vacancy (555|777, DV2), and finally (f) adatom, called hollow site [184].

In the defect-free structure of h-As (h-Sb) of Fig. 3.17(a) the stable configuration gives a buckle of 0.139 (0.164) nm with an As–As (Sb-Sb) bond length of 0.251 (0.288) nm [157,184]. The defects of SW (55|77) type, formed by two opposite pentagons and heptagons rings with adjacent shortened and elongated As-As bonds of a few hundreds of nanometers, are obtained by rotating two adjacent atoms, Fig. 3.17(b). Bright-dark STM spots are visible, representing high-low atoms locations [184]. The missing As or Sb atom produces only one type of SV defect for both h-As and h-Sb (Fig. 3.17c). In the simulated STM the three atoms around the defective site with dangling bonds appear as bright spots [184]. The DVs type defects are formed without any dangling bond left, by removing a pair of adjacent atoms. For both h-As and h-Sb, there are two defects in DVs, named DV1 (5-8-5) and DV2 (555|777) as shown in Fig. 3.17(d) and (e), where the 555|777 the defective structure is obtained by the rotation bonds of the 5-8-5 structure [184]. Furthermore, there are three other defects for adatom types (Fig. 3.17f), including the top, bridge, and hollow sites. Among them, the hollow site is the most stable with an energy of 0.16 eV (0.23 eV) lower than the other two types for h-As and h-Sb. The extra As (Sb) atom in the hollow site bonded to two adjacent atoms has the shortest bond length 0.238 nm for As-As and 0.279 nm for Sb-Sb [184]. This type of defect could be easily detected in the STM image because the position of the adatom is higher than the position of atom in the basal plane, 0.172 nm for As and 0.216 nm for Sb, respectively [184].

When speaking of point defects, one must keep in mind that theoretical predictions are extremely important for obtaining their formation energy, migration probability, effect on band gap and their effect on the electronic and magnetic properties.

The defects formation energies give access to the average equilibrium areal density n (m^{-2}) of defects, at a finite temperature T, via the Arrhenius equation [184,185] $n = N \exp(-E_f/k_B T)$, where N is the areal density of atoms in 2D materials and k_B is the Boltzmann constant [185]. The calculated formation energies E_f (eV per atom) depend on the defect type, ranging from 1.33 eV (SW) to 2.28 eV (adatom) for h-As, and lower, from 1.09 eV (SW) to 1.99 eV (adatom) for h-Sb [184], being considerably lower than the values obtained for graphene [66,68] and silicene [63,68,69]. In particular, the SVs defects can diffuse very quickly towards the edges with a lower diffusion (migration) barrier of 0.75 eV and 0.86 eV in h-As and h-Sb, respectively [184].

The various point defects can significantly influence the electronic structures of 2D materials. This is the case for ML h-As (h-Sb), which are indirect band gap semiconductors, where the band gap (E_g) is reduced to 0.1 (0) eV from 1.59 eV, respectively, for h-As; and to 0 eV from 0.99 eV for h-Sb under the effect of SV and adatom defects [184]. The bandgap values between these extremes are obtained for SW, DV1 (5-8-5) and DV2 (555|777) for both h-As and h-Sb [184], noting that, in general, the defective structures significantly reduce the bandgaps. Finally, SVs and adatom type defects can carry magnetic moments due to the dangling bonds, resulting from the absent or extra atom [184]. A clear example of experimentally observed defects is shown in large area STM images of monolayer arsenene epitaxial grown on Ag(111) reported in Fig. 3.18, where (a) is the filled-state STM (254 × 254) nm^2 and (b) the empty-state STM image (31 × 31) nm^2 [163]. These STM images show that the arsenene layer is uniformly grown upon the surface, forming a 4 × 4 reconstruction, whose features appeared much weaker at low bias (+10 meV), as displayed by the high-resolution empty-state STM image in the inset (11.7 × 9.2) nm^2 of Fig. 3.18(b) [163]. The hexagonal structure, consistent with buckled arsenene, as well as the Moiré character of the 4 × 4 periodicity due to the difference in lattice constants between Ag(111) and arsenene, is clearly visible [163]. Interestingly, the white and blue arrows point to two types of defects, triangular vacancies and straight lines [163]. These are the only two types of defects observed up to now on the buckled surface of arsenene on Ag(111) [163]. The few dark triangular defects are attributed to missing As atoms [161]. The other type, indicated in Fig. 3.18 (a) by a blue arrow, appears as dark lines. These lines shown mainly in three different orientations are consistent with a hexagonal structure [163].

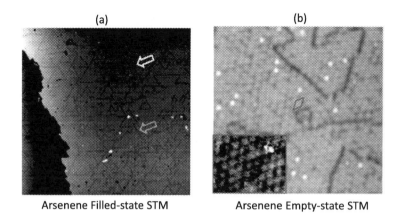

Arsenene Filled-state STM   Arsenene Empty-state STM

FIGURE 3.18 Filled-state STM image of arsenene (−1 V) STM image (254 × 254) nm^2 (a) and empty state (+2 V) image (31 × 31) nm^2 (b), while the inset is the high-resolution STM image (10 meV (11.7 × 9.2) nm^2. The white and blue arrows indicated the two types of defects, namely triangular vacancies and straight lines. The red 4 × 4 unit cell is also marked. Adapted from Fig. 1 of Ref. [163]. "Reprinted Fig. 2 with permission from [J. Shah et al., 2D Mater., 7, 025013, 2020] Copyright (2014) licensed by Creative Commons Attribution 3.0 Unported License https://creativecommons.org/licenses/by/3.0/."

3.9 Bismuthene

3.9.1 Synthesis and atomic structure

The heaviest atom, Bi (Z=83), is the last of the pnictogen group-VA of the periodic table. Ultrathin Bi(111) was predicted by first-principles calculations to be a stable not trivial Z_2 topological insulator, independent of the film thickness [186]. The Bi films with 1–4 bilayers (BLs) were found to be intrinsic 2D topological insulators (TI) [186], as also previously predicted by Murakami et al. [187] for a single free-standing BL Bi (111), while Bi films with 5-8 BLs are 2D TIs sandwiched with trivial metallic surfaces [186]. Two-dimensional bismuth exhibited the quantum spin-Hall effect, both by calculating the helical edge states, and by showing the non-triviality of the Z_2 topological number [188]. Among various possible bismuthene allotropes (α, β, ζ), which have a similar energy configuration, the β phase with a buckled atomic form is the most stable structure [189]. The β-type bismuth has the same crystal structure as gray arsenic under T=300 K and P = 1 atm conditions with the same space group ($R3 = m$) [189].

Experimentally, a thin bismuth film has been grown on Si(111) and Si(001) [190], on an insulating α-Al$_2$O$_3$(0001) substrate, obtaining stable islands of Bi(110) and Bi(111) up to 400 K [191] and on a superconducting 2H-NbSe$_2$ substrate, finding that the band structures of the ultrathin Bi was a quasi-freestanding film [192]. Reis et al., [193] have synthesized monolayer bismuthene by epitaxial growth on 4H-SiC(0001) substrates. A fully planar graphene-like film was obtained, instead of buckled bismuthene, due to the contribution of the SiC substrate strain [193]. Bi flakes were studied by STM, where a clear honeycomb structure of bismuthene was shown [193]. These flakes have a typical diameter of ~ 25 nm. They smoothly cover the whole surface and are separated by phase-slip domain boundaries, including occasional localized defects [193].

A new strategy that combined acid-interaction and liquid exfoliation was successfully used to transform bulk metal Bi into bilayers 2D Bi [194]. The ultrasonication-assisted electrochemical exfoliation method was also proposed to prepare ultrathin few layered bismuthene nanosheets [195].

Other authors reported free-standing buckled bismuthene by compressing (hot-pressing method) Bi nanoparticles [196] and/or by using the sonochemical exfoliation method to produce high-quality bismuthene with enhanced stability [197]. The 2D bismuthene α-phase with the puckered structure [198] can be obtained using suitable substrates such as highly oriented pyrolytic graphite, and Walker et al., [199] reported the first direct dry transfer by using a double cantilever beam fracture technique to transfer epitaxial from 2D bismuth nanosheets grown on silicon (111) to silicon strips coated with epoxy.

3.9.2 Defects in bismuthene

Since 2D Bi development is in its early infancy, there is little experimental and theoretical information. An example on the effect of point defects on the electronic structure, magnetic structure, and topological phase of 2D bismuth monolayers was recently reported by Kadioglu et al., [200] through first-principles calculations. Single vacancies and divacancies can be readily formed at room temperature in h-bismuthene structures since their calculated formation energies are 1.10 and 1.44 eV respectively and remain stable at high temperatures. The Bismuth atoms surrounding the vacancy have sp^2-type dangling bonds emerging from the vacancy. Dangling bonds give rise to localized gap states and resonant states locally modifying the electronic structure of the underlying bismuthene and promoting the vacancy as a chemically active site. Furthermore, the formation energy of a divacancy is unexpectedly small owing to the rebondings of Bi atoms surrounding the divacancy. These re-bondings lead to the removal of localized gap states because the dangling bonds are saturated [200]. In addition, in h-Bi, the divacancies transform into Stone-Wales-type defects consisting of two pentagonal rings and one octagonal ring. At high temperature, these defects transform into larger rings consisting of fourteen Bi atoms [200]. Depending on the type and coupling between the point defects, the topologically non trivial phase of the bare bismuthene can be destroyed.

Fig. 3.19(a) and (c) show the optimized atomic structures for a single vacancy and a divacancy, respectively, while the corresponding electronic structures are presented in (b) and (d). After the structure optimization, three Bi atoms surrounding the single vacancy collapse towards the center, while the overall character of the honeycomb structure remains the same. Fig. 3.19(c) shows the Bi atoms surrounding the divacancy which form one octagonal ring and two adjacent pentagonal rings transforming the surrounding atoms threefold coordinated [200]. This makes the divacancy of h-Bi have a larger hole than a single vacancy, but it is chemically less active. The single vacancy on the other hand modifies the electronic properties by closing the band gap (b). In the case of divacancy, the localized gap states are expelled from the fundamental band gap, but they appear near the band edges due to dangling bond saturation (d) [200].

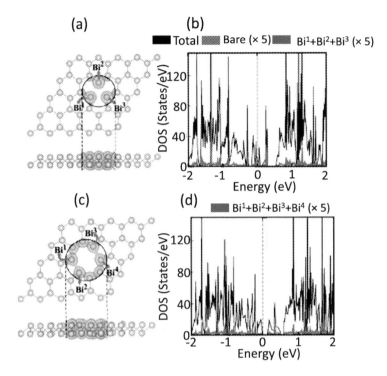

FIGURE 3.19 Top- and side-view of optimized atomic configuration of a single vacancy in free-standing single layer h-Bi and charge density isosurfaces of surrounding three Bi_1, Bi_2 and Bi_3 atoms (a). Total (TDOS) densities of states of single vacancy +h-Bi in black; DOS projected onto three Bi atoms surrounding the vacancy in red; and density of states projected onto a Bi atom farthest from the single vacancy representing extended and bare h-Bi, in blue (b). Top- and side-view of optimized atomic configuration of a divacancy in free-standing single layer h-Bi and charge density isosurfaces of surrounding three Bi_1, Bi_2, Bi_3 and Bi_4 atoms (c). (d) DOS projected onto four Bi atoms surrounding the vacancy in red. The zero of energy is set at the Fermi level shown by the vertical dashed line. The fundamental band gap of the extended and bare h-Bi band is shaded in yellow. Adapted from Fig. 7 of Ref. [200].

3.10 Selenene and tellurene

At ambient temperature and pressure, bulk tellurium and selenium have a trigonal crystal structure, the so-called Te-I and Se-I structures, with the $P3_121$-D_3^4 or $P3_221$-D_3^6 space group, consisting of helical chains with three atoms per turn along the c-axis (the screw axis), arranged in a hexagonal array which spirals around the axes parallel to the crystalline c axis. Depending on the chirality of the screw axis, and sense of rotation of the helical chains (right- or left-handed screw), the space group can be $P3_121$ (right-handed) or $P3_221$ (left-handed). [201–203].

To date, there have been very few reports on monolayered Se and Te and much less is known about their defect chemistry of these newborn 2D Se and Te nanosheets, the growth processes, and their properties. Here, we present the synthesis and theoretical predictions of selenene and tellurene, giving a broad reconstruction of these emerging 2D elemental materials of the group-VIA elements of the periodic table.

Zhu et al., [204] predicted a new category of 2D monolayers comprised by the element Te named tellurene, extending the realm of 2D materials to group-VI. Three types of structures are found for monolayer Te, α-Te (tetragonal – 1T-MoS$_2$-like), β-Te (tetragonal), and γ-Te (2H-MoS$_2$-like). Of these, α-Te and β-Te are semiconducting with the band gap of 0.75 and 1.47 eV respectively as determined by the particle-swarm optimization method [205] in combination with first-principles density functional theory calculations [204]. All these arrangements are composed of trilayers, which are driven by the multivalence behavior of tellurium, with the central-layer being more metal-like, and the two outer layers more semiconductor-like. It was calculated that α-Te, which has the structure similar to 1T-MoS$_2$, contains three Te atoms per unit cell [204]. Here, there are two distinct types of Te atom with different coordination numbers (n_c), a central Te atom located at the Mo site has $n_c = 6$, while the Te atom in the upper or lower layer at the S sites has $n_c = 3$. β-Te is composed of the planar four-membered and chair-like six-membered rings arranged alternately with the lattice constants a = 0.417 and b = 0.549 nm, where a central Te atom has $n_c = 4$, and an upper or lower Te atom has $n_c = 3$. γ-Te have the 2H-MoS$_2$-like structure, with smaller lattice constants a = b = 0.392 nm than those (a = b = 0.415 nm) of α-Te [204]. Recent theoretical calculations suggest that the α-phase, derived from the bulk trigonal structure, is the most stable phase for few-layer tellurene [206], and the tetragonal β-phase is more stable for monolayer tellurene, due to structural relaxation [204]. Indeed, in the structural transformation from the helical chains to the β-Te phase, the geometric structures of the Te slab are truncated along the [100] direction of bulk Te-I and optimized [204, Suppl. Inf.]. Before relaxation, each atom was covalently bonded with two neighboring atoms within a helical chain. After relaxation, the helical chains become closer to each other, forming the β-Te phase, in which all the Te atoms become threefold- or fourfold-coordinated [204, Suppl. Inf.].

Fig. 3.20 reports the three-layers structure of 2D tellurene (α, β and γ), whereas (d) shows the first STM image of the β-Te phase synthesized on highly oriented pyrolytic graphite [204].

The β-Te phase was also confirmed by Chen et al., [207], presenting an STM study on the ultrathin β-tellurium layers grown in HOPG by MBE from a standard Knudsen cell operated at 540 K where the substrate temperature was kept at 400 K. Tellurium (Te) films with monolayer and few-layer thickness were also grown by MBE on graphene/6H-SiC(0001) substrates and investigated by STM and STS. It was deduced that the Te films were composed of parallel-arranged helical Te chains flat-anchored on the graphene surface, exposing the (1×1) facet of (101̄0) of the bulk crystal and the band gap of Te films increases monotonically, up to 0.92 eV for the monolayer Te, with decreasing thickness, reaching in fact the near-infrared band for the monolayer Te [208]. The arrangement of parallel-helical Te chains was retained in a way different from the three structural models proposed theoretically [204]. As occurred for the majority of 2D synthesized materials, the substrate is expected to play also a critical role indicating the structure and properties of the supported tellurene.

Ultrathin 2D non-layered Te nanosheets were formed by a liquid-phase exfoliation method with lateral dimension ranging from 41.5 to 177.5 nm and a thickness ranging from 5.1 to 6.4 nm [209], and ultrathin 2D tellurium–polymer membranes were fabricated using a facile and cost-effective liquid-phase exfoliation method [210]. A review on the physical properties, scalable manufacturing, and device applications of tellurene was recently reported [211], highlighting the prototypal field-effect transistors made from 2D tellurene flakes obtained by liquid-solution, showing an interesting figure of merit [212].

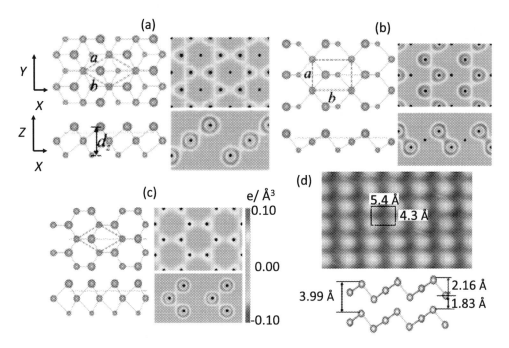

FIGURE 3.20 Top and side views of the optimized structures of the different tellurene phases; α-Te (a), β-Te (b), and γ-Te (c). The red dashed lines indicate the unit cell of each structure. The total charge density of each structure is also reported along with the horizontal and vertical cross sections marked by the blue dotted lines. (3.5 × 2.2) nm² high-resolution STM image with drawn the rectangular unit cell (d) and drawn side view of a bilayer film in the β-Te phase. The calculated monolayer thickness and interlayer spacing are shown. Adapted from Figs. 1 and 4 of Ref. [204]. "Reprinted Fig. 1 and with permission from [Z. Zhu, et al., Phys. Rev. Lett. 119, 106101, 2017] Copyright (2017) by the American Physical Society."

Finally, a novel stable 2D layered structure for group-VI elements Se and Te, square selenene and tellurene, was predicted by first-principles calculations [213]. It was found that this special structure gives rise to anisotropic band dispersions near the Fermi level, described by a generalized semi-Dirac Hamiltonian. These films have a band gap of ∼ 0.1 eV opened by spin-orbit coupling, making square selenene and tellurene topological insulators, hosting non-trivial edge states [213]. The structure of the 2D monolayers of Se has been theoretically stabilized, but only with the 1T-MoS$_2$-like (α-Se) structure (see Fig. 3.20a, α-Te) [214]. Although selenene was less investigated than tellurene, some theoretical work, based on first-principles calculations, reported interesting physical properties, such as the abnormally low thermal conductivity for the 2D Se α-phase [214], high bipolar conductivity and robust spontaneous electric polarization in the plane [215]. Growth of 2D selenium nanosheets was recently reported by Qin et al., [216] where they used physical vapor deposition of pure Se powder kept at 480 K in a multizone furnace and deposited on a freshly cleaned Si(111) substrate held at 370 K in pure Ar gas (100 mbar) for 60 min. After deposition, a black needle-like material was observed that covered the Si substrate. Fig. 3.21(a) shows an optical microscopy image of Se nanosheets, which, shows that most of the Se nanosheets have an irregular quad-

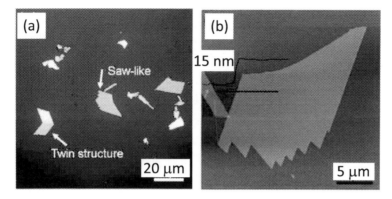

FIGURE 3.21 Optical microscopy image of selenium nanosheets after being transferred to the SiO$_2$/Si substrate, using the Scotch tape method (a) and AFM topography image and saw-like height profile Se nanosheet (minimum thickness was of nm) (b). Adapted from Figs. 2 of Ref. [216]. "Reprinted Fig. 2 with permission from [J. Qin, et al., ACS Nano, 11, 10222, 2017] Copyright (2017) by the American Chemical Society."

rangular shape with zigzag edges on one side, while the remaining have a twin structure with mirror symmetry [216]. Atomic force microscopy (AFM) was employed to determine the thickness and topography of the nanosheets, as displayed in Fig. 3.21b, where a saw-like height profile of 15 nm was measured [216].

3.11 Gallenene

A superconducting phase, with a T$_c$ as high as 5.4 K, was recently found in two monolayers of hexagonal Ga, gallenene, film grown epitaxially on a 3 µm thick film of GaN(0001) grown at T$_{substrate}$ = 650 °C from a high-purity solid Ga source [217]. The high-resolution atomic STM image of 2 MLs Ga on GaN(0001) is illustrated in Fig. 3.22(a), where an atomically flat Ga film with hexagonal symmetry, evidenced by its FFT pattern in the inset, is clearly visible [217]. Furthermore, the matching between the Ga film and the GaN substrate is almost perfect, showing that the hexagonal Ga film has a lattice constant of 0.318 nm, a value that is very close to that of the underlying GaN(0001) substrate (0.319 nm). On the other hand, Fig. 3.22(b) shows the TEM image collected *ex situ*, after the deposition of 80 nm thick Ag capping layer, showing a sharp interface between the 2 MLs of Ga and GaN(0001), with a stacking height (distance between the two Ga layers and GaN substrate) of 0.276 nm, as drafted in (c) [217].

Recently, gallenene, an analogue of graphene, composed of a one-atom layer of gallium with a honeybomb lattice was synthesized [218]. The $\sqrt{3} \times \sqrt{3}$-Ga grown on Si(111)7 × 7, obtained by depositing 1/3 ML Ga atoms on the 7 × 7 reconstructed Si surface at room temperature and subsequently annealing at ~ 820 K for 30 min, was used as a template for gallenene growth at substrate temperature of ~ 320 K [218]. At 1.4 ML Ga coverage, the $\sqrt{3} \times \sqrt{3}$-Ga structure on Si(111) is completely buried, and the growth of the Ga monolayer

76　　　　　　　　　　3. Defects in two-dimensional elemental materials beyond graphene

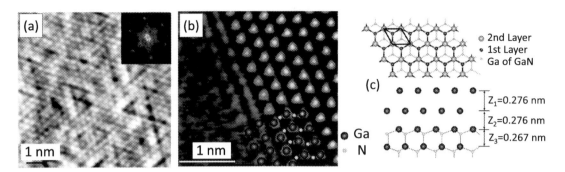

FIGURE 3.22 High resolution STM image (0.22 V, 0.05 nA, 8 × 8 nm^2) of 2 MLs of Ga film (a) and its corresponding FFT pattern shown in the inset. Cross section TEM image from Ag/Ga/GaN heterostructure (b), and top- and side-view drawing of 2 MLs Ga on GaN(0001) (c). The unit cell is marked in blue, as well as the distances between the Ga layers. Adapted from Fig. 1 of Ref. [217]. "Reprinted Fig. 1 with permission from [H.-M. Zhang, et al., Phys. Rev. Lett. 114, 107003, 2015] Copyright (2015) by the American Physical Society."

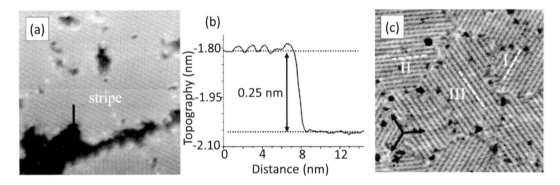

FIGURE 3.23 STM image (70 nm × 70 nm, 2.8 V, 30 pA) of 1.2 ML Ga deposited on $\sqrt{3} \times \sqrt{3}$ -Ga grown on Si(111)7 × 7 (a); (b) is the height profile along the black line traced in (a); and (c) shows the STM image (50 nm ×50 nm, 2.0 V, 30 pA) of the Ga monolayer, where the three kinds of striped domains, are visible, marked by the white dashed lines, and aligned along the main Si(111) symmetry. Adapted from Figs. 1 and 2 of Ref. [218]. "Reprinted Figs. 1 and 2 with permission from [M.-L. Tao, et al., 2D Mater., 5, 035009, 2018] Copyright (2018) by the IOP."

is complete. Lower surface coverage of 1.2 ML Ga, whose atomic resolution STM image, collected at 78 K, is displayed in Fig. 3.23(a), shows a Ga monolayer height of ∼ 0.25 nm, by tracing the line profile along the black line in (a), see Fig. 3.23(b), in nice agreement with the thickness of a Ga atomic layer obtained in Ref. [217]. As can be clearly observed from the STM image of Fig. 3.23(c) the Ga monolayer on Si(111) has many defects, most of which are located along domain boundaries of the Ga monolayer [218] as indicated by the black arrows. Additionally, the Ga atoms are aligned along the Si(111) crystallographic directions.

It is interesting to note that, very recently, atomically thin gallium layers were also obtained from solid-melt exfoliation [219], and the investigation on the thermal stability of freestanding bilayer and trilayer gallenene by DFT simulations resulted in a thermal stability enhancement when the bulk Ga is reduced [220].

3.12 Hafnene

Hafnium (Hf) is a transition metal of the group-IVB of the Periodic Table after zirconium, with the atomic number Z=72. Its discovery was announced in 1923 after the seminal work of Coster and Hevesy [221], "On the Missing Element of Atomic Number 72", where they succeeded in detecting six lines, L $\alpha_1, \alpha_2, \beta_1, \beta_2, \beta_3$ and γ_1, in zirconium minerals through the x-ray spectra, ascribed to the new element 72 [221], according to the theoretical prediction of the Bohr theory of spectra and atomic constitution [222]. This new element was found to be chemically analogous to zirconium and was found to be an impurity in Zr, from 0,01 to 0.1% Hf in the first samples [221]. Shortly after they investigated a great number of Zr minerals from different parts of the world [223], where some contained a higher percentage of Hf, between 5 and 10%, which was later increased by means of the chemical separation method and increased to about 50% Hf in Zr [223]. For the new discovered element they proposed the name Hafnium (Hafniae = Copenhagen) [221], which historically was at the center of a strong diatribe [224–230] on its erroneous attribution to an element named, by Urbain in 1911, Celtium [226], claiming to have chemical properties typical of rare earths.

In 2013, hafnene 2D honeycomb single layer, was synthesized, for the first time, on the surface of Ir (111), by depositing Hf in UHV at RT and post-annealing it at a temperature of 573 K, by Li et al., [231,232]. The 2D Hf, hafnene, layer exhibited a LEED pattern with a (2×2) reconstruction, and a well-ordered continuous honeycomb lattice [231], as can easily be observed in the HR STM images shown in Fig. 3.24 [231]. At the same time hafnene was also obtained on Rh(111) ([231] Supp. Inf., [232]). Just after, in 2014 the review of Gao's group already had exhaustively reported on the 2D atomic crystal and transition metal, such as graphene, silicene, and the TM 2D hafnene, synthesized until then [233], and subsequently enlarging the assessment to an ever increasing number of 2D materials [234]. Interestingly, the electronic and magnetic properties of d-electron-based transition metals such as Ti, Zr, Hf, V, Nb, or Ta, monolayer arranged in honeycomb structure were studied by combining first-principles with mean field theory and Monte Carlo approaches, demonstrating, as in the case of p-electron-based materials graphene and its 2D materials family, that they could also host the Dirac cone structures in both spin-up and spin-down states [235]. In addition, first-principles calculations from hafnene-graphene van der Waals heterostructure have predicted exciting electronic and magnetic properties directly deriving by the interfacial coupling between these two different p- and d-electron-based 2D honeycomb materials, envisioning, furthermore, the possibility to fulfill a spin-polarized field-effect transistor [236]. Lastly, the transition from non-magnetic to FM behavior was predicted by adding BN as a spacer layer between the Ir(111) substrate and the hafnene lattice, which has no magnetic states of the 2D hafnene due to its strong hybridization with the Ir(111) substrate [237].

Finally, this mention has been granted to other emerging metal-enes, such as Rh nanosheets [238,239] and goldene single layers embedded in between couples of graphene layers [240,241], as well as to Fe membranes suspended in graphene pores [242].

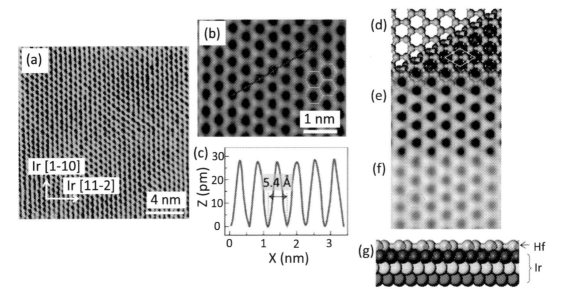

FIGURE 3.24 The STM image of Hafnene on Ir(111) (U = −1.0 V and I = 0.8 nA) is reported in (a) with the main directions of the crystal [1–10] and [11,12], while the HR-STM image (U = −1.0 V and I = 0.8 nA) image is shown in (b), where the blue line traces the distance between the dark hollows of the honeycomb Hf pattern, whose line profile is traced in (c). A lattice periodicity of 0.54 nm was found corresponding to the Hf-Hf atomic distance of 0.312 nm ($0.54/\sqrt{3}$), in nice agreement with the Hf-Hf atomic distance of 0.319 nm on the bulk Hf (0001). Panels (d–g) report the atomic configuration of the Hf honeycomb lattice on Ir(111). The calculated top view of the Hf atomic structure of Hf calculated by DFT is shown in (d), with the Ir(111)-(2 × 2) white rhombus unit cell. The simulated STM image (−1.5 V) is displayed in (e), while the corresponding experimental atomic STM image (−1.5 V, 0.1 nA) is showed in (f). In both panels (d–f) the green hexagons remark the same atomic features, showing the obtaining of planar honeycomb Hf layer on an Ir(111) surface. The side view of the relaxed Hf/Ir(111) atomic structure of (d) is drawn in (g). Adapted from Figs. 2 and 3 of Ref. [231]. "Reprinted Figs. 2 and 3 with permission from [L. Li, et al., ACS Nano, 13, 4671, 2013] Copyright (2013) by the American Chemical Society."

3.13 Conclusions and outlook

In this chapter, we report a comprehensive review of two-dimensional elemental materials beyond graphene, such as borophene, silicene, germanene, phosphorene, stanene, arsenene, antimonene, bismuthene, selenene and tellurene and the effects of point defects on their 2D honeycomb structure. Different types of structural point defects, including Stone-Wales, single vacancy and bivacancies, and grain boundaries (lines) and adatoms have been described on the basis of theoretical and experimental published reports. Several theoretical works have concentrated their assessment on the probability of formation and thermodynamic stability of point defects in the lattice structure of 2D materials. The identification of these structural defects, diffusion behavior, and impact on electronic/magnetic properties is believed to be crucial for future applications of 2D materials. We also highlighted some emerging 2D material properties with non-trivial topological behavior due to the presence of defects that could be of great interest for potential electronic applications. As already mentioned in the fabrication and processing of 2D materials, structural defects are inevitable, and whether they are native or intentional their existence can strongly affect their fundamental physical properties. As in the case of existing semiconductor materials, defects can also be intentionally introduced by physical methods such as stress, irradiation, annealing or chemical treatments. Although two-dimensional materials are still quite young and many new physical phenomena have yet to be discovered, there are many challenges that have to be addressed where the impact of defects in 2D materials will play an enormous role in the improvement of the desired properties for many applications. Finally, the synthesis of the new 2D elemental metal-enes, such as gallenene, group-IIIA, and hafnene, group IVB, has also been reported.

References

[1] https://www.nobelprize.org/prizes/physics/2010/.

[2] K.S. Novoselov, A.K. Geim, S.V. Morozov, D. Jiang, Y. Zhang, S.V. Dubonos, I.V. Grigorieva, A.A. Firsov, Electric field effect in atomically thin carbon films, Science 306 (2004) 666–669.

[3] K.S. Novoselov, A.K. Geim, S.V. Morozov, D. Jiang, M.I. Katsnelson, I.V. Grigorieva, S.V. Dubonos, A.A. Firsov, Two-dimensional gas of massless Dirac fermions in graphene, Nature 438 (2005) 197–200.

[4] A.K. Geim, K.S. Novoselov, The rise of graphene, Nature Materials 6 (2007) 183–191.

[5] R.E. Peierls, Quelques proprieties typiques des corpses solides, Annales Henri Poincaré 5 (1935) 177–222.

[6] L.D. Landau, Phys. Z. Sowjetunion, Zur Theorie der phasenumwandlungen II. 11 (1936) 26–35.

[7] L.D. Landau, E.M. Lifshitz, Statistical Physics, Part I, Pergamon, Oxford, 1980.

[8] N.D. Mermin, Crystalline order in two dimensions, Physical Review 176 (1968) 250–254.

[9] T. Ohta, A. Bostwick, T. Seyller, K. Horn, E. Rotenberg, Controlling the electronic structure of bilayer graphene, Science 313 (2006) 951–953.

[10] I. Pletikosic´, M. Kralj, P. Pervan, R. Brako, J. Coraux, A.T. N'Diaye, C. Busse, T. Michely, Dirac cones and minigaps for graphene on Ir(111), Physical Review Letters 102 (2009) 056808.

[11] P. Vogt, P. De Padova, C. Quaresima, J. Avila, E. Frantzeskakis, M.C. Asensio, A. Resta, B. Ealet, G. Le Lay, Silicene: compelling experimental evidence for graphenelike two-dimensional silicon, Physical Review Letters 108 (2012) 155501.

[12] A.J. Stone, D.J. Wales, Theoretical studies of icosahedral C60 and some related species, Chemical Physics Letters 124 (1986) 501.

[13] B.E. Douglas, S.-M. Ho, Structure and Chemistry of Crystalline Solids, Springer, 2006.

[14] D.E. Sands, J. Hoard, Rhombohedral elemental boron, Journal of the American Chemical Society 79 (1957) 5582.

[15] L.V. McCarty, J.S. Kasper, F.H. Horn, B.F. Decker, A.E. Newkirk, A new crystalline modification of boron, Journal of the American Chemical Society 80 (1958) 2592.

[16] A.R. Oganov, J. Chen, C. Gatti, Y. Ma, Y. Ma, C.W. Glass, Z. Liu, T. Yu, O.O. Kurakevych, V.L. Solozhenko, Ionic high-pressure form of elemental boron, Nature 460 (2009) 292.

[17] P. Ball, Why is boron so hard?, Nature Materials 9 (2010) 6.

[18] M.J. Van Setten, M.A. Uijttewaal, G.A. de Wijs, R.A. de Groot, Thermodynamic stability of boron: the role of defects and zero point motion, Journal of the American Chemical Society 129 (2007) 2458.

[19] M. Widom, M. Mihalkovic, Symmetry-broken crystal structure of elemental boron at low temperature, Physical Review B 77 (2008) 064113.

[20] T. Ogitsu, F. Gygi, J. Reed, Y. Motome, E. Schwegler, G. Galli, Imperfect crystal and unusual semiconductor: boron, a frustrated element, Journal of the American Chemical Society 2009 (1903) 131.

[21] B. Kiran, S. Bulusu, H.-J. Zhai, S. Yoo, X.C. Zeng, L.-S. Wang, Planar-to-tubular structural transition in boronclusters: B20 as the embryo of single-walled boron nanotubes, PNAS 102 (2005) 4961.

[22] D. Ciuparu, R.F. Klie, Y. Zhu, L. Pfefferle, Synthesis of pure boron single-wall nanotubes, Journal of Physical Chemistry. B 108 (2004) 3967.

[23] H. Tang, S. Ismail-Beigi, Novel precursors for boron nanotubes: the competition of two-center and three-center bonding in boron sheets, Physical Review Letters 99 (2007) 115501.

[24] N.G. Szwacki, A. Sadrzadeh, B.I. Yakobson, B80 fullerene: an ab initio prediction of geometry, stability, and electronic structure, Physical Review Letters 98 (2007) 166804.

[25] R. Ettl, I. Chao, F. Diederich, R.L. Whetten, Isolation of C76, a chiral (D2) allotrope of carbon, Nature 353 (1991) 149.

[26] J. Kunstmann, A. Quandt, Broad boron sheets and boron nanotubes: an ab initio study of structural, electronic, and mechanical properties, Physical Review 74 (2006) 035413.

[27] X. Yang, Y. Ding, J. Ni, Ab initio prediction of stable boron sheets and boron nanotubes: structure, stability, andelectronic properties, Physical Review 77 (2008), 041402 (R).

[28] A.K. Singh, A. Sadrzadeh, B.I. Yakobson, Probing properties of boron α-tubes by ab initio calculations, Nano Letters 8 (2008) 1314.

[29] H. Tang, S. Ismail-Beigi, First-principles study of boron sheets and nanotubes, Physical Review B 82 (2010) 115412.

[30] Z. Zhang, Y. Yang, G. Gao, B.I. Yakobson, Two-dimensional boron monolayers mediated by metal substrates, Angewandte Chemie. International Edition in English 54 (2015) 13022.

[31] L.Z. Zhang, Q.B. Yan, S.X. Du, G. Su, H.J. Gao, Boron sheet adsorbed on metal surfaces: structures and electronic properties, Journal of Physical Chemistry C 116 (2012) 18202.

[32] Z. Zhang, E.S. Penev, B.I. Yakobson, Two-dimensional boron: structures, properties and applications, Chemical Society Reviews 46 (2017) 6746.

[33] E.S. Penev, V.I. Artyukhov, F. Ding, B.I. Yakobson, Unfolding the fullerene: nanotubes, graphene and poly-elemental varieties by simulations, Advanced Materials 24 (2012) 4956.

[34] C. Özdogan, S. Mukhopadhyay, W. Hayami, Z.B. Güvenç, R. Pandey, I. Boustani, The unusually stable B100 fullerene, structural transitions in boron nanostructures, and a comparative study of α- and γ-boron and sheets, Journal of Physical Chemistry C 114 (2010) 4362.

[35] X. Wu, J. Dai, Y. Zhao, Z. Zhuo, J. Yang, X.C. Zen, Two-dimensional boron monolayer sheets, ACS Nano 6 (2012) 7443.

[36] E.S. Penev, S. Bhowmick, A. Sadrzadeh, B.I. Yakobson, Polymorphism of two-dimensional boron, Nano Letters 12 (2012) 2441.

[37] Y. Liu, E.S. Penev, B.I. Yakobson, Probing the synthesis of two-dimensional boron by first principles computations, Angewandte Chemie. International Edition in English 52 (2013) 3156.

[38] A.J. Mannix, X.-F. Zhou, B. Kiraly, J.D. Wood, Di. Alducin, B.D. Myers, X. Liu, B.L. Fisher, U. Santiago, J.R. Guest, M.J. Yacaman, A. Ponce, A.R. Oganov, M.C. Hersam, N.P. Guisinger, Synthesis of borophenes: anisotropic, two-dimensional boron polymorphs, Science 350 (2015) 1512.

[39] B. Feng, J. Zhang, Q. Zhong, W. Li, S. Li, H. Li, P. Cheng, S. Meng, L. Chen, K. Wu, Experimental realization of two-dimensional boron sheets, Nature Chemistry 8 (2016) 563.

[40] G. Bhattacharyya, A. Mahata, I. Choudhuri, B. Pathak, Semiconducting phase in borophene: role of defect and strain, Journal of Physics. D, Applied Physics 50 (2017) 405103.

[41] Z.-D. Sha, Q.-X. Pei, K. Zhou, Z. Dong, Y.-W. Zhang, Temperature and strain-rate dependent mechanical properties of single-layer borophene, Extreme Mechanics Letters 19 (2018) 39.

[42] V. Wang, W.T. Geng, Lattice defects and the mechanical anisotropy of borophene, Journal of Physical Chemistry C 121 (2017) 10224.

[43] Z. Zhang, A.J. Mannix, Z. Hu, B. Kiraly, N.P. Guisinger, M.C. Hersam, B.I. Yakobson, Substrate-induced nanoscale undulations of borophene on silver, Nano Letters 16 (2016) 6622.

[44] P. Hohenberg, W. Kohn, Inhomogeneous electron gas, Physical Review 136 (1964) B864.

[45] W. Kohn, L.J. Sham, Self-consistent equations including exchange and correlation effects, Physical Review 140 (1965) A1133.

[46] X. Liu, Z. Zhang, L. Wang, B.I. Yakobson, M.C. Hersam, Intermixing and periodic self-assembly of borophene line defects, Nature Materials 17 (2018) 783.

[47] A.V. Krasheninnikov, When defects are not defects, Nature Materials 17 (2018) 757, https://doi.org/10.1038/s41563-018-0153-y.

[48] K. Takeda, K. Shiraishi, Theoretical possibility of stage corrugation in Si and Ge analogs of graphite, Physical Review B 50 (1994) 14916.

[49] G. Guzmán-Verri, L.C. Lew Yan Voon, Electronic structure of silicon-based nanostructures, Physical Review B 76 (2007) 075131.

References   **81**

[50] S. Lebègue, O. Eriksson, Electronic structure of two-dimensional crystals from ab initio theory, Physical Review B 79 (2009) 115409.

[51] S. Cahangirov, M. Topsakal, E. Aktürk, H. Şahin, S. Ciraci, Two- and one-dimensional honeycomb structures of silicon and germanium, Physical Review Letters 102 (2009) 236804.

[52] D. Jose, A. Datta, Understanding of the buckling distortions in silicene, Journal of Physical Chemistry C 116 (2012) 24639.

[53] B. Feng, Z. Ding, S. Meng, Y. Yao, X. He, P. Cheng, L. Chen, K. Wu, Evidence of silicene in honeycomb structures of silicon on Ag(111), Nano Letters 12 (2012) 3507–3511.

[54] C.L. Lin, R. Arafune, K. Kawahara, N. Tsukahara, E. Minamitami, Y. Kim, N. Takagi, M. Kawai, Structure of silicene grown on Ag(111), Applied Physics Express 5 (2012) 045802.

[55] A. Fleurence, R. Friedlein, T. Osaki, H. Kawai, T. Wang, Y. Yamada-Takamura, Experimental evidence for epitaxial silicene on diboride thin films, Physical Review Letters 108 (2012) 245501.

[56] L. Meng, Y. Wang, L. Zhang, S. Du, R. Wu, L. Li, Y. Zhang, G. Li, H. Zhou, W.A. Hofer, H.-J. Gao, Buckled silicene formation on Ir(111), Nano Letters 13 (2013) 685–690.

[57] Y. Du, J. Zhuang, H. Liu, X. Xu, S. Eilers, K. Wu, P. Cheng, J. Zhao, X. Pi, K.W. See, G. Peleckis, X. Wang, S.X. Dou, Tuning the band gap in silicene by oxidation, ACS Nano 8 (2014) 10019.

[58] L. Huang, Y.F. Zhang, Y.Y. Zhang, W.Y. Xu, Y.D. Que, E. Li, J.B. Pan, Y.L. Wang, Y.Q. Liu, S.X. Du, S.T. Pantelides, H.J. Gao, Sequence of silicon monolayer structures grown on a ru surface: from a herringbone structure to silicene, Nano Letters 17 (2017) 1161.

[59] M. Krawiec, Functionalization of group-14 two-dimensional materials, Journal of Physics. Condensed Matter 30 (2018) 233003.

[60] A. Stępniak-Dybala, M. Krawiec, Formation of silicene on ultrathin Pb(111) films, Journal of Physical Chemistry C 123 (2019) 17019.

[61] A. Stępniak-Dybala, P. Dyniec, M. Kopciuszyński, R. Zdyb, M. Jałochowski, M. Krawiec, Planar silicene: a new silicon allotrope epitaxially grown by segregation, Advanced Functional Materials 29 (2019) 1906053.

[62] J. Tersoff, D.R. Hamann, Theory of the scanning tunneling microscope, Physical Review B 31 (1985) 805.

[63] J. Gao, J. Zhang, H. Liu, Q. Zhang, J. Zhao, Structures, mobilities, electronic and magnetic properties of point defects in silicene, Nanoscale 5 (2013) 9785.

[64] P.E. Blöchl, Projector augmented-wave method, Physical Review B 50 (1994) 17953.

[65] J. Zhao, H. Liu, Z. Yu, R. Quhe, S. Zhou, Y. Wang, C.C. Liu, H. Zhong, N. Han, J. Lu, Y. Yao, K. Wu, Rise of silicene: a competitive 2D material, Progress in Materials Science 83 (2016) 24.

[66] F. Banhart, J. Kotakoski, A.V. Krasheninnikov, Structural defects in graphene, ACS Nano 5 (2011) 26.

[67] M. Ali, X. Pi, Y. Liu, D. Yang, Electronic and magnetic properties of graphene, silicene and germanene with varying vacancy concentration, AIP Advances 7 (2017) 045308.

[68] V.O. Özçelik, H.H. Gurel, S. Ciraci, Self-healing of vacancy defects in single-layer graphene and silicene, Physical Review B 88 (2013) 045440.

[69] H. Sahin, J. Sivek, S. Li, B. Partoens, F.M. Peeters, Stone-Wales defects in silicene: formation, stability, and reactivity of defect sites, Physical Review B 88 (2013) 045434.

[70] S. Li, Y. Wu, Y. Tu, Y. Wang, T. Jiang, W. Liu, Y. Zhao, Defects in silicene: vacancy clusters, extended line defects, and di-adatoms, Scientific Reports 5 (2015) 7881, https://doi.org/10.1038/srep07881.

[71] S. Wang, C. Ren, Y. Li, H. Tian, W. Lu, M. Sun, Spin and valley filter across line defect in silicene, Applied Physics Express 11 (2018) 053004.

[72] C.D. Ren, B. Zhou, S. Zhang, W. Lu, Y. Li, H. Tian, J. Liu, Controllable valley polarization using silicene double line defects due to Rashba spin-orbit coupling, Nanoscale Research Letters 14 (2019) 350, https://doi.org/10.1186/s11671-019-3196-3.

[73] H. Liu, H. Feng, Y. Du, J. Chen, K. Wu, J.J. Zhao, Point defects in epitaxial silicene on Ag(111) surfaces, 2D Materials 3 (2016) 025034.

[74] Z.-L. Liu, M.-X. Wang, C. Liu, J.-F. Jia, P. Vogt, C. Quaresima, C. Ottaviani, B. Olivieri, P. De Padova, G. Le Lay, The fate of the $2\sqrt{3} \times 2\sqrt{3}R(30\circ)$ silicene phase on Ag(111), Applied Physics Letters Materials 2 (2014) 092513.

[75] L. Li, S.-Z.L.J. Pan, Z. Qin, Y.-Q. Wang, Y. Wang, G.-Y. Cao, S. Du, H.-J. Gao, Buckled germanene formation on Pt(111), Advanced Materials 26 (2014) 4820–4824.

[76] M.E. Dávila, L. Xian, S. Cahangirov, A. Rubio, G. Le Lay, Germanene: a novel two-dimensional germanium allotrope akin to graphene and silicene, New Journal of Physics 16 (2014) 095002.

[77] P. Bampoulis, L. Zhang, A. Safaei, R. van Gastel, B. Poelsema, H.J.W. Zandvliet, Germanene termination of Ge_2Pt crystals on Ge(110), Journal of Physics. Condensed Matter 26 (2014) 442001.

[78] M. Derivaz, D. Dentel, R. Stephan, M.-C. Hanf, A. Mehdaoui, P. Sonnet, C. Pirri, Continuous germanene layer on Al(111), Nano Letters 15 (2015) 2510–2516.

[79] L. Zhang, P. Bampoulis, A.N. Rudenko, Q. Yao, A. van Houselt, B. Poelsema, M.I. Katsnelson, H.J.W. Zandvliet, Structural and electronic properties of germanene on MoS_2, Physical Review Letters 116 (2016) 256804.

[80] Z.H. Qin, J.B. Pan, S.Z. Lu, S. Yan, Y.L. Wang, S.X. Du, H.J. Gao, G.Y. Cao, Direct evidence of Dirac signature in bilayer germanene islands on Cu(111), Advanced Materials 29 (2017) 1606046.

[81] C.H. Lin, A. Huang, W.W. Pai, W.C. Chen, T.Y. Chen, T.R. Chang, R. Yukawa, C.M. Cheng, C.Y. Mou, I. Matsuda, T.C. Chiang, H.T. Jeng, S.J. Tang, Single-layer dual germanene phases on Ag(111), Physical Review Materials 2 (2018) 024003.

[82] E. Bianco, S. Butler, S. Jiang, O.D. Restrepo, W. Windl, J.E. Goldberger, Stability and exfoliation of germanane: a germanium graphane analogue, ACS Nano 5 (2013) 4414.

[83] S. Jiang, S. Butler, E. Bianco, O.D. Restrepo, W. Windl, J.E. Goldberger, Improving the stability and optical properties of germanane via one-step covalent methyl-termination, Nature Communications 5 (2014) 3383.

[84] J.E. Padilha, R.B. Pontes, Electronic and transport properties of structural defects in monolayer germanene: an ab initio investigation, Solid State Communications 225 (2016) 38–43.

[85] M. Saito, K. Yamashita, T. Oda, Magic numbers of graphene multivacancies, Japanese Journal of Applied Physics 46 (2007) L1185.

[86] O.V. Yazyev, L. Helm, Defect-induced magnetism in graphene, Physical Review B 75 (2007) 125408.

[87] P. Jamdagni, A. Kumar, M. Sharma, A. Thakur, P.K. Ahluwalia, Electronic properties and STM images of vacancy clusters and chains in functionalized silicene and germanene, Physica E 85 (2017) 65–73.

[88] L. Zhu, X. Chang, D. He, Q. Xue, X. Li, Y. Jin, H. Zheng, C. Ling, Defective germanene as a high-efficiency helium separation membrane: a first principles study, Nanotechnology 28 (2017) 135703.

[89] R. Paul, T. Tasnim, S. Saha, M. Motalab, Atomistic analysis to characterize the impact of temperature and defects on the mechanical properties of germanene sheet, Materials Research Express 5 (2018) 015062.

[90] Y. Xu, B. Yan, H.-J. Zhang, J. Wang, G. Xu, P. Tang, W. Duan, S.-C. Zhang, Large-gap quantum spin Hall insulators in tin films, Physical Review Letters 111 (2013) 136804.

[91] C.C. Liu, H. Jiang, Y. Yao, Low-energy effective Hamiltonian involving spin-orbit coupling in silicene and two-dimensional germanium and tin, Physical Review B 84 (2011) 195430.

[92] S.-C. Wu, G. Shan, B. Yan, Prediction of near-room-temperature quantum anomalous Hall effect on honeycomb materials, Physical Review Letters 113 (2014) 256401.

[93] S. Rachel, M. Ezawa, Giant magnetoresistance and perfect spin filter in silicene, germanene, and stanene, Physical Review B 89 (2014) 195303.

[94] J. Wang, Y. Xu, S.-C. Zhang, Two-dimensional time-reversal-invariant topological superconductivity in a doped quantum spin-Hall insulator, Physical Review B 90 (2014) 054503.

[95] H. Zhou, Y. Cai, G. Zhang, Y.-W. Zhang, Quantum thermal transport in stanene, Physical Review B 94 (2016) 045423.

[96] F.-F. Zhu, W.-J. Chen, Y. Xu, C.-L. Gao, D.-D. Guan, C.-H. Li, D. Qian, S.-C. Zhang, J.-F. Jia, Epitaxial growth of two-dimensional stanene, Nature Materials 14 (2015) 1020.

[97] J. Gou, L. Kong, H. Li, Q. Zhong, W. Li, P. Cheng, L. Chen, K. Wu, Strain-induced band engineering in monolayer stanene on Sb(111), Physical Review Materials 1 (2017) 054004.

[98] C.-Z. Xu, Y.-H. Chan, P. Chen, X. Wang, D. Flötotto, J.A. Hlevyack, G. Bian, S.-K. Mo, M.-Y. Chou, T.-C. Chiang, Gapped electronic structure of epitaxial stanene on InSb(111), Physical Review B 97 (2018) 035122.

[99] Y. Zang, T. Jiang, Y. Gong, Z. Guan, C. Liu, M. Liao, K. Zhu, Z. Li, L. Wang, W. Li, C. Song, D. Zhang, Y. Xu, K. He, X. Ma, S.-C. Zhang, Q.-K. Xue, Realizing an epitaxial decorated stanene with an insulating bandgap, Advanced Functional Materials 28 (2018) 1802723.

[100] J. Yuhara, Y. Fujii, K. Nishino, N. Isobe, M. Nakatake, L. Xian, A. Rubio, G. Le Lay, Large area planar stanene epitaxially grown on Ag(111), 2D Materials 5 (2018) 025002.

[101] J. Gao, Ga. Zhang, Y.-W. Zhang, Exploring Ag(111) substrate for epitaxially growing monolayer stanene: a first-principles study, Scientific Reports 6 (2016) 29107, https://doi.org/10.1038/srep29107.

[102] J. Li, T. Lei, J. Wang, R. Wu, H. Qian, K. Ibrahim, In-plane crystal field constrained electronic structure of stanene, Applied Physics Letters 116 (2020) 101601.

References

[103] M. Liao, Y. Zang, Z. Guan, H. Li, Y. Gong, K. Zhu, X.-P. Hu, D. Zhang, Y. Xu, Y.-Y. Wang, K. He, X.-C. Ma, S.-C. Zhang, Q.-K. Xue, Superconductivity in few-layer stanene, Nature Physics 14 (2018) 344.

[104] T. Eguchi, J. Nakamura, Structure and electronic states of the α-Sn(111)-(2 $\times$ 2) surface. & T. Osaka, Journal of the Physical Society of Japan 67 (1998) 381.

[105] W. Xiong, C. Xia, T. Wang, J. Du, Y. Peng, X. Zhaoa, Y. Jia, Tuning electronic structures of the stanene monolayer via defects and transition-metal-embedding: spin–orbit coupling, Physical Chemistry Chemical Physics 18 (2016) 28759.

[106] L. Shen, M. Lan, X. Zhang, G. Xiang, The structures and diffusion behaviors of point defects and their influences on the electronic properties of 2D stanene, RSC Advances 7 (2017) 9840.

[107] S. Das, S. Mojumder, T. Rakib, M.M. Islam, M. Motalab, Atomistic insights into mechanical and thermal properties of stanene with defects, Physical Review. B, Condensed Matter 553 (2019) 127.

[108] J. Deng, B. Xia, X. Ma, H. Chen, H. Shan, X. Zhai, B. Li, A. Zhao, Y. Xu, W. Duan, S.-C. Zhang, B. Wang, J.G. Hou, Epitaxial growth of ultraflat stanene with topological band inversion, Nature Materials 17 (2018) 1081.

[109] Z.-Q. Huang, C.-H. Hsu, F.-C. Chuang, Y.-T. Liu, H. Lin, W.-S. Su, V. Ozolins, A. Bansil, Strain driven topological phase transitions in atomically thin films of group IV and V elements in the honeycomb structures, New Journal of Physics 16 (2014) 105018.

[110] H. Zhao, C. Zhang, W. Ji, R. Zhang, S. Li, S. Yan, B. Zhang, P. Li, P. Wang, Unexpected giant-gap quantum spin Hall insulator in chemically decorated plumbene monolayer, Scientific Reports 6 (2016) 20152.

[111] X.-L. Yu, L. Huang, J. Wu, From a normal insulator to a topological insulator in plumbene, Physical Review B 95 (2017) 125113.

[112] D.K. Das, J. Sarkar, S.K. Singh, Effect of sample size, temperature and strain velocity on mechanical properties of plumbene by tensile loading along longitudinal direction: a molecular dynamics study, Computational Materials Science 151 (2018) 196.

[113] L. Zhang, H. Zhao, W. Ji, C. Zhang, P. Li, P. Wang, Discovery of a new quantum spin Hall phase in bilayer plumbene, Chemical Physics Letters 712 (2018) 78.

[114] J. Yuhara, B. He, N. Matsunami, M. Nakatake, G. Le Lay, Graphene's latest cousin: plumbene epitaxial growth on a "Nano water cube", Advanced Materials 31 (2019) 1901017.

[115] G. Bihlmayer, J. Sassmannshausen, A. Kubetzka, S. Blügel, Plumbene on a magnetic substrate: a combined scanning tunneling microscopy and density functional theory study, Physical Review Letters 124 (2020) 126401.

[116] A. Brown, S. Rundqvist, Refinement of the crystal structure of black phosphorus, Acta Crystallographica 19 (1965) 684.

[117] R. Hultgren, N.S. Gingrich, B.E. Warren, The atomic distribution in red and black phosphorus and the CrystalStructure of black phosphorus, Journal of Chemical Physics 3 (1935) 351.

[118] H. Thurn, H. Kerbs, Crystal structure of violet phosphorus, Angewandte Chemie. International Edition in English 5 (1966) 1047.

[119] S. Zhang, S. Guo, Z. Chen, Y. Wang, H. Gao, J. Gómez-Herrero, P. Ares, F. Zamora, Z. Zhu, H. Zeng, Recent progress in 2D group-VA semiconductors: from theory to experiment, Chemical Society Reviews 47 (2018) 982.

[120] L. Pauling, M. Simonetta, Bond orbitals and bond energy in elementary phosphorus, Journal of Chemical Physics 20 (1952) 29.

[121] R.W. Keyes, The electrical properties of black phosphorus, Physical Review 92 (1953) 580.

[122] P.W. Bridgman, Two new modifications of phosphorus, Journal of the American Chemical Society 36 (1914) 1344.

[123] L.-Q. Sun, M.-J. Li, K. Sun, S.-H. Yu, R.-S. Wang, H.-M. Xie, Electrochemical activity of black phosphorus as an anode material for lithium-ion batteries, Journal of Physical Chemistry C 116 (2012) 14772.

[124] J.C. Jamieson, Crystal structures adopted by black phosphorus at high pressures, Science 139 (1963) 1291.

[125] L. Li, Y. Yu, G.J. Ye, Q. Ge, X. Ou, H. Wu, D. Feng, X.H. Chen, Y. Zhang, Black phosphorus field-effect transistors, Nature Nanotechnology 9 (2014) 372.

[126] Z. Zhu, D. Tománek, Semiconducting layered blue phosphorus: a computational study, Physical Review Letters 112 (2014) 176802.

[127] M. Wu, H. Fu, L. Zhou, K. Yao, X.C. Zeng, Nine new phosphorene polymorphs with non-HoneycombStructures: a much extended family, Nano Letters 15 (2015) 3557.

[128] J. Guan, Z. Zhu, D. Tománek, Phase coexistence and metal-insulator transition in few-layer phosphorene: a computational study, Physical Review Letters 113 (2014) 046804.

[129] S. Zhang, M. Xie, F. Li, Z. Yan, Y. Li, E. Kan, W. Liu, Z. Chen, H. Zeng, Semiconducting group 15 monolayers: a broad range of band gaps and high carrier mobilities, Angewandte Chemie 128 (2016) 1698.

[130] X. Wang, A.M. Jones, K.L. Seyler, V. Tran, Y. Jia, H. Zhao, H. Wang, L. Yang, X. Xu, F. Xia, Highly anisotropic and robust excitons in monolayer black phosphorus, Nature Nanotechnology 10 (2015) 517.

[131] M. Buscema, D.J. Groenendijk, S.I. Blanter, G.A. Steele, H.S.J. van der Zant, A. Castellanos-Gomez, Fast and broadband photoresponse of few-layer black phosphorus field-effect transistors, Nano Letters 14 (2014) 3347.

[132] H. Liu, A.T. Neal, Z. Zhu, Z. Luo, X. Xu, D. Tománek, P.D. Ye, Phosphorene: an unexplored 2D semiconductor with a high hole mobility, ACS NANO 8 (2014) 4033.

[133] R. Galceran, E. Gaufres, A. Loiseau, M. Piquemal-Banci, F. Godel, A. Vecchiola, O. Bezencenet, M.-B. Martin, B. Servet, F. Petroff, B. Dlubak, P. Seneor, Stabilizing ultra-thin black phosphorus within-situ-grown 1 nm-Al_2O_3 barrier, Applied Physics Letters 111 (2017) 243101.

[134] J.R. Brent, N. Savjani, E.A. Lewis, S.J. Haigh, D.J. Lewis, P. O'Brien, Production of few-layer phosphorene by liquid exfoliation of black phosphorus, Chemical Communications 50 (2014) 13338.

[135] W. Zhao, Z. Xue, Ji. Wang, J. Jiang, X. Zhao, T. Mu, Large-scale, highly efficient, and green liquid-exfoliation of black phosphorus in ionic liquids, ACS Applied Materials & Interfaces 7 (2015) 27608.

[136] W. Lu, H. Nan, J. Hong, Y. Chen, C. Zhu, Z. Liang, X. Ma, Z. Ni, C. Jin, Z. Zhang, Plasma-assisted fabrication of monolayer phosphorene and its Raman characterization, Nano Research 7 (7) (2014) 853.

[137] Z. Yang, J. Hao, S. Yuan, S. Lin, H.M. Yau, J. Dai, S.P. Lau, Field-effect transistors based on amorphous black phosphorus ultrathin films by pulsed laser deposition, Advanced Materials 27 (2015) 3748.

[138] J.B. Smith, D. Hagaman, H.F. Ji, Growth of 2D black phosphorus film from chemical vapor deposition, Nanotechnology 27 (2016) 215602.

[139] J.L. Zhang, S. Zhao, C. Han, Z. Wang, S. Zhong, S. Sun, R. Guo, X. Zhou, C.D. Gu, K.D. Yuan, Z. Li, W. Chen, Epitaxial growth of single layer blue phosphorus: a new phase of two-dimensional phosphorus, Nano Letters 16 (2016) 4903.

[140] C. Gu, S. Zhao, J.L. Zhang, S. Sun, K. Yuan, Z. Hu, C. Han, Z. Ma, L. Wang, F. Huo, W. Huang, Z. Li, W. Chen, Growth of quasi-free-standing single-layer blue phosphorus on tellurium monolayer functionalized Au(111), ACS Nano 11 (2017) 4943.

[141] W. Hu, J. Yang, Defects in phosphorene, Journal of Physical Chemistry C 119 (2015) 20474.

[142] Brian Kiraly, Nadine Hauptmann, Alexander N. Rudenko, Mikhail I. Katsnelson, Alexander A. Khajetoorians, Probing single vacancies in black phosphorus at the atomic level, Nano Letters 17 (2017) 3607.

[143] M. Sun, J.-P. Chou, A. Hu, U. Schwingenschlögl, Point defects in blue phosphorene, Chemistry of Materials 31 (2019) 8129.

[144] Andrew Cupo, Paul Masih Das, Chen-Chi Chien, Gopinath Danda, Neerav Kharche, Damien Tristant, Marija Drndić, Vincent Meunier, Periodic arrays of phosphorene nanopores as antidot lattices with tunable properties, ACS Nano 11 (2017) 7494.

[145] J.V. Riffle, C. Flynn, B.St. Laurent, C.A. Ayotte, C.A. Caputo, S.M. Hollen, Impact of vacancies on electronic properties of black phosphorus probed by STM, Journal of Applied Physics 123 (2018) 044301.

[146] Abhishek Kumar, F. Telesio, S. Forti, A. Al-Temimy, C. Coletti, M. Serrano-Ruiz, M. Caporali, M. Peruzzini, F. Beltram, S. Heun, STM study of exfoliated few layer black phosphorus annealed in ultrahigh vacuum, 2D Materials 6 (2019) 015005.

[147] Z.-D. Sha, Q.-X. Pei, Y.-Y. Zhang, Y.-W. Zhang, Atomic vacancies significantly degrade the mechanical properties of phosphorene, Nanotechnology 27 (2016) 315704.

[148] Y. Chen, H. Xiao, Y. Liu, X. Chen, Effects of temperature and strain rate on mechanical behaviors of stone-Wales defective monolayer black phosphorene, Journal of Physical Chemistry C 122 (2018) 6368.

[149] H. Qin, V. Sorkin, Q-X. Pei, Y. Liu, Y-W. Zhang, Failure in two-dimensional materials: defect sensitivity and failure criteria, Journal of Applied Mechanics 87 (2020) 030802.

[150] N.C. Norman, Chemistry of Arsenic, Antimony and Bismuth, Springer Science & Business Media, 1997.

[151] A.F. Welles, Structural Inorganic Chemistry, fifth edition, Clarendon Press, Oxford University Press, United Kingdom, 1982.

[152] H. Krebs, W. Holz, K.H. Worms, Über die Struktur und die Eigenschaften der Halbmetalle, X. Eine Neue Rhombische Arsenmodifikation und Ihre Mischkristallbildung mit Schwarzem Phosphor, Chem., Ber 90 (1957) 103.

[153] P.M. Smith, A.J. Leadbetter, A.J. Apling, The structures of orthorhombic vitreous arsenic, Philosophical Magazine 31 (1975) 57.

[154] P.F. Zhang, Z. Liu, W. Duan, F. Liu, J. Wu, Topological and electronic transitions in a Sb(111) nanofilm: The interplay between quantum confinement and surface effect, Physical Review B 85 (2012), 201410(R).

[155] F.-C. Chuang, C.-H. Hsu, C.-Y. Chen, Z.-Q. Huang, V. Ozolins, H. Lin, A. Bansil, Tunable topological electronic structures in Sb(111) bilayers: a first-principles study, Applied Physics Letters 102 (2013) 022424.

[156] Z. Zhu, J. Guan, D. Tománek, Strain-induced metal-semiconductor transition in monolayers and bilayers of gray arsenic: a computational study, Physical Review B 91 (2015), 161404(R).

[157] C. Kamal, M. Ezawa, Arsenene: two-dimensional buckled and puckered honeycomb arsenic systems, Physical Review B 91 (2015) 085423.

[158] G. Wang, R. Pandey, S.P. Karna, Atomically thin group V elemental films: theoretical investigations of antimonene allotropes, ACS Applied Materials & Interfaces 7 (2015) 11490.

[159] S.K. Gupta, Y. Sonvane, G. Wang, Ravindra pandey, size and edge roughness effects on thermal conductivity of pristine antimonene allotropes, Chemical Physics Letters 641 (2015) 169.

[160] O. Madelung, Semiconductors: Data Handbook, 3rd ed., Springer, Berlin, 2004.

[161] D. Schiferl, C.S. Barrett, The crystal structure of arsenic at 4.2, 78 and 299 °K, Journal of Applied Crystallography 2 (1969) 30.

[162] S. Zhang, Z. Yan, Y. Li, Z. Chen, H. Zeng, Atomically thin arsenene and antimonene: semimetal–semiconductor and indirect–direct band-gap transitions, Angewandte Chemie. International Edition in English 54 (2015) 3112.

[163] J. Shah, W. Wang, H.M. Sohail, R.I.G. Uhrberg, Experimental evidence of monolayer arsenene: an exotic 2D semiconducting material, 2D Materials 7 (2020) 025013.

[164] H.-S. Tsai, S.-W. Wang, C.-H. Hsiao, C.-W. Chen, H. Ouyang, Y.-L. Chueh, H.-C. Kuo, J.-H. Liang, Direct synthesis and practical bandgap estimation of multilayer arsenene nanoribbons, Chemistry of Materials 28 (2016) 425.

[165] R. Gusmao, Z. Sofer, D. Bousa, M. Pumera, Pnictogen (As, Sb, Bi)nanosheets for electrochemical applications are produced by shear exfoliation using kitchen blenders, Angewandte Chemie. International Edition in English 56 (2017) 14417.

[166] P. Vishnoi, M. Mazumder, S.K. Pati, C.N.R. Rao, Arsenene nanosheets and nanodots, New Journal of Chemistry 42 (2018) 14091.

[167] T. Lei, C. Liu, J.-L. Zhao, J.-M. Li, Y.-P. Li, Ji.-O. Wang, R. Wu, H.-J. Qian, H.-Q. Wang, K. Ibrahim, Electronic structure of antimonene grown on Sb_2Te_3 (111) and Bi_2Te_3 substrates, Journal of Applied Physics 119 (2016) 015302.

[168] H.-S. Tsai, C.-W. Chen, C.-H. Hsiao, H. Ouyang, J.-H. Liang, The advent of multilayer antimonene nanoribbons with room temperature orange light emission, Chemical Communications 52 (2016) 8409.

[169] P. Ares, F. Aguilar-Galindo, D. Rodríguez-San-Miguel, D.A. Aldave, S. Díaz-Tendero, M. Alcamí, F. Martín, J. Gómez-Herrero, F. Zamora, Mechanical isolation of highly stable antimonene under ambient conditions, Advanced Materials 28 (2016) 6332.

[170] P. Ares, F. Zamora, J. Gomez-Herrero, Optical identification of few-layer antimonene crystals, ACS Photonics 4 (2017) 600.

[171] C. Gibaja, D. Rodriguez-San-Miguel, P. Ares, J. Gomez- Herrero, M. Varela, R. Gillen, J. Maultzsch, F. Hauke, A. Hirsch, G. Abellan, F. Zamora, Few-layer antimonene by liquid-phase exfoliation, Angewandte Chemie. International Edition in English 55 (2016) 14345.

[172] L. Lu, X. Tang, R. Cao, L. Wu, Z. Li, G. Jing, B. Dong, S. Lu, Y. Li, Y. Xiang, J. Li, D. Fan, H. Zhang, Broadband nonlinear optical response in few-layer antimonene and antimonene quantum dots: a promising optical Kerr media with enhanced stability, Advanced Optics Materials 5 (2017) 1700301.

[173] J. Ji, X. Song, Jizi Liu, Z. Yan, C. Huo, S. Zhang, M. Su, L. Liao, W. Wang, Z. Ni, Y. Hao, H. Zeng, Two-dimensional antimonene single crystals grown by van der Waals epitaxy, Nature Communications 7 (2016) 13352, https://doi.org/10.1038/ncomms13352.

[174] X. Wu, Y. Shao, H. Liu, Z. Feng, Y.-L. Wang, J.-T. Sun, C. Liu, J.-O. Wang, Z.-L. Liu, S.-Y. Zhu, Y.-Q. Wang, S.-X. Du, Y.-G. Shi, K. Ibrahim, H.-J. Gao, Epitaxial growth and air-stability of monolayer antimonene on $PdTe_2$, Advanced Materials 29 (2017) 1605407.

[175] Synthesis of Antimonene on Germanium M. Fortin-Deschênes, O. Waller, T.O. Mentes, A. Locatelli, S. Mukherjee, F. Genuzio, P.L. Levesque, A. Hébert, R. Martel, O. Moutanabbir, Synthesis of antimonene on germanium, Nano Letters 17 (2017) 174970.

[176] Y. Shao, Z.-L. Liu, C. Cheng, X. Wu, H. Liu, C. Liu, J.-O. Wang, S.-Y. Zhu, Y.-Q. Wang, D.-X. Shi, K. Ibrahim, J.-T. Sun, Y.-L. Wang, H.-J. Gao, Epitaxial growth of flat antimonene monolayer: a new honeycomb analogue of graphene, Nano Letters 18 (2018) 2133.

[177] M. Jałochowski, M. Krawiec, Antimonene on Pb quantum wells, 2D Materials 6 (2019) 045028.

[178] Y. Hu, Y. Wu, S. Zhang, Influences of Stone–Wales defects on the structure, stability and electronic properties of antimonene: a first principle study, Physica B 503 (2016) 126.

[179] A. Bafekry, M. Ghergherehchi, S.F. Shayesteh, Tuning the electronic and magnetic properties of antimonene nanosheets via point defects and external fields: first-principles calculations, Physical Chemistry Chemical Physics 21 (2019) 10552.

[180] Z. Hu, J. Gao, S. Zhang, J. Zhao, W. Zhou, H. Zeng, Topologically protected states and half-metal behaviors: defect-strain synergy effects in two-dimensional antimonene, Physical Review. Matter 3 (2019) 074005.

[181] X. Liang, S.-P. Ng, N. Ding, C.-M.L. Wu, Characterization of point defects in monolayer arsenene, Applied Surface Science 443 (2018) 74.

[182] K. Iordanidou, J. Kioseoglou, V.V. Afanas'ev, A. Stesmansa, M. Houssa, Intrinsic point defects in buckled and puckered arsenene: a first-principles study, Physical Chemistry Chemical Physics 19 (2017) 9862.

[183] Y. Liu, T. Wang, J. Robertson, J. Luo, Y. Guo, D. Liu, Journal of Physical Chemistry C 124 (2020) 7441.

[184] X. Sun, Y. Liu, Z. Song, Y. Li, W. Wang, H. Lin, L. Wang, Y. Li, Structures, mobility and electronic properties of point defects in arsenene, antimonene and an antimony arsenide alloy, Journal of Materials Chemistry 5 (2017) 4159.

[185] C. Kittel, Introduction to Solid State Physics, Wiley, 2005.

[186] Z. Liu, C.-X. Liu, Y.-S. Wu, W.-H. Duan, F. Liu, J. Wu, Stable nontrivial Z_2 topology in ultrathin Bi(111) films: a first-principles study, Physical Review B 107 (2011) 136805.

[187] Shuichi Murakami, Quantum spin Hall effect and enhanced magnetic response by spin-orbit coupling, Physical Review Letters 97 (2006) 236805.

[188] M. Wada, S. Murakami, F. Freimuth, G. Bihlmayer, Localized edge states in two-dimensional topological insulators: ultrathin Bi films, Physical Review B 83 (2011) 121310(R).

[189] S. Zhang, M. Xie, F. Li, Z. Yan, Y. Li, E. Kan, W. Liu, Z. Chen, H. Zeng, Semiconducting group 15 monolayers: a broad range of band gaps and high carrier mobilities, Angewandte Chemie. International Edition in English 55 (2016) 1666.

[190] T. Nagao, S. Yaginuma, M. Saito, T. Kogure, J.T. Sadowski, T. Ohno, S. Hasegawa, T. Sakurai, Strong lateral growth and crystallization via two-dimensional allotropic transformation of semi-metal Bi film, Surface Science 590 (2005) L247.

[191] M. Jankowski, D. Kamiński, K. Vergeer, M. Mirolo, F. Carla, G. Rijnders, T.R.J. Bollmann, Controlling the growth of Bi(110) and Bi(111) films on an insulating substrate, Nanotechnology 28 (2017) 155602.

[192] H. -Hua Sun, M.-X. Wang, F. Zhu, G.-Y. Wang, H.-Y. Ma, Z.-A. Xu, Q. Liao, Y. Lu, C.-L. Gao, Y.-Y. Li, C. Liu, D. Qian, D. Guan, J.-F. Jia, Coexistence of topological edge state and superconductivity in bismuth ultrathin film, Nano Letters 17 (2017) 3035.

[193] F. Reis, G. Li, L. Dudy, M. Bauernfeind, S. Glass, W. Hanke, R. Thomale, J. Schäfer, R. Claessen, Bismuthene on a SiC substrate: a candidate for a high-temperature quantum spin Hall material, Science 357 (2017) 287.

[194] Q.-Q. Yang, R.-T. Liu, C. Huang, Y.-F. Huang, L.-F. Gao, B. Sun, Z.-P. Huang, L. Zhang, C.-X. Hu, Z.-Q. Zhang, C.-L. Sun, Q. Wang, Y.-L. Tang, H.-L. Zhang, 2D bismuthene fabricated via acid-intercalated exfoliation showing strong nonlinear near-infrared responses for mode-locking lasers, Nanoscale 10 (2018) 21106.

[195] C. Shen, T. Cheng, C. Liu, L. Huang, M. Cao, G. Song, D. Wang, B. Lu, J. Wang, C. Qin, X. Huang, P. Peng, X. Lia, Y. Wu, Bismuthene from sonoelectrochemistry as a superior anode for potassium-ion batteries, Journal of Materials Chemistry A 8 (2020) 453.

[196] N. Hussain, T. Liang, Q. Zhang, T. Anwar, Y. Huang, J. Lang, K. Huang, H. Wu, Ultrathin Bi nanosheets with superior photoluminescence, Small 13 (2017) 1701349.

[197] F. Xing, D. Fan, H. Zhang, Few-layer bismuthene: sonochemical exfoliation, nonlinear optics and applications for ultrafast photonics with enhanced stability, Laser & Photonics Reviews (2017) 1700221.

[198] Y. Lu, W. Xu, M. Zeng, G. Yao, L. Shen, M. Yang, Z. Luo, F. Pan, K. Wu, T. Das, P. He, J. Jiang, J. Martin, Y.P. Feng, H. Lin, X.-S. Wang, Topological properties determined by atomic buckling in self-assembled ultrathin Bi(110), Nano Letters 15 (2015) 80.

[199] E.S. Walker, S.R. Na, D. Jung, S.D. March, J.S. Kim, T. Trivedi, W. Li, L. Tao, M.L. Lee, K.M. Liechti, D. Akinwande, S.R. Bank, Large-area dry transfer of single-crystalline epitaxial bismuth thin films, Nano Letters 16 (2016) 6931.

[200] Y. Kadioglu, S.B. Kilic, S. Demirci, O.Ü. Aktürk, E. Aktürk, S. Ciraci, Modification of electronic structure, magnetic structure, and topological phase of bismuthene by point defects, Physical Review B 96 (2017) 245424.

References

[201] J. Li, A. Ciani, J. Gayles, D.A. Papaconstantopoulos, N. Kioussis, C. Grein, F. Aqariden, Non-orthogonal tight-binding model for tellurium and selenium, Philosophical Magazine 93 (23) (2013) 3216–3230, https://doi.org/10.1080/14786435.2013.801569.

[202] P. Cherin, P. Unger, The crystal structure of trigonal selenium, Inorganic Chemistry 6 (1967) 1589.

[203] R. Keller, W.B. Holzapfel, H. Schulz, Effect of pressnre on the atom positions in Se and Te, Physical Review B 16 (1977) 4404.

[204] Z. Zhu, X. Cai, S. Yi, J. Chen, Y. Dai, C. Niu, Z. Guo, M. Xie, F. Liu, J.-H. Cho, Y. Jia, Z. Zhang, Multivalency-driven formation of te-based monolayer materials: a combined first-principles and experimental study, Physical Review Letters 119 (2017) 106101.

[205] Y. Wang, J. Lv, L. Zhu, Y. Ma, Crystal structure prediction via particle-swarm optimization, Physical Review B 82 (2010) 0941162010.

[206] C. Wang, X. Zhou, J. Qiao, L. Zhou, X. Kong, Y. Pan, Z. Cheng, Y. Chai, W. Ji, Charge-governed phase manipulation of few-layer tellurium, Nanoscale 10 (2018) 22263.

[207] J. Chen, Y. Dai, Y. Ma, X. Dai, W. Ho, M. Xie, Ultrathin β-tellurium layers grown on highly oriented pyrolytic graphite by molecular-beam epitaxy, Nanoscale 9 (2017) 15945.

[208] X. Huang, J. Guan, Z. Lin, B. Liu, S. Xing, W. Wang, J. Guo, Epitaxial growth and band structure of Te film on graphene, Nano Letters 17 (2017) 4619.

[209] Z. Xie, C. Xing, W. Huang, T. Fan, Z. Li, J. Zhao, Y. Xiang, Z. Guo, J. Li, Z. Yang, B. Dong, J. Qu, D. Fan, H. Zhang, Ultrathin 2D nonlayered tellurium nanosheets: facile liquid-phase exfoliation, characterization, and photoresponse with high performance and enhanced stability, Advanced Functional Materials 28 (2018) 1705833.

[210] J. Guo, J. Zhao, D. Huang, Y. Wang, F. Zhang, Y. Ge, Y. Song, aChenyang Xing, D. Fana, H. Zhang, Two-dimensional tellurium–polymer membrane for ultrafast photonics, Nanoscale 11 (2019) 6235.

[211] W. Wu, G. Qiu, Y. Wang, R. Wang, P. Ye, Tellurene: its physical properties, scalable nanomanufacturing, and device applications, Chemical Society Reviews 47 (2018) 7203.

[212] Y. Wang, G. Qiu, R. Wang, S. Huang, Q. Wang, Y. Liu, Y. Du, W.A. Goddard III, M.J. Kim, X. Xu, P.D. Yeand, W. Wu, Field-effect transistors made from solution-grown two-dimensional tellurene, Nature Electronics 1 (2018) 228.

[213] L. Xian, A.P. Paz, El. Bianco, P.M. Ajayan, A. Rubio, Square selenene and tellurene: novel group VI elemental 2D materials with nontrivial topological properties, 2D Materials 4 (2017) 041003.

[214] G. Liu, Z. Gao, G.-L. Li, H. Wang, Abnormally low thermal conductivity of 2D selenene: an ab initio study, Journal of Applied Physics 127 (2020) 065103.

[215] D. Wang, L.-M. Tang, X.-X. Jiang, J.-Y. Tan, M.-D. He, X.-J. Wang, K.-Q. Chen, High bipolar conductivity and robust in-plane spontaneous electric polarization in selenene, Advanced Electronic Materials 5 (2019) 1800475.

[216] J. Qin, G. Qiu, J. Jian, H. Zhou, L. Yang, A. Charnas, D.Y. Zemlyanov, C.-Y. Xu, X. Xu, W. Wu, H. Wang, P.D. Ye, Controlled growth of a large-size 2D selenium nanosheet and its electronic and optoelectronic applications, ACS Nano 11 (2017) 10222.

[217] H.-M. Zhang, et al., Detection of a superconducting phase in a two-atom layer of hexagonal Ga film grown on semiconducting GaN(0001), Physical Review Letters 114 (2015) 107003.

[218] M.-L. Tao, Y.-B. Tu, K. Sun, Y.-L. Wang, Z.-B. Xie, L. Liu, M.-X. Shi, J.-Z. Wang, Gallenene epitaxially grown on Si(111), 2D Materials 5 (2018) 035009.

[219] V. Kochat, et al., Atomically thin gallium layers from solid-melt exfoliation, Science Advances 4 (2018) e1701373.

[220] K.G. Steenbergen, N. Gaston, Thickness dependent thermal stability of 2D gallenene, Chemical Communications 55 (2019) 8872.

[221] D. Coster, G. Hevesy, On the missing element of atomic number 72, Nature 111 (1923) 79, https://doi.org/10.1038/111079a0.

[222] N. Bohr, The Theory of Spectra and Atomic Constitution, Cambridge University Press, 1922, C.F. Clay, Manager London: Fetter Lane, E.C.

[223] D. Coster, G. Hevesy, On the new element hafnium, Nature 111 (1923) 182, https://doi.org/10.1038/111182a0.

[224] D. Coster, G. Hevesy, On celtium and hafnium, Nature 111 (1993) 462, https://doi.org/10.1038/111462a0.

[225] M.G. Urbain, A. Dauvillier, On the element of atomic number 72, Nature 111 (1923) 218.

[226] M.G. Urbain, Sur un nouvel élément qui accompagne le lutécium e le scandinium dans les terres de la gadolinite: le celtium, Comptes Rendus 152 (1911) 141–143, https://www.biodiversitylibrary.org/item/31511#page/9/mode/1up.

[227] M.A. Dauvillier, Sur les series L du lutécium et de l'ytterbium et sur l'identification du celtium avec l'élément de numbre atomique, Comptes Rendus 72 (174) (1922) 1347–1349, https://www.biodiversitylibrary.org/item/31193#page/9/mode/1up.

[228] Les numéros atomique du néo-ytterbium, du lutecium et du celtium, Comptes Rendus 174 (1922) 1349–1351, https://www.biodiversitylibrary.org/item/31511#page/9/mode/1up.

[229] H.M. Hansen, S. Werner, The optical spectrum of hafnium, Nature 111 (1923) 322.

[230] H.M. Hansen, S. Werner, On urbain's celtium lines, Nature 111 (1923) 461.

[231] L. Li, Y. Wang, S. Xie, X.-B. Li, Y.-Q. Wang, R. Wu, H. Sun, S. Zhang, H.-J. Gao, Two-dimensional transition metal honeycomb realized: Hf on Ir(111), Nano Letters 3 (2013) 4671.

[232] L. Li, Springer theses recognizing outstanding Ph.D. research, fabrication and physical properties of novel two-dimensional crystal materials beyond graphene: germanene, hafnene and PtSe2, https://www.springer.com/gp/book/9789811519628, 2020.

[233] Y. Pan, L. Zhang, L. Huang, L. Li, L. Meng, M. Gao, Q. Huan, X. Lin, Y. Wang, S. Du, H.-J. Freund, H.-J. Gao, Construction of 2D atomic crystals on transition metal surfaces: graphene, silicene, and hafnene, Small 3 (2014) 2215.

[234] G. Li, Y.-Y. Zhang, H. Guo, L. Huang, H. Lu, X. Lin, Y.-L. Wang, S. Du, H.-J. Gao, Epitaxial growth and physical properties of 2D materials beyond graphene: from monatomic materials to binary compounds, Chemical Society Reviews 47 (2018) 6073.

[235] X. Li, Y. Dai, Y. Ma, B. Huang, Electronic and magnetic properties of honeycomb transition metal monolayers: first-principles insights, Physical Chemistry Chemical Physics 16 (2014) 13383.

[236] B. Zhou, X. Wang, S. Dong, K. Zhang, W. Mi, Tunable gap opening and spin polarization of two dimensional graphene/hafnene van der Waals heterostructures, Carbon 120 (2017) 121.

[237] A. Hashmi, M.U. Farooq, I. Khan, J. Hong, Two-dimensional honeycomb hafnene monolayer: stability and magnetism by structural transition, Nanoscale 9 (2017) 10038.

[238] H. Duan, N. Yan, R. Yu, C.-R. Chang, G. Zhou, H.-S. Hu, H. Rong, Z. Niu, J. Mao, H. Asakura, T. Tanaka, P.J. Dyson, J. Li, Y. Li, Ultrathin rhodium nanosheets, Nature Communications 5 (2014) 3093, https://doi.org/10.1038/ncomms4093.

[239] L. Zhao, C. Xu, H. Su, J. Lian, S. Lin, L. Gu, X. Wang, M. Chen, N. Zheng, Single-crystalline rhodium nanosheets with atomic thickness, Advanced Science 2 (2015) 1500100.

[240] X. Wang, C. Wang, C. Chen, H. Duan, K. Du, Free-standing monatomic thick two-dimensional gold, Nano Letters 19 (2019) 4560.

[241] W. Yuan, Z. Deng, Z. Ren, Y. Shen, W. Xi, J. Luo, Monolayer goldene intercalated in graphene layers, Applied Physics Letters 117 (2020) 233102, https://doi.org/10.1063/5.0019487.

[242] J. Zhao, Q. Deng, A. Bachmatiuk, G. Sandeep, A. Popov, J. Eckert, M.H. Rümmeli, Free-standing single-atom-ThickIron membranes suspended in graphene pores, Science 343 (2014) 1228.

CHAPTER 4

Defects in transition metal dichalcogenides

Stephen McDonnell and Petra Reinke

Department of Materials Science and Engineering, University of Virginia, Charlottesville, VA, United States

4.1 Introduction

This chapter discusses defects in transition metal dichalcogenides (TMDs). We employ a simple definition of defect as being anything that deviates from a perfect periodic structure of the pure material and separate out our subsections by dimensionality. This includes point defects (0D) such as vacancies, anti-sites as well as impurities. Impurities are given their own section in this chapter and include both contaminants (substitution, interstitial or intercalated impurities as well as inclusions (3D)), and deliberate impurities such as substitutional dopants, intercalation functional ions, or alloying elements. Grain boundaries are expressed in 2D materials as "line defects", in contrast to their 2D defect designation in 3D materials. This is a natural result of the lack of primary bonding out-of-plane. Finally, we briefly summarize the chapter by discussing process control of defects and engineering of defects for targeted functionality.

4.2 Point defects

Point defects in TMDs can decisively impact performance in electronic, photonic, and optoelectronic devices [1–9]. These defects can also provide active sites for catalytic reactions [10–12], or anchor sites for chemical functionalization. In general, uncontrolled point defects are seen as detrimental to device performance [13,14]. Recent studies have taken advantage of defect engineering, which is well established in the "traditional" semiconductor industry, and defects with unique electronic or photonic signatures are created and add functionality to the TMD layer [15–17]. Establishing the link between the "point defect inventory" in a given

90   4. Defects in transition metal dichalcogenides

TMD layer and its performance is a highly active area of research and impacts nearly all uses of TMD materials [1,18].

The intrinsic defect density is generally larger in TMDs compared to graphene, where densities as low as 10^9 cm^{-2} have been reported in exfoliated films, and it includes defects in the chalcogen as well as the transition metal sublattice. Point defects can act as scattering centers, induce local changes in band structure as well as band bending [14,19,20], cause recombination, and localized photon emission which all rely on the local electronic structure of specific defects [21–23]. The current interest in the community is on isolated point defects that involve fewer than 5 atoms. The next challenge is the study of interactions between closely spaced defects, perhaps within the active region of a device, and geometrically complex and more extended defects which can form during annealing, or irradiation [24,25].

4.2.1 Defect inventory

The "point defect inventory" as defined here includes information about spatial distribution, density, geometric, and electronic structure of all point defects across a TMD layer. Initially, growth of TMDs was directed towards minimizing the defect density, and approach the very low defect concentrations reported in high quality "bulk samples", which can be grown utilizing a flux-growth method starting from the melt [26]. Most growth methods such as CVD (chemical vapor deposition), MOCVD (metal organic CVD), and MBE (molecular beam epitaxy) are now able (with optimized growth parameters for a given reactor) to reach, or come at least close to bulk sample values, which are reported as having as low as 7×10^{10} cm^{-2} point defects per unit area by Edelberg *et al.* [26] for flux grown MoS$_2$ crystals. This result shifts the challenge from "counting" defects to understanding their geometric and electronic structure, and ultimately correlating the defect type with device performance [5,27]. Our understanding of point defects in sulfides and selenides has advanced in the last decade owing to careful imaging and spectroscopic studies. However, while interest in tellurides is growing, the data on their defect inventory remains scarce. Pandey *et al.* [28] introduced an interesting distinction between defect-tolerant and defect-sensitive TMDs, whereas defect states which are far away from the band edges compared to $k_B T$, where k_B is the Boltzmann constant, introduce localized states with a significant impact on material properties, while shallow states tend to modify mostly conductivity values. Hence, the exact nature of the electronic structure of defects and their effects on the electronic and/or photonic properties is indeed critically important to understand.

Experimentally, the characteristics of defects can be measured by spatially integrated methods such as photoluminescence (PL), which is highly sensitive to the electronic signatures of optically active defects, and microscopy with atomic resolution such as transmission electron microscopy (TEM) [3,29–31], scanning tunneling microscopy and spectroscopy (STM/STS), and atomic force microscopy (AFM). Other methods are sensitive to specific types of defects, or reflect changes in the overall electronic structure, and include photoelectron and Raman spectroscopy. [19,32–36] Spatially integrative methods lack information about the defect density but can capture electronic and optical signatures rapidly, and recent technical advances have significantly improved spatial resolution of several spectroscopic techniques. The PL signature is usually related to the Wannier-Mott exciton, which delivers two states due to the spin orbit splitting, and the intensity ratio of these two peaks [37] can be used as a signature of the defects in MoS$_2$ and likely also in other TMDs [22]. The localization of an exciton is strongly temperature dependent in correlation with its binding energy to a defect, and it was shown that excitons tend to be trapped in the vicinity of defect-rich edges of TMD flakes, where a higher density of vacancies is present as shown in WS$_2$ flakes using scanning electron microscopy [22,38].

Scanning probe methods, on the other hand, tend to be slow, but offer a more complete picture of the geometric, and often also the local electronic structure of point defects. The limited area which can be studied with scanning probe techniques (STM, and AFM) remains a challenge: a defect density of 10^{12} cm^{-2} offers only 20 defects for imaging in a 20×20 nm^2 image. STM/STS still gives the most direct insight into the correlation between geometric and electronic structure, albeit it can be influenced by tip conditions, or tip induced band bending. All studies of point defects should be ideally combined with density functional theory (DFT) calculations, which offer insight in the energy of formation as a function of chemical potential, defect electronic states within the band gap, and distortions of the local symmetry [39,40]. The defect inventory is not only controlled/influenced by the specific choice of growth conditions or method but frequently modified by after-growth processing including annealing, exposure to air, which can cause oxygen substitution of chalcogen vacancies and oxidation at step edges and can lead to the introduction of light elements (H, O, OH, CO) as unintentional dopants [17,41].

4.2.2 Defect classification

Point defects are defined as zero dimensional (0D), localized defects, and the following types of point defects have been observed experimentally or have been studied with DFT [42]:

(i) **Vacancies** – a missing atom in either the chalcogen (V$_{S,Se,Te}$) or metal sub-lattice (V$_m$), as well as multi-atom vacancies, for example, V$_{MoS3}$ where an entire trigonal unit with a metal at the center and three chalcogen atoms, is missing. Larger vacancy units are mostly studied with DFT methods and are very challenging to assign unambiguously based on imaging methods.

(ii) **Substitutional** – intentional on either the chalcogen or metal sublattice with a chalcogen (O, S, Se, Te) or metal (any of the transition metals) respectively, or unintentional dopants on either of the sublattices.

(iii) **Antisite defects** – the chalcogen and transition metal atom switch places.

(iv) **Interstitials** are mostly relevant in multilayer TMDs and consist of an atom trapped in the intercalation space between layers. Sometimes chemisorbed adatoms on the TMD surface are described as interstitials due to the similarity in bonding states.

(v) **Geometrically complex defects** extending over several atoms, which emerge during annealing and are the initial reaction steps in TMD degradation.

Intentional doping with electrically active elements to introduce, enhance, or convert the conduction of films to p- or n-type is not considered in this section, despite their formally correct designation point defects. These are instead discussed as impurities in Section 4.3. Defects (and especially vacancies) can develop in the top or bottom (substrate facing) chalcogen sub-lattice and the energy of formation, as well as the electronic signature of the defects will depend on the nature of the substrate, the specific interfacial bonding and structure, and the dielectric constant of substrate and TMD. [43,44] A recent experiment, which observed defect formation as a function of annealing temperature showed that the defect density on the 1st WSe$_2$ layer in contact with a graphite substrate is higher than the defect density on the 2nd WSe$_2$ layer, which is only in contact with the 1st TMD layer [24]. The electronic signature of point defects is critical to many aspects of TMD materials uses: (i) catalytic activity frequently depends on the filling and emptying of specific gap states to facilitate chemical reactions, and (ii) the presence of gap states controls local n-or p-doping. However, creating a vacancy (or any other point defect) distorts the local symmetry leading, for example, to Jahn-Teller distortions, which in turn impact photonic processes, excitation as well as recombination.

An extensive DFT study by Haldar *et al.* [40] compares the energy of formation of chalcogen and metal vacancies across the spectrum for the most frequently used TMDs: MoS$_2$, MoSe$_2$, MoTe$_2$, WS$_2$, WSe$_2$, and WTe$_2$ under metal and chalcogen-rich conditions. Metal and chalcogen vacancy energies of formation decrease from sulfides to tellurides when prepared under chalcogen-rich conditions, while the energy of formation increases from sulfides to tellurides when prepared under metal-rich conditions, and the metal vacancy energy of formation decreases significantly from sulfides to tellurides. However, the thermodynamic stability of a given defect does not consider kinetic aspects; this has been shown for WSe$_2$, where antisite defects dominate despite their larger energy of formation compared to the chalcogen vacancy [26]. Our understanding of the defect kinetics, and how they play out during growth processes far from equilibrium is not well developed but is clearly important for defect control and engineering efforts.

4.2.3 The nature of vacancies

A particularly interesting example of defect characterization is the study of chalcogen vacancies, which are expressed in STM topography images as a "trefoil" shaped defects with an apparent "missing" atom, or depression at the center [16,26,45–47]. The appearance of a depression lead to the interpretation of this defect, which is seen prominently in many STM topography images, as a chalcogen vacancy. Fig. 4.1 shows the direct comparison between "true" S-vacancies and O-terminated ones from the work by Schuler *et al.* [45] with additional discussion reported by Barja et al. [47]. However, the electronic signature of a chalcogen vacancy includes states that are located deep in the bandgap closer to the conduction band minimum (n-type), and one additional state, which overlaps with the valence band maximum. The presence of the gap states distinguishes a chalcogen vacancy from a substitutional oxygen atom, which replaces any other group VI elements in the sublattice. Calculations of the local density of states demonstrate that such a substitutional oxygen atom does not introduce the same gap states but creates near identical STM topography images. STM topography images always represent a convolution of information from topography (tip-sample distance), local density of states, and transition matrix elements between the wavefunctions, which contribute to the tunneling process. The combination of STM imaging and spectroscopy studies with non-contact AFM [45], which afford a different contrast mechanism, allowed for the re-interpretation of this defect signature. Non-oxygen chalcogen vacancies can, on the other hand, be created in abundance by annealing in vacuum, confirming the difference between vacancy signature and oxygen-substitutional defects. Similar

4.2. Point defects   93

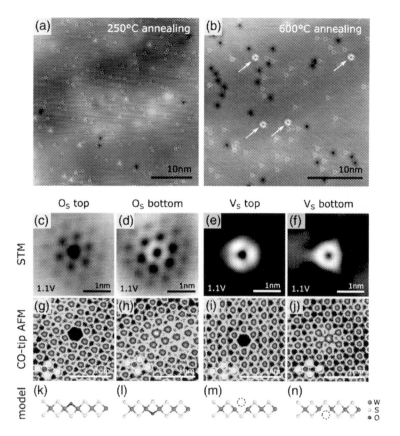

FIGURE 4.1 Illustrates the topography images measured with STM [a–f] and non-contact AFM (CO-terminated tip) [g–j] at 4 K. (a), (b) STM topography images with $I_t = 20$ pA, $V_t = 1.1$ V of CVD-grown monolayer WS_2 graphene/SiC. (a) is measured after low-temperature sample annealing and Oxygen substituents at sulfur sites are most abundant. (b) By in-vacuum annealing at about 600 °C, sulfur vacancies can be generated (white arrows). (c)–(f) STM topography ($I_t = 20$ pA, $V_t = 1.1$ V) of an oxygen substituting sulfur (OS) in the (c) top and (d) bottom sulfur plane as well as a sulfur vacancy in the (e) top and (f) bottom sulfur sublattice. (g)–(j) Corresponding CO-tip NCAFM images of the same defects as in (c)–(f). The unit cell has been indicated as a guide for the eye. Yellow, S atom; blue, W atom. (k)–(n) DFT calculated defect geometry. Adapted with permission from B. Schuler et al. *Phys Rev Lett* **123**, 076801 (2019). Copyright (2019) by the American Physical Society.

contrast mechanisms and defect representations are likely to be also present in MoS_2 albeit a directly comparable study using STM and AFM has not yet been done [48].

4.2.4 Complex defects created by annealing of WSe_2

The majority of defect studies assess as grown material but rarely consider the impact of annealing at higher temperatures on the defect inventory [24,47,49,50]. However, while TMD-based devices are not likely to experience temperatures in excess of about 250 °C under normal device operation, their use as catalysts or sensors might exceed these temperatures.

Further, the deposition of another TMD layer to create van der Waals heterostructures necessitates significant heating during the synthesis of the next layer. It is therefore of interest to understand TMD degradation as a function of temperatures, up to the growth temperature. TMD degradation is usually initiated by chalcogen loss, which is readily identified in photoelectron spectroscopy, as seen for example in the increase of sulfur vacancies in MoS_2 and WS_2; it is expected that similar reactions occur in all TMDs albeit to a different degree[50]. WSe_2 when annealed at a temperature of 600 °C can degrade significantly leading to detrimental modifications in the electronic structure [24]. Fig. 4.2 includes the distribution of bandgaps measured locally with STS across a 70×70 nm^2 area, which can be used as a measure of electronic structure modulation initiated by annealing and shows several complex defects created by annealing. The distribution of band gaps (and concurrently the CBM and VBM positions) is significantly broadened, and while there are still sizeable areas within the layer with only a few defects, the TMD layer overall has changed quite dramatically owing to a near 10-fold increase in the defect density. In addition, the small point defects which involve only a few atoms are substituted by structurally complex distortions of the material. Measurements at the atomic scale cannot be achieved in most instances suggesting that if defects are observed there is substantial loss of chalcogen and breakdown of the TMD structure. It is still possible to classify defects by geometry and electronic structure, but the details of the atomic structure remain unresolved at present.

4.3 Impurities

While substitutional impurities are strictly a point defect, it is worth discussing impurities separately. After all, while many of the point defects discussed in the preceding section can be considered intrinsic to the materials, impurities are by definition extrinsic defects. Existing in both geological and synthetic materials these impurities are usually unwanted and often unknown defects in a material. But when intentionally introduced they can be used to control optical and electrical properties as well as other properties of the material when introduced in small quantities, much less than 1 at%, and can play a more significant role in the band structure when introduced at concentrations higher than 1 at%.

4.3.1 Contaminants

Contaminants within TMDs come from a number of sources. For synthetic TMDs, impurities could be introduced from the precursor or source materials, as well as from the growth system, post growth processes and handling. However, for geological or naturally occurring materials, the impurities are either introduced from the handling or from the high temperature formation process of nature. A review article by Wilson and Yoffe [51] shows that much of the early work on determining the electronic properties was carried out on geological crystals, compact powders, and crystals grown by vapor transport. Vapor transport methods would continue to be used for generating samples to determine the photoelectric properties [52,53], reactivity [54], and photoelectrochemical properties [55,56] of TMDs in the latter half of the twentieth century. After the graphene inspired resurgence of interest in 2D materials much of

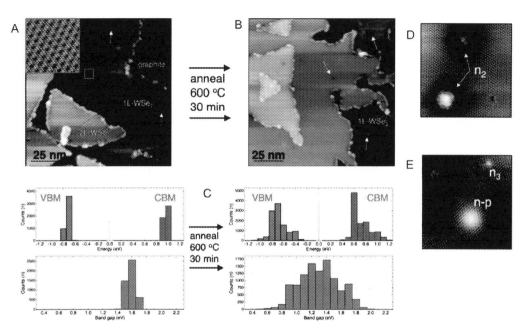

FIGURE 4.2 STM image of WSe$_2$ grown on HOPG by molecular beam epitaxy prior to (A) and after the 30 min anneal (B) at 600 °C. (C) summarizes the CBM, VBM, and band gap values extracted from STS maps recorded before and after the anneal over the 1st layer across an area of 70 × 70 nm^2 and step edges are excluded from the analysis. Images D and E show a selection of geometrically complex point defects found on the first layer of WSe$_2$ after the 600 °C anneal. Each unique point defect is labeled according to both its geometric and electronic characteristics (not shown here). STM conditions for image A: $V_t = 1.5$ V, $I_t = 50$ pA, and the 10 × 10 nm^2 image inset: $V_t = 1.0$ V, $I_t = 100$ pA. Images G and H: 12 × 12 nm^2 at $V_t = -1.0$ V, $I_t = 100$ pA. Adapted from Blades et al. *J. of Phys. Chem C* 124, 15337-15346 (2020) with permission.

the work in the early 21st century utilized flakes from geological crystals and many discrepancies in reported properties resurfaced. In particular, while the overwhelming majority of work reported n-type behavior of geological MoS$_2$ flakes [57–59], it was also shown by McDonnell *et al.* that single crystals could exhibit both n and p behavior depending on the probe location [60]. In that work, p-type behavior was found to coincide with areas that showed high concentrations of morphological defects and accounted for very large percentages of the analyzed area of such crystals. However, this may still be consistent with the electrical reports of crystal that showed almost entirely n-type behavior. This is so because device fabrication would typically only be carried out on flakes that appeared to be 'high quality' based on optical microscopy, thereby ignoring the p-type flakes and skewing the statistics. It is worth noting that in the work of McDonnell *et al.* [60] the n-type regions were not defect free, instead, metal-like defects were observed on 1 to 3% of the surface area.

In their review, Wilson and Yoffe [51] reported that for MoS$_2$, as an example, three prior papers reporting the electronic properties, one reported p-type behavior [61], another reported 'mainly p' [62], and another n+p [63]. Also in the one cited paper that investigated the properties of vapor transport grown MoS$_2$, the authors reported all n-type characteristics [64].

Therefore, it may be inferred that when the macroscopic properties of geological MoS_2 crystals are measured, a p-type character might be expected. This leads us to consider the origins of carrier type determination in MoS_2.

Some have attempted to explain the doping behavior of TMDs by considering external factors such as the influence of the substrate. For example, Kang & Han [65], considered the role of hydrogen in an SiO_2 substrate (which is commonly employed when studying TMD flakes), and calculated the impact of hydrogen impurities near the SiO_2 surface on the Fermi level position of TMDs. They showed that this could predict n-type behavior in MoS_2, bipolar behavior in $MoSe_2$, and p-type behavior in WSe_2 [65]. While it is interesting to consider that even pure TMDs could be doped by substrate impurities or interfacial defects, it seems unlikely that TMD flakes are indeed pure as will now be discussed.

It has been suggested that mined MoS_2 is intrinsically n-type [66], with the possible origin being the Cl and Br impurities detected by Yin *et al.* [67]. In addition to this, Re has been shown [68] to exist in geological MoS_2 and has been shown [69] to behave as an n-type dopant when added to synthetic MoS_2. Addou *et al.* [70] employed inductively coupled ion mass spectroscopy (ICP-MS) to investigate the composition of geological and CVT grown MoS_2 crystal. They showed that a large number of impurities could be detected. In fact, of the 35 elements investigated, 22 were found as shown for the CVT samples (see Fig. 4.3). The highest concentration impurities in geological samples were Fe, Ag, Cd, Sb, W, Pb, and Bi. For the CVT samples the trends were qualitatively similar, although low levels of Li and Zn were also observed while being below the limit of detection in geological crystals. Low levels of In and Sn were also detected in geological samples but not in CVT samples; such differences could be within the error of the measurement.

The large number of impurities found in these samples result in a wide range of complex doping and counter doping scenarios in the case of electrical active impurities. If the impurities are present as simple inclusions (in the intercalated regions), then they could impact the electronic properties of the crystal. However, if a monolayer flake is exfoliated from such a crystal, the electrical properties of the monolayer might not be impacted by such inclusions, provided that there is no bonding between the intercalant and the monolayer. At this time, we cannot be certain as to the location of all the impurities present. However, one recent study of Bi impurities [71], which are the highest concentration impurity reported by Addou *et al.* [70], showed that large amounts of the Bi impurities are likely present as inclusions (clusters). In that work, XPS mapping revealed that the Bi could be detected in 15% of the areas mapped, as shown in Fig. 4.4. By varying the size of the analysis area, the authors were able to estimate the size of individual clusters, as well as their depth below the surface MoS_2 layers in a bulk crystal. It was noted that the observation of Bi impurities as clusters does not preclude the presence as substitutional dopants, but merely emphasizes that a detectable portion of them exist as inclusions. To our knowledge, such studies on the other impurities in MoS_2 have not yet been carried out yet.

We now consider the impact of impurities on functional properties. Without identifying the impurities, Toh *et al.* compared ultrapure synthetic MoS_2 to commercial powders with comparable particle sizes and found superior catalytic performance which they attributed to the absence of unintentional dopants [72]. This highlighted that not all dopants/impurities are beneficial, and the authors commented on the importance of the intentional introduction of specific dopants to achieve desired properties. This work was followed by a detailed re-

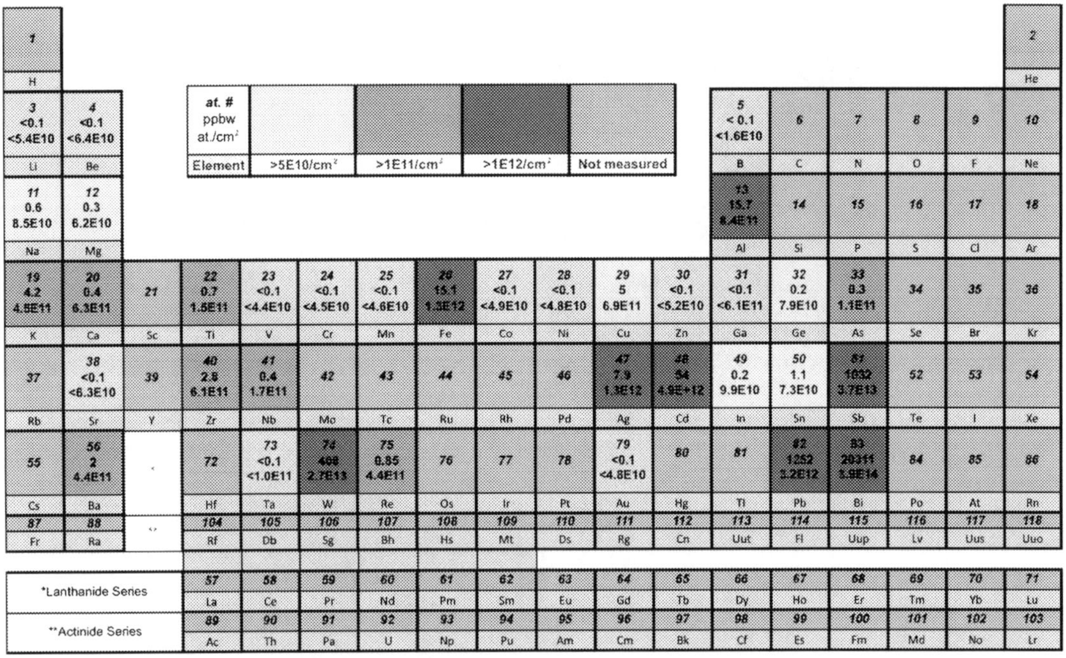

4.3. Impurities

Legend: at. # / ppbw / at./cm² / Element — shading scale: >5E10/cm², >1E11/cm², >1E12/cm², Not measured

1	2		3	4	5	6	7	8	9	10	11	12	13	14	15	16	17	18
1 H																		**2** He
3 <0.1 <5.4E10 Li	**4** <0.1 <6.4E10 Be												**5** <0.1 <1.6E10 B	**6** C	**7** N	**8** O	**9** F	**10** Ne
11 0.6 8.5E10 Na	**12** 0.3 6.2E10 Mg												**13** 15.7 8.4E11 Al	**14** Si	**15** P	**16** S	**17** Cl	**18** Ar
19 4.2 4.5E11 K	**20** 0.4 6.3E11 Ca	**21** Sc	**22** 0.7 1.5E11 Ti	**23** <0.1 <4.4E10 V	**24** <0.1 <4.5E10 Cr	**25** <0.1 <4.6E10 Mn	**26** 15.1 1.3E12 Fe	**27** <0.1 <4.9E10 Co	**28** <0.1 <4.8E10 Ni	**29** 5 6.9E11 Cu	**30** <0.1 <5.2E10 Zn	**31** <0.1 <6.1E11 Ga	**32** 0.2 7.9E10 Ge	**33** 8.3 1.1E11 As	**34** Se	**35** Br	**36** Kr
37 Rb	**38** <0.1 <6.3E10 Sr	**39** Y	**40** 2.8 6.1E11 Zr	**41** 0.4 1.7E11 Nb	**42** Mo	**43** Tc	**44** Ru	**45** Rh	**46** Pd	**47** 7.9 1.3E12 Ag	**48** 54 4.9E+12 Cd	**49** 0.2 9.9E10 In	**50** 1.1 7.3E10 Sn	**51** 1032 3.7E13 Sb	**52** Te	**53** I	**54** Xe
55 Cs	**56** 2 4.4E11 Ba	*	**72** Hf	**73** <0.1 <1.0E11 Ta	**74** 408 2.7E13 W	**75** 0.85 4.4E11 Re	**76** Os	**77** Ir	**78** Pt	**79** <0.1 <4.8E10 Au	**80** Hg	**81** Tl	**82** 1252 3.2E12 Pb	**83** 20311 1.9E14 Bi	**84** Po	**85** At	**86** Rn
87 Fr	**88** Ra	**	**104** Rf	**105** Db	**106** Sg	**107** Bh	**108** Hs	**109** Mt	**110** Ds	**111** Rg	**112** Cn	**113** Uut	**114** Fl	**115** Uup	**116** Lv	**117** Uus	**118** Uuo

*Lanthanide Series: **57** La, **58** Ce, **59** Pr, **60** Nd, **61** Pm, **62** Sm, **63** Eu, **64** Gd, **65** Tb, **66** Dy, **67** Ho, **68** Er, **69** Tm, **70** Yb, **71** Lu

Actinide Series: **89 Ac, **90** Th, **91** Pa, **92** U, **93** Np, **94** Pu, **95** Am, **96** Cm, **97** Bk, **98** Cf, **99** Es, **100** Fm, **101** Md, **102** No, **103** Lr

FIGURE 4.3 Periodic Table of Elements showing the impurity concentration of 30 elements in MoS_2 as determined by ICPMS. Al, Fe, Ag, Cd, Sb, W, Pb, and Bi are detected in the largest amounts. Reprinted with permission from Addou et al. ACS Nano, 9(9) 9124-9133. Copyright 2015 American Chemical Society.

view of the role of impurities in the catalytic properties of 2D materials [73]. The authors also challenged the misconception that doping generally leads to enhanced electrocatalytic properties. They considered four reactions, hydrogen evolution reaction, oxygen reduction reaction, oxygen evolution reaction, and carbon dioxide reduction reaction. They showed that in many situations dopants inhibit rather than enhance the performance. For field effect transistor applications, Mirabelli *et al.* [74] have demonstrated using continuum-based TCAD (technology computer aided design) device models that the low field effect mobility observed in MoS_2 based devices can be explained once impurity concentrations are considered. Vialla *et al.* [75] studied $MoSe_2/WSe_2$ heterostructures with temporally resolved photoluminescence and concluded that impurities can dominate the interlayer luminescence spectrum, adding to the argument that achieving ultrahigh purity materials is vital to for obtaining adequate control of optoelectronic properties.

At the present time it appears that contaminants in the form of inclusions are a likely contributor to the difference between macroscopic properties of crystal and the properties of the monolayer flakes exfoliated from them. However, the knowledge that such inclusions exist does not rule out the likelihood that such impurities may also exist as substitutional dopants. In fact, the work comparing synthetic powders to commercial powders, as well as photoluminescence studies of $MoSe_2/WSe_2$ interfaces and the TCAD simulations of MoS_2 based FETs all suggest that impurities do behave as unintentional dopants in exfoliated TMDs.

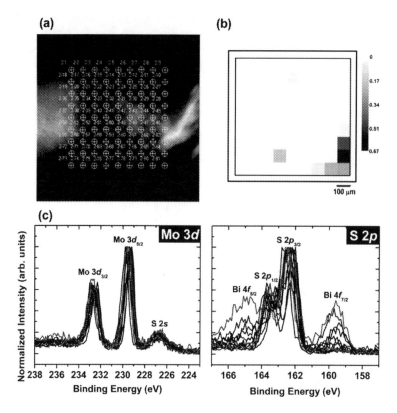

FIGURE 4.4 (a) Set-up of the array of 81 points spaced 100 µm apart for XPS mapping on an 800 µm × 800 µm surface of geological MoS$_2$. (b) Map of normalized Bi intensities obtained through XPS mapping. (c) Normalized spectra of the 12 analysis spots that were found to have some amount of Bi. Reprinted from Applied Surface Science, 508, M. G. Sales et al., MoS$_2$ impurities: Chemical identification and spatial resolution of bismuth impurities in geological material, 145256. Copyright (2020), with permission from Elsevier.

4.3.2 Intercalants

We begin the discussion of intercalants into TMDs with the work of Woollam and Somboano [76] where the properties of A$_x$:MoS$_2$ (where "A" are the intercalants Na, Ca, Sr, K, Rb, and Cs) were reported. They group these intercalants into two groups, with different structures. One group with A= K, Rb, Cs have hexagonal structures and have an approximately fixed A$_x$:MoS$_2$ stoichiometry of x∼0.3 (actually reporting x=0.4 for K). These materials exhibited superconducting behavior at 6.9 K. The second group (A=Na, Ca, Sr) where more varied, exhibiting superconductivity at 3.6, 4.0, and 5.6 K respectively. Their structures were found to be tetragonal (3.6 K), orthorhombic (4.0 K), and tetragonal (5.6 K), and they had stoichiometry ranges of $0.3 < x < 0.6$ (Na), $0.05 < x < 0.07$ (Ca) and $0.06 < x < 0.1$ (Sr). They also report that in all cases, the intercalation results in transformation from semiconducting to metallic behavior. This may seem obvious given the presence of electron donors, however, the structural change itself is enough to induce the metallic behavior.

Famously, it was the intercalation of Li into MoS_2 crystals that was the method first employed for isolating monolayer MoS_2 in the 1980s [77]. It has been reproducibly shown that such a process induces a phase change of 2H-MoS_2 to the 1T'-MoS_2 polytypes. The 1T' is similar to the 1T in that it is based on an octahedral coordination of the Mo rather than the tetrahedral of the 2H [51]. However the 1T' is a distortion of the 1T. While many articles report that the alkali intercalation of 2H-MoS_2 induces a phase transformation to the 1T structure, this structure is known to be unstable and decays into the 1T' phase [78,79]. In fact, while the J_1, J_2 and J_3 features that appear in the Raman spectra, at 156 cm^{-1}, 226 cm^{-1}, and 333 cm^{-1} respectively, have previously been used as evidence of the 1T-MoS_2, it is actually the superlattice formed by the distortion of the 1T that enables Raman scattering events to involve zone-edge phonons without violating wave-vector conservations [80].

This distorted 1T' structure has been well characterized. Using combinations of XRD and STM, Wypych $et~al.$ [81–83] describe potassium intercalated MoS_2 resulting in an $a \times a\sqrt{3}$ superstructure which is distinct from the $2a \times a$ superlattice that is observed [84] after Li based intercalation. Specifically, the $a \times a\sqrt{3}$ superstructure results from the formation of rows of sulfur atoms along the b-axis. This is illustrated in Fig. 4.5 where every second row cannot be seen because they lie 0.13 nm deeper than those observed in the image [83]. Increased oxidation of this compound can lead to the observation of a $a\sqrt{3} \times a\sqrt{3}$ superstructure that is evidence of the undistorted 1T-MoS_2 [82,83]. In the case of Li intercalation, for Li_x:MoS_2 compounds when x is greater than 1, the transformation from Mo in a trigonal prismatic site to an octahedral coordinated site is achieved. This distortion was originally assumed to be due to the Li in octahedral and tetrahedral interstices but was later explained to be due to the trigonal clusters of Mo distorted from their 2H-MoS_2 positions [85,86]. This suggests that while alkali intercalation typically results in a transformation to a distorted 1T phase, the details of the intercalation can modify the observed distortion.

In more recent work, this intercalation process was optimized with respect to exfoliation and subsequent functional properties of the material. For example, in the work of Ambrosi $et~al.$ [87], the authors studied the exfoliation of Li intercalated MoS_2 using methyllithium (Me-Li), n-butyllithium (n-Bu-Li) and $tert$-butyllithium (t-Bu-Li) and found that the exfoliation efficiency was reduced for Me-Li. In a detailed study of Na intercalation in MoS_2 ($Na_x MoS_2$), Wang $et~al.$ [88] determined that reversibility was only possible up to x=1.5. Beyond this, the material irreversibly formed NaS and metallic molybdenum.

Beyond the intercalation of alkali metal, it has also been shown that alkali intercalation followed by exfoliation and flocculation can enable the intercalation of other ions. For example, Dungey $et~al.$ [86] used this method to intercalate both Co and Fe. Zubavichus $et~al.$ [89] explored the intercalation of Mn, Co, Ni, and Ru. In that work they claimed that the Co and Mn intercalation did not alter the Mo coordination from trigonal prismatic while Ni and Ru did yield the previously reported change to octahedral. Li and Li [90] reported on the intercalation and deintercalation of Mg ions in MoS_2, finding that their synthesized MoS_2 enable more reversible processing than MoS_2 powders. Interestingly, separate work also compared 'ultrapure' synthetic samples to commercial powders and showed the ultrapure sample had superior catalytic properties when compared to unintentionally doped, impure commercial MoS_2 powders [72].

In summary, the intercalation of MoS_2 has been studied extensively. The majority of the intercalants induce a structural transition to 1T or some form of distorted 1T structure in the

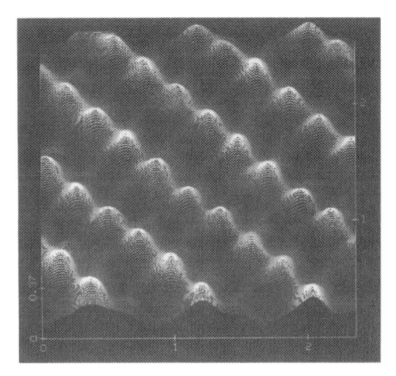

FIGURE 4.5 Topographical representation of an STM image of $K_x(H_2O)_yMoS_2$ in a region of 2.4 x 2.4 nm obtained with a bias voltage of -3.7 mV and an average current of 4.3 nA. Reprinted from Surface Science, 380, F. Wypych et al., Scanning Tunneling Microscopic investigation of $K_x(H_2O)_yMoS_2$, L474-L478., Copyright (1997), with permission from Elsevier.

MoS_2. This yields metallic behavior and the resulting material can be exfoliated to produce monolayer material [77] which can be annealed to recover its semiconducting behavior [91]. Beyond being a vehicle to achieve monolayers, the materials can be restacked with foreign intercalants to obtain a range of functionalities. Moreover, even without introducing additional ions, the properties of the material can depend on the exfoliation process [87]. The intercalation and deintercalation mechanism in these materials are interesting for alkali battery applications, however the other property changes such as enhanced catalytic activity and conductivity are also of interest to the energy conversion and nanoelectronic communities.

4.3.3 Dopants

As discussed in Section 4.3.1, impurities may play an important role in determining the transport properties of both geological and synthetic TMDs. This current section focuses on the deliberate doping of TMDs with impurities, while also briefly highlighting other doping mechanism that do not depend on impurities.

While MoS_2 has been considered to be intrinsically n-type [66] we avoid that terminology here because the word intrinsic, as a counter to extrinsic, implies that the doping is not due to impurities. Instead, we acknowledge that such doping may be inherent to geological samples due to specific impurities that modify the transport properties. Many of the first studies of monolayer TMD based devices relied on this inherent doping of geological materials [59]. However, other strategies have also been employed. For example, Das *et al.* used electrostatic doping induced by top gates to achieve PFET and NFET characteristics for WSe_2 based CMOS inverter [92,93]. This concept had previously been used for graphene [94] and has been adopted by other groups for their work on WSe_2 [95–98]. Electrostatic doping has been employed in a range of 2D-based devices including: (1) Controllable switching between ferromagnetic and antiferromagnetic states in CrI_3 [99], (2) reconfigurable p-n junction diodes in MoS_2 [100], (3) controlling exciton-plasmon couple in MoS_2 [101], and (4) inducing phase changes in $MoTe_2$. The concept of electrostatic doping is not unique to 2D materials and a number of reviews have discussed the general application of the field effect principle for doping [102–104]. Pertinent to these discussions, Lu *et al.* [105] have carried out a first principle study of electrostatic doping specifically in van der Waals heterostructures. Their predictions suggest the potential for tunable transistors enabled by electrostatic control. It has also been proposed that modulation doping can avoid the negative impact of ionized impurities such as reduced mobility. In the work by Xu *et al.* [106], it was shown that by selecting oxides with appropriate charge neutrality levels both n-type and p-type doping can be induced in MoS_2. In that work, TiO_2 was found to donate electrons leading to n-type behavior, while MoO_3 was found to extract electrons leading to p-type behavior with doping levels of 7.8×10^{11} cm^{-2} and 6.0×10^{11} cm^{-2} respectively. Another novel method of doping is exposure to dopant vapor such as potassium or NO_2 [107,108]. In these reports the contact regions were degenerately doped by surface charge transfer to achieve a performance enhancement of 1000x in on current with no increase in off current.

A few first principles studies have considered dopants in 2D materials. For example, Li *et al.* [109] calculated the properties of WSe_2 with Nb, Ta, and V dopants. They considered both the substitutional replacement of W as well as intercalation of the dopant at the octahedral site. They found that the substitutional position was energetically favorable, the WSe_2 structure was preserved despite some lattice distortion and that all three dopants were acceptors (p-type dopants). In a separate work, Yang *et al.* [110] studied the impact of substitutional dopants where Se was the element being replaced. In that work they predicted that N, P, and As would all be acceptors, while F, Cl, Br, and I would all be donors. They also noted that N, P, F, and Br all introduced a magnetic character to the WSe_2 monolayer with a prediction that F-doped WSe_2 exhibit a magnetic moment of 0.846 μ_B. A comprehensive study has been carried out by Onofrio *et al.* [111]. In that work, the authors considered MoS_2, $MoSe_2$, and $MoTe_2$, with 54 different dopant elements, their formation energies for substitutional replacement of the metal or the chalcogen site as well being on an interstitial site. They report on the formation energies always as Fermi energy position within the band gap. An excerpt of the results for MoS_2 is shown in Fig. 4.6; the data for other TMDs, formation energies, and intercalants sites are included in this publication. Their predictions show that while many dopants clearly prefer to substitute either the metal or the chalcogen, there are some, such as Mg, Ca, Sr, Ni, Pd, Cu, Ag, Zn, Cd, Ga, In, Si, Sn, and Ge can occupy either site by starting with chalcogen rich (to promote metal substitution) or metal rich (to promote chalcogen substitution). This suggests that for many dopants, the substitutional site can be engineered through stoichiometric control of the TMD.

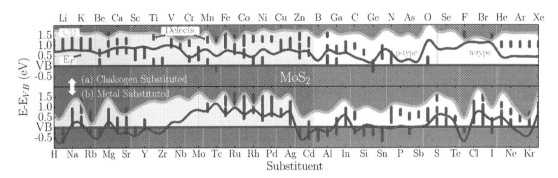

FIGURE 4.6 Energy of the conduction band (CB, green), Fermi level (E_F, red), and defects in the bandgap (BG, blue) with respect to the energy of the valence band (VB, blue line x = 0) of S- [(a): top] and Mo- [(b): bottom] substituted monolayer MoS_2 as a function of substituents. Reprinted from N. Onofrio et al. "Novel doping alternatives for single-layer transition metal dichalcogenides" J. Appl. Phys. 122 (2017) 185102, with the permission of AIP Publishing.

To begin our discussion of experimental substitutional doping we will first acknowledge the continuation of the previously discussed surface charge transfer doping by NO_2 to an air stable doped WSe_2. This air stability was achieved by covalently attaching N through the exposure of WSe_2 to NO_2 at 150 °C [112]. The authors considered a number of adsorption positions and determined that only NO, where N is bonded to the W at the place of Se vacancies, would result in the p-type doping. In another work, N_2 plasma was employed to achieve lower doping concentrations of 2.5×10^{18} cm^{-3} to 1.5×10^{19} cm^{-3} [113,114]. Importantly, these sub-degenerate doping levels promise good control over the doping concentration controlled by plasma exposure time. The authors noted that if the strain imposed by the incorporation of nitrogen is considered, then the band gap increases by 100 meV.

It has also been demonstrated that Mn doping of up to 2% in MoS_2 could be achieved during growth [102]. In that work, doped MoS_2 was grown by sulfurization of the metal oxide powders at high temperature; MoO_3 and $Mn_2(CO)_{10}$, with Mn/Mo precursor ratio of 1:50. A key finding of that work was a marked substrate dependence in the concentration and localization of Mn. On SiO_2 and sapphire substrates, no Mn incorporation was detected, while Mn doping/alloying was achieved on graphene. The authors speculate that the Mn may have been incorporated into the SiO_2 or sapphire substrate instead of the MoS_2 at high temperature. Interestingly the authors also reported that their attempts to exceed a 2% doping concentration resulted in film degradation and the loss of 2D character or phase transformation. In a separate work, WSe_2 grown by CVT was doped during synthesis with Re for n-type doping and Nb for p-type doping [115]. Nb concentrations of nominally 0.5% where sufficient to result in degenerate doping, and p-n junctions made from these materials were shown to exhibit rectifying behavior.

4.3.4 Alloys

Alloying is one of the most powerful tools that can be wielded by a materials scientist and is applicable to a wide range of fields. In the field of electronic materials, alloying is used to tune the properties of III-Vs by creating ternary alloys, enhancing Si mobility through strain by the addition of Ge, and lowering the Ni-Si Schottky barrier by forming a nickel silicide alloy. Alloying TMDs takes nominally two forms: (1) mixed chalcogen (anion) alloys of the form $MX^1_xX^2_{2-x}$, (2) mixed cations of the form $M^1_xM^2_{1-x}X_2$. Of course, one could envision more complex four element alloys as well. The goal of alloying is generally application driven and we will briefly discuss some of these applications in Section 4.5. Here we focus on the reports of properties dependent on alloy composition as well as the potential for phase transition as a result of alloying to a different phase rather than just forming a solid solution.

Some of the earliest examples of alloying focus on the combination Mo, W, S, and Se. For example, Agarwal *et al.* [116] reported on the resistivity, Hall mobility, and temperature dependent Seebeck coefficient in MoS_xSe_{2-x} alloys. They found that the resistivity increased with increasing S content. The authors report an analogous resistance increase with Se content in $MoSe_xTe_{2-x}$ alloys. With increasing S content, it was also reported that both the band gap and the Hall coefficient increased. With respect to $Mo_xW_{1-x}Se_2$ alloys, Hofmann *et al.* [117] claim to be the first to synthesize these by vapor transport. Their work included a detailed characterization of the materials crystal structure, chemistry, and valance band structure as well as reporting on the materials photoactivity in a range of electrolytes. While they reported similar onset potentials for $Mo_{0.5}W_{0.5}Se_2$ and WSe_2, they suggest that the band alignment difference between $MoSe_2$ and WSe_2 could still lead to marked differences in contact behavior with varying composition.

The above examples are for TMDs that form complete solid solutions and have been studied for their electrochemical and optical properties for some time [118–121]. However, many TMDs can be found in both the 2H or 1T structure; they are often meta-stable in the other structure, suggesting that there may be a limited solubility to form one or the other structure. As such, alloying across these phase transitions has been of recent interest. An example of such a system is the $WSe_{2-x}Te_x$ alloy. WSe_2 is a semiconductor with a 2H crystal structure, while WTe_2 is a semimetal exhibiting a distorted 1T crystal structure. Yu *et al.* [122] synthesized such an alloy with varying compositions. They demonstrated that the compositional range where both the 2H and 1T phases are stable is very narrow, $x\sim$ 0.5–0.6. This is shown graphically in Fig. 4.7. They showed that within the 2H phase the bandgap can be tuned from 1.67 to 1.44 eV before becoming semi-metallic. FETs fabricated with bilayer WSe_2 and bilayer $WSe_{1.4}Te_{0.6}$ yielded calculated mobilities of 10 $cm^2V^{-1}s^{-1}$ and 46 $cm^2V^{-1}s^{-1}$ showing enhanced mobility despite the expected increase in impurity scattering as well as on/off ratios of about 10^6 for both. This study was somewhat analogous to the predictive work of Duerloo and Reed [123] who considered the phases in a $Mo_xW_{1-x}Te$ alloy system. They discussed two extreme cases: 1) by considering only thermodynamic stability, this would yield a phase diagram with a wide two-phase region between two terminal phases (i.e. limited solubility); 2) by considering the limit of quenched W diffusion, this would yield a metastable system with a transition from H to T' at a specific composition (i.e. no two phase region). Paired with the work of Yu *et al.* [122] we can see that while there is a miscibility gap, it is quite narrow, suggesting some combination of the two scenarios considered by Duerloo and Reed [123].

For 2D materials there exists a novel type of alloy where one surface of the layer is different from the other. These are commonly referred to as Janus structures, analogous to Janus nanoparticles [124], and can lead to characteristics that are unique from homogenous alloys. The origin of Janus structures dates back at least to graphene research when Zhou et al. [125]

104   4. Defects in transition metal dichalcogenides

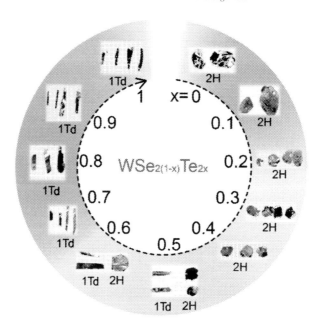

FIGURE 4.7 Photographs of as-grown WSe$_{2(1-x)}$Te$_{2x}$ $_{(x=0-1)}$ single crystals on the millimeter-grid paper, and 2H and 1Td types TMD alloys show circularly platy and bar shapes, which are in agreement with hexagonal and orthorhombic crystal systems, respectively. Adapted from P. Yu et al. Adv. Mater. 2017, 29, 1603991, © 2016 WILEY-VCH Verlag GmbH & Co. KGaA, Weinheim.

predicted the properties of hydrogenated graphene (graphane) with the hydrogen removed from one side (graphone). Other would then consider other functional groups such as halogens to replace the hydrogen on one side [126]. For TMDs of the form MX$_2$, (x = S, Se, Te) a Janus structure takes the form MYX. This is sometimes written as YMX, to indicate that the differing chalcogens are on opposite sides of the metal, rather than homogenously distributed. Lu et al. [127] created an SeMoS Janus structure by first synthesizing MoS$_2$ by CVD, subsequently removed the surface sulphur atoms with a hydrogen plasma, and then without breaking vacuum, thermally selenizing the sample. The reverse structure has been formed by sulfurizing MoSe$_2$ and does not require the prior removal of the top Se layer [128]. However, the temperature must be carefully controlled between 750 and 850 °C to avoid sulfurization of the bottom layer. Li, Cheng, and Huang have provided a more detailed review of Janus structures, their synthesis, characterization, and applications [129]], also discussed in Chapter 8.

4.4 Line defects

Line defects emerge in TMD layers mostly during the growth process and arise from the fusion of individual islands to form a continuous film. Ideally, the epitaxial relation with the substrates promotes an epitaxial relation to the TMD which in turns leads to islands in the first layer, which are aligned and only rotated with respect to the three-fold symmetry axis. The growth of TMD islands and their shape is controlled by the relative growth rates of the metal and chalcogen terminated island edges, respectively, and the metal and chalcogen fluxes [130]. The island edges are catalytically active, and their local chemistry can be compared to those of specific point defects. At some point in the growth process the edges of the islands meet, which determines the specific grain boundary orientation, and atomic structure. A grain-boundary free graphene layer can be grown on a liquid Cu-substrate, from a single graphene seed, the individual domains "wiggle" and adjust their orientation to access the minimum energy configuration, which is to avoid formation of a grain boundary all together if there are no residues on the surface of the liquid Cu [131–134].

Line defects impact electron transport, and generally reduce charge carrier mobility, but can at the same time introduce unique 1D electronic states in the form of metallic or semi-metallic nanowires [135–137]. Considering the relatively large distances and extension of the grain boundaries and line defects, they will also modulate the phonon density of states, and hence offer a path to the control electron-phonon interactions. It has been proposed that the metastable 1T structure can be considered as a 'nucleus' for line defects [138]. As shown schematically in Fig. 4.8, such a defect can reorganize itself into a line defect that subsequently would decay again into a more stable state. While many potential pathways were considered, the two most favorable begin with a line of either 4-fold coordinated (Fig. 4.8 I) or 8-fold coordinated (Fig. 4.8 II) Mo atoms. These subsequently reduce their energy by reorganizing to include Mo-Mo bonding, or -S- bridging atoms respectively.

A grain boundary of intense interest is the twin boundary with metallic character, which has been observed in $MoSe_2$ grown by MBE and MOCVD. Triangular grain boundaries are formed during growth and present metallic properties, which have been confirmed with STM/STS [136,137]. It is even possible to form 1D charge density waves in $MoSe_2$, which are confined within the grain boundaries [139]. Komsa and Krasheninnikov [140], also Chapter 2 of this book, describe the atomic scale structure of three commonly observed grain boundaries which are shown in Fig. 4.9 along with representative TEM images from other works [141–143]. These are 4|4E (two four-fold rings connected at an edge), 4|4P (two four-fold rings connected at a chalcogen point), and 55|8 (two five-fold rings connected to a octagon) and the calculated formation energies for these defects appear to agree well with experimental observations. Interestingly, while the 4|4E has been observed in as grown MoS_2 and $MoSe_2$, it has not been observed in as grown WS_2 or WSe_2 which is an indication of the subtle but significant variation in defect stability between different TMDs. The subsequent deposition of Nb onto WSe_2 has been shown to induce similar twin grain boundaries.

It is worth noting that while we have chosen to define grain boundaries as 1D defects in 2D materials, 1D dislocations have also been discussed in relation to 2D materials. Yazyev and Louie [144] state that in 2D material, only edge dislocations are possible and define the dislocation as a combination of positive and negative disclinations such as the 5|7 structure where each carbon atom retains its three-fold coordination. Grain boundaries are on the other hand periodic arrays of dislocations [144,145]. Liu and Yakobson [145] have shown that an armchair to zig-zag grain boundary can be described entirely with 5|7 dislocations. Using first principles calculations, Liu et al. have shown that in h-BN the 4|8 structure has a lower energy [146]. McGuigan et al. considered the h-BN and graphene lateral heterojunctions and evalu-

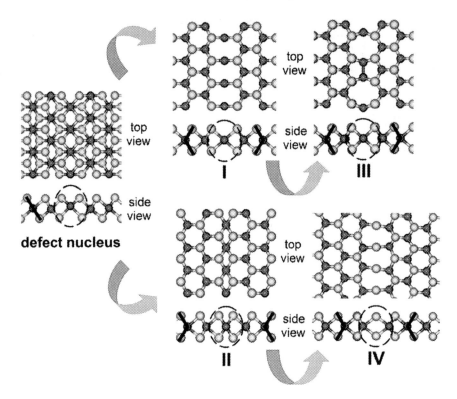

FIGURE 4.8 Models of MoS$_2$ layer fragments containing line defects and optimized using the density functional based tight binding method. The defect nucleus of 1T-layer is composed of MoS$_6$ octahedra interlinked with a line of MoS$_6$ prisms (on the left). The defect nucleus rearranges to a line defect of 4-fold coordinated Mo atoms (I) and then to a line defect containing Mo-Mo metallic bonds. Alternatively, the defect nucleus may rearrange to a line defect of 8-fold coordinated Mo atoms (II) and then into a line defect composed of S bridging atoms (IV). All the line defects are marked by a dashed circle on the side view, and black chevrons are overlaid on the layers to guide the eye to the layer direction. Reprinted with permission from A. N. Enyashin et al. "Line Defects in Molybdenum Disulfide" J. of Phys. Chem. C. 2013, 117, 10842-10848. Copyright (2013) American Chemical Society..

ated the 2D critical thickness analogy of heteroepitaxy, i.e., the thickness beyond which strain energy is sufficient to induce dislocations [147]. In the case of WS$_2$, Azizi et al. [148] noted that the grain boundaries were dominated by 6|8 structures which are derived from 5|7 by the insert of an S column between the otherwise W-W bond. The motion of these dislocations, captured by scanning transmission electron microscopy, is mediated by the highly mobile S atoms, which interact with the dislocation cores and form different derivative structures [148]. Beyond dislocations within a layer, Pochet *et al.* [149] have interpreted moiré structures resulting from lattice mismatch or rotation between two dissimilar 2D material layers in the framework of dislocation theory. Screw dislocations, which are not defined by Yazyev and Louie [144] in a 2D layer, can emerge under specific growth conditions, where the growth progresses in such a way as to produce a spiral arrangement. This unique growth morphology is supported by the nucleation rate and S-concentration and visualizes in the corkscrew

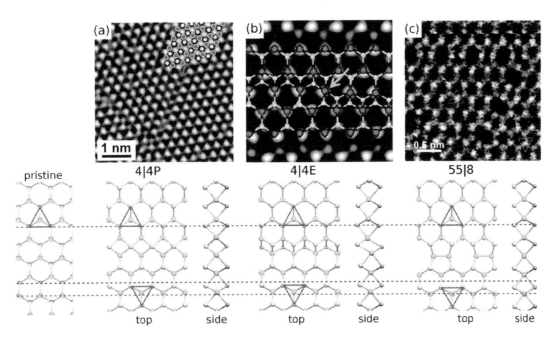

FIGURE 4.9 Transmission electron microscopy images for the three Mirror Twin Boundary (MTB) structures. a) 4|4P MTB in MoSe$_2$ (bright field TEM image). b) 4|4E MTB in MoS$_2$ (annular dark field STEM image). c) 55|8 MTB in WSe$_2$ (annular dark field STEM image). Corresponding atomic structures are show below each panel. The triangles highlight the registry of the lattice at the two sides of the MTB. The dashed horizontal lines highlight strain in the perpendicular direction, when the lattices are aligned at the top. Reproduced from H.-P. Komsa and A. Krasheninnikov. Adv. Electron. Mater. 2017, 3, 1600468, © 2017 WILEY-VCH Verlag GmbH & Co. KGaA, Weinheim. Part (a) Reproduced with permission.[131] Copyright 2015, American Chemistry Society. Part (b) Reproduced with permission.[130] Copyright 2013, American Chemical Society. (c) Reproduced with permission [132]. Copyright 2015, Nature Publishing Group.

like layering of the TMD material [150–152]. Such structures have been shown to have weakened interlayer coupling and as such may prove to be another route to manipulating the properties or unveiling new properties, in van der Waals materials.

4.5 Control of defects and their applications

We will now summarize our prior discussions by considering how defects can be controlled during growth or by post growth modification, and how they can be used in applications. Much of the defect engineering in TMDs is achieved post synthesis, but some groups have demonstrated control during growth. The most common is the work on growth of TMD alloys, but other examples include doping and grain or domain control. Dopants have been added during powder vaporization [102], vapor transport [115,153], and MBE [154] Similarly, a number of groups have demonstrated control of grain size during growth [155–157]. He *et al.* [155] achieved ultra-small grain sizes (<10 nm) by seeding the growth of MoS_2 using Au quantum dots, while Yue *et al.* [156] employed higher temperature and low metal flux conditions to achieve an order of magnitude increase in grain size of MBE grown WSe_2. There also exists a full review on the characterization and control of grain boundaries [157].

The modification of TMDs after growth tends to be a more common route to defect control. Vacancies can be generated by vacuum annealing or ion irradiation, as also described in Chapter 2 [158,159]. Doping can be achieved by NO_2 or N_2 plasma exposure [112–114]. Some TMDs such as $MoSe_2$ and $MoTe_2$ are known to be susceptible to metallic twin boundary formation [137,160,161]. This has been demonstrated to occur spontaneously during MBE and MOCVD growth of $MoSe_2$, but can also be promoted by creating a Mo-rich environment, either by annealing (to remove Se) or by depositing additional Mo [137]. One would expect similar results could be obtained by deposition in a Se deficient environment. A similar result was observed for Nb doped WSe_2 which is particularly interesting given that pure WSe_2 does not exhibit these boundaries in contrast to $MoSe_2$ [160]. Similar structures can be formed from the low temperature ($< 400\,°C$) vacuum annealing of $2H-MoTe_2$ [161], however vacuum annealing at $400–500\,°C$ can result in the conversion of $2H-MoTe_2$ to metallic Mo_6Te_6 nanoribbons [162]. Another very common modification is the intercalation of alkali metals which can be used to obtain monolayer materials and even to restack TMD layers with other intercalants [77,86].

Defect engineering has been shown to be a powerful tool for TMDs. The intercalation of Li and other alkali metals, discussed earlier, can be used for a number of applications beyond the obvious battery field. The induced structural changes modify the electronic properties and thereby the functional properties of the MoS_2. For example, the optical absorption is found to be markedly reduced for Li intercalated MoS_2 even while the conductivity increases [163]. The catalytic properties are also seen to improve upon Li intercalation [164]. We now discuss some of the potential applications enabled by these changes in functional properties.

One such application is in photoelectrochemical water splitting. The improved performance of $1T-MoS_2$ over $2H-MoS_2$ with respect to the hydrogen evolution reaction (HER) has been shown experimentally, also discussed in Chapter 10 [164,165]. First principles study by Tang and Jiang [166] predict that the 1T phase of MoS_2 has a catalytically active basal plane. In the work of Ambrosi *et al.* [87] the authors found that the details of the exfoliation process of MoS_2 impacted the resultant electrochemical properties. Specifically, when comparing HER for $1T'-MoS_2$ obtained from exfoliation using methyllithium (Me-Li), *n*-butyllithium (*n*-Bu-Li) and *tert*-butyllithium (*t*-Bu-Li), they found that the catalytic activity was markedly superior for materials synthesized with *n*-Bu-Li and *t*-Bu-Li than for Me-Li.

Focusing on the enhanced conductivity afforded by Li-intercalation, it has also been demonstrated that locally inducing a transition to metallic MoS_2 can facilitate reduced contact resistance between metals and MoS_2 [167]. Given the difficulties and importance of achieving low contact resistance in TMD based nanoelectronics, such a process has significant potential if it can remain stable through a full process flow. This enhanced conductivity is of course related to stabilizing the distort 1T phase of MoS_2; however, Li intercalation is not the only mechanism for doing this. Both argon plasma exposure and electron beam irradiation have been shown to induce this phase change which offers flexibility for device processing [168,169].

Ataca *et al.* [170] employed density functional theory to predict the role of defects in tuning the electronic and magnetic properties of MoS_2 nanoribbons. Their calculations suggest that zigzag nanoribbons which have undergone an energy lowering edge reconstruction are half-metallic, i.e. metal for minority spins, but metallic for majority spins. This half-metallic nature is destroyed by edge passivation with hydrogen. The magnetic dipole moment, however, is found to vary with the nature of hydrogen passivation, specifically, the number of hydrogen atoms involved in the passivation. For armchair nanoribbons are predicted to be direct bandgap semiconductors with a band gap that increases with hydrogen termination of the edges. They find this structure to have no magnetic dipole moment. It is worth noting that early work by Mendez *et al.* predicted that bare armchair nanoribbons are metallic, and that the magnetic state is 14 meV. Mendez *et al.* [171] also predicted that for hydrogen passivated zigzag nanoribbons the antiferromagnetic state is 15 meV more favorable than the ferromagnetic state. Ataca *et al.* [170] extended their study to consider vacancies and found that only an MoS_2 triple vacancy resulted in a significant dipole moment.

There are also many advantages to be gained by alloying TMDs. By controlling the composition of each layer of a multilayer thin film it is possible to generate broadband absorbers. For example, a multilayer photodetector with a graded composition from MoS_2 to WS_2 was shown to generate a photocurrent four times greater than MoS_2 or WS_2 devices by themselves [172]. This enhancement was attributed to the gradual transition from the MoS_2 indirect and direct bandgaps at 1.2 and 1.9 eV to the WS_2 bandgaps and 1.3 and 2.1 eV leading to broadband absorption. Janus structures that show large Rashba splitting may have promise for spintronic applications [173].

Kosmala *et al.* [158] observed a boost in the HER hydrogen evolution reaction (HER) of $MoTe_2$ and $MoSe_2$ solid solutions and speculated that this may be due to the increased concentration of mirror twin grain boundaries (MTBs), however a systematic study of the correlation between defect type, concentration and HER activity is lacking. In their study of the hydrogen evolution reaction using $1T$-MoS_2 Tang and Jiang found that substitutional doping with Mn, Cr, Cu, Ni, or Fe could make the $1T$ MoS_2 a better catalyst. Improvements were also predicted were by Attanayake *et al.* [174], for the intercalation of Na^+, Ca^{2+}, Ni^{2+} and Co^{2+} into the interlayer region of $1T$-MoS_2. Janus structures have also been considered for HER. It was found that SeMoS, and SMoSe should enhance HER behavior when compared to MoS_2 and $MoSe_2$. However, a comparison to homogenous MoSSe alloys was not included [128].

4.6 Summary

In summary, like all materials, many of the functional properties of TMDs are dominated by their defects. Numerous reports have shown that defect engineering can be used to control catalytic properties, conductivity, optical properties, and magnetic properties. However, we are still a long way from synthesizing truly high-purity films comparable to silicon. Therefore, we must continue to consider the important role of non-intentional and random defects in determining the final properties of our material. Both geological and synthetic material are impacted by the presence of impurities, vacancies, vacancies clusters, and grain boundaries. In the long term, achieving high purity material will significantly simplify our efforts in engineering the properties of TMDs, but in the near term, we must continue to thoroughly characterize our materials and continue to consider both random/uncontrolled and engineered defects when interpreting the functional properties of these materials.

References

[1] S. Das, J.A. Robinson, M. Dubey, H. Terrones, M. Terrones, Beyond graphene: progress in novel two-dimensional materials and van der Waals solids, Annual Review of Materials Research 45 (2015) 1.

[2] Y.M. He, G. Clark, J.R. Schaibley, Y. He, M.C. Chen, Y.J. Wei, X. Ding, Q. Zhang, W. Yao, X. Xu, Single quantum emitters in monolayer semiconductors, Nature Nanotechnology 10 (2015) 497.

[3] J. Hong, C. Jin, J. Yuan, Z. Zhang, Atomic defects in two-dimensional materials: from single-atom spectroscopy to functionalities in opto-/electronics, nanomagnetism, and catalysis, Advanced Materials 29 (2017) 1606434.

[4] M. Koperski, K. Nogajewski, A. Arora, V. Cherkez, P. Mallet, J.Y. Veuillen, J. Marcus, P. Kossacki, M. Potemski, Single photon emitters in exfoliated WSe2 structures, Nature Nanotechnology 10 (2015) 503.

[5] Z. Lin, Y. Lei, S. Subramanian, N. Briggs, Y. Wang, C.-L. Lo, E. Yalon, D. Lloyd, S. Wu, K. Koski, R. Clark, S. Das, R.M. Wallace, T. Kuech, J.S. Bunch, X. Li, Z. Chen, E. Pop, V.H. Crespi, J.A. Robinson, M. Terrones, Research update: recent progress on 2D materials beyond graphene: from ripples, defects, intercalation, and valley dynamics to straintronics and power dissipation, APL Materials 6 (8) (2018) 080701.

[6] K.F. Mak, J. Shan, Photonics and optoelectronics of 2D semiconductor transition metal dichalcogenides, Nature Photonics 10 (2016) 216.

[7] S. Manzeli, D. Ovchinnikov, D. Pasquier, O.V. Yazyev, A. Kis, 2D transition metal dichalcogenides, Nature Reviews Materials 2 (2017) 17033.

[8] K. Novoselov, A. Mishchenko, A. Carvalho, A.C. Neto, 2D materials and van der Waals heterostructures, Science 353 (2016), aac9439.

[9] J.R. Schaibley, H. Yu, G. Clark, P. Rivera, J.S. Ross, K.L. Seyler, W. Yao, X. Xu, Valleytronics in 2D materials, Nature Reviews Materials 1 (2016) 16055.

[10] H. Li, C. Tsai, A.L. Koh, L. Cai, A.W. Contryman, A.H. Fragapane, J. Zhao, H.S. Han, H.C. Manoharan, F. Abild-Pedersen, Activating and optimizing MoS2 basal planes for hydrogen evolution through the formation of strained sulphur vacancies, Nature Materials 15 (2016) 48.

[11] Y. Ji, J.K. Nørskov, K. Chan, Scaling relations on basal plane vacancies of transition metal dichalcogenides for CO2 reduction, The Journal of Physical Chemistry C 123 (7) (2019) 4256–4261.

[12] G. Li, D. Zhang, Q. Qiao, Y. Yu, D. Peterson, A. Zafar, R. Kumar, S. Curtarolo, F. Hunte, S. Shannon, All the catalytic active sites of MoS2 for hydrogen evolution, Journal of the American Chemical Society 138 (2016) 16632.

[13] I. Delac Marion, D. Capeta, B. Pielic, F. Faraguna, A. Gallardo, P. Pou, B. Biel, N. Vujicic, M. Kralj, Atomic-scale defects and electronic properties of a transferred synthesized MoS2 monolayer, Nanotechnology 29 (30) (2018) 305703, https://www.ncbi.nlm.nih.gov/pubmed/29726400.

[14] K. Kaasbjerg, T. Low, A.-P. Jauho, Electron and hole transport in disordered monolayer MoS2: atomic vacancy induced short-range and Coulomb disorder scattering, Physical Review B 100 (11) (2019) 115409.

[15] Z. Lin, B.R. Carvalho, E. Kahn, R. Lv, R. Rao, H. Terrones, M.A. Pimenta, M. Terrones, Defect engineering of two-dimensional transition metal dichalcogenides, 2D Materials 3 (2) (2016) 022002.

[16] C. Zhang, C. Wang, F. Yang, J.K. Huang, L.J. Li, W. Yao, W. Ji, C.K. Shih, Engineering point-defect states in monolayer WSe2, ACS Nano 13 (2) (2019) 1595–1602, https://www.ncbi.nlm.nih.gov/pubmed/30689361.

[17] B. Schuler, J.H. Lee, C. Kastl, K.A. Cochrane, C.T. Chen, S. Refaely-Abramson, S. Yuan, E. van Veen, R. Roldan, N.J. Borys, R.J. Koch, S. Aloni, A.M. Schwartzberg, D.F. Ogletree, J.B. Neaton, A. Weber-Bargioni, How substitutional point defects in two-dimensional WS2 induce charge localization, spin-orbit splitting, and strain, ACS Nano 13 (9) (2019) 10520–10534, https://www.ncbi.nlm.nih.gov/pubmed/31393700.

[18] M. Aghajanian, A.A. Mostofi, J. Lischner, Tuning electronic properties of transition-metal dichalcogenides via defect charge, Scientific Reports 8 (2018) 13611.

[19] C. Kastl, R.J. Koch, C.T. Chen, J. Eichhorn, S. Ulstrup, A. Bostwick, C. Jozwiak, T.R. Kuykendall, N.J. Borys, F.M. Toma, S. Aloni, A. Weber-Bargioni, E. Rotenberg, A.M. Schwartzberg, Effects of defects on band structure and excitons in WS2 revealed by nanoscale photoemission spectroscopy, ACS Nano 13 (2) (2019) 1284–1291, https://www.ncbi.nlm.nih.gov/pubmed/30645100.

[20] T. Le Quang, K. Nogajewski, M. Potemski, M.T. Dau, M. Jamet, P. Mallet, J.Y. Veuillen, Band-bending induced by charged defects and edges of atomically thin transition metal dichalcogenide films, 2D Materials 5 (3) (2018).

[21] S. Tongay, J. Suh, C. Ataca, W. Fan, A. Luce, J.S. Kang, J. Liu, C. Ko, R. Raghunathanan, J. Zhou, Defects activated photoluminescence in two-dimensional semiconductors: interplay between bound, charged, and free excitons, Scientific Reports 3 (2013) 2657.

[22] T. Verhagen, V.L.P. Guerra, G. Haider, M. Kalbac, J. Vejpravova, Towards the evaluation of defects in MoS2 using cryogenic photoluminescence spectroscopy, Nanoscale 12 (5) (2020) 3019–3028, https://www.ncbi.nlm.nih.gov/pubmed/31834348.

[23] S. Yuan, R. Roldán, M. Katsnelson, F. Guinea, Effect of point defects on the optical and transport properties of MoS2 and WS2, Physical Review. B, Condensed Matter and Materials Physics 90 (2014) 041402.

[24] W.H. Blades, N.J. Frady, P.M. Litwin, S.J. McDonnell, P. Reinke, Thermally induced defects on WSe2, The Journal of Physical Chemistry C 124 (28) (2020) 15337–15346.

[25] E. Mitterreiter, B. Schuler, K.A. Cochrane, U. Wurstbauer, A. Weber-Bargioni, C. Kastl, A.W. Holleitner, Atomistic positioning of defects in helium ion treated single-layer MoS2, Nano Letters 20 (6) (2020) 4437–4444, https://www.ncbi.nlm.nih.gov/pubmed/32368920.

[26] D. Edelberg, D. Rhodes, A. Kerelsky, B. Kim, J. Wang, A. Zangiabadi, C. Kim, A. Abhinandan, J. Ardelean, M. Scully, D. Scullion, L. Embon, R. Zu, E.J.G. Santos, L. Balicas, C. Marianetti, K. Barmak, X. Zhu, J. Hone, A.N. Pasupathy, Approaching the intrinsic limit in transition metal diselenides via point defect control, Nano Letters 19 (7) (2019) 4371–4379, https://www.ncbi.nlm.nih.gov/pubmed/31180688.

[27] K.K.H. Smithe, S.V. Suryavanshi, M. Munoz Rojo, A.D. Tedjarati, E. Pop, Low variability in synthetic monolayer MoS2 devices, ACS Nano 11 (8) (2017) 8456–8463, https://www.ncbi.nlm.nih.gov/pubmed/28697304.

[28] M. Pandey, F.A. Rasmussen, K. Kuhar, T. Olsen, K.W. Jacobsen, K.S. Thygesen, Defect-tolerant monolayer transition metal dichalcogenides, Nano Letters 16 (4) (2016) 2234–2239, https://www.ncbi.nlm.nih.gov/pubmed/27027786.

[29] J. Hong, Z. Hu, M. Probert, K. Li, D. Lv, X. Yang, L. Gu, N. Mao, Q. Feng, L. Xie, J. Zhang, D. Wu, Z. Zhang, C. Jin, W. Ji, X. Zhang, J. Yuan, Z. Zhang, Exploring atomic defects in molybdenum disulphide monolayers, Nature Communications 6 (2015) 6293, https://www.ncbi.nlm.nih.gov/pubmed/25695374.

[30] H.P. Komsa, J. Kotakoski, S. Kurasch, O. Lehtinen, U. Kaiser, A.V. Krasheninnikov, Two-dimensional transition metal dichalcogenides under electron irradiation: defect production and doping, Physical Review Letters 109 (2012) 035503.

[31] W. Zhou, X. Zou, S. Najmaei, Z. Liu, Y. Shi, J. Kong, J. Lou, P.M. Ajayan, B.I. Yakobson, J.C. Idrobo, Intrinsic structural defects in monolayer molybdenum disulfide, Nano Letters 13 (6) (2013) 2615–2622, https://www.ncbi.nlm.nih.gov/pubmed/23659662.

[32] C. Lee, B.G. Jeong, S.J. Yun, Y.H. Lee, S.M. Lee, M.S. Jeong, Unveiling defect-related Raman mode of monolayer WS2 via tip-enhanced resonance Raman scattering, ACS Nano 12 (10) (2018) 9982–9990, https://www.ncbi.nlm.nih.gov/pubmed/30142265.

[33] R. Addou, R.M. Wallace, Surface analysis of WSe2 crystals: spatial and electronic variability, ACS Applied Materials & Interfaces 8 (39) (2016) 26400–26406, https://www.ncbi.nlm.nih.gov/pubmed/27599557.

[34] P.K. Chow, R.B. Jacobs-Gedrim, J. Gao, T.M. Lu, B. Yu, H. Terrones, N. Koratkar, Defect-induced photoluminescence in monolayer semiconducting transition metal dichalcogenides, ACS Nano 9 (2015) 1520.

[35] C. Kastl, C. Chen, R.J. Koch, B. Schuler, T. Kuykendall, A. Bostwick, C. Jozwiak, T. Seyller, E. Rotenberg, A. Weber-Bargioni, Multimodal spectromicroscopy of monolayer WS2 enabled by ultra-clean van der Waals epitaxy, 2D Materials 5 (2018) 045010.

[36] D.Y. Qiu, F.H. da Jornada, S.G. Louie, Optical spectrum of MoS2: many-body effects and diversity of exciton states, Physical Review Letters 111 (2013) 216805.

[37] K.M. McCreary, A.T. Hanbicki, S.V. Sivaram, B.T. Jonker, A- and B-exciton photoluminescence intensity ratio as a measure of sample quality for transition metal dichalcogenide monolayers, APL Materials 6 (11) (2018).

[38] V. Carozo, Y. Wang, K. Fujisawa, B.R. Carvalho, A. McCreary, S. Feng, Z. Lin, C. Zhou, N. Perea-Lopez, A.L. Elias, B. Kabius, V.H. Crespi, M. Terrones, Optical identification of sulfur vacancies: bound excitons at the edges of monolayer tungsten disulfide, Science Advanced 3 (2017) e1602813.

[39] K. Kaasbjerg, Atomistic T-matrix theory of disordered two-dimensional materials: bound states, spectral properties, quasiparticle scattering, and transport, Physical Review B 101 (2020) 4.

[40] S. Haldar, H. Vovusha, M.K. Yadav, O. Eriksson, B. Sanyal, Systematic study of structural, electronic, and optical properties of atomic-scale defects in the two-dimensional transition metal dichalcogenidesMX2(M=Mo, W,X=S, Se, Te), Physical Review B 92 (23) (2015).

[41] X. Zhang, Z.Y. Al Balushi, F. Zhang, T.H. Choudhury, S.M. Eichfeld, N. Alem, T.N. Jackson, J.A. Robinson, J.M. Redwing, Influence of carbon in metalorganic chemical vapor deposition of few-layer WSe2 thin films, Journal of Electronic Materials 45 (12) (2016) 6273–6279.

[42] H.-P. Komsa, A.V. Krasheninnikov, Native defects in bulk and monolayer MoS2 from first principles, Physical Review B 91 (12) (2015).

[43] M.H. Naik, M. Jain, Substrate screening effects on the quasiparticle band gap and defect charge transition levels in MoS2, Physical Review Materials 2 (2018) 084002.

[44] Y. Li, A. Chernikov, X. Zhang, A. Rigosi, H.M. Hill, A.M. van der Zande, D.A. Chenet, E.M. Shih, J. Hone, T.F. Heinz, Measurement of the optical dielectric function of monolayer transition-metal dichalcogenides: MoS2, MoSe2, WS2, and WSe2, Physical Review. B, Condensed Matter and Materials Physics 90 (2014) 205422.

[45] B. Schuler, D.Y. Qiu, S. Refaely-Abramson, C. Kastl, C.T. Chen, S. Barja, R.J. Koch, D.F. Ogletree, S. Aloni, A.M. Schwartzberg, J.B. Neaton, S.G. Louie, A. Weber-Bargioni, Large spin-orbit splitting of deep in-gap defect states of engineered sulfur vacancies in monolayer WS_{2}, Physical Review Letters 123 (7) (2019) 076801, https://www.ncbi.nlm.nih.gov/pubmed/31491121.

[46] Y.J. Zheng, Y. Chen, Y.L. Huang, P.K. Gogoi, M.Y. Li, L.J. Li, P.E. Trevisanutto, Q. Wang, S.J. Pennycook, A.T.S. Wee, S.Y. Quek, Point defects and localized excitons in 2D WSe2, ACS Nano 13 (5) (2019) 6050–6059, https://www.ncbi.nlm.nih.gov/pubmed/31074961.

[47] S. Barja, S. Refaely-Abramson, B. Schuler, D.Y. Qiu, A. Pulkin, S. Wickenburg, H. Ryu, M.M. Ugeda, C. Kastl, C. Chen, C. Hwang, A. Schwartzberg, S. Aloni, S.K. Mo, D. Frank Ogletree, M.F. Crommie, O.V. Yazyev, S.G. Louie, J.B. Neaton, A. Weber-Bargioni, Identifying substitutional oxygen as a prolific point defect in monolayer transition metal dichalcogenides, Nature Communications 10 (1) (2019) 3382, https://www.ncbi.nlm.nih.gov/pubmed/31358753.

[48] R. Addou, L. Colombo, R.M. Wallace, Surface defects on natural MoS_2, ACS Applied Materials & Interfaces 7 (22) (2015) 11921–11929, http://pubs.acs.org/doi/abs/10.1021/acsami.5b01778.

[49] R. Addou, C.M. Smyth, J.-Y. Noh, Y.-C. Lin, Y. Pan, S.M. Eichfeld, S. Fölsch, J.A. Robinson, K. Cho, R.M. Feenstra, R.M. Wallace, One dimensional metallic edges in atomically thin WSe2 induced by air exposure, 2D Materials 5 (2) (2018).

[50] M. Donarelli, F. Bisti, F. Perrozzi, L. Ottaviano, Tunable sulfur desorption in exfoliated MoS2 by means of thermal annealing in ultra-high vacuum, Chemical Physics Letters 588 (2013) 198–202.

[51] J.A. Wilson, A.D. Yoffe, Transition metal dichalcogenides discussion and interpretation of observed optical, electrical and structural properties, Advances in Physics 18 (73) (1969) 193, <Go to ISI>://WOS:A1969E605200001.

[52] W. Kautek, H. Gerischer, H. Tributsch, The role of carrier diffusion and indirect optical transitions in the photoelectrochemical behavior of layer type d-band semiconductors, Journal of The Electrochemical Society 127 (11) (1980) 2471–2478.

[53] K. Kam, B. Parkinson, Detailed photocurrent spectroscopy of the semiconducting group VIB transition metal dichalcogenides, The Journal of Physical Chemistry 86 (4) (1982) 463–467.

[54] W. Jaegermann, D. Schmeisser, Reactivity of layer type transition metal chalcogenides towards oxidation, Surface Science 165 (1) (1986) 143–160.

[55] W. Hofmann, H. Lewerenz, C. Pettenkofer, Mixed cation group VI layered transition metal dichalcogenides: preparation, characterization and photoactivity of $MoxW_{1-x}Se_2$, Solar Energy Materials 17 (3) (1988) 165–178.

[56] W. Jaegermann, H. Tributsch, Interfacial properties of semiconducting transition metal chalcogenides, Progress in Surface Science 29 (1) (1988) 1–167.

[57] S. Kim, A. Konar, W.S. Hwang, J.H. Lee, J. Lee, J. Yang, C. Jung, H. Kim, J.B. Yoo, J.Y. Choi, Y.W. Jin, S.Y. Lee, D. Jena, W. Choi, K. Kim, High-mobility and low-power thin-film transistors based on multilayer MoS2 crystals, Nature Communications 3 (2012) 1011, <Go to ISI>://WOS:000308801100027.

[58] S. Das, H.-Y. Chen, A.V. Penumatcha, J. Appenzeller, High performance multi-layer MoS_2 transistors with scandium contacts, Nano Letters 13 (2013) 100–105.

[59] B. Radisavljevic, A. Radenovic, J. Brivio, V. Giacometti, A. Kis, Single-layer MoS_2 transistors, Nature Nanotechnology 6 (3) (2011) 147–150, https://www.nature.com/articles/nnano.2010.279.

[60] S. McDonnell, R. Addou, C. Buie, R.M. Wallace, C.L. Hinkle, Defect dominated doping and contact resistance in MoS_2, ACS Nano 8 (3) (2014) 2880–2888, http://pubs.acs.org/doi/abs/10.1021/nn500044q.

[61] J. Lagrenaudie, Comparaison des composés de la famille de MoS2 (structure et propriétés optiques et électriques), Journal de Physique Et Le Radium 15 (4) (1954) 299–300, https://doi.org/10.1051/jphysrad:01954001504029900.

[62] R. Mansfield, S. Salam, Electrical properties of molybdenite, Proceedings of the Physical Society. Section B 66 (5) (1953) 377.

[63] F. Regnault, P. Aigrain, C. Dugas, B. Jancovivi, Sur Les Proprietes semi-conductrices de la molybdenite, Numéro Comptes Rendus Hebdomadaires des Séances de l'Académie des Sciences 235 (1952) 31–32.

[64] R. Fivaz, E. Mooser, Mobility of charge carriers in semiconducting layer structures, Physical Review 163 (3) (1967) 743.

[65] Y. Kang, S. Han, An origin of unintentional doping in transition metal dichalcogenides: the role of hydrogen impurities, Nanoscale 9 (12) (2017) 4265–4271.

[66] B. Baugher, H.O. Churchill, Y. Yang, P. Jarillo-Herrero, Intrinsic electronic transport properties of high quality monolayer and bilayer MoS_2, Nano Letters (2013) 4212–4216.

[67] Z. Yin, H. Li, H. Li, L. Jiang, Y. Shi, Y. Sun, G. Lu, Q. Zhang, X. Chen, H. Zhang, Single-layer MoS_2 phototransistors, ACS Nano 6 (1) (2012) 74–80.

[68] J. Golden, M. McMillan, R.T. Downs, G. Hystad, I. Goldstein, H.J. Stein, A. Zimmerman, D.A. Sverjensky, J.T. Armstrong, R.M. Hazen, Rhenium variations in molybdenite (MoS_2): evidence for progressive subsurface oxidation, Earth and Planetary Science Letters 366 (2013) 1–5.

[69] T. Hallam, S. Monaghan, F. Gity, L. Ansari, M. Schmidt, C. Downing, C.P. Cullen, V. Nicolosi, P.K. Hurley, G.S. Duesberg, Rhenium-doped MoS2 films, Applied Physics Letters 111 (20) (2017) 203101.

[70] R. Addou, S. McDonnell, D. Barrera, Z. Guo, A. Azcatl, J. Wang, H. Zhu, C.L. Hinkle, M. Quevedo-Lopez, H.N. Alshareef, L. Colombo, J.W.P. Hsu, R.M. Wallace, Impurities and electronic property variations of natural MoS_2 crystal surfaces, ACS Nano 9 (9) (2015) 9124–9133, https://doi.org/10.1021/acsnano.5b03309, http://pubs.acs.org/doi/pdfplus/10.1021/acsnano.5b03309.

[71] M.G. Sales, L. Herweyer, E. Opila, S. McDonnell, MoS2 impurities: chemical identification and spatial resolution of bismuth impurities in geological material, Applied Surface Science (2020) 145256.

[72] R.J. Toh, Z. Sofer, J. Luxa, M. Pumera, Ultrapure molybdenum disulfide shows enhanced catalysis for hydrogen evolution over impurities-doped counterpart, ChemCatChem 9 (7) (2017) 1168–1171.

[73] S.M. Tan, M. Pumera, Two-dimensional materials on the rocks: positive and negative role of dopants and impurities in electrochemistry, ACS Nano 13 (3) (2019) 2681–2728.

[74] G. Mirabelli, F. Gity, S. Monaghan, P.K. Hurley, R. Duffy, In impact of impurities, interface traps and contacts on MoS_2 MOSFETs: modelling and experiments, in: 2017 47th European Solid-State Device Research Conference (ESSDERC), IEEE, 2017, pp. 288–291.

[75] F. Vialla, M. Danovich, D.A. Ruiz-Tijerina, M. Massicotte, P. Schmidt, T. Taniguchi, K. Watanabe, R.J. Hunt, M. Szyniszewski, N.D. Drummond, Tuning of impurity-bound interlayer complexes in a van der Waals heterobilayer, 2D Materials 6 (3) (2019) 035032.

[76] J.A. Woollam, R.B. Somoano, Physics and chemistry of MoS_2 intercalation compounds, Materials Science and Engineering 31 (1977) 289–295.

[77] P. Joensen, R. Frindt, S.R. Morrison, Single-layer MoS_2, Materials Research Bulletin 21 (4) (1986) 457–461.

[78] H. Huang, X. Fan, D.J. Singh, W. Zheng, First principles study on 2H–1T′ transition in MoS 2 with copper, Physical Chemistry Chemical Physics 20 (42) (2018) 26986–26994.

[79] G. Eda, T. Fujita, H. Yamaguchi, D. Voiry, M. Chen, M. Chhowalla, Coherent atomic and electronic heterostructures of single-layer MoS2, ACS Nano 6 (8) (2012) 7311–7317.

[80] S.J. Sandoval, D. Yang, R. Frindt, J. Irwin, Raman study and lattice dynamics of single molecular layers of MoS_2, Physical Review B 44 (8) (1991) 3955.

[81] F. Wypych, T. Weber, R. Prins, Scanning tunneling microscopic investigation of 1T-MoS_2, Chemistry of Materials 10 (3) (1998) 723–727.

[82] F. Wypych, R. Schöllhorn, 1T-MoS_2, a new metallic modification of molybdenum disulfide, Journal of the Chemical Society, Chemical Communications 19 (1992) 1386–1388.

[83] F. Wypych, T. Weber, R. Prins, Scanning tunneling microscopic investigation of $K_x(H_2O)_yMoS_2$, Surface Science 380 (1) (1997) L474–L478.

[84] X. Qin, D. Yang, R. Frindt, J. Irwin, Real-space imaging of single-layer MoS_2 by scanning tunneling microscopy, Physical Review B 44 (7) (1991) 3490.

[85] K. Chrissafis, M. Zamani, K. Kambas, J. Stoemenos, N. Economou, I. Samaras, C. Julien, Structural studies of MoS_2 intercalated by lithium, Materials Science and Engineering: B 3 (1) (1989) 145–151.

[86] K.E. Dungey, M.D. Curtis, J.E. Penner-Hahn, Structural characterization and thermal stability of MoS2 intercalation compounds, Chemistry of Materials 10 (8) (1998) 2152–2161.

[87] A. Ambrosi, Z. Sofer, M. Pumera, Lithium intercalation compound dramatically influences the electrochemical properties of exfoliated MoS_2, Small 11 (5) (2015) 605–612.

[88] X. Wang, X. Shen, Z. Wang, R. Yu, L. Chen, Atomic-scale clarification of structural transition of MoS2 upon sodium intercalation, ACS Nano 8 (11) (2014) 11394–11400.

[89] Y.V. Zubavichus, A. Golub, Y.N. Novikov, Y.L. Slovokhotov, A. Nesmeyanov, P. Schilling, R. Tittsworth, XAFS study of MoS_2 intercalation compounds, Le Journal de Physique IV 7 (C2) (1997), C2-1057–C2-1059.

[90] X.-L. Li, Y.-D. Li, MoS_2 nanostructures: synthesis and electrochemical Mg^{2+} intercalation, The Journal of Physical Chemistry B 108 (37) (2004) 13893–13900.

[91] G. Eda, H. Yamaguchi, D. Voiry, T. Fujita, M. Chen, M. Chhowalla, Photoluminescence from chemically exfoliated MoS_2, Nano Letters 11 (12) (2011) 5111–5116, http://pubs.acs.org/doi/abs/10.1021/nl201874w.

[92] S. Das, A. Roelofs, in: Electrostatically Doped WSe_2 CMOS Inverter, 72nd Device Research Conference, IEEE, 2014, pp. 185–186.

[93] S. Das, M. Dubey, A. Roelofs, High gain, low noise, fully complementary logic inverter based on bi-layer WSe2 field effect transistors, Applied Physics Letters 105 (8) (2014) 083511.

[94] A. Ossipov, M. Titov, C.W. Beenakker, Reentrance effect in a graphene n-p-n junction coupled to a superconductor, Physical Review B 75 (24) (2007) 241401.

[95] G.V. Resta, Y. Balaji, D. Lin, I.P. Radu, F. Catthoor, P.-E. Gaillardon, G. De Micheli, In doping-free complementary inverter enabled by 2D WSe_2 electrostatically-doped reconfigurable transistors, in: 2018 76th Device Research Conference (DRC), IEEE, 2018, pp. 1–2.

[96] C.-S. Pang, N. Thakuria, S.K. Gupta, Z. Chen, First demonstration of WSe_2 based CMOS-SRAM, in: 2018 IEEE International Electron Devices Meeting (IEDM), IEEE, 2018, pp. 22.2.1–22.2.4.

[97] C.-S. Pang, Z. Chen, First demonstration of WSe_2 CMOS inverter with modulable noise margin by electrostatic doping, in: 2018 76th Device Research Conference (DRC), IEEE, 2018, pp. 1–2.

[98] L. Yu, A. Zubair, E.J. Santos, X. Zhang, Y. Lin, Y. Zhang, T. Palacios, High-performance WSe_2 complementary metal oxide semiconductor technology and integrated circuits, Nano Letters 15 (8) (2015) 4928–4934.

[99] S. Jiang, L. Li, Z. Wang, K.F. Mak, J. Shan, Controlling magnetism in 2D CrI_3 by electrostatic doping, Nature Nanotechnology 13 (7) (2018) 549–553.

[100] S. Sutar, P. Agnihotri, E. Comfort, T. Taniguchi, K. Watanabe, J. Ung Lee, Reconfigurable pn junction diodes and the photovoltaic effect in exfoliated MoS_2 films, Applied Physics Letters 104 (12) (2014) 122104.

[101] B. Lee, W. Liu, C.H. Naylor, J. Park, S.C. Malek, J.S. Berger, A.C. Johnson, R. Agarwal, Electrical tuning of exciton–plasmon polariton coupling in monolayer MoS_2 integrated with plasmonic nanoantenna lattice, Nano Letters 17 (7) (2017) 4541–4547.

[102] K. Zhang, S. Feng, J. Wang, A. Azcatl, N. Lu, R. Addou, N. Wang, C. Zhou, J. Lerach, V. Bojan, Manganese doping of monolayer MoS_2: the substrate is critical, Nano Letters 15 (10) (2015) 6586–6591.

[103] G. Gupta, B. Rajasekharan, R.J. Hueting, Electrostatic doping in semiconductor devices, I.E.E.E. Transactions on Electron Devices 64 (8) (2017) 3044–3055.

[104] S. Cristoloveanu, K.H. Lee, H. Park, M.S. Parihar, The concept of electrostatic doping and related devices, Solid-State Electronics 155 (2019) 32–43.

[105] A.K.A. Lu, M. Houssa, I.P. Radu, G. Pourtois, Toward an understanding of the electric field-induced electrostatic doping in van der Waals heterostructures: a first-principles study, ACS Applied Materials & Interfaces 9 (8) (2017) 7725–7734.

[106] K. Xu, Y. Wang, Y. Zhao, Y. Chai, Modulation doping of transition metal dichalcogenide/oxide heterostructures, Journal of Materials Chemistry C 5 (2) (2017) 376–381.

[107] H. Fang, S. Chuang, T.C. Chang, K. Takei, T. Takahashi, A. Javey, High-performance single layered WSe_2 p-FETs with chemically doped contacts, Nano Letters 12 (7) (2012) 3788–3792, <Go to ISI>://WOS:000306296200073.

[108] H. Fang, M. Tosun, G. Seol, T.C. Chang, K. Takei, J. Guo, A. Javey, Degenerate n-doping of few-layer transition metal dichalcogenides by potassium, Nano Letters 13 (5) (2013) 1991–1995, <Go to ISI>://WOS:000318892400019.

Reference list:

[109] H. Li, S. Liu, S. Huang, Q. Zhang, C. Li, X. Liu, J. Meng, Y. Tian, Metallic impurities induced electronic transport in WSe$_2$: first-principle calculations, Chemical Physics Letters 658 (2016) 83–87.

[110] C. Yang, X. Zhao, T. Wang, S. Wei, Characters of group V and VII atoms doped WSe2 monolayer, Journal of Alloys and Compounds 699 (2017) 291–296.

[111] N. Onofrio, D. Guzman, A. Strachan, Novel doping alternatives for single-layer transition metal dichalcogenides, Journal of Applied Physics 122 (18) (2017) 185102, https://aip.scitation.org/doi/abs/10.1063/1.4994997.

[112] P. Zhao, D. Kiriya, A. Azcatl, C. Zhang, M. Tosun, Y.-S. Liu, M. Hettick, J.S. Kang, S. McDonnell, S. KC, Air stable p-doping of WSe$_2$ by covalent functionalization, ACS Nano 8 (10) (2014) 10808–10814, https://pubs.acs.org/doi/10.1021/nn5047844.

[113] A. Azcatl, X. Qin, A. Prakash, C. Zhang, L. Cheng, Q. Wang, N. Lu, M.J. Kim, J. Kim, K. Cho, Covalent nitrogen doping and compressive strain in MoS2 by remote N$_2$ plasma exposure, Nano Letters 16 (9) (2016) 5437–5443.

[114] A. Khosravi, R. Addou, C.M. Smyth, R. Yue, C.R. Cormier, J. Kim, C.L. Hinkle, R.M. Wallace, Covalent nitrogen doping in molecular beam epitaxy-grown and bulk WSe2, APL Materials 6 (2) (2018) 026603, https://aip.scitation.org/doi/abs/10.1063/1.5002132.

[115] R. Mukherjee, H. Chuang, M. Koehler, N. Combs, A. Patchen, Z. Zhou, D. Mandrus, Substitutional electron and hole doping of WSe$_2$: synthesis, electrical characterization, and observation of band-to-band tunneling, Physical Review Applied 7 (3) (2017) 034011.

[116] M. Agarwal, L. Talele, Transport properties of molybdenum sulphoselenide (MoSxSe2-x, $0 \leqslant x \leqslant 2$) single crystals, Solid State Communications 59 (8) (1986) 549–551.

[117] W. Hofmann, H. Lewerenz, C. Pettenkofer, Mixed cation group VI layered transition metal dichalcogenides: preparation, characterization and photoactivity of $Mo_xW_{1-x}Se_2$, Solar Energy Materials 17 (3) (1988) 165–178.

[118] S.M. Delphine, M. Jayachandran, C. Sanjeeviraja, Review of material properties of (Mo/W)Se$_2$-layered compound semiconductors useful for photoelectrochemical solar cells, Crystallography Reviews 17 (4) (2011) 281–301.

[119] S.M. Delphine, M. Jayachandran, C. Sanjeeviraja, A. Almusallam, Study on (Mo/W) Se2 layered compound semi conductors useful for photoelectrochemical solar cells, International Journal of ChemTech Research 3 (2011) 846–852.

[120] D. Dolakia, G. Solanki, S. Patel, M. Agarwal, Optical band gap studies of tungsten sulphoselenide single crystals grown by DVT technique, Scientia Iranica 10 (4) (2003) 373–382.

[121] D. Gujarathi, G. Solanki, M. Deshpande, M. Agarwal, Band gap in tungsten sulphoselenide single crystals determined by the optical absorption method, Materials Science in Semiconductor Processing 8 (5) (2005) 576–586.

[122] P. Yu, J. Lin, L. Sun, Q.L. Le, X. Yu, G. Gao, C.H. Hsu, D. Wu, T.R. Chang, Q. Zeng, Metal–semiconductor phase-transition in $WSe_{2(1-x)}te_{2x}$ monolayer, Advanced Materials 29 (4) (2017) 1603991.

[123] K.-A.N. Duerloo, E.J. Reed, Structural phase transitions by design in monolayer alloys, ACS Nano 10 (1) (2016) 289–297.

[124] M. Lattuada, T.A. Hatton, Synthesis, properties and applications of Janus nanoparticles, Nano Today 6 (3) (2011) 286–308.

[125] J. Zhou, Q. Wang, Q. Sun, X. Chen, Y. Kawazoe, P. Jena, Ferromagnetism in semihydrogenated graphene sheet, Nano Letters 9 (11) (2009) 3867–3870.

[126] F. Li, Y. Li, Band gap modulation of Janus graphene nanosheets by interlayer hydrogen bonding and the external electric field: a computational study, Journal of Materials Chemistry C 3 (14) (2015) 3416–3421.

[127] A.-Y. Lu, H. Zhu, J. Xiao, C.-P. Chuu, Y. Han, M.-H. Chiu, C.-C. Cheng, C.-W. Yang, K.-H. Wei, Y. Yang, Janus monolayers of transition metal dichalcogenides, Nature Nanotechnology 12 (8) (2017) 744–749.

[128] J. Zhang, S. Jia, I. Kholmanov, L. Dong, D. Er, W. Chen, H. Guo, Z. Jin, V.B. Shenoy, L. Shi, Janus monolayer transition-metal dichalcogenides, ACS Nano 11 (8) (2017) 8192–8198.

[129] R. Li, Y. Cheng, W. Huang, Recent progress of Janus 2D transition metal chalcogenides: from theory to experiments, Small 14 (45) (2018) 1802091.

[130] X. Zhang, T.H. Choudhury, M. Chubarov, Y. Xiang, B. Jariwala, F. Zhang, N. Alem, G.C. Wang, J.A. Robinson, J.M. Redwing, Diffusion-controlled epitaxy of large area coalesced WSe2 monolayers on sapphire, Nano Letters 18 (2) (2018) 1049–1056, https://www.ncbi.nlm.nih.gov/pubmed/29342357.

[131] I.N.M. Groot, Personal communication – AVS 66 – invited presentation, 2019.

[132] M. Saedi, J.M. de Voogd, A. Sjardin, A. Manikas, C. Galiotis, M. Jankowski, G. Renaud, F. La Porta, O. Konovalov, G.J.C. van Baarle, I.M.N. Groot, Development of a reactor for the in situ monitoring of 2D materials growth on liquid metal catalysts, using synchrotron x-ray scattering, Raman spectroscopy, and optical microscopy, Review of Scientific Instruments 91 (1) (2020) 013907, https://www.ncbi.nlm.nih.gov/pubmed/32012586.

[133] X. Li, C.W. Magnuson, A. Venugopal, R.M. Tromp, J.B. Hannon, E.M. Vogel, L. Colombo, R.S. Ruoff, Large-area graphene single crystals grown by low-pressure chemical vapor deposition of methane on copper, Journal of the American Chemical Society 133 (9) (2011) 2816–2819.

[134] Z. Yan, J. Lin, Z. Peng, Z. Sun, Y. Zhu, L. Li, C. Xiang, E.L. Samuel, C. Kittrell, J.M. Tour, Toward the synthesis of wafer-scale single-crystal graphene on copper foils, Acs Nano 6 (10) (2012) 9110–9117.

[135] D. Le, T.S. Rahman, Joined edges in MoS2: metallic and half-metallic wires, Journal of Physics. Condensed Matter 25 (31) (2013) 312201, https://www.ncbi.nlm.nih.gov/pubmed/23835417.

[136] T. Kosmala, H. Coy Diaz, H.-P. Komsa, Y. Ma, A.V. Krasheninnikov, M. Batzill, S. Agnoli, Metallic twin boundaries boost the hydrogen evolution reaction on the basal plane of molybdenum selenotellurides, Advanced Energy Materials 8 (20) (2018) 8.

[137] Y. Ma, S. Kolekar, H. Coy Diaz, J. Aprojanz, I. Miccoli, C. Tegenkamp, M. Batzill, Metallic twin grain boundaries embedded in MoSe$_2$ monolayers grown by molecular beam epitaxy, ACS Nano 11 (5) (2017) 5130–5139, https://pubs.acs.org/doi/abs/10.1021/acsnano.7b02172.

[138] A.N. Enyashin, M. Bar-Sadan, L. Houben, G. Seifert, Line defects in molybdenum disulfide layers, The Journal of Physical Chemistry C 117 (20) (2013) 10842–10848.

[139] S. Barja, S. Wickenburg, Z.-F. Liu, Y. Zhang, H. Ryu, Miguel M. Ugeda, Z. Hussain, Z.-X. Shen, S.-K. Mo, E. Wong, Miquel B. Salmeron, F. Wang, M.F. Crommie, D.F. Ogletree, Jeffrey B. Neaton, A. Weber-Bargioni, Charge density wave order in 1D mirror twin boundaries of single-layer MoSe2, Nature Physics 12 (8) (2016) 751–756.

[140] H.P. Komsa, A.V. Krasheninnikov, Engineering the electronic properties of two-dimensional transition metal dichalcogenides by introducing mirror twin boundaries, Advanced Electronic Materials 3 (6) (2017) 1600468.

[141] W. Zhou, X.L. Zou, S. Najmaei, Z. Liu, Y.M. Shi, J. Kong, J. Lou, P.M. Ajayan, B.I. Yakobson, J.C. Idrobo, Intrinsic structural defects in monolayer molybdenum disulfide, Nano Letters 13 (6) (2013) 2615–2622, <Go to ISI>://WOS:000320485100048.

[142] O. Lehtinen, H.-P. Komsa, A. Pulkin, M.B. Whitwick, M.-W. Chen, T. Lehnert, M.J. Mohn, O.V. Yazyev, A. Kis, U. Kaiser, A.V. Krasheninnikov, Atomic scale microstructure and properties of Se-deficient two-dimensional MoSe2, ACS Nano 9 (3) (2015) 3274–3283, https://doi.org/10.1021/acsnano.5b00410, http://pubs.acs.org/doi/pdfplus/10.1021/acsnano.5b00410.

[143] Y.-C. Lin, T. Björkman, H.-P. Komsa, P.-Y. Teng, C.-H. Yeh, F.-S. Huang, K.-H. Lin, J. Jadczak, Y.-S. Huang, P.-W. Chiu, Three-fold rotational defects in two-dimensional transition metal dichalcogenides, Nature Communications 6 (2015) 6736.

[144] O.V. Yazyev, S.G. Louie, Topological defects in graphene: dislocations and grain boundaries, Physical Review B 81 (19) (2010) 195420.

[145] Y. Liu, B.I. Yakobson, Cones, pringles, and grain boundary landscapes in graphene topology, Nano Letters 10 (6) (2010) 2178–2183.

[146] Y. Liu, X. Zou, B.I. Yakobson, Dislocations and grain boundaries in two-dimensional boron nitride, ACS Nano 6 (8) (2012) 7053–7058.

[147] B.C. McGuigan, P. Pochet, H.T. Johnson, Critical thickness for interface misfit dislocation formation in two-dimensional materials, Physical Review B 93 (21) (2016) 214103.

[148] A. Azizi, X. Zou, P. Ercius, Z. Zhang, A.L. Elías, N. Perea-López, G. Stone, M. Terrones, B.I. Yakobson, N. Alem, Dislocation motion and grain boundary migration in two-dimensional tungsten disulphide, Nature Communications 5 (1) (2014) 1–7.

[149] P. Pochet, B.C. McGuigan, J. Coraux, H.T. Johnson, Toward moiré engineering in 2D materials via dislocation theory, Applied Materials Today 9 (2017) 240–250.

[150] P. Kumar, B. Viswanath, Effect of sulfur evaporation rate on screw dislocation driven growth of MoS2 with high atomic step density, Crystal Growth & Design 16 (12) (2016) 7145–7154.

[151] L. Zhang, K. Liu, A.B. Wong, J. Kim, X. Hong, C. Liu, T. Cao, S.G. Louie, F. Wang, P. Yang, Three-dimensional spirals of atomic layered MoS2, Nano Letters 14 (11) (2014) 6418–6423.

[152] X. Wang, H. Yang, R. Yang, Q. Wang, J. Zheng, L. Qiao, X. Peng, Y. Li, D. Chen, X. Xiong, Weakened interlayer coupling in two-dimensional $MoSe_2$ flakes with screw dislocations, Nano Research 12 (8) (2019) 1900–1905.

[153] H.-J. Chuang, B. Chamlagain, M. Koehler, M.M. Perera, J. Yan, D. Mandrus, D. Tomanek, Z. Zhou, Low-resistance 2D/2D ohmic contacts: a universal approach to high-performance WSe_2, MoS_2, and $MoSe_2$ transistors, Nano Letters 16 (3) (2016) 1896–1902.

[154] Y. Xia, J. Zhang, Z. Yu, Y. Jin, H. Tian, Y. Feng, B. Li, W. Ho, C. Liu, H. Xu, A shallow acceptor of phosphorous doped in $MoSe_2$ monolayer, Advanced Electronic Materials 6 (1) (2020) 1900830.

[155] Y. He, P. Tang, Z. Hu, Q. He, C. Zhu, L. Wang, Q. Zeng, P. Golani, G. Gao, W. Fu, Engineering grain boundaries at the 2D limit for the hydrogen evolution reaction, Nature Communications 11 (1) (2020) 1–12.

[156] R. Yue, Y. Nie, L.A. Walsh, R. Addou, C. Liang, N. Lu, A.T. Barton, H. Zhu, Z. Che, D. Barrera, Nucleation and growth of WSe_2: enabling large grain transition metal dichalcogenides, 2D Materials 4 (4) (2017) 045019.

[157] W. Yao, B. Wu, Y. Liu, Growth and grain boundaries in 2D materials, ACS Nano (2020).

[158] T. Kosmala, H. Coy Diaz, H.P. Komsa, Y. Ma, A.V. Krasheninnikov, M. Batzill, S. Agnoli, Metallic twin boundaries boost the hydrogen evolution reaction on the basal plane of molybdenum selenotellurides, Advanced Energy Materials (2018) 1800031, https://onlinelibrary.wiley.com/doi/abs/10.1002/aenm.201800031.

[159] D. Kim, H. Du, T. Kim, S. Shin, S. Kim, M. Song, C. Lee, J. Lee, H. Cheong, D.H. Seo, The enhanced low resistance contacts and boosted mobility in two-dimensional p-type WSe_2 transistors through Ar^+ ion-beam generated surface defects, AIP Advances 6 (10) (2016) 105307.

[160] B. Wang, Y. Xia, J. Zhang, H.-P. Komsa, M. Xie, Y. Peng, C. Jin, Niobium doping induced mirror twin boundaries in MBE grown WSe_2 monolayers, Nano Research (2020) 1–8.

[161] H. Zhu, Q. Wang, L. Cheng, R. Addou, J. Kim, M.J. Kim, R.M. Wallace, Defects and surface structural stability of MoTe2 under vacuum annealing, ACS Nano 11 (11) (2017) 11005–11014.

[162] H. Zhu, Q. Wang, C. Zhang, R. Addou, K. Cho, R.M. Wallace, M.J. Kim, New Mo6Te6 sub-nanometer-diameter nanowire phase from 2H-MoTe2, Advanced Materials 29 (18) (2017) 1606264.

[163] F. Xiong, H. Wang, X. Liu, J. Sun, M. Brongersma, E. Pop, Y. Cui, Li intercalation in MoS_2: in situ observation of its dynamics and tuning optical and electrical properties, Nano Letters 15 (10) (2015) 6777–6784.

[164] M.A. Lukowski, A.S. Daniel, F. Meng, A. Forticaux, L. Li, S. Jin, Enhanced hydrogen evolution catalysis from chemically exfoliated metallic MoS_2 nanosheets, Journal of the American Chemical Society 135 (28) (2013) 10274–10277.

[165] D. Voiry, M. Salehi, R. Silva, T. Fujita, M. Chen, T. Asefa, V.B. Shenoy, G. Eda, M. Chhowalla, Conducting MoS2 nanosheets as catalysts for hydrogen evolution reaction, Nano Letters 13 (12) (2013) 6222–6227, http://pubs.acs.org/doi/abs/10.1021/nl403661s.

[166] Q. Tang, D.-e. Jiang, Mechanism of hydrogen evolution reaction on 1T-MoS_2 from first principles, ACS Catalysis 6 (8) (2016) 4953–4961, https://pubs.acs.org/doi/10.1021/acscatal.6b01211.

[167] R. Kappera, D. Voiry, S.E. Yalcin, B. Branch, G. Gupta, A.D. Mohite, M. Chhowalla, Phase-engineered low-resistance contacts for ultrathin MoS_2 transistors, Nature Materials 13 (2014) 1128–1134, http://www.nature.com/nmat/journal/v13/n12/full/nmat4080.html.

[168] S. Kretschmer, H.-P. Komsa, P. Bøggild, A.V. Krasheninnikov, Structural transformations in two-dimensional transition-metal dichalcogenide MoS2 under an electron beam: insights from first-principles calculations, Journal of Physical Chemistry Letters 8 (13) (2017) 3061–3067.

[169] J. Zhu, Z. Wang, H. Yu, N. Li, J. Zhang, J. Meng, M. Liao, J. Zhao, X. Lu, L. Du, Argon plasma induced phase transition in monolayer MoS_2, Journal of the American Chemical Society 139 (30) (2017) 10216–10219.

[170] C. Ataca, H. Sahin, E. Akturk, S. Ciraci, Mechanical and electronic properties of MoS_2 nanoribbons and their defects, Journal of Physical Chemistry C 115 (10) (2011) 3934–3941, <Go to ISI>://WOS:000288113400015.

[171] A.R. Botello-Méndez, F. Lopez-Urias, M. Terrones, H. Terrones, Metallic and ferromagnetic edges in molybdenum disulfide nanoribbons, Nanotechnology 20 (32) (2009) 325703.

[172] J.-G. Song, G.H. Ryu, S.J. Lee, S. Sim, C.W. Lee, T. Choi, H. Jung, Y. Kim, Z. Lee, J.-M. Myoung, Controllable synthesis of molybdenum tungsten disulfide alloy for vertically composition-controlled multilayer, Nature Communications 6 (1) (2015) 1–10.

[173] Y. Cheng, Z. Zhu, M. Tahir, U. Schwingenschlögl, Spin-orbit–induced spin splittings in polar transition metal dichalcogenide monolayers, Europhysics Letters 102 (5) (2013) 57001.

[174] N.H. Attanayake, A.C. Thenuwara, A. Patra, Y.V. Aulin, T.M. Tran, H. Chakraborty, E. Borguet, M.L. Klein, J.P. Perdew, D.R. Strongin, Effect of intercalated metals on the electrocatalytic activity of 1T-MoS_2 for the hydrogen evolution reaction, ACS Energy Letters 3 (1) (2017) 7–13.

CHAPTER

5

Realization of electronic grade graphene and h-BN

Vitaliy Babenko and Stephan Hofmann

Department of Engineering, University of Cambridge, Cambridge, United Kingdom

5.1 Challenges overview: growth, transfer, and integration

Two-dimensional (2D) materials at the atomically thin mono- or few-layer limit are essentially "all-surface". As a result, their highly sought after opto-electronic and other properties are not only dependent on the overall structural quality, but also to an unprecedented level on extrinsic influences such as substrate interactions and substrate-2D interface contaminants. A monolayer graphene film across a 4″ wafer corresponds to an aspect ratio in excess of 3×10^8, highlighting the challenges faced in wrinkle- and fold-free growth and processing. The new material process technology has been driven by two different streams, which are starting to come together: (1) top-down exfoliation of flakes from bulk 2D crystals [1] and (2) bottom-up growth of mono- and few-layer 2D films [2]. While the first process has seen tremendous progress in ultra-clean transfer and interfacing of 2D layers, the second has seen the emergence of a whole new framework of crystal growth for 2D layers, particularly by catalytic enhancement of chemical vapor deposition (CVD) for graphene and hexagonal boron nitride (h-BN) [3,4]. The progress in the field is reflected in the drastic material cost reduction from \$530 for a 3″ wafer of CVD graphene [2] on Cu in 2013 (Graphenea) to \$220 for a 6″ square in 2020 (Grolltex), representing an order of magnitude reduction in the price per area in 7 years. Yet, the potential of 2D material based opto-electronic devices to outperform current technology remains in many cases limited. Radically different approaches and mindsets are required to drive their large-scale integration, which is why 2D technology faces many challenges towards commercial adoption by mainstream electronics manufacturers, particularly for any device architecture of higher complexity as common in the semiconductor industry and nanoelectromechanical systems.

Taking graphene as the most developed 2D material, a crucial technological challenge is to bring the many recent scientific advancements in its growth, transfer and integration together to reach the performance requirements for the most demanding applications. State-of-the-art roadmaps in 2D electronics target not only the device channel to be a single crystal, but also

Defects in Two-Dimensional Materials
https://doi.org/10.1016/B978-0-12-820292-0.00011-2

Copyright © 2022 Elsevier Inc. All rights reserved.

suggest the development of single crystal dielectrics and encapsulants either as 2D materials or compatible 3D atomic interfaces [5–7]. Amorphous oxides that work well in silicon technology have ill-defined and yet poorly understood interfaces with 2D materials [5]. All high-performance graphene devices to date have been reliant on h-BN encapsulation, and while the dielectric properties and band offsets for h-BN are not ideally suited across the semiconductor roadmap, h-BN is widely agreed as a highly promising dielectric material for analog 2D devices, photodetectors and sensors employing graphene channels. This motivates us here to discuss challenges towards scalable process technology for both single crystal graphene and h-BN layers.

Crystal growth of monolayer graphene by CVD offers a good starting point for the more complex growth of h-BN. This is particularly pertinent as CVD has without a doubt become the method of choice for the large-area, cost-effective production of graphene and h-BN, with the parameter space involving conventional CVD and heterogeneous catalysis, which is notoriously large and complex. Catalytic CVD revealed, for instance, the immense sensitivity of graphene and h-BN to minute substrate impurities like oxygen, to a level far below the detection limit of conventional techniques like X-ray photoelectron spectroscopy (XPS) [8,9]. Challenges such as the control of foreign impurities during CVD highlight the required interdisciplinarity of advanced 2D material growth, with significant efforts being dedicated to re-assess some topics in metallurgy and epitaxial thin film growth for cost-efficient and scalable substrate preparation (Section 5.4) [10].

Most of the existing literature and technology roadmaps point to the need for the 2D layer transfer from the growth substrate, and such transfer and handling technology in many ways is the key "silent partner" to CVD, especially where catalytic metals are involved. Mechanical "dry" exfoliation approaches of sub-100 μm flakes have been applied to CVD graphene enabling the transfer without the use of polymers leading to cleaner interfaces and demonstrating high charge carrier mobilities when embedded between two multilayer h-BN films [11]. However, such approach is currently not scalable. Matched to the application need, a multitude of variations of "wet" and "semi-dry" transfer routes have emerged that were optimized in conjunction with the CVD substrate, where liquids and additional polymer scaffolds are used to support and handle the atomically thin films [2]. Albeit being the current standard and used by most commercial suppliers [12], arguably, wet transfer has received relatively little scientific attention due to the vast parameter space spanned by 5–15 interdependent steps, some of which with large variability. Such complexities are driving the introduction of (semi)automated and robotic processing to achieve improved yield and reliability in industrial settings [13].

The heterogeneous integration of 2D materials with common 3D materials used in the semiconductor industry is another prominent challenge, starting with basic aspects of electrical contacts, and device encapsulation using dielectrics. Atomic layer deposition (ALD) of dielectrics, for instance, not only shows a very distinct nucleation behavior on a 2D material layer, but due to intercalation phenomena can also lead to significant changes underneath the 2D layer [14–16]. These exemplar topics highlight the complexity of achieving a reliable multi-process flow with 2D layers and current challenges toward their full integrated processing. When discussing graphene "quality", it is usually quantified by the charge carrier mobility, and it is important to consider that such quality metrics include the complete process flow, with growth only being a starting point, and post-growth transfer and processing being important "quality-limiting" factors [11,17]. In this chapter we focus on recent progress in 2D crystal growth while considering such process dependencies and technological needs for material integration. We start with synthetic bulk crystal growth as a reference, in the context of basic pyrolytic and catalytic 2D crystal growth.

5.2 Apparatus and methodology overview

5.2.1 Bulk crystal production and layer exfoliation

There is a long history of graphite mining and synthetic bulk graphite growth, most importantly the process for the widely used highly oriented pyrolytic graphite (HOPG). HOPG production involves the pyrolysis of a carbonaceous precursor, e.g. methane at 2200 °C, to pyrolytic graphite (PG), followed by high temperature (3200 °C) stress annealing [18]. Such developments were driven and optimized for applications mainly in composites, moulds and lubricants. The key addition for more recent applications in nanoelectronics has been the controlled and clean exfoliation of layers from such bulk graphite. This progress reflects the situation for many common 2D materials (e.g. TMDs), where there have been historic growth studies of layered bulk crystals often driven by niche applications [19], which are now used as source material for exfoliation of "electronic-grade" 2D layers. For some materials such as h-BN there are no known natural deposits, thus the focus is entirely on synthetic bulk crystals [20,21].

The starting point of crystal growth is to achieve phase selectivity with suitable pressure-temperature conditions, which is generally more complex for compound materials like BN compared to elemental carbon. Recently high temperature (1500–1750 °C) growth of h-BN crystals from metal solvents have been developed, either at high temperature and high pressure (HTHP, 4.0–5.5 GPa) [20] or high temperature and atmospheric pressure (HTAP) [22,23] with combinations of pressure-temperature to select the hexagonal phase of the material. Such methodologies for avoiding defects require rigorous parameter optimization to obtain a low nucleation density for the growth of large crystals. This is done by choosing a suitable metal "solvent" system to optimize the growth rate, avoid dendrite formation, decrease the point defect density and minimize impurity incorporation [24]. By using high temperature solvents, oxygen and carbon impurities can be reduced by two and three orders of magnitude [25], respectively, in the as-grown h-BN bulk crystals compared to the starting BN material (powder), as confirmed by the observation of a very intense cathodoluminescence signature [20].

Bulk layered crystals can be used to produce exfoliated flakes with thicknesses from a monolayer to tens of nanometres via mechanical or chemical exfoliation of the weakly-bound atomic planes. The former is the current standard technique to produce reference model mono- and few-layered materials and devices with the highest charge carrier mobility. The layer dimensions and quality achievable by exfoliation are ultimately limited by the crystal size and quality of available bulk crystals. The natural graphite of highest known crystallinity is vein or lump graphite [26], believed to form pyrolytically from subterranean fluid, and flake graphite formed in metamorphic or calcareous sedimentary rocks. Average crystal dimensions for such natural graphite are of the order of tens of micrometres, while rare, isolated

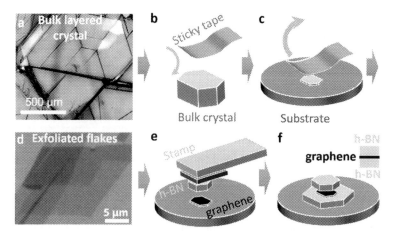

FIGURE 5.1 Overview of mechanical exfoliation. (a) An optical image of metal-melt synthesized bulk h-BN crystal; adapted from [23] under CC BY 4.0 license. (b–c) A schematic of bulk crystal exfoliated with tape. (d) An optical image of graphene flakes on a SiO$_2$ substrate. Reprinted from [29], with the permission of AIP Publishing. (e–f) Stamp-based encapsulation of graphene in multilayer h-BN flakes used in high-quality devices.

flakes can reach centimetres in plane crystal dimensions [27]. Synthetic HTHP h-BN bulk crystals have been reported up to millimetre sizes [23] (Fig. 5.1a). Mechanical exfoliation in principle is based on peeling layers from a bulk crystal using sticky tape, followed by pressing them onto the desired substrate, such as SiO$_2$/Si wafers [28]. Successful transfer relies on stronger 2D layer adhesion to the target substrate than to the stamp, as shown schematically in Fig. 5.1b–c. The use of an optimized SiO$_2$ thickness allows convenient optical observation of the resulting flakes down to the monolayer limit due to interference effects [29], as shown in Fig. 5.1d for graphene. However, SiO$_2$ and many other widely available substrates do not result in the best possible electronic performance due to charge traps, roughness and surface phonon-scattering effects [30,31]. Additionally, water or other adsorbed species from ambient exposure [32–34] can also degrade the pristine properties of 2D materials.

While conceptually simple, mechanical exfoliation is non-trivial and is associated with a large parameter space, including design of polymeric stamps, control of local atmosphere and temperature to achieve clean surfaces. State-of-the-art approaches utilize a custom polymer-based stamp that allows the pick-up of a thick h-BN flake (pre-exfoliated to SiO$_2$/Si), which in turn is used to pick up a pre-exfoliated graphene monolayer and the resulting h-BN/graphene stack is finally placed on top of another thick pre-exfoliated h-BN flake [31,35] (Fig. 5.1e–f). Such processing avoids any direct contact with polymers, etchants, or solvents. Additionally, specific stamp angles and heating are applied to move trapped adsorbates and contaminants, that typically accumulate in blisters [36] achieving localized atomically clean interfaces. Combined with the developed "edge contact" geometry such structures have been demonstrated to support room temperature charge carrier mobilities in excess of 120,000 cm^{2}V^{-1}s^{-1} (5×10^{11} cm^{-2} carrier density in monolayer graphene [31,35]). Such prototyping platform, consisting of high-quality crystals, processing and integration, has been expanded to many other 2D materials. However, while avenues for robotic scaling of such flake peeling processes have recently been proposed to increase throughput and reproducibility [37], its scalability is ultimately limited by the small crystal sizes and difficulty of exfoliating controlled number of layers. The industrial demand for large-area, continuous "electronic grade" films of controlled layer number has driven the alternative stream of bottom-up growth of mono- and few-layer 2D films.

5.2.2 Chemical vapor deposition and related methods overview

CVD is a material production technique where a solid film is deposited from reactants supplied in the gas phase. CVD and related approaches have been utilized for the production of coatings from the late 18th century and greatly expanded in the 19th–20th centuries with notable examples of the Mond process for Ni purification (1890) [38]; metal carbide, nitride and oxide coatings (1920–1970) and later diamond, carbon nanotube and BCN coatings (1980-ies), and a plethora of thin film processes of many metals and dielectrics used by the semiconductor industry [39]. Ultra-thin film growth on catalytic substrates was also undertaken in the nineties, including monolayer graphene, monolayer h-BN and monolayer MoS_2 [40–44] however, the deposits were thought to be stabilized by the substrate and were never isolated, being considered integral part of the larger 3D structure [45]. Successful isolation from bulk crystals by exfoliation and measurements of the properties of such monolayer graphene [1,28] led to the award of the Novel Prize in 2010.

The biggest advantage of CVD is its inherent scalability, but also low cost of production, large choice of parameters and, as has been demonstrated, the possibility to produce high quality 2D materials that rival exfoliated layers [17]. CVD can be characterized by various factors, such as the apparatus design, for example, cold-wall or hot-wall chambers, operating conductions, e.g. ultra-high vacuum (UHV) or ambient pressure, precursor chemistry, e.g. metal-organic CVD (MOCVD), substrate type, e.g. catalytic or non-catalytic (Sections 5.3.2 and 5.3.1) and, whether plasma-enhanced decomposition is utilized (PECVD). Each variation offers its advantages and disadvantages. For example, the hot-wall or tube furnace apparatus (Fig. 5.2a) allows good control and measurement of the temperature distribution but suffers from precursor decomposition with distance traveled within the hot zone and deposition non-uniformity due to precursor gas concentration depletion along the tube. On the other hand, it is more difficult to establish temperature uniformity in cold-wall systems (Fig. 5.2b) due to steep gradients with water-cooled walls, but control of what gas species arrive at the substrate is significantly improved. Advanced engineering solutions have been developed in industry to overcome temperature uniformity challenges needed for large wafer processing. All types of CVD reactors have been successfully adapted to the growth of monolayer graphene and h-BN. Furthermore, large-scale commercial production CVD equipment is already available. For example, polycrystalline graphene was achieved on 200–300 mm wafer sizes by CVD Equipment Corporation using hot wall systems [46] (Fig. 5.2c) and by Aixtron with a cold wall CVD system (Fig. 5.2d) [47]. Roll-to-roll growth systems using Cu foils as the catalyst have also been in development [48,49]. While the roll-to-roll growth processes can produce large quantities of material, it faces great challenges that involve material transfer at a given speed, and the need for quality monitoring techniques that can match the speed of production [47,50–52]. Catalytic CVD and its complex underpinning mechanisms in particular, have been the subject of decades of interdisciplinary fundamental research, including surface science and catalysis studies, such as in operando scanning tunneling microscopy

124　　5. Realization of electronic grade graphene and h-BN

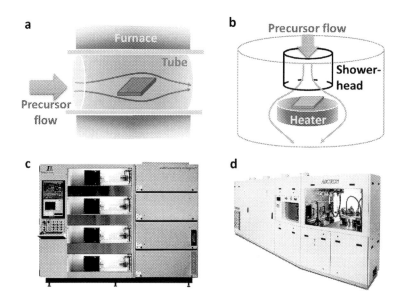

FIGURE 5.2 CVD apparatus configurations. (a–b) Hot wall and cold wall systems. Production examples of such systems include (c) EasyTube 6000; picture courtesy of CVD Equipment Corporation, and (d) CCS MOCVD; picture courtesy of Aixtron.

(STM), low-energy electron diffraction (LEED), X-ray photoelectron spectroscopy (XPS) and other diffraction measurements, combined with atomistic modeling [53–57].

Hybrid CVD approaches, such as film growth by precipitation on cooling also exist and rely on the property of some metals (such as Ni, Co, Fe) to dissolve constituent elements after the gaseous precursor decomposition on the surface [58]. The precursor source can also be a solid pre-deposited layer [59–62]. In this process, upon cooling, the solubility is gradually reduced forcing the dissolved species to the surface. Nevertheless, challenges for controlling such processes include faster precipitation via metal grain boundaries that lead to non-uniformities, lack of layer control, and precipitation of impurities. While these alternative processes have not yielded any improvements on material quality over gas-based CVD on optimized substrates, they are of interest for scaled, integrated device and microelectromechanical systems fabrication.

5.3 Scalable growth by chemical vapor deposition

5.3.1 Pyrolytic growth

In its pure form, pyrolytic growth follows a vapor-solid (VS) growth model where supersaturation is not dependent on the underlying substrate. As a simple guideline for VS type growth, crystal quality is dependent on the balance between arrival (or thermal decomposition rate) and incorporation rate requiring a sufficiently high temperature that facilitates

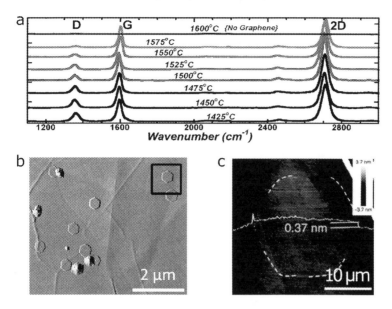

FIGURE 5.3 Non-catalytic growth. (a) Dependence of graphene Raman spectra vs. temperature on Al_2O_3. Reprinted with permission from ref. [64]. Copyright 2021 American Chemical Society. (b–c) Direct growth of graphene on exfoliated h-BN without and with silane gas, showing ca. 0.5 and 20 µm domains respectively. A line height profile is shown as the yellow trace. Reprinted by permission from Springer Nature: ref. [73], Copyright (2021); and from ref. [74] under CC BY 4.0 license.

mobility of relevant species and healing of defects. Due to the strong covalent bonding and high activation energies, both for graphene and h-BN, extremely high temperatures are required for pyrolytic growth. This requirement restricts possible substrates to a select range, for example, sapphire (Al_2O_3), gallium nitride (GaN), silicon nitride (Si_3N_4) and quartz (SiO_2). Furthermore, gaseous precursors need to have suitable thermal stability and purity at optimized concentrations to achieve good crystalline quality. Similarly to PG growth developed in the 19th century [18,63], typical reported deposition temperatures for graphene range from 1200 to 1700 °C [64]. Lower temperatures or plasma enhancement generally led to highly defective deposits [65,66]. Particularly for compatibility with CMOS-technology, metal-free graphene and h-BN growth is of high interest to prevent the incorporation of impurities that are detrimental to CMOS devices. Si wafer substrates are the most widely-available crystalline substrates due to their use in electronics now as large as 300 mm diameter, and the graphene integration in a Si device flow is highly desired. It is important to note that above 900 °C unwanted formation of Si carbide (SiC) occurs [67], in addition to the surface instability of single-crystal Si in hydrocarbon environments [65]. A related approach of growing graphene is via the surface decomposition of SiC wafers [68,69], which has its own complexities including the surface preparation, interphase structures, high substrate costs and limited quality of resulting graphene [70–72].

On the other hand, sapphire wafers are widely available at low costs due to their high-volume production and use as substrates for light-emitting diodes, which is highlighted by a large body of literature on pyrolytic graphene and h-BN growth on sapphire. Fig. 5.3a shows the dependence of graphene quality on the growth temperature (1425 °C to 1575 °C) on sapphire seen from Raman spectra, [64] exhibiting a decrease in D peak intensity (1350 cm^{-1}) with increased temperature. The Raman D/G peaks ratio is related to the defect density in graphene and a low value of 0.05 has been achieved at ca. 1575 °C, while the production of 50 mm wafers and device fabrication was also demonstrated, reaching mobilities of around 3000 cm^{2}V^{-1}s^{-1} at 5×10^{11} cm^{-2} carrier density (room temperature, in vacuum). More recent efforts on the CVD parameter optimization point to the critical dependence on the sapphire pre-treatment before the growth step. Hydrogen annealing of sapphire led to surface reconstruction to the Al-rich termination that allowed to lower the growth temperature to 1200 °C and obtain epitaxial alignment of the as-grown graphene [47]. Large area uniform production on 150 mm sapphire wafers has also been demonstrated, however, the electronic properties have not significantly improved; the charge carrier mobility measurements directly on sapphire reached values between 2000 to 3000 cm^{2}V^{-1}s^{-1} [64,66].

Pyrolytic graphene growth has also been reported on h-BN and vice versa (h-BN on HOPG) [73,75], but in limited size and crystallinity. For example, the growth of graphene domains was demonstrated on exfoliated, flat h-BN flakes (on a quartz support) using methane at 1200 °C, obtaining domain sizes of hundreds of nm (Fig. 5.3b), occasionally reaching 2.5 μm [73,76]. The graphene domains exhibited good alignment with the h-BN lattice and a superstructure visible in an atomic force microscope (AFM). The growth rate was very slow, on the order of 0.01 μm min^{-1}, which is up to 4 orders of magnitude lower than for catalytic CVD on metals [77]. The growth methodology was further refined in follow-on works by utilizing a different reaction chemistry with silane gas as a possible catalyst in addition to acetylene and argon [74]. Such precursor gas modification allowed faster growth rate of ∼1 μm min^{-1} and also larger domains, up to 20 μm (Fig. 5.4c). The as-grown graphene on multilayer h-BN exhibited low Raman D-peak, and a narrow Raman 2D-peak width of 20 cm^{-1}, but only for domains aligned to the h-BN lattice at a specific angle. Such domains exhibited high charge carrier mobilities of up to 23,000 cm^{2}V^{-1}s^{-1} at room temperature. It is evident that direct growth of graphene on atomically flat h-BN demonstrated much better performance as a substrate than for sapphire, however, scalable growth of such h-BN materials is currently not available, limiting the method to small, exfoliated flakes (<0.1 mm) as the growth substrate. In a related strategy, attempts have been made to grow multilayer h-BN on sapphire, followed by graphene transfer from Cu [78], but have not resulted in improved performance, achieving charge carrier mobilities of ca. 2200 cm^{2}V^{-1}s^{-1} at ∼5×10^{12} cm^{-2} carrier density at room temperature, which is one order of magnitude lower than the direct growth on exfoliated h-BN discussed above. Controlling the roughness of the as-grown films remains a major challenge, which for exfoliated flakes is dictated by the atomically smooth crystal planes, while for pyrolytic CVD h-BN, much rougher nanocrystalline deposits are typically observed [78]. In summary, non-catalytic growth has been implemented in various settings on select high-temperature resistant substrates, however, growth on large area substrates have not achieved suitable material performance. There are other obstacles, such as the lack of integration routes for pre-patterned device substrates and low growth rates compared to catalytic CVD. Further developments or novel approaches are critically required.

5.3.2 Catalytic CVD: substrate and catalyst effects

Heterogeneous catalysis concepts underpin the growth of graphene and h-BN on transition metals and some post-transition metals where the processes are fundamentally different to decomposition in the gas phase or pyrolysis (Section 5.3.1). Analogous to the growth of 1D nanomaterials such as nanotubes and nanowires with prevalent vapor-liquid-solid (VLS) and vapor-solid-solid (VSS) growth mechanisms [79–81], the overall growth process can be considered catalytic if any of the constituting processes has its activation energy lowered by the presence of a catalyst. Notably, this concept encompasses not just catalytically enhanced gaseous precursor dissociation but includes 2D crystal nucleation and growth interface propagation. The use of suitable catalysts allows not only to lower the required process temperature (e.g., down to ca. 450 °C for polycrystalline graphene monolayer [82], or ca. 1035 °C for mm-sized crystalline graphene domains [83]), but also to gain nucleation and phase control and increase the growth rate. The choice of a catalytic substrate, however, can narrow the accessible parameter space and bring additional complexity due to, for example, their melting points, excessive sublimation, or phase changes when exposed to different process gases or vacuum at high temperatures. In order to provide an overview on the efficiency of transition metals for catalytic dissociation, analogies can be made to "volcano plots" used in catalysis plotting efficiency vs. the metal d-orbital energy [84]. For example, for graphene CVD, early transition metals (e.g. Ti) are seen as unsuitable catalysts for graphene growth because they have too strongly bound carbides, whereas for noble metals like Au, Cu, Pt (Fig. 5.4a) the precursor molecule is only weakly adsorbed and its dissociation rate is slow. Ni, Co and Fe are in that aspect most efficient as carbon precursor catalysts (Fig. 5.4a). However, another important consideration is the solubility of growth species in the catalyst that has a large effect on the supersaturation and therefore the growth of graphene and h-BN. A high bulk solubility of growth species can be considered to give rise to a reservoir, and typically leads to kinetic processes dominating the growth [85], and a possibility of growth during cooling from high temperature after saturation [86]. It is often the case that there is a link between the catalytic ability and solubility in the commonly used transition metal catalysts.

Low solubility catalysts under optimized conditions, facilitate self-limiting monolayer growth (Fig. 5.4b, case 1) [87–89]. Hence, Cu, which has negligible carbon bulk solubility, offers a facile parameter window for monolayer graphene growth by CVD [2]. While the need to go beyond 1000 °C (and often to near the melting point of Cu) highlights that it is a weaker catalyst, the wide availability and low cost of Cu foils allowed the development of many substrate processing methodologies such as roughness control via electropolishing or chemical mechanical polishing [90], supplier differentiation and cleaning methods [91], annealing [92], and bulk interstitial control [9,93,94]. Additionally, the possibility of facile transfer (Section 5.4.2) has made Cu one of the most widely used catalysts particularly for graphene CVD [2]. While h-BN domain sizes on Cu are typically limited to <50 µm (Fig. 5.4g) even after prolonged growth times of 4 hours [94], graphene domain sizes of hundreds of microns are commonly achieved [83], including on polycrystalline Cu foils as shown in Fig. 5.4d. Platinum can be considered as a similarly attractive weakly interacting catalyst for h-BN growth with some B solubility and low N solubility resulting in typical domain sizes of only a few microns [95] at 1000 °C, but due to its high melting point and low vapor pressure it is possible to utilize temperatures in excess of 1200 °C, where h-BN domains of up to 0.5 mm were obtained [96]. Interestingly, at high temperatures (>1200 °C), the addition of borazine promoted

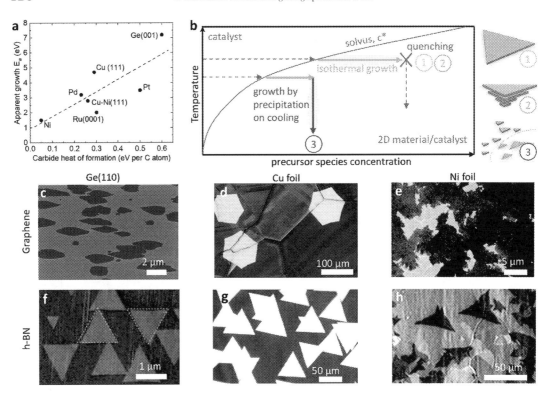

FIGURE 5.4 (a) Exemplar catalytic effects of different substrates on graphene CVD from CH₄ represented by the apparent growth activation energy vs. carbide heat of formation per C atom.[*] (b) A schematic of routes for graphene and h-BN growth via isothermal growth of monolayer and multilayer deposits (cases 1–2), and via precipitation upon cooling (case 3). (c–h) Scalable substrate examples for graphene and h-BN CVD. Domains are shown on Ge (c,f), Cu (d,g), Ni (e,h). (e) Reprinted with permission from ref. [82]. Copyright (2021) American Chemical Society. (f),(h) Reprinted with permission from refs. [103,104]. © 2015 WILEY-VCH Verlag GmbH & Co. KGaA, Weinheim. * Data adapted from refs. [84,97–102]. The growth rate is defined as in ref. [105]. The extracted growth rates do not differentiate the time for nucleation. For the Cu-Ni alloy, the carbide heat of formation was calculated as the average of 6 Cu and 1 Ni atoms according to the surface composition identified in ref. [100]. For Ge(001), the nanoribbon domain areas from ref. [102] were used to estimate the growth rate at different temperatures.

recrystallization of Pt foils to a single crystal with the Pt(111) surface orientation, caused by the enhanced Pt grain boundary mobility either due to the removal of pinning impurities or the formation of eutectic species. This approach achieved continuous epitaxial h-BN films on such surfaces in centimetre sizes [96].

Catalysts with higher C or B, N solubilities show bulk-facilitated growth mechanisms [85,87–89], and depending on the CVD conditions, isothermal growth of monolayer (Fig. 5.4b, case 1) or multilayer deposits (Fig. 5.4b, case 2 and Fig. 5.4e) [106–108], and growth via precipitation upon cooling (Fig. 5.4b, case 3) [85,86,89,109,110] can occur. The bulk solubility of growth species causes reservoir effects, and typically leads to kinetic processes including time-dependent bulk concentration gradients and continuous permeation of growth species through the bulk. Process parameters such as catalyst thickness and cooling rates become accordingly more pertinent [111]. "Kinetically stabilized" growth mechanisms explain the possibility of monolayer formation on high solubility catalysts [85], while precise tweaking of such solubility, e.g., by alloying (Section 5.4.1), could potentially be used for controlled growth of bilayer [112] or multilayer films [113], under rigorously optimized conditions. For transfer-free process technology, catalysts like Ni allow for much lower growth temperatures than Cu and the stronger Ni-graphene interaction can be exploited to achieve long term interface stability [114] which is highly interesting e.g., for magnetic tunnel junctions or corrosion resistant contacts [115,116].

Catalytic CVD is not restricted to transition metallic substrates, for example, oxide substrates such as hafnia, zirconia, and germanium may be considered as weakly-catalytic substrates (Fig. 5.4a). For the latter, graphene can be obtained at temperatures just below the melting point of Ge [103,117], at around 930 °C. Small domain sizes of around 1 µm are typically obtained (Fig. 5.4c,f). Significant efforts have been dedicated to this substrate due to its semiconducting properties and potential for metal-free and/or transfer-free integration. However, it is difficult to avoid defect formation in such films as seen by the frequently reported high Raman D/G peaks ratios [117]. This highlights the covalent bonding nature of Ge surfaces, and makes such substrates significantly more challenging to utilize due to the very narrow window of suitable CVD parameters compared to metal substrates that generally are much more forgiving, error and impurity tolerant, as discussed above. If properly optimized, including a substrate cost reduction, the use of Ge is promising, but it remains to be seen if it can go beyond niche applications.

Beyond simple first order growth models, there is a lot of complexity with regards to catalyst facets [57,118–120] and reaction dependent (dynamic) catalyst surface states and how these influence the underpinning chemical reactions. The simplest practical graphene CVD example can be considered as methane (CH_4) dissociation into elemental C, hydrogenated CH_x species, atomic (H) and molecular (H_2) hydrogen. It was shown that on Cu abundant atomic and molecular hydrogen chemisorption competes with CH_4 dehydrogenation, overall inhibiting graphene growth, while on Ni, atomic H rapidly recombines and desorbs from the surface, thus creating available surface sites for catalyzed graphene growth [121]. Analogously, on Cu, ammonia borane decomposition proceeds via the formation of deleterious oligomers, which can results in unwanted amorphous BN deposits, while on Ni-Cu alloys, no such deposits are observed due to the enhanced oligomer dissociation [94,122].

For compound materials like h-BN the growth mechanisms are inherently more complex, with the supersaturation, for instance, dependent on the solubility of both B and N in the catalyst. The basic growth models developed for catalytic graphene CVD, however, can serve as a useful basis and be readily adopted to capture the added complexity of h-BN CVD. Such understanding has enabled progress in optimizing elemental catalysts such as Ge, Cu, Ni (Fig. 5.4f–h), Pt, Rh, Ir, Fe and alloy catalysts such as Cu/Ni, for h-BN CVD. For some catalysts, e.g. for Fe, there is an added challenge of accommodating a bcc/fcc phase change upon cooling within the CVD parameter range. It should not be forgotten that for most applications transfer from the catalyst substrate is required (Section 5.4.2), i.e. the ease of subsequent 2D material release from the process catalyst becomes a quintessential catalyst attribute, linked to the challenge of cost-efficient and scalable catalyst substrate preparation and its re-usability. Analogously to graphene, for h-BN CVD the optimal choice of process catalyst requires holistic consideration of application needs, specifically if post-growth transfer is required. For catalyst materials that have been widely used in surface science literature for h-BN growth, including Rh(111) and Ir(111) [54], post-growth h-BN transfer is extremely difficult. Hence less strongly interacting catalysts like Cu and Pt have been optimized for monolayer h-BN production, including transfer (Section 5.4.2) [96,123]. In the following we expand in more detail on the parameter space of catalytic CVD as basis for emergent developments to achieve higher material quality combined with scalability.

5.3.3 Catalytic CVD: growth parameters and process optimization

5.3.3.1 Overview

Beyond the substrate and process catalyst considerations for CVD, there are multiple growth parameters and choices in process and system implementation that can critically affect the quality and properties of the resulting 2D material. These include controls of the temperature, heating and cooling rates, chamber pressure, precursor choice and properties, precursor partial pressure, carrier or functional gas choice, their flow rates (Fig. 5.5b). In a typical CVD process the steps include substrate preparation, such as electropolishing or more advanced treatments (Section 5.4.1), followed by annealing in an auxiliary gas or mixture, 2D material growth by decomposition of the precursor gas and cooling. A schematic process diagram for the model system of graphene CVD on Cu [90] is shown in Fig. 5.5a with growth temperature of 1065 °C and a constant total process pressure of 50 mbar, but varying gas mixtures for the annealing, growth, and cooling stages. While the influence of core CVD parameters is discussed below within specific examples, it should be highlighted that the parameters and their effects closely connect to the choice of substrate (Section 5.3.1) or catalyst (Section 5.3.2). The resulting large number of variables are often convoluted in ways that are not always obvious. Furthermore, there are parameters such as impurity levels in reactors that are not easy to capture, an issue that particularly flares up when for instance quartz tubing is changed in a hot-walled CVD reactor or graphite heaters are used in cold-walled reactors in conjunction with reactive gases such as NH_3. Hence, recipes reported across the literature are difficult to reproducibly transfer between different reactors, and each system requires its own detailed optimization.

5.3.3.2 Precursor choice

The choice of CVD precursors is typically motivated by their thermodynamic stability matched to the reaction conditions.

For bulk material growth, advanced thickness control can be achieved, for example, by atomic layer deposition (ALD) or molecular beam epitaxy (MBE) techniques. For the former such control is achieved by alternating precursor pulse and purge cycles, which relies on surface saturation and thereby allows a self-limited reaction regime without being reliant on exact precursor dosage, while for the latter, UHV systems and low growth rates are employed. In contrast, for 2D materials like graphene and h-BN the self-limited reaction regime originates from their anisotropic, layered crystal structure and related growth mechanisms, allowing atomic monolayer deposition control with CVD precursors even under extended continuous exposure.

Graphene is a single element material, only requiring widely-available C for the growth, which can be selected from a plethora of organic compounds, starting from simple methane

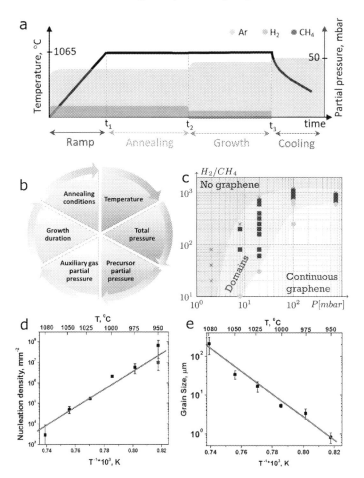

FIGURE 5.5 (a) A schematic process diagram for an exemplar graphene growth process on Cu, following ref. [90]. (b) CVD parameters optimization cycle crucial for graphene and h-BN growth. (c) The effect of the H$_2$:CH$_4$ gas ratio on graphene deposition. (d–e) The dependence of the nucleation density and grain size of graphene grown on Cu on the temperature. (c–e) Adapted with permission from refs. [124,125]. Copyright (2021) American Chemical Society.

[2] and other gaseous hydrocarbons, such as ethylene and acetylene [74], to liquids such as alcohols [126], and even soybean oil [127] or chicken fat [128]. While oils or fats are unlikely to result in high-quality crystalline graphene growth they have been proposed as low-cost, renewable natural precursors or as a route for recycling of waste materials. Precursor stability or composition can influence the evolution of graphene domains, for example, ethane allows higher growth rate than methane for the same temperature [129], while ethanol could provide a consistent oxygen balance in CVD reactions [126]. With modern growth approaches that utilize optimized CVD parameters and catalysts (Section 5.4.1) it is possible to achieve crystalline graphene domains with low defect density with a variety of C precursors [2,126,129].

For h-BN however, only around six precursors or gas combinations are typically available, including borazine [130], ammonia borane (AB) [131], triethylborane (TEB) and ammonia [78], boron chloride and ammonia [132], trichloroborazine [133], diborane and ammonia.[134,135] TEB, BCl$_3$ and B$_2$H$_6$ have significant associated safety concerns, being either pyrophoric or toxic, which is why borazine and AB are currently the two most common precursors for h-BN CVD. The B and N feed for stoichiometric h-BN growth is highly dependent on the substrate and its B and N solubilities, and a stoichiometric precursor does not warrant a precursor flux balance [136]. While borazine is flammable, it is a liquid at room temperature with a high vapor pressure that can be easily controlled by standard means [136], AB is a solid with low vapor pressure that is considered safer in its pristine form, but requires heating to its decomposition point. Uncontrolled AB decomposition releases different volatile species simultaneously, some of which lead to the formation of h-BN while other to amorphous deposits and is thus complicated to control [94,122]. High quality monolayer h-BN has been produced with dedicated solutions for these precursors in simple CVD reactors and chemistries owning to the self-limiting 2D growth mode. As such there is no advantage in devising, for example, specific ALD precursors in terms of monolayer h-BN control specifically on catalytic substrates. For h-BN CVD, carrier gases such as N$_2$ or NH$_3$ cannot be considered inert, as they can affect the ratio between B and N species in the catalyst bulk and at the growth front and thus control the growth supersaturation, termination of the domains [137] and consequently their shape and stitching [138].

5.3.3.3 Process pressure

In CVD technology the process pressure is usually used to control the flow regime and traditionally links to control of the overall rate-limiting reaction step, with film growth typically either under a surface reaction controlled or mass transfer limited regime. For catalytic 2D growth this situation is more complex, and it has been shown for graphene CVD that the rate-limiting step can differ for each growth facet and be reaction stage dependent [57]. This is further convoluted with catalyst specific effects. Surface science studies looking at graphene and h-BN growth on surfaces such as Pt, Ni, etc. [53,54] typically utilize ultrahigh vacuum, UHV and low precursor pressure ($<10^{-4}$ mbar) without any additional gases. More recently, for industry-oriented growth on Cu, low pressure CVD (<1 mbar, LPCVD) was developed as the early route towards large-area monolayer graphene growth [2] with a mixture of a precursor (CH$_4$) and an auxiliary gas (such as H$_2$). Such LPCVD conditions offered suitable parameter controls and improved safety due to the low volume of pure hydrogen in the reactor. However, excessive Cu sublimation and roughening occurred, which led to the development of Cu enclosures and related confinement approaches [83], which are somewhat cumbersome. More recent studies successfully utilized intermediate pressures (ca. 50 mbar, Fig. 7.5a) [90] or atmospheric pressures (ca. 1000 mbar) with diluted gases, such as 0.1% CH$_4$ in Ar, nanomolar flows of B$_x$N$_y$H$_z$ species in Ar, or $\sim$2% H$_2$ in Ar, which are intrinsically safe as the concentrations are below the lower explosivity limits [94,139]. Such gas mixtures allow sufficiently low precursor partial pressures for monolayer graphene or h-BN growth, while also establishing suitable boundary layers for suppressed substrate surface sublimation. The gas mixtures adding up to the total chamber pressure are then thoroughly optimized.

5.3.3.4 *Precursor and auxiliary gas pressures*

The choice of a suitable partial pressure of precursors and auxiliary/carrier gases is the most versatile parameter that can be readily controlled through the CVD reactor design, dedicated pressure controls and gas mixture controls (e.g. via electronic mass flow controllers) allowing the formulation of custom gas ambients. The detailed mechanistic understanding and experimental methods of these parameters frequently change when better processes are identified. In particular, the role of H_2 gas has received significant attention as modern CVD methods typically employ large flows of H_2, especially for high quality graphene and h-BN growth. From one perspective, this is because CVD precursor molecules for graphene and h-BN growth almost always contain hydrogen where their chemical decomposition reaches dynamic equilibrium, which is why introducing an additional H_2 gas flow can be used to control the reaction kinetics and edge termination. Even for the simplest CH_4 molecule decomposition, multiple steps [140] and surface species can be considered, many of which are influenced by the hydrogen gas partial pressure (pp) in the vessel. For example, for graphene growth specifically on Cu, increasing the H_2 pp (P_{H_2}) or lowering the CH_4 pp (P_{CH_4}) often has a similar effect [105,121,125,140,141], allowing to consider their ratio P_{H_2}/P_{CH_4} [125], or the $(P_{H_2})^2/P_{CH_4}$ ratio [140], as a representative parameter. A high ratio leads to no deposition or very slow growth rates, while a very low ratio leads to high growth rates, higher defect densities and more multilayer formation, depending on the type of catalyst substrate. Furthermore, a high H_2:CH_4 ratio results in reduced nucleation density but requires longer times to grow sufficiently large domains or films, thus, it is also possible to consider the P_{H_2}/P_{CH_4} and time as connected parameters. Several other complications typically need to be considered, such as the absolute partial pressure of CH_4; too low value could result in faster growth of intrusions along the edges and thus dendrite formation, akin to snowflake growth [142]. Such deposits are typically avoided due to the unfavorable film stitching. Consequently, the ratio required to obtain 2D material domains is itself dependent on the total chamber pressure as shown in Fig. 5.5c and other specific conditions, such as the process temperature, presenting a multi-dimensional parameter space. It must be noted, for example, on Ni the role of H_2 is considered to be completely different [121].

From a different perspective, it was also proposed that common research grade gases, for example, 99.9995% Ar or H_2, have oxidative impurities such as O_2 and H_2O in the parts per million (ppm) and tens of ppm level, respectively [143,144]. Because these impurities are much more reactive towards carbonaceous deposits than H_2, even low impurity levels can have a decisive effect on the evolution of graphene domains during growth. In particular, it was shown that graphene is etched in un-purified gas environments at high temperatures, while utilizing O_2 and H_2O filters that reduce O_2 and H_2O level to <0.001 ppm does not result in graphene etching [145]. Other studies developed further insight that actually, some level of oxidative impurities could be beneficial, as their full removal resulted in increased nucleation density and an increased number of adlayer patches for graphene growth [9,143, 146]. To attain reliable growth and improved graphene quality, some oxidative species were controllably added to purified gases [143,146].

Early works in large graphene crystal growth on Cu relied on meticulously optimizing the gas mixtures, for example, a H_2:CH_4 ratio of 6000 combined with a 48-hour growth stage for the first time demonstrated graphene domain sizes of 5 mm [147]. For comparison, latest state-of-the-art approaches after the full catalyst, parameter space and CVD flow geometry

134   5. Realization of electronic grade graphene and h-BN

optimizations can reach 5 mm domains in 30 mins [88] presenting a two orders of magnitude improvement.

5.3.3.5 Temperature

As a generic trend for many crystal growth processes, including graphene and h-BN, higher temperatures allow larger crystal size and lower defect densities [141], but are typically limited by substrate degradation (e.g. sublimation, roughening or melting), a high precursor desorption regime or full precursor decomposition before it reaches the surface (for highly reactive precursors, e.g. C_2H_2). On simple catalytic metals the growth rate is significantly increased while the nucleation density is reduced when the process temperature is increased (Fig. 5.5d–e), allowing larger domains to form. For example, the growth rate of graphene on Cu increases by an order of magnitude when the temperature is increased from 1025 °C to near the melting point at 1080 °C (from 0.2 to 2 μm min^{-1}), while the nucleation density is lowered fifty times [125]. Pt metal has a higher melting point where growth rates of up to 40 μm min^{-1} were also achieved at 1150 °C [77]. For widely used poly-crystalline metal foil substrates, a high process temperature is additionally beneficial to drive normal and abnormal metal grain growth, foil purification, dispersion of localized impurities, and surface restructuring [90,92,96,148].

In contrast, transfer-free process integration and process sustainability motivate the lowering of the overall process temperature. A significant lowering of the temperature can be achieved by choice of an efficient process catalyst, for example Ni for graphene CVD [82,149], yet such efficiency typically links to a stronger coupling of the 2D material to the substrate, and thus is not suited for transfer-based process design. Plasmas can promote precursor dissociation at low temperatures, so can the choice of more reactive precursors [74,150]. Neither approach results in low defect density or the growth of large crystal domains, with the general consensus pointing to the need of temperatures closer to 1000 °C.

5.3.3.6 Time-dependent controls

CVD process control extends beyond the simple process stages and constant set parameters. The resulting material quality can be improved if different reactor types, levels of supersaturation and partial pressures are matched to dynamic parameters, such as varying gas composition. One reason being that the precursor gets consumed with distance in long tube-furnaces, as was shown for 91 cm × 43 cm Cu foils [151], as such creating spatial variation with length: from multilayer deposits near the tube inlet, to a monolayer film in the middle, to isolated domains at the outlet; a known historical issue with hot-wall CVD. However, specifically for graphene (or h-BN), due to its self-limiting monolayer growth, the spatial problem can be circumvented by finely increasing the precursor partial pressure with time. Here, the low initial precursor dose allows monolayer nucleation at the inlet of the tube, then later in the process a higher precursor dose can flow over the partially covered regions without causing multilayer formation. Furthermore, the precursor flow of such increased concentration will be partially consumed with distance in the tube, thus creating suitable growth conditions for monolayer graphene on the catalyst near the outlet. Other, more fundamental reasoning includes the enlargement of the domain perimeter of the 2D material and thus the need to supply more precursor species to compensate the decreasing

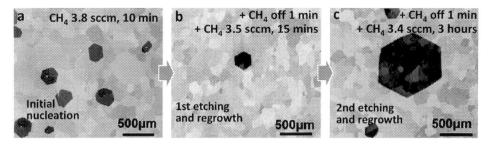

FIGURE 5.6 Time-dependent CVD optimization with the growth–etching–regrowth approach. (a) Primary nuclei are formed with the initial CH$_4$ dose, (b) smaller secondary nuclei are etched when methane flow is stopped and the remaining seeds are regrown again, (c) larger graphene domains grow from primary seeds upon further iterations of the etching-regrowth steps. Adapted with permission from ref. [153]. Copyright (2021) American Chemical Society.

active uncovered catalyst area [151,152]. The most advanced time-dependent precursor controls also aim to separate the nucleation stage of graphene or h-BN and their growth. A long incubation period is often present if 2D materials are grown at ultra-low precursor partial pressures or high H$_2$:Precursor ratios, thus two different sets of parameters could be used to first trigger nucleation rapidly, followed by the domain expansion with different conditions. Furthermore, if the precursor supply is stopped, the larger domains are partially etched away from the edges, but crucially, small deleterious nucleation sites can be fully removed allowing a more precise control of the initial seeding and homogenization stages, followed by controlled growth [96]. Such methodologies, for instance, allowed to obtain monolayer h-BN domain sizes of up to 600 μm after only 23 mins within the modified growth stage on Pt foils at 1200 °C [96]. Furthermore, it is possible to continuously regulate the nucleation density via a pulsed-type precursor flow, or the "growth–etching–regrowth" mechanism [153]. Controlled lowering of the precursor concentration fully etches small nucleation sites, while the larger domains marginally shrink. Restoring the precursor supply resumes the growth of the larger domains only (Fig. 5.6a–c), with sizes up to 3 mm demonstrated on Pt foils [153]. Such methods are beneficial but require a precise and careful calibration of precursor pressures and gas ratios at specific times within a CVD process.

For a typical CVD process all the discussed parameters are simultaneously optimized to achieve highest crystallinity and growth rate [154]. However, it is worth noting that several studies reversed this general understanding of the parameter optimization direction discussed in this section and instead optimized towards highly defective thin film deposits. Glass-like carbon monolayers [155] and amorphous BN [156] have recently been demonstrated with CVD parameters aimed to optimize for amorphous film homogeneity. Potential new applications for such amorphous 2D films include buffer layers, diffusion or interfacial barriers, membranes or ultra-low dielectric constant materials (BN) [157]. The wealth of knowledge for graphene and h-BN CVD from the variety of studies has led to the establishment of core methodologies that allow the formation of good quality material, with remaining key challenges being in the transfer (section 5.4.2) and other optimization steps to bridge scalable commercial production and outstanding structure quality (Section 5.4.3).

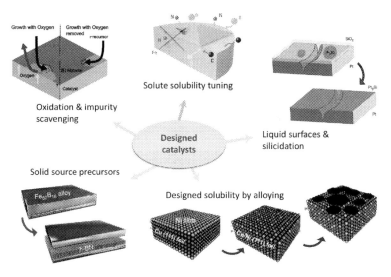

FIGURE 5.7 An overview of various routes to designed catalysts for graphene and h-BN CVD. Adapted with permission from refs. [8,100]. Copyright (2021) American Chemical Society. Adapted from refs. [77,89,160] under CC BY 4.0 license.

5.4 Material optimization

5.4.1 Designed catalysts

There are many transition metals in the periodic table but only several are suitable for graphene or h-BN growth by CVD (Section 5.3.2), nevertheless, as well known from heterogeneous catalysis, elemental catalysts can be modified or alloyed to yield more advantageous properties for 2D material growth. An overview of such modifications is shown in Fig. 5.7.

5.4.1.1 Oxidation & impurity scavenging

Catalyst pre-treatment is of paramount importance specifically to control the nucleation density of graphene and h-BN. Taking the example of Cu catalyzed graphene CVD, one widely used catalyst pre-growth treatment is the intentional overoxidation of the Cu surface, followed by Ar annealing to allow oxygen diffusion in the bulk, followed by a reducing treatment (often with H_2) and, subsequently, graphene growth. Such a process is designed specifically for facile growth of ultra-large graphene domains [9,147]. On such substrates the graphene nucleation density is drastically reduced, while a higher precursor dose can be used to increase the growth rate. Early reports have suggested that catalytic surface passivation with Cu-oxides was the reason for the reduced nucleation density, while also identifying the role of O to reduce the C edge attachment barrier, and accelerating graphene growth [9]. Follow-on studies identified a C-impurity scavenging mechanism, effectively suppressing graphene nucleation at C-impurity sites [90]. Latest results also point to the existence of a minute oxygen reservoir in the bulk of the catalyst that is only detectable with very highly sensitive instrumentation such as time-of-flight secondary ion mass spectrometry (ToF-SIMS).

The diffusion of O from the Cu bulk to the surface during graphene CVD leads to continuous nucleation site etching and thus the observed low nucleation density. However, another critical finding of the study suggests that excessive concentration of oxygen in the bulk causes point defect formation in graphene. By introducing a H_2 annealing step after oxidation and Ar annealing, the level of O in the catalyst bulk can be decreased to a much lower level leading to a reduction in the defect density as evidenced by a lower D/G Raman peaks ratio and a reduced propensity to graphene etching. Such catalyst oxidation was also attempted for h-BN [93,94,158], but did not result in a significant improvement when compared to graphene, with some reports suggesting negative oxygen interference in h-BN growth chemistry [8,94,159].

5.4.1.2 Catalyst bulk solubility tuning

Another catalyst pre-treatment approach employs the deliberate introduction of foreign species via secondary CVD gases into specific catalysts to create interstitial solutes that allow the tuning of the bulk solubilities of relevant growth species. This approach allows to minimize the bulk reservoir effects and is most relevant to h-BN growth, where, for example, the bulk pre-fill with C or N can lower the B and N solubilities in Fe [89,136]. In-situ XRD and XPS measurements of Fe foils annealed in H_2 and NH_3 environments showed that the former results in pristine Fe spectra, while the latter leads to a nitride-containing composition, Fe_3N. Consequently, when borazine is introduced, iron boride formation is observed in the former case, due to B dissolution in Fe bulk leading to multilayer h-BN pyramids via a bulk-mediated growth mechanism (Fig. 5.4b). For NH_3 pre-growth annealing, however, the study showed that the N species from the Fe bulk are consumed upon contact with B-species on the surface resulting in a purely monolayer coverage. In a different approach, the use of acetylene (C_2H_2) to add C to the catalyst prior to h-BN growth (carburisation) was also investigated for Fe, Co, and Ni [89,161,162]. For Co and Ni, the addition of C resulted in an increase in the size of h-BN crystals, achieving lateral dimensions of the order of 600 μm, and the suppression of multilayer formation, leading to large-area monolayer h-BN deposits. Further mechanistic studies on the improved growth of h-BN with the addition of C to Ni showed that a minute quantity of oxygen on the surface of the catalyst forms deleterious, passivating BO_x species that are very stable and require C to be reduced to B and form BN [159]. For Fe, the addition of C also resulted in some suppression of multilayer formation and marginally larger domain sizes, however, a combination of oxidation and carburisation to control the balance of oxygen and carbon species resulted in a drastic suppression of the nucleation density and full monolayer formation leading to h-BN domain sizes of 1.1 mm [89]. Notably, time-dependent bulk composition changes were also reported due to O and C diffusion away from the bulk, requiring a complex optimization of multiple convoluted parameters.

5.4.1.3 Designed solubility by alloying

Custom alloys can be tailored to attain high catalytic performance, while preserving layer control by tweaking the overall precursor solubility through the alloy composition optimization. One of the early studies of such an approach explored the differences in graphene growth on Ni and gold-coated Ni films [82]. A 5 nm Au/550 nm Ni film resulted in the formation of purely monolayer graphene deposits at low temperature (450 °C), rationalized by the Au passivation of nucleation sites such as catalyst step edges and the lower C bulk solubility. While the crystallinity was not assessed and the Raman D/G ratio was relatively high at 0.2, the low growth temperature suggested process aims towards CMOS compatibility. On pristine Ni films, only patchy multilayer graphene was formed at the same conditions.

The use of gallium as a co-alloying element has been also demonstrated for the iron-group metals (Fe, Co, Ni) via the mechanism of forming an anti-perovskite structure, $GaCNi_3$, with a significantly reduced bulk C fraction compared to pure Ni [163]. A wide range of CVD conditions resulted in uniform monolayer graphene formation suggesting a robust self-limiting mechanism.

Recently, the use of Cu-Ni alloys has also gained momentum [88,164–167], driven by the advantages of lower cost compared to e.g. Au and being less cumbersome than the utilization of liquid Ga droplets, and especially due to the apparent advantages for the growth of both graphene and h-BN [122,167]. Cu-Ni alloys can be prepared by Ni electroplating of Cu foils [100], for example, and a Ni atomic composition ranging from 1.3 at% to 7.8 at% was investigated for graphene growth with a value of 6 at% chosen for the monolayer, while a value of 7.8 at% resulted in some bilayer and multilayer island formation [100]. The growth rate reached 70 µm/min, which was 17 times higher than for pristine Cu at 1075 °C. A related study investigated Cu-Ni alloys with Ni concentration from 10 at% to 20 at% and found that the growth of graphene on $Cu_{90}Ni_{10}$ was dominantly isothermal and surface mediated, while a mixed surface and bulk mechanism was observed for $Cu_{85}Ni_{15}$ [88]. One of the additional advantages of using Cu-Ni alloys is the somewhat increased melting temperature allowing graphene growth at temperatures higher than the melting point of elemental Cu (1085 °C). The graphene growth rate is strongly depended on the temperature (Fig. 5.5e) [88], reaching 170 µm/min at 1100 °C with domain sizes of ca. 3 mm after only 15 mins. Further improvements were demonstrated by devising a specific flow geometry (Section 5.4.3). Most studies point to a Ni composition value between 10 at% to 15 at% for the highest monolayer graphene growth rate and reduced nucleation density [166,168], while values between 17–22 at% have been shown to be useful for the segregation-driven formation of continuous bilayer and trilayer graphene under specific growth and cooling conditions [112,167]. A variety of alloys were also developed to facilitate the growth of a controlled number of layers, including Cu-Si alloy for 1–4 layer graphene [169], Pt-Si for AB-stacked bilayer graphene [170], Co-Cu alloy for 2 and 3 layer AB-stacked graphene [171].

The Cu-Ni alloy has also been used to form h-BN [122]. The composition of around 15–20 at% Ni was found to result in monolayer h-BN formation with growth rates of 1.2 µm/min and domain sizes up to 90 µm at 1050 °C. Increasing the Ni fraction resulted in the reduction of the monolayer h-BN domain size and at 50 at% Ni, for example, no deposits were observed after 90 min, suggesting that a higher precursor dose is required to reach supersaturation. Additionally, ammonia borane was found to decompose to more suitable surface species on such alloy surfaces, avoiding polymeric oligomer formation and thus deleterious amorphous deposits. For monolayer h-BN CVD the overall benefits of such Cu-Ni alloys are arguably more limited compared to graphene CVD, in particular considering the available other pathways [104,133,161,172,173] to monolayer control (5.4.1.2) and the possibility to grow directly on Ni [89,161,162].

A different strategy to advanced catalyst design for graphene CVD is highlighted by the use of Ni-Mo alloys [174], where Mo forms carbidic species in the bulk and acts as a high-capacity C "sink", allowing the kinetically-stabilized monolayer growth [85]. Fully monolayer graphene coverage was obtained under a wide range of conditions, however, the Raman

D/G peaks ratio ($\sim$0.2) was much higher than what can be obtained on pristine Cu ($\sim$0.02) [2,8], indicative of a higher defect density. The crystallinity of the resulting material was not assessed, highlighting the difficulties of domain size control on such high C-solubility substrates, and possibly explaining why the methodology is not widely utilized.

Conversely, the catalyst alloy composition can also be tailored to attain high precursor species solubilities with the aim of producing few-layer films, which is technologically important especially for h-BN. Recent research efforts aim to scalably produce multilayer h-BN, matched to the application demand for ultra-flat dielectric support and encapsulation for graphene and TMD channels. As an example, the Fe-Ni alloy system was investigated in that respect, specifically to tune and balance both N and B solubilities [113]. An alloy of 30 at% Fe and 70 at% Ni was epitaxially deposited on a $MgAl_2O_4$ (100) support, and slow cooling was employed to allow effective diffusion out of the catalyst bulk to form a h-BN film of ca. 3 nm in thickness. However, the crystallinity of such multi-layer CVD h-BN remains compromised as seen from the relatively wide Raman E_{2g} peak width of 17 cm^{-1}, and further work is required, including a better understanding of underlying growth mechanisms, to achieve the quality set by exfoliated h-BN flakes (Section 5.2.1).

Catalyst alloying can also be used to tailor the coupling between the catalyst and 2D material. This is specifically pertinent to optimize post-growth transfer or to tune 2D layer properties if left on the catalyst. A few published papers have reported the study of metal intercalation underneath already grown graphene or h-BN layers. Examples include Sn [175], Al [175], Au [176], Ag [177] or Cu [178] intercalation underneath graphene on Ni. While this opens a vast additional parameter space, the mechanisms of such intercalation remain unclear, specifically if successful effective intercalation depends on the presence of defects in the 2D layer which may preclude high quality material.

5.4.1.4 Growth on liquid surfaces

The use of liquid catalysts for 2D material CVD is motivated by the potential of having an ultra-smooth, uniform growth surface that is free of grain boundaries and free of influences from the metal lattice; high precursor diffusion rates, and the possibility of high growth temperatures facilitating enhanced growth rates, as well as reported possibilities of 2D domain self-alignment on clean surfaces [179]. The disadvantages or challenges include 2D material damage due to (re)solidification of the catalyst, rapid surface evaporation especially at low pressures, and difficulties to upscale due to de-wetting and droplet mobility (thickness and wetting dependent). Growth on molten surfaces has been demonstrated in various growth configurations, from e.g., molten pure metals, such as Cu [179] to eutectic alloys, such as metal silicides [77]. In most cases a suitable wetting high temperature support, such as W, must be also optimized for the specific liquid metal. Self-aligned graphene growth has been demonstrated on molten Cu [179] where mobile hexagonal domains took the most compact packing arrangement with growth rates between 10–50 µm/min at typical process temperatures between 1120–1180 °C. A related study focused on the less discussed aspect of Cu recrystallization and macroscopic restructuring upon cooling that caused graphene cracks. A solution was developed where graphene was first grown on molten Cu (1090 °C) and then cooled to the point of Cu re-solidification and graphene crack formation (1060 °C), followed by a second growth process at different conditions to heal the cracks [180]. Other examples of pristine liquid catalysts include graphene domain expansion on liquid Ga [181] at low temperatures (50–300 °C) and single-crystal h-BN growth (section 5.4.3) on liquid Au [182] – metals with melting points of 30 °C and 1064 °C, respectively. More advanced liquid surfaces were also designed by utilizing binary alloys, for example a Si film deposited on Pt foils. Upon annealing the two elements reacted forming a eutectic platinum silicide (Pt_3Si) film with a lowered melting point (830 °C) than either Pt or Si. This liquid "coating" on the surface allowed large hexagonal graphene domains to grow at rates of up to 120 μm/min obtaining 1 mm domains at 1150 °C [77]. Follow-on studies showed that such a Pt_3Si liquid surface can also be used to grow wafer-scale uniform bilayer graphene [170]. Furthermore, Fe-Si and Cu-Si alloys have been utilized for h-BN and bi/tri/tetra-layer graphene, respectively [169,183].

5.4.1.5 Solid source precursors

A conceptually-different, yet related strategy to purely gas-fed CVD is graphene and h-BN growth via precursors that come from either a solid film, or pre-dissolved species in the bulk, or via a combination of a solid precursor and a gas-based precursor. For graphene, simple examples include the use of an amorphous carbon film [60,184], polymer [59], or self-assembled monolayers (SAMs) [61] under or above a metal catalyst with adequate catalytic properties and carbon bulk solubility, such as Ni. Upon high temperature annealing the carbonaceous species dissolve into the metal, which can then be converted into graphene either isothermally or by precipitation upon cooling. In a related study a solid Fe_2B alloy combined with pure N_2 gas flow led to the formation of 30 nm thick h-BN deposits [185] at 1300 °C. The width of the E_{2g} Raman peak was ca. 15 cm^{-1} after transfer from the growth substrate to SiO_2. The latter value is typical for CVD h-BN but is still quite far from values of 8–9 cm^{-1} typically achieved by mechanically exfoliated bulk h-BN [21,22]. Interestingly, a different ratio of the elements namely $Fe_{82}B_{18}$ generated a eutectic alloy with a melting point of ca. 1180 °C, lower than either Fe or B, thus providing an avenue to also investigate a liquid form of such a substrate and growth mediated by bulk diffusion [160]. Controlled h-BN multilayer growth was obtained with 5–50 nm thickness and a Raman E_{2g} peak width of 10.4 cm^{-1} after transfer to SiO_2.

5.4.2 Transfer routes overview

Given the need to transfer the 2D materials from the catalyst substrate for most applications and the flexibility this endows, a variety of methodologies have been demonstrated in literature. Specifically for catalytic CVD, the ease of subsequent 2D material release from the process catalyst becomes a quintessential catalyst attribute. For many transfer methodologies the field developed through ad-hoc experiments and materials (e.g., polymer choice) that are readily available, requiring a careful evaluation of their suitability from a more fundamental perspective. Further experimental developments and detailed understanding of all the steps and convoluted influences in the transfer processes are critically required. The transfer process builds on important fundamental surface science and many interdisciplinary aspects such as intercalation chemistries, wetting, adhesion, polymer science, graphene-polymer interactions, solvent effects, interactions with all the transfer media and environment [186]. While some aspects, such as the graphene-catalyst interface has been extensively studied, other important factors received less attention than they deserve.

The most widely used, large-area transfer approach utilizes a polymeric film as a temporary scaffold that is applied to the as-grown 2D material, which supports the delicate 2D layer for further chemical treatments and handling. The transfer approach needs to match application needs, particularly scalability, environmental sustainability and cost efficiency as demanded for industrial use. The large variety of approaches developed for 2D material transfer can be broadly grouped into "wet transfer" and "semi-dry transfer"; the former can also be categorized as destructive or non-destructive with respect to the catalyst. We distinguish the semi-dry transfer from the fully "dry transfer" (Fig. 5.1) because a polymer removal step is typically required for scalable processes that involve the use of an organic solvent. Currently fully dry transfer processes using a h-BN stamp are only applicable to mechanical exfoliation (Section 5.2.1) or small CVD graphene flakes on post-growth oxidized Cu [17] due to the lack of high-quality large-area multilayer h-BN and strong attachment to the growth substrates. In this section we focus only on the scalable methods.

Fig. 5.8a outlines the three most common routes to 2D material transfer. Briefly, a support template is first applied, either a polymer coating or a specially designed stamp, followed by the detachment of the template/2D material by either (1) chemical etching of the catalyst, (2) electrochemical delamination, or (3) peeling. The wet routes require further cleaning e.g. in pure water to remove ionic residues. Finally, the template together with the 2D material is lifted onto the target substrate by "fishing out" from water (1–2) or by applying a stamp to the substrate surface (3). The contacting template is then dissolved in an organic solvent. Every step of the process has been meticulously investigated, starting from the earliest etchant optimization for Cu catalysts, such as the use of the more soluble ammonium persulfate, $(NH_4)_2S_2O_8$ instead of iron(III) chloride, $FeCl_3$ and various metallic contamination cleaning approaches [187,188]. Furthermore, a plethora of polymers and/or their combinations have been investigated leading to various transfer quality improvements with reduced doping or better continuity [189]. Some approaches were developed to reduce organic contamination by avoiding such templates altogether – including the deposition of a secondary metal film (e.g. Ni) to peel graphene [190] and utilization of an interface between two immiscible liquids to stabilize CVD graphene [191]. The catalyst etching approach remains very topical as most studies that rely on strong catalysts for graphene and h-BN CVD, such as Ni, Cu-Ni, Fe, Co [53,54] etc. utilize catalyst etching as the surface attachment is too strong for the other transfer methods to work. Additionally, such catalyst etching is a gentler approach where no pulling forces are involved, which makes it more suitable for applications where minute cracks can be detrimental. Alternatives to the full catalyst etching have also been developed where only the surface in immediate contact to the 2D material was oxidized or intercalated followed by etching of the converted surface layer, with the possibility to reuse the remainder of the catalyst [89,192].

For the electrochemical delamination approach, a voltage is applied between the polymer/2D material/substrate stack, and a second inert electrode is submerged into an electrolyte solution. The polymer/2D material stack can be detached and typically floats on the surface of the electrolyte. A variety of fundamental studies has been performed to identify the mechanism of graphene detachment, which was originally proposed to be the mechanical action of H_2 bubbles generated at the interface between the graphene and metal [193]. Excessive bubble generation, however, was shown to damage graphene [194]. Recent studies indicate that ion intercalation from the electrolyte is the primary cause for the delamination

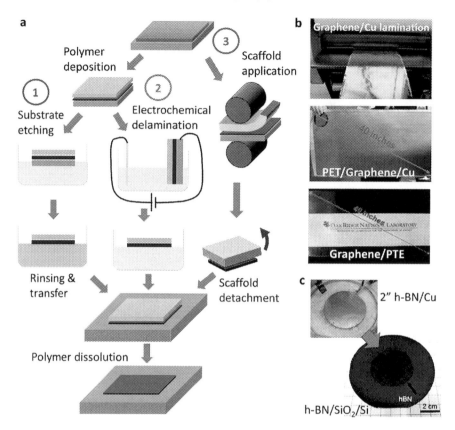

FIGURE 5.8 (a) A diagram of the three most common routes for graphene and h-BN transfer: (1) catalyst etching; (2) electrochemical "bubbling" transfer; (3) stamp-assisted "semi-dry" transfer or "peeling". (b) Exemplar lamination-based transfer of large area graphene from Cu foil. Reprinted from ref. [151], Copyright (2021), with permission from Elsevier. (c) Electrochemical delamination of a h-BN from a 2" Cu film wafer. Reprinted by permission from Springer Nature: ref. [123], Copyright (2021).

from metal surfaces which is possible even without H_2 generation at the electrode [195]. It is therefore necessary to have very weakly interacting interfaces where ions can be inserted, which means that the method is limited to metals with weak graphene or h-BN interactions, such as Cu or Pt. Similar to electrochemical delamination, semi-dry "peeling" can only be applied to such substrates. The procedure involves an application of a robust, but flexible scaffold to the 2D material surface, for example by lamination or simple heating above the glass temperature of a polymer, such as polyvinyl alcohol (PVA), followed by applying a force at a suitable angle to facilitate the 2D material detachment from the substrate (Fig. 5.8) [96,196–198]. Furthermore, pre-transfer processing procedures were developed to lower the interaction between the 2D material and Cu or Pt. Water intercalation has also shown to result in Cu oxide layer formation [199,200] and a layer of interfacial water, respectively [194], resulting in lower voltages required for the electrochemical delamination and better continuity for mechanical peeling. Large-area transfer of graphene and h-BN has been achieved with all

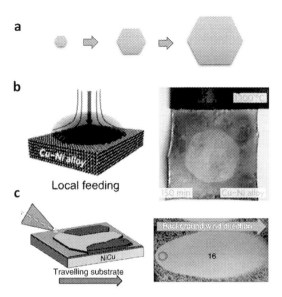

FIGURE 5.9 (a) A diagram of the single domain growth approach. (b) A schematic of a local precursor injection through a nozzle over a Cu-Ni alloy, and an optical image of the resulting inch-sized graphene domain. (c) A schematic of the graphene growth approach utilizing a local nozzle and a moving substrate, allowing seed selection and expansion; with an optical image of the resulting graphene domain. (b,c) Reprinted by permission from Springer Nature: refs. [51,88], Copyright (2021).

of the above methods, for example for 50–300 mm wafer-like samples and foils up to ca. 90 cm × 40 cm [123,151,200] (Fig. 5.8b,c). Additionally, roll-to-roll apparatus and methodologies for ultra-large samples have also been demonstrated [49].

5.4.3 State-of-the-art: large area single 2D crystal production

For a variety of applications, including simple heaters [201] or sensors, graphene and h-BN layers of limited crystallinity may be adequate, with the functionality not requiring high performance from a specific property, but possibly a combination of unusual properties, such as transparency and flexibility. The main focus being large area and cost-effective production, e.g., via a roll-to-roll process. Nevertheless, analogous to the history of bulk materials, large-area single crystal production is an important step for 2D materials, not only for fundamental research but towards future technology and higher value-added applications, e.g., in integrated optoelectronics and quantum devices where such defects are clearly detrimental or where detailed control is required at atomic level. To date, two principal routes to single crystal graphene and h-BN production have emerged, each with their advantages and disadvantages: (1) expansion of a single graphene or h-BN seed into a large crystal (Fig. 5.9a), and (2) nucleation of multiple aligned domains that seamlessly merge into a single crystal (Fig. 5.10a).

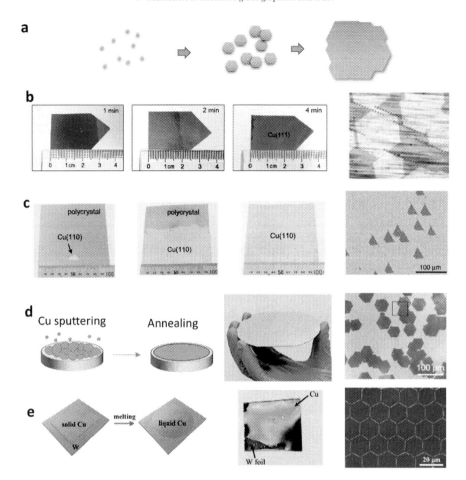

FIGURE 5.10 (a) A schematic of single crystal growth via domain merging their seamless stitching. (b–c) Examples of single crystal foil formation via grain growth from a narrow tip or from a seed Cu piece, respectively, with exemplar aligned graphene and h-BN domains. Reprinted from ref. [50], Copyright (2021), with permission from Elsevier. Reprinted by permission from Springer Nature: ref. [205], Copyright (2021). (d) A schematic of the single crystal metal thin film formation after PVD and high temperature annealing, with an optical image of the resulting 4″ Cu(111) wafer and exemplar aligned graphene domains. Reprinted with permission from ref. [206]. Copyright (2021) American Chemical Society. (e) Aligned graphene grown on liquid copper on a tungsten support. Reprinted with permission from ref. [179].

5.4.3.1 Single domain growth

The single domain growth approach is based on an initial selection of a single seed and its controlled expansion while avoiding secondary nucleation. To date such approach has been implemented via localized, confined precursor injection into a CVD system to locally drive supersaturation and limit growth to expansion of a single monolayer crystal domain. In one study a small quartz nozzle was placed above a Cu-Ni alloy foil (Fig. 5.10b) with the supplied gas mixture of $Ar/H_2/CH_4$ suitable for graphene nucleation and growth just below the nozzle, while globally a different gas composition with only Ar and H_2 was utilized [88]. The dilution of the CH_4 away from the nozzle and the increase of the H_2:CH_4 ratio (Section 5.3.3) meant that nucleation away from the nozzle was suppressed. While the nucleation directly under the nozzle was still a probabilistic process, it was possible to slowly nucleate and expand just a single domain with optimized gas ratios and pressures. Due to the expansion of the crystal growth edge (perimeter) with time it was necessary to increase the flow of CH_4 with time. The size of the single crystal graphene domain reached up to 38 mm at a high growth rate of ca. 250 μm/min, boosted by the use of a high growth temperature of 1100 °C and an optimized $Cu_{85}Ni_{15}$ alloy composition. The need to increase the required precursor flow in this configuration, however, could present a limitation as eventually the concentration of CH_4 directly under the nozzle is likely to cause undesired deposit formation, such as amorphous carbon or multilayers [202]. Electrical transport measurements after wet transfer onto exfoliated h-BN flakes showed charge carrier mobility up to 20000 $cm^2V^{-1}s^{-1}$ at room temperature.

In an alternative implementation of the localized precursor flow strategy, a quartz nozzle was positioned inside a tube system with a movable Cu roll (Fig. 5.9c, Fig. 5.11d) and a carbon precursor flow was introduced through the nozzle with a background, global "wind" consisting of just H_2/Ar [51]. The idea behind such optimized flow geometries and concentrations was to create conditions whereby multiple nucleation seeds are initially formed, but eventually the fastest growing seed will dominate and expand into a single crystal. High growth rates of ca. 400 μm/min were obtained, again boosted by a designed catalytic alloy ($Cu_{83}Ni_{17}$), less stable precursor (C_2H_6) and a high growth temperature (1100 °C) [51].

The biggest advantage of the single domain expansion approaches is a reduction in catalyst single crystal requirement allowing processed polycrystalline foils to be used. While it is still not fully understood, a growing graphene domain can cross a metal grain boundary and continue to grow as a single crystal [9,51]. Similar reports also exist for h-BN where triangular domains crossed different metal grains without being affected in their crystallinity [203]. However, only specific catalysts under specific conditions and a limited number of grains have been explored thus far, requiring further studies in the future. Some disadvantages include the need for the development of new CVD system parts, including movement of hot parts inside reactor chambers and meticulous optimization of the CVD conditions.

5.4.3.2 *Domain stitching*

The domain stitching approach for the formation of graphene and h-BN films is based on the ability to nucleate aligned domains that seamlessly merge, which is typically based on their epitaxial registry to the underlying substrate or self-alignment on molten surfaces (discussed in Section 5.4.1). The pre-requisite for the former is the need for high-quality single crystal substrates with suitable crystallographic structures. For typical transition metal catalysts, two principal approaches exist: epitaxial thin film growth or crystallization of conventional metal foils or films. For either approach, the complexity of metal template formation overshadows in practical difficulty the actual CVD of 2D materials. The catalyst surface crystal symmetry (not to mention surface reconstructions) must be compatible with the graphene or h-BN crystal symmetry, so that the domain edges are perfectly aligned, including potential additional considerations due to the registry to metal step edges. Early reports within surface science studies focused on a variety of catalytic metals, including Pt, Rh, Re, Ir etc. [53,54]

as epitaxial surfaces for graphene, but more recent studies expand to widely available and industrially applicable metals, such as Cu [204] and Ni [172] or their alloys [165].

Driven by cost and ease of scalability, a range of reports has focused on the crystallization of commercial cold-rolled metal foils, particularly Cu. For example, applying tension to polycrystalline Cu foils resulted in large-area recrystallization into a single Cu(111) surface orientation due to promoted abnormal grain growth [204], with other studies also pointing to the importance of how the foil is supported – improved recrystallization was obtained by the design of a holder that minimized the contact with the foil surface. More advanced mechanisms are based on the adaptation of a "zone refining" method to metal foils, which relies on the controlled expansion of a select facet into a single crystal. In one implementation a Cu foil was cut in such a way as to create a pointed edge, where one grain would rapidly consume other grains and create a single crystal [50]. Upon annealing in a temperature gradient and by pulling the foil through a hot furnace, the initial seed was expanded into a very large single crystal as shown in Fig. 5.10b; the principle of this process is similar to the Bridgman process or seeded Czochralski single growth. Single crystal Cu foil with the (111) fcc surface orientation was demonstrated in sizes of up to 50 cm $\times$ 5 cm [50]. Aligned hexagonal graphene domains were also demonstrated on such foils (Fig. 5.10b). Assuming perfect stitching of such domains, the growth rate of each individual domain is not a suitable parameter for the overall speed of crystal production, instead the surface coverage rate can be used as such a parameter, and depends on the nuclei density and the edge growth rate of individual domains. Full coverage of the maximum foil size was achieved in 20 mins, corresponding to an effective linear coverage rate of 8000 µm/min – which is significantly faster than rates achievable by the single domain expansion approach. Furthermore, large area crystallinity was confirmed by controlled etching of holes in the film, followed by their orientation analysis with 99% of the geometrical shapes exhibiting alignment. However, the assumption of perfect stitching is extremely difficult to rigorously confirm over such large areas and requires the development of large area rapid characterization techniques for in-line quality monitoring.

In another implementation of the seed expansion process, which, in addition to the Cu(111) facet allowed the production of various Cu orientations such as (110) [205], (112), (233), (355) etc. [207], a piece of Cu foil with the desired orientation, for example Cu(110) was placed onto a polycrystalline commercial Cu foil. Upon annealing the seed expanded faster than any of the polycrystalline grains (abnormal growth) resulting in large size single crystal Cu foils (Fig. 5.10c) with with dimensions up to 20 cm $\times$ 30 cm [207]. The annealing treatment required for the conversion of such foils was around 5 hours, which would become the limiting factor for production considering the much faster times for CVD growth of tens of minutes. The possibility of substrate re-use is thus an important requirement. h-BN growth was demonstrated on such substrates with good alignment (Fig. 5.10c).

An emerging focus is the utilization of thin catalytic films deposited epitaxially on suitable high temperature substrates. This builds on a rich body of literature on heteroepitaxial systems and thin film deposition of transition metals by MBE and physical vapor deposition (PVD) approaches such as magnetron sputtering that have historically been driven by surface science and applications such as spintronics which are sensitive to the film structure and require ultra-flat surfaces. Epitaxial film growth of metals, including Cu, Fe, Ni, Co, Cr, Pt, V, etc. on substrates such as MgO and Al_2O_3 has been evolving for several decades [208], and is now being adopted to the needs of catalytic graphene and h-BN growth by CVD, with a focus on high throughput, scalability, and low cost. Cu(111), Ni(111) and Cu-Ni(111) alloy films were among the first to be deposited at the wafer-scale [141,165,166,206,209,210]. The process typically involves the PVD of the catalyst metal with a thickness of ca. 500 nm onto a specially-prepared sapphire substrate, followed by high temperature annealing, resulting in large-area single crystal catalysts (Fig. 5.10d). The reported sapphire treatment methods include high temperature oxygen annealing [206] or acid treatments [141,209] to create suitable surface termination, to avoid twinned grain formation and promote metal-sapphire epitaxy, while avoiding de-wetting of the deposited film. As produced epitaxial films have a significantly reduced (nm scale) overall surface roughness compared to metal foils (tens to hundreds of nm RMS roughness). Some studies also report suppression of graphene wrinkle formation, rationalized by a lower overall thermal expansion of the Cu(111)/sapphire substrate and a higher coupling of graphene to the Cu(111) surface, causing compressive strain in the graphene rather than its relaxation via wrinkle formation [206].

Domain stitching approaches for h-BN growth are more complex due to its polarity. There is a possibility of antiparallel alignment on (111) surfaces which could cause domain boundaries and polycrystallinity even on single crystal fcc metal (111) faces [210]. For example, on Ni(111) it was shown that 180°–rotated h-BN domains cannot merge without a defect line because of their different registry to the substrate [211]. A number of reports therefore suggested the necessity to utilize substrates that have a unique direction, such as Cu(102) or (103) – with one edge of h-BN triangles being perpendicular to this direction. Also, Cu(110) [205], Cu(410), Cu(013), Cu(025), Cu(122), etc. [207] grains seeded on Cu foils were shown to produce fully-aligned h-BN domains as shown in Fig. 5.10c. However, a recent report where h-BN growth was performed on a Cu(111) film demonstrated h-BN of nearly single orientation with 99.6% alignment [123]. The reasons for such striking differences are not yet clear.

While the above approaches on foil crystallization and epitaxial metal PVD require not only time intensive preparation, but also separate equipment and (protected) transfer to the CVD reactor, the latest research targets a combined process flow of epitaxial catalyst metalization and 2D CVD in a standard single reactor [212]. Using a sandwich substrate holder, a close-spaced sublimation (CSS) approach to epitaxial, ultra-flat Cu deposition from a commercial Cu foil source onto an opposing sapphire wafer has been demonstrated, seamlessly combined with subsequent graphene CVD process flow. The CSS-like metalization shows innate advantages similar to prior reports on CSS of polycrystalline CdTe thin films [213], namely, simplicity, low source material wastage, and fast deposition rates.

An interesting comparison to epitaxial growth substrates is attempts to produce single crystal graphene and h-BN on molten surfaces, driven by the idea of mobile 2D material domains that are able to rotate and self-align to form most energetically favorable interactions [182] or aided by the gas flow [214]. For example, aligned hexagonal graphene domains were observed on liquid Cu when such domains were of similar size (Fig. 5.10e) [179]. Similarly, circular h-BN domains grown on liquid Au showed alignment over 4 mm × 4 mm areas measured by LEEM [182]. Due to the mobility of the liquid metal and recrystallization on cooling (Section 5.4.1), various practical challenges need to be solved for the scalable expansion of CVD on liquid surfaces.

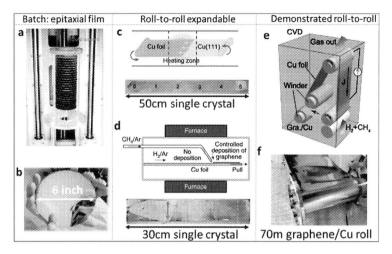

FIGURE 5.11 Opportunities for large-area production of graphene and h-BN. (a–b) Exemplar epitaxial Cu-Ni thin films for 2D material growth produced as 25 four inch wafers or a single six inch wafer, respectively. (c–d) Roll-to-roll expandable approaches based on Cu foil recrystallization with domain stitching and single domain growth via localized precursor flows, respectively. (e–f) Demonstrated polycrystalline graphene production in roll-to-roll geometry: apparatus overview and an exemplar 70 m ×12 cm graphene/Cu roll, respectively. (a,c) Reprinted from refs. [50,166], Copyright (2021), with permission from Elsevier. (b) Reprinted with permission from ref. [165]. © 2019 WILEY-VCH Verlag GmbH & Co. KGaA, Weinheim. (d) Reprinted by permission from Springer Nature: ref. [51], Copyright (2021). (e) Reprinted from ref. [215], with the permission of AIP Publishing. (f) Reprinted from ref. [204]. © IOP Publishing. Reproduced with permission. All rights reserved.

5.4.3.3 Large area production

A number of studies already demonstrate large area single crystal 2D material production or can be extended, for example, to the roll-to-roll approach [44]. Alternatively, volume production of batches of wafers is another route that is well-established, for example, in the semiconductor industry. Recently, such production route has been demonstrated for graphene CVD on single crystal metallic film catalysts, where 25 four-inch Cu-Ni alloy wafers were simultaneously used for graphene growth [166] (Fig. 5.11a). However, only basic characterization was performed to assess the uniformity across each wafer, echoing the need for rapid and large-area quality monitoring. Furthermore, production on 6" Cu-Ni alloy wafers has also been demonstrated [165] (Fig. 5.11b). Tension-assisted recrystallization, the "evolutionary selection" and metal seeding-based methods are potentially expandable to roll-to-roll processes (Fig. 5.11c,d), but so far have not been utilized in this configuration. A schematic for the large-area polycrystalline graphene growth apparatus is shown in Fig. 5.11e with a continuous flow of gas and electrical joule heating of the foil [204]. Polycrystalline foil sizes as large as 70 m × 12 cm (Fig. 5.11f) were covered with graphene with complimentary roll-to-roll transfer equipment designs being in development [215]. Scalability for 2D materials is currently achieved at a compromise, e.g. in material quality, while delivering the necessary throughput and cost reduction – a balance will be required in future manufacturing to match application needs.

5.5 Conclusions and outlook

Extraordinary efforts have been dedicated to the advancement of scalable and defect-free production of graphene and h-BN, built upon the vast knowledge accumulated over the last decades across a multitude of disciplines and applied research fields. From surface science studies and modeling at atomic scales to heterogeneous catalysis and crystal growth, from metallurgical processing to epitaxial single crystal film growth, from atomistic hetero-interfacial studies to industry-relevant processing to name a few. Availability of bulk layered crystals and their exfoliation will continue to help driving fundamental 2D materials discovery and to facilitate small-scale proof of concept devices. On the other hand, bottom-up growth will be critical for the transition to commercial applications and the challenge now is to reach the quality set by exfoliated materials or the quality demanded by high-value-added applications, while preserving scalability. Methodologies applicable to roll-to-roll processes have reached ultra-fast growth rates and greatly improved performance compared to the early studies, while smooth epitaxial thin film catalysts have resulted in significantly improved graphene and h-BN crystallinity on wafer scale. Such single crystal 2D materials will be able to satisfy applications with stringent quality requirements in the future, with the growth arguably not being the limiting factor owning to the recent advances in the CVD field. The focus is gradually shifting from narrow dedicated studies to holistic full-cycle integrated material development, where the growth, processing and integration cannot be considered independent. The discussed graphene and monolayer h-BN CVD crystals are the required building blocks for future materials engineering at the atomic layer limit, such as twisted bilayers and designer heterostructures. In parallel, crystalline multilayer exfoliated h-BN has played a critical role in boosting the performance of almost every 2D-material based device and the lack of matching synthetically grown h-BN materials requires further development of the growth methodologies or alternative materials.

The ultimate quality and performance requirement is typically tied to an application, with a wide range of applications that can tolerate some imperfections, such as sensors or photodetectors, currently driving 2D materials commercialization. 2D materials are natural candidates for monolithic 3D integration for advanced functionality, however, while 2D material films are reaching maturity, the next research challenge is not only scalable, clean 2D-2D layer interfacing, but also heterogeneous 2D-3D materials integration, i.e., moving from basic material control at atomic scale to heterogeneous complementary technology and bringing together many materials. Deeper integration levels of 2D materials, especially in CMOS technology, will require focused research efforts and adaptation of the manufacturing approaches that are at the core of the semiconductor industry. Understanding the role of defects and their control is a crucial vector, with a topical example being the recently discovered single phonon emission from atomic defects in h-BN [216]. Such atomic engineering requires high-quality 2D material to start with, which are now within reach of modern scalable growth technologies.

References

[1] K.S. Novoselov, et al., Electric field effect in atomically thin carbon films, Science 306 (2004) 666–669.

[2] X.S. Li, et al., Large-area synthesis of high-quality and uniform graphene films on copper foils, Science 324 (2009) 1312–1314.

[3] J. Sun, et al., Recent progress in the tailored growth of two-dimensional hexagonal boron nitride via chemical vapour deposition, Chemical Society Reviews 47 (2018) 4242–4257.

[4] L. Lin, et al., Bridging the gap between reality and ideal in chemical vapor deposition growth of graphene, Chemical Reviews 118 (2018) 9281–9343.

[5] Y.Y. Illarionov, et al., Insulators for 2D nanoelectronics: the gap to bridge, Nature Communications 11 (2020) 3385.

[6] K.S. Novoselov, et al., A roadmap for graphene, Nature 490 (2012) 192–200.

[7] A.C. Ferrari, et al., Science and technology roadmap for graphene, related two-dimensional crystals, and hybrid systems, Nanoscale 7 (2015) 4598–4810.

[8] O.J. Burton, et al., The role and control of residual bulk oxygen in the catalytic growth of 2D materials, Journal of Physical Chemistry C 123 (2019) 16257–16267.

[9] Y.F. Hao, et al., The role of surface oxygen in the growth of large single-crystal graphene on copper, Science 342 (2013) 720–723.

[10] S. Xu, et al., Chemical vapor deposition of graphene on thin-metal films, Cell Reports Physical Science 2 (2021) 100372.

[11] D. De Fazio, et al., High-mobility, wet-transferred graphene grown by chemical vapor deposition, ACS Nano 13 (2019) 8926–8935.

[12] L.-P. Ma, et al., Transfer methods of graphene from metal substrates: a review, Small Methods 3 (2019) 1900049.

[13] A. Boscá, et al., Automatic graphene transfer system for improved material quality and efficiency, Scientific Reports 6 (2016) 21676.

[14] A.I. Aria, et al., Parameter space of atomic layer deposition of ultrathin oxides on graphene, ACS Applied Materials & Interfaces 8 (2016) 30564–30575.

[15] S. Kim, et al., Realization of a high mobility dual-gated graphene field-effect transistor with Al2O3 dielectric, Applied Physics Letters 94 (2009).

[16] B. Lee, et al., Characteristics of high-k Al2O3 dielectric using ozone-based atomic layer deposition for dual-gated graphene devices, Applied Physics Letters 97 (2010) 043107.

[17] L. Banszerus, et al., Ultrahigh-mobility graphene devices from chemical vapor deposition on reusable copper, Science Advances 1 (2015) e1500222.

[18] M.R. Null, et al., Thermal diffusivity and thermal conductivity of pyrolytic graphite from 300 to 2700°K, Carbon 11 (1973) 81–87.

[19] R.M.A. Lieth, Preparation and Crystal Growth of Materials with Layered Structures, D. Reidel Pub. Co., Dordrecht, Holland, Boston, 1977.

[20] K. Watanabe, et al., Direct-bandgap properties and evidence for ultraviolet lasing of hexagonal boron nitride single crystal, Nature Materials 3 (2004) 404–409.

[21] S. Liu, et al., Large-scale growth of high-quality hexagonal boron nitride crystals at atmospheric pressure from an Fe–Cr flux, Crystal Growth & Design 17 (2017) 4932–4935.

[22] Y. Kubota, et al., Deep ultraviolet light-emitting hexagonal boron nitride synthesized at atmospheric pressure, Science 317 (2007) 932–934.

[23] J. Sonntag, et al., Excellent electronic transport in heterostructures of graphene and monoisotopic boron-nitride grown at atmospheric pressure, 2D Materials 7 (2020) 031009.

[24] T. Taniguchi, S. Yamaoka, Spontaneous nucleation of cubic boron nitride single crystal by temperature gradient method under high pressure, Journal of Crystal Growth 222 (2001) 549–557.

[25] T. Taniguchi, K. Watanabe, Synthesis of high-purity boron nitride single crystals under high pressure by using Ba–BN solvent, Journal of Crystal Growth 303 (2007) 525–529.

[26] A.D. Jara, et al., Purification, application and current market trend of natural graphite: a review, International Journal of Mining Science and Technology 29 (2019) 671–689.

[27] S. Hofmann, et al., CVD-enabled graphene manufacture and technology, Journal of Physical Chemistry Letters 6 (2015) 2714–2721.

[28] K.S. Novoselov, et al., Two-dimensional atomic crystals, Proceedings of the National Academy of Sciences of the United States of America 102 (2005) 10451–10453.

[29] P. Blake, et al., Making graphene visible, Applied Physics Letters 91 (2007) 063124.

References 151

[30] J. Xue, et al., Scanning tunnelling microscopy and spectroscopy of ultra-flat graphene on hexagonal boron nitride, Nature Materials 10 (2011) 282–285.

[31] L. Wang, et al., One-dimensional electrical contact to a two-dimensional material, Science 342 (2013) 614–617.

[32] A.S. Mayorov, et al., Micrometer-scale ballistic transport in encapsulated graphene at room temperature, Nano Letters 11 (2011) 2396–2399.

[33] A. Pirkle, et al., The effect of chemical residues on the physical and electrical properties of chemical vapor deposited graphene transferred to SiO_2, Applied Physics Letters 99 (2011).

[34] J. Chan, et al., Reducing extrinsic performance-limiting factors in graphene grown by chemical vapor deposition, ACS Nano 6 (2012) 3224–3229.

[35] D.G. Purdie, et al., Cleaning interfaces in layered materials heterostructures, Nature Communications 9 (2018) 5387.

[36] F. Pizzocchero, et al., The hot pick-up technique for batch assembly of van der Waals heterostructures, Nature Communications 7 (2016) 11894.

[37] S. Masubuchi, et al., Autonomous robotic searching and assembly of two-dimensional crystals to build van der Waals superlattices, Nature Communications 9 (2018) 1413.

[38] L. Mond, et al., L.—action of carbon monoxide on nickel, Journal of the Chemical Society, Transactions 57 (1890) 749–753.

[39] R. Haubner, The history of hard CVD coatings for tool applications at the university of technology Vienna, International Journal of Refractory Metals & Hard Materials 41 (2013) 22–34.

[40] A. Nagashima, et al., 2-dimensional plasmons in monolayer graphite, Solid State Communications 83 (1992) 581–585.

[41] T.A. Land, et al., STM investigation of single layer graphite structures produced on Pt(111) by hydrocarbon decomposition, Surface Science 264 (1992) 261–270.

[42] A. Nagashima, et al., Electronic-structure of monolayer hexagonal boron-nitride physisorbed on metal-surfaces, Physical Review Letters 75 (1995) 3918–3921.

[43] A. Nagashima, et al., Electronic dispersion-relations of monolayer hexagonal boron-nitride formed on the Ni(111) surface, Physical Review B 51 (1995) 4606–4613.

[44] S.J. Sandoval, et al., Raman-study and lattice-dynamics of single molecular layers of MoS_2, Physical Review B 44 (1991) 3955–3962.

[45] A.K. Geim, K.S. Novoselov, The rise of graphene, Nature Materials 6 (2007) 183–191.

[46] K. Strobl, L. Rosenbaum, Scalable 2D-Film CVD Synthesis (U.S. Patent Application US201414547362A), U.S. Patent and Trademark Office, 2013.

[47] N. Mishra, et al., Wafer-scale synthesis of graphene on sapphire: toward fab-compatible graphene, Small 15 (2019) 1904906.

[48] H. Xin, W. Li, A review on high throughput roll-to-roll manufacturing of chemical vapor deposition graphene, Applied Physics Reviews 5 (2018) 031105.

[49] S. Bae, et al., Roll-to-roll production of 30-inch graphene films for transparent electrodes, Nature Nanotechnology 5 (2010) 574–578.

[50] X. Xu, et al., Ultrafast epitaxial growth of metre-sized single-crystal graphene on industrial Cu foil, Science Bulletin 62 (2017) 1074–1080.

[51] I.V. Vlassiouk, et al., Evolutionary selection growth of two-dimensional materials on polycrystalline substrates, Nature Materials 17 (2018) 318–322.

[52] P. Braeuninger-Weimer, et al., Fast, noncontact, wafer-scale, atomic layer resolved imaging of two-dimensional materials by ellipsometric contrast micrography, ACS Nano 12 (2018) 8555–8563.

[53] M. Batzill, The surface science of graphene: metal interfaces, CVD synthesis, nanoribbons, chemical modifications, and defects, Surface Science Reports 67 (2012) 83–115.

[54] W. Auwärter, Hexagonal boron nitride monolayers on metal supports: versatile templates for atoms, molecules and nanostructures, Surface Science Reports 74 (2019) 1–95.

[55] R.S. Weatherup, et al., Interdependency of subsurface carbon distribution and graphene–catalyst interaction, Journal of the American Chemical Society 136 (2014) 13698–13708.

[56] P.R. Kidambi, et al., Observing graphene grow: catalyst–graphene interactions during scalable graphene growth on polycrystalline copper, Nano Letters 13 (2013) 4769–4778.

[57] R.S. Weatherup, et al., Situ Graphene Growth Dynamics on Polycrystalline Catalyst Foils Nano Letters 16 (2016) 6196–6206.

[58] G. Wang, et al., Lattice selective growth of graphene on sapphire substrate, Journal of Physical Chemistry C 119 (2015) 426–430.

[59] Z. Peng, et al., Direct growth of bilayer graphene on SiO2 substrates by carbon diffusion through nickel, ACS Nano 5 (2011) 8241–8247.

[60] K. Banno, et al., Transfer-free graphene synthesis on insulating substrates via agglomeration phenomena of catalytic nickel films, Applied Physics Letters 103 (2013) 082112.

[61] Z. Yan, et al., Growth of bilayer graphene on insulating substrates, ACS Nano 5 (2011) 8187–8192.

[62] R.S. Weatherup, et al., Introducing carbon diffusion barriers for uniform, high-quality graphene growth from solid sources, Nano Letters 13 (2013) 4624–4631.

[63] O. Vohler, et al., Deposition of pyrolytic carbon in the pores of graphite bodies — I. Introduction to and results of deposition experiments using methane, Carbon 6 (1968) 397–403.

[64] M.A. Fanton, et al., Characterization of graphene films and transistors grown on sapphire by metal-free chemical vapor deposition, ACS Nano 5 (2011) 8062–8069.

[65] L. Tai, et al., Direct growth of graphene on silicon by metal-free chemical vapor deposition, Nano-Micro Letters 10 (2017) 20.

[66] D. Wei, et al., Critical crystal growth of graphene on dielectric substrates at low temperature for electronic devices, Angewandte Chemie. International Edition in English 52 (2013) 14121–14126.

[67] A. Khan, et al., New insight into the metal-catalyst-free direct chemical vapor deposition growth of graphene on silicon substrates, Journal of Physical Chemistry C 125 (2021) 1774–1783.

[68] C. Berger, et al., Ultrathin epitaxial graphite: 2D electron gas properties and a route toward graphene-based nanoelectronics, The Journal of Physical Chemistry B 108 (2004) 19912–19916.

[69] A.J. Van Bommel, et al., LEED and Auger electron observations of the SiC(0001) surface, Surface Science 48 (1975) 463–472.

[70] M. Ruan, et al., Epitaxial graphene on silicon carbide: introduction to structured graphene, MRS Bulletin 37 (2012) 1138–1147.

[71] K.V. Emtsev, et al., Towards wafer-size graphene layers by atmospheric pressure graphitization of silicon carbide, Nature Materials 8 (2009) 203–207.

[72] N. Mishra, et al., Graphene growth on silicon carbide: a review, Physica Status Solidi A 213 (2016) 2277–2289.

[73] S. Tang, et al., Precisely aligned graphene grown on hexagonal boron nitride by catalyst free chemical vapor deposition, Scientific Reports 3 (2013).

[74] S. Tang, et al., Silane-catalysed fast growth of large single-crystalline graphene on hexagonal boron nitride, Nature Communications 6 (2015) 6499.

[75] Y.-J. Cho, et al., Hexagonal boron nitride tunnel barriers grown on graphite by high temperature molecular beam epitaxy, Scientific Reports 6 (2016) 34474.

[76] S. Tang, et al., Nucleation and growth of single crystal graphene on hexagonal boron nitride, Carbon 50 (2012) 329–331.

[77] V. Babenko, et al., Rapid epitaxy-free graphene synthesis on silicidated polycrystalline platinum, Nature Communications 6 (2015) 7536.

[78] S. Vangala, et al., Wafer scale BN on sapphire substrates for improved graphene transport, Scientific Reports 8 (2018) 8842.

[79] L. Chen, et al., in: Semiconductor Nanowires: from Next-Generation Electronics to Sustainable Energy 1–53, The Royal Society of Chemistry, 2015.

[80] S. Hofmann, et al., Ledge-flow-controlled catalyst interface dynamics during Si nanowire growth, Nature Materials 7 (2008) 372–375.

[81] F.M. Ross, Controlling nanowire structures through real time growth studies, Reports on Progress in Physics 73 (2010) 114501.

[82] R.S. Weatherup, et al., In situ characterization of alloy catalysts for low-temperature graphene growth, Nano Letters 11 (2011) 4154–4160.

[83] X.S. Li, et al., Large-area graphene single crystals grown by low-pressure chemical vapor deposition of methane on copper, Journal of the American Chemical Society 133 (2011) 2816–2819.

[84] J. Robertson, Heterogeneous catalysis model of growth mechanisms of carbon nanotubes, graphene and silicon nanowires, Journal of Materials Chemistry 22 (2012) 19858–19862.

[85] R.S. Weatherup, et al., Kinetic control of catalytic CVD for high-quality graphene at low temperatures, ACS Nano 6 (2012) 9996–10003.

References   153

[86] J.C. Shelton, et al., Equilibrium segregation of carbon to a nickel (111) surface: a surface phase transition, Surface Science 43 (1974) 493–520.

[87] X. Li, et al., Evolution of graphene growth on Ni and cu by carbon isotope labeling, Nano Letters 9 (2009) 4268–4272.

[88] T.R. Wu, et al., Fast growth of inch-sized single-crystalline graphene from a controlled single nucleus on Cu-Ni alloys, Nature Materials 15 (2016) 43.

[89] V. Babenko, et al., Oxidising and carburising catalyst conditioning for the controlled growth and transfer of large crystal monolayer hexagonal boron nitride, 2D Materials 7 (2020).

[90] P. Braeuninger-Weimer, et al., Understanding and controlling Cu-catalyzed graphene nucleation: the role of impurities, roughness, and oxygen scavenging, Chemistry of Materials 28 (2016) 8905–8915.

[91] A.T. Murdock, et al., Targeted removal of copper foil surface impurities for improved synthesis of CVD graphene, Carbon 122 (2017) 207–216.

[92] S. Jin, et al., Colossal grain growth yields single-crystal metal foils by contact-free annealing, Science 362 (2018) 1021–1025.

[93] R.-J. Chang, et al., Growth of large single-crystalline monolayer hexagonal boron nitride by oxide-assisted chemical vapor deposition, Chemistry of Materials 29 (2017) 6252–6260.

[94] V. Babenko, et al., Time dependent decomposition of ammonia borane for the controlled production of 2D hexagonal boron nitride, Scientific Reports 7 (2017) 14297.

[95] Y. Gao, et al., Repeated and controlled growth of monolayer, bilayer and few-layer hexagonal boron nitride on Pt foils, Acs Nano 7 (2013) 5199–5206.

[96] R. Wang, et al., A peeling approach for integrated manufacturing of large monolayer h-BN crystals, ACS Nano 13 (2019) 2114–2126.

[97] Y.-P. Hsieh, et al., Promoter-assisted chemical vapor deposition of graphene, Carbon 67 (2014) 417–423.

[98] E. Loginova, et al., Evidence for graphene growth by C cluster attachment, NJPh 10 (2008) 093026.

[99] X.H. An, et al., Large-area synthesis of graphene on palladium and their Raman spectroscopy, Journal of Physical Chemistry C 116 (2012) 16412–16420.

[100] M. Huang, et al., Highly oriented monolayer graphene grown on a Cu/Ni(111) alloy foil, ACS Nano 12 (2018) 6117–6127.

[101] H.-S. Jang, et al., Toward scalable growth for single-crystal graphene on polycrystalline metal foil, ACS Nano 14 (2020) 3141–3149.

[102] R.M. Jacobberger, et al., Direct oriented growth of armchair graphene nanoribbons on germanium, Nature Communications 6 (2015) 8006.

[103] J. Yin, et al., Aligned growth of hexagonal boron nitride monolayer on germanium, Small 11 (2015) 5375–5380.

[104] H. Wang, et al., Synthesis of large-sized single-crystal hexagonal boron nitride domains on nickel foils by ion beam sputtering deposition, Advanced Materials 27 (2015) 8109–8115.

[105] J. Kraus, et al., Understanding the reaction kinetics to optimize graphene growth on cu by chemical vapor deposition, AnP 529 (2017) 1700029.

[106] P. Solís-Fernández, et al., Isothermal growth and stacking evolution in highly uniform Bernal-stacked bilayer graphene, ACS Nano 14 (2020) 6834–6844.

[107] H.Q. Ta, et al., Stranski–Krastanov and Volmer–Weber CVD growth regimes to control the stacking order in bilayer graphene, Nano Letters 16 (2016) 6403–6410.

[108] Q. Li, et al., Growth of adlayer graphene on cu studied by carbon isotope labeling, Nano Letters 13 (2013) 486–490.

[109] G. Odahara, et al., In-situ observation of graphene growth on Ni(111), Surface Science 605 (2011) 1095–1098.

[110] M. Li, et al., Equilibrium shape of graphene domains on Ni(111), Physical Review B 88 (2013) 041402.

[111] A. Cabrero-Vilatela, et al., Towards a general growth model for graphene CVD on transition metal catalysts, Nanoscale 8 (2016) 2149–2158.

[112] Y. Takesaki, et al., Highly uniform bilayer graphene on epitaxial Cu–Ni(111) alloy, Chemistry of Materials 28 (2016) 4583–4592.

[113] Y. Uchida, et al., Controlled growth of large-area uniform multilayer hexagonal boron nitride as an effective 2D substrate, ACS Nano 12 (2018) 6236–6244.

[114] R.S. Weatherup, et al., Long-term passivation of strongly interacting metals with single-layer graphene, Journal of the American Chemical Society 137 (2015) 14358–14366.

[115] M. Piquemal-Banci, et al., Spin filtering by proximity effects at hybridized interfaces in spin-valves with 2D graphene barriers, Nature Communications 11 (2020) 5670.

[116] D.D. Nuzzo, et al., Graphene-passivated nickel as an efficient hole-injecting electrode for large area organic semiconductor devices, Applied Physics Letters 116 (2020) 163301.

[117] L. Persichetti, et al., Driving with temperature the synthesis of graphene on Ge(110), Applied Surface Science 499 (2020) 143923.

[118] P. Braeuninger-Weimer, et al., Crystal orientation dependent oxidation modes at the buried graphene–Cu interface, Chemistry of Materials 32 (2020) 7766–7776.

[119] A.T. Murdock, et al., Controlling the orientation, edge geometry, and thickness of chemical vapor deposition graphene, ACS Nano 7 (2013) 1351–1359.

[120] N. Mendelson, et al., Grain dependent growth of bright quantum emitters in hexagonal boron nitride, Advanced Optical Materials 9 (2021) 2001271.

[121] M. Losurdo, et al., Graphene CVD growth on copper and nickel: role of hydrogen in kinetics and structure, Physical Chemistry Chemical Physics 13 (2011) 20836–20843.

[122] G. Lu, et al., Synthesis of large single-crystal hexagonal boron nitride grains on Cu–Ni alloy, Nature Communications 6 (2015) 6160.

[123] T.-A. Chen, et al., Wafer-scale single-crystal hexagonal boron nitride monolayers on Cu (111), Nature 579 (2020) 219–223.

[124] B. Huet, J.-P. Raskin, Pressure-Controlled Chemical vapor deposition of single-layer graphene with millimeter-size domains on thin copper film, Chemistry of Materials 29 (2017) 3431–3440.

[125] I. Vlassiouk, et al., Graphene nucleation density on copper: fundamental role of background pressure, Journal of Physical Chemistry C 117 (2013) 18919.

[126] X. Chen, et al., Chemical vapor deposition growth of 5 mm hexagonal single-crystal graphene from ethanol, Carbon 94 (2015) 810–815.

[127] D.H. Seo, et al., Single-step ambient-air synthesis of graphene from renewable precursors as electrochemical genosensor, Nature Communications 8 (2017) 14217.

[128] M.S. Rosmi, et al., Synthesis of uniform monolayer graphene on re-solidified copper from waste chicken fat by low pressure chemical vapor deposition, Materials Research Bulletin 83 (2016) 573–580.

[129] X. Sun, et al., Low-temperature and rapid growth of large single-crystalline graphene with ethane, Small 14 (2018) 1702916.

[130] Y.M. Shi, et al., Synthesis of few-layer hexagonal boron nitride thin film by chemical vapor deposition, Nano Letters 10 (2010) 4134.

[131] K.K. Kim, et al., Synthesis of monolayer hexagonal boron nitride on cu foil using chemical vapor deposition, Nano Letters 12 (2012) 161–166.

[132] J. Lee, et al., Atomic layer deposition of layered boron nitride for large-area 2D electronics, ACS Applied Materials & Interfaces 12 (2020) 36688–36694.

[133] W. Auwarter, et al., Synthesis of one monolayer of hexagonal boron nitride on Ni(111) from B-trichloroborazine (ClBNH)$_{(3)}$, Chemistry of Materials 16 (2004) 343–345.

[134] A. Ismach, et al., Toward the controlled synthesis of hexagonal boron nitride films, Acs Nano 6 (2012) 6378–6385.

[135] S. Sonde, et al., Ultrathin, wafer-scale hexagonal boron nitride on dielectric surfaces by diffusion and segregation mechanism, 2D Materials 4 (2017) 025052.

[136] S. Caneva, et al., Controlling catalyst bulk reservoir effects for monolayer hexagonal boron nitride CVD, Nano Letters 16 (2016) 1250.

[137] Y. Stehle, et al., Synthesis of hexagonal boron nitride monolayer: control of nucleation and crystal morphology, Chemistry of Materials 27 (2015) 8041–8047.

[138] B.C. Bayer, et al., Introducing overlapping grain boundaries in chemical vapor deposited hexagonal boron nitride monolayer films, ACS Nano 11 (2017) 4521–4527.

[139] X. Wu, et al., Growth of continuous monolayer graphene with millimeter-sized domains using industrially safe conditions, Scientific Reports 6 (2016) 21152.

[140] H. Kim, et al., Modeling of the self-limited growth in catalytic chemical vapor deposition of graphene, NJPh 15 (2013).

[141] K. Verguts, et al., Single-layer graphene synthesis on a Al2O3(0001)/Cu(111) template using chemical vapor deposition, ECS Journal of Solid State Science and Technology 5 (2016), Q3060-Q3066.

[142] B. Wu, et al., Self-organized graphene crystal patterns, NPG Asia Materials 5 (2013).

[143] W. Guo, et al., Controlling fundamental fluctuations for reproducible growth of large single-crystal graphene, ACS Nano 12 (2018) 1778–1784.

[144] S. Choubak, et al., Graphene CVD: interplay between growth and etching on morphology and stacking by hydrogen and oxidizing impurities, Journal of Physical Chemistry C 118 (2014) 21532–21540.

[145] S. Choubak, et al., No graphene etching in purified hydrogen, Journal of Physical Chemistry Letters 4 (2013) 1100–1103.

[146] W. Guo, et al., Oxidative-etching-assisted synthesis of centimeter-sized single-crystalline graphene, Advanced Materials 28 (2016) 3152–3158.

[147] H. Zhou, et al., Chemical vapour deposition growth of large single crystals of monolayer and bilayer graphene, Nature Communications 4 (2013) 2096.

[148] L. Zhan, et al., Preparation of ultra-smooth Cu surface for high-quality graphene synthesis, Nanoscale Research Letters 13 (2018) 340.

[149] R. Addou, et al., Monolayer graphene growth on Ni(111) by low temperature chemical vapor deposition, Applied Physics Letters 100 (2012).

[150] K.-J. Peng, et al., Hydrogen-free PECVD growth of few-layer graphene on an ultra-thin nickel film at the threshold dissolution temperature, Journal of Materials Chemistry C 1 (2013) 3862–3870.

[151] I. Vlassiouk, et al., Large scale atmospheric pressure chemical vapor deposition of graphene, Carbon 54 (2013) 58–67.

[152] I. Vlassiouk, et al., Role of hydrogen in chemical vapor deposition growth of large single-crystal graphene, ACS Nano 5 (2011) 6069–6076.

[153] T. Ma, et al., Repeated growth–etching–regrowth for large-area defect-free single-crystal graphene by chemical vapor deposition, ACS Nano 8 (2014) 12806–12813.

[154] T. Wu, et al., Triggering the continuous growth of graphene toward millimeter-sized grains, Advanced Functional Materials 23 (2013) 198–203.

[155] W.-J. Joo, et al., Realization of continuous Zachariasen carbon monolayer, Science Advances 3 (2017) e1601821.

[156] S. Hong, et al., Ultralow-dielectric-constant amorphous boron nitride, Nature 582 (2020) 511–514.

[157] A. Antidormi, et al., Emerging properties of non-crystalline phases of graphene and boron nitride based materials, Nano Materials Science (2021).

[158] L. Wang, et al., Water-assisted growth of large-sized single crystal hexagonal boron nitride grains, Materials Chemistry Frontiers 1 (2017) 1836–1840.

[159] A. Ismach, et al., Carbon-assisted chemical vapor deposition of hexagonal boron nitride, 2D Materials 4 (2017) 025117.

[160] Z. Shi, et al., Vapor–liquid–solid growth of large-area multilayer hexagonal boron nitride on dielectric substrates, Nature Communications 11 (2020) 849.

[161] H. Tian, et al., Role of carbon interstitials in transition metal substrates on controllable synthesis of high-quality large-area two-dimensional hexagonal boron nitride layers, Nano Letters 18 (2018) 3352–3361.

[162] H. Tian, et al., Growth dynamics of millimeter-sized single-crystal hexagonal boron nitride monolayers on secondary recrystallized Ni (100) substrates, Advanced Materials Interfaces 6 (2019) 1901198.

[163] L. Chen, et al., Growth of uniform monolayer graphene using iron-group metals via the formation of an antiperovskite layer, Chemistry of Materials 27 (2015) 8230–8236.

[164] S. Chen, et al., Synthesis and characterization of large-area graphene and graphite films on commercial Cu–Ni alloy foils, Nano Letters 11 (2011) 3519–3525.

[165] X. Zhang, et al., Epitaxial growth of 6 in. single-crystalline graphene on a Cu/Ni (111) film at 750 °C via chemical vapor deposition, Small 15 (2019) 1805395.

[166] B. Deng, et al., Scalable and ultrafast epitaxial growth of single-crystal graphene wafers for electrically tunable liquid-crystal microlens arrays, Science Bulletin 64 (2019) 659–668.

[167] M. Huang, et al., Large-area single-crystal AB-bilayer and ABA-trilayer graphene grown on a Cu/Ni(111) foil, Nature Nanotechnology 15 (2020) 289–295.

[168] Y. Liu, et al., How low nucleation density of graphene on CuNi alloy is achieved, Advanced Science 5 (2018) 1700961.

[169] V.L. Nguyen, et al., Layer-controlled single-crystalline graphene film with stacking order via Cu–Si alloy formation, Nature Nanotechnology (2020).

[170] W. Ma, et al., Interlayer epitaxy of wafer-scale high-quality uniform AB-stacked bilayer graphene films on liquid Pt3Si/solid Pt, Nature Communications 10 (2019) 2809.

[171] T. Lin, et al., Self-regulating homogenous growth of high-quality graphene on Co–Cu composite substrate for layer control, Nanoscale 5 (2013) 5847–5853.

[172] J. Meng, et al., Controlled growth of unidirectionally aligned hexagonal boron nitride domains on single crystal Ni (111)/MgO thin films, Crystal Growth & Design 19 (2019) 453–459.

[173] Y. He, et al., Growth of high-quality hexagonal boron nitride single-layer films on carburized Ni substrates for metal–insulator–metal tunneling devices, ACS Applied Materials & Interfaces 12 (2020) 35318–35327.

[174] B.Y. Dai, et al., Rational design of a binary metal alloy for chemical vapour deposition growth of uniform single-layer graphene, Nature Communications 2 (2011).

[175] R. Addou, et al., Graphene on ordered Ni-alloy surfaces formed by metal (Sn, Al) intercalation between graphene/Ni(111), Surface Science 606 (2012) 1108–1112.

[176] A. Varykhalov, et al., Electronic and magnetic properties of quasifreestanding graphene on Ni, Physical Review Letters 101 (2008) 157601.

[177] D. Farías, et al., Synthesis of a weakly bonded graphite monolayer on Ni(111) by intercalation of silver, Journal of Physics. Condensed Matter 11 (1999) 8453–8458.

[178] Y.S. Dedkov, et al., Intercalation of copper underneath a monolayer of graphite on Ni(111), Physical Review B 64 (2001) 035405.

[179] D.C. Geng, et al., Uniform hexagonal graphene flakes and films grown on liquid copper surface, Proceedings of the National Academy of Sciences of the United States of America 109 (2012) 7992–7996.

[180] Y. Fan, et al., Crack-free growth and transfer of continuous monolayer graphene grown on melted copper, Chemistry of Materials 26 (2014) 4984–4991.

[181] J.-i. Fujita, et al., Near room temperature chemical vapor deposition of graphene with diluted methane and molten gallium catalyst, Scientific Reports 7 (2017) 12371.

[182] J.S. Lee, et al., Wafer-scale single-crystal hexagonal boron nitride film via self-collimated grain formation, Science 362 (2018) 817–821.

[183] S. Caneva, et al., Nucleation control for large, single crystalline domains of monolayer hexagonal boron nitride via Si-doped Fe catalysts, Nano Letters 15 (2015) 1867–1875.

[184] M. Zheng, et al., Metal-catalyzed crystallization of amorphous carbon to graphene, Applied Physics Letters 96 (2010) 063110.

[185] Z. Shi, et al., Controlled synthesis of uniform multilayer hexagonal boron nitride films on Fe_2B alloy, RSC Advances 9 (2019) 10155–10158.

[186] G. Zhao, et al., The physics and chemistry of graphene-on-surfaces, Chemical Society Reviews 46 (2017) 4417–4449.

[187] J.W. Suk, et al., Transfer of CVD-grown monolayer graphene onto arbitrary substrates, ACS Nano 5 (2011) 6916–6924.

[188] X. Liang, et al., Toward clean and crackless transfer of graphene, ACS Nano 5 (2011) 9144–9153.

[189] M. Chen, et al., Advances in transferring chemical vapour deposition graphene: a review, Materials Horizons 4 (2017) 1054–1063.

[190] A.V. Zaretski, et al., Metal-assisted exfoliation (MAE): green, roll-to-roll compatible method for transferring graphene to flexible substrates, Nanotechnology 26 (2015).

[191] G. Zhang, et al., Versatile polymer-free graphene transfer method and applications, ACS Applied Materials & Interfaces 8 (2016) 8008–8016.

[192] R. Wang, et al., Catalyst interface engineering for improved 2D film lift-off and transfer, ACS Applied Materials & Interfaces 8 (2016) 33072–33082.

[193] L.B. Gao, et al., Repeated growth and bubbling transfer of graphene with millimetre-size single-crystal grains using platinum, Nature Communications 3 (2012) 699.

[194] K. Verguts, et al., Controlling water intercalation is key to a direct graphene transfer, ACS Applied Materials & Interfaces 9 (2017) 37484–37492.

[195] K. Verguts, et al., Graphene delamination using 'electrochemical methods': an ion intercalation effect, Nanoscale 10 (2018) 5515–5521.

[196] S.Y. Yang, et al., Metal-etching-free direct delamination and transfer of single-layer graphene with a high degree of freedom, Small 11 (2015) 175–181.

References

157

[197] T. Yoon, et al., Direct measurement of adhesion energy of monolayer graphene as-grown on copper and its application to renewable transfer process, Nano Letters 12 (2012) 1448–1452.

[198] A. Shivayogimath, et al., Do-it-yourself transfer of large-area graphene using an office laminator and water, Chemistry of Materials 31 (2019) 2328–2336.

[199] F. Pizzocchero, et al., Non-destructive electrochemical graphene transfer from reusable thin-film catalysts, Carbon 85 (2015) 397–405.

[200] P.R. Whelan, et al., Raman spectral indicators of catalyst decoupling for transfer of CVD grown 2D materials, Carbon 117 (2017) 75–81.

[201] Y. Zhang, et al., Properties of undoped few-layer graphene-based transparent heaters, Materials 13 (2020) 104.

[202] L. Sun, et al., A force-engineered lint roller for superclean graphene, Advanced Materials 31 (2019) 1902978.

[203] J. Li, et al., Growth of polar hexagonal boron nitride monolayer on nonpolar copper with unique orientation, Small 12 (2016) 3645–3650.

[204] I. Jo, et al., Tension-controlled single-crystallization of copper foils for roll-to-roll synthesis of high-quality graphene films, 2D Materials 5 (2018) 024002.

[205] L. Wang, et al., Epitaxial growth of a 100-square-centimetre single-crystal hexagonal boron nitride monolayer on copper, Nature 570 (2019) 91–95.

[206] B. Deng, et al., Wrinkle-free single-crystal graphene wafer grown on strain-engineered substrates, Acs Nano 11 (2017) 12337–12345.

[207] M. Wu, et al., Seeded growth of large single-crystal copper foils with high-index facets, Nature 581 (2020) 406–410.

[208] G.R. HARP, Thin Films: Heteroepitaxial Systems, pp. 117–166.

[209] K. Verguts, et al., Epitaxial Al2O3(0001)/Cu(111) template development for CVD graphene growth, Journal of Physical Chemistry C 120 (2016) 297–304.

[210] J. Meng, et al., Aligned growth of millimeter-size hexagonal boron nitride single-crystal domains on epitaxial nickel thin film, Small 13 (2017) 1604179.

[211] W. Auwärter, et al., Defect lines and two-domain structure of hexagonal boron nitride films on Ni(111), Surface Science 545 (2003) L735–L740.

[212] O.J. Burton, et al., Integrated wafer scale growth of single crystal metal films and high quality graphene, ACS Nano 14 (2020) 13593–13601.

[213] S.N. Alamri, The growth of CdTe thin film by close space sublimation system, Physica Status Solidi A 200 (2003) 352–360.

[214] X. Xue, et al., Gas-flow-driven aligned growth of graphene on liquid copper, Chemistry of Materials 31 (2019) 1231–1236.

[215] T. Kobayashi, et al., Production of a 100-m-long high-quality graphene transparent conductive film by roll-to-roll chemical vapor deposition and transfer process, Applied Physics Letters 102 (2013) 023112.

[216] T.T. Tran, et al., Robust multicolor single photon emission from point defects in hexagonal boron nitride, ACS Nano 10 (2016) 7331–7338.

CHAPTER

6

Realization of electronic-grade two-dimensional transition metal dichalcogenides by thin-film deposition techniques

Yu-Chuan Lin, Riccardo Torsi, Nicholas A. Simonson, Azimkhan Kozhakhmetov, and Joshua A. Robinson

The Pennsylvania State University, University Park, PA, United States

6.1 Current challenges in transition metal dichalcogenide synthesis

"Smaller would be better" perhaps describes the general trend of the semiconductor technology development the best. The advances of semiconductor technology have been driven by device dimensional scaling toward improved device performance and reduced cost. The intense miniaturization of a transistor's physical dimensions and the adoption of equivalent scaling that utilizes strained silicon [1], advanced fin field effect transistor (FinFET) structure [2], and high-k gate dielectrics and metal gates have managed to keep Moore's law validated for more than five decades [3] and improve the key metrics in a power performance area analysis, as well as reduced cost [4,5]. Currently, the 10 nm technology node in Intel's production line has a silicon fin width of about 7 nm and a gate length of 18 nm [5]. As the scaling endeavor continues into sub-10 nm technology node, several challenges in three-dimensional (3D) semiconductor arise such as surface roughness, thickness variation, and the influence of surface dangling bonds. These imperfections will unavoidably scatter charge carriers within the transistor channel and worsen with smaller transistor body thickness (t_b), resulting in degraded device performance. For example, with t_b below 5 nm, the carrier mobility (μ) is scaled with the sixth power of t_b ($\mu \propto t_b^6$) [6] due to thickness variation-induced scattering in ultrathin 3D semiconductors. On the other hand, semiconducting two-dimensional (2D) materials like transition metal dichalcogenides (TMD) address material-related issues that ultrathin 3D semiconductors are encountering. They have near zero surface roughness and have the

Defects in Two-Dimensional Materials
https://doi.org/10.1016/B978-0-12-820292-0.00012-4

Copyright © 2022 Elsevier Inc. All rights reserved.

promise of much lower dangling bonds on the films surfaces [7]. Their room-temperature μ can range from tens to hundreds of cm^2/Vs and do not show a strong correlation with the thickness variation when t_b is < 3 nm [4,5]. Additionally, the subthreshold swing of 2D is predicted to be sufficiently small with the gate length below 10 nm [8], which can conveniently compensate for the limitations of Si electronics in this ultrathin regime [5]. Therefore, the use of mono- and few-layer (< 3 nm) semiconducting TMD could open a pathway towards 5 nm technology node and beyond.

So far, the state-of-the-art electronic properties of 2D transistors cited in the literature or future device roadmaps typically come from exfoliated TMD bulk crystals [4,5,9]. Bulk crystals exhibit low intrinsic defects or interfacial disorder and hence are popular options for the proof-of-the-concept experiments. Even though exfoliated 2D crystals have no hope for wafer-scale production, their material quality and device performance has set "electronic-grade" standards for atomically thin synthetic TMDs. Taking WSe_2 crystals as an example, the state-of-the-art bulk crystals exhibit a surface defect density below 10^{11} cm^{-2} [10], which is one order lower than the density of WSe_2 monolayer grown by state-of-the-art metal-organic chemical vapor deposition (MOCVD) [11]. Compared to bulk crystal synthesis, CVD and other non-equilibrium synthesis techniques would generate other types of defects seldomly seen in the bulk TMD such as antisite defects [12] (e.g., chalcogen constituent sites substituted by transition metal atoms) that would cause carrier scattering in a transistor channel. As a result, the electronic properties of 2D TMD films made by thin-film deposition methods including MOCVD, molecular beam epitaxy (MBE), and atomic layer deposition (ALD) differ from the properties of the exfoliated crystals significantly.

Significant progress is being made in scalable growth of high-quality TMD by thin-film deposition techniques with *in situ* growth control. High-quality, large-area 2D crystals are essential for the next-generation semiconductor technology and a broad range of practical applications, therefore it is important to tackle the following significant challenges encountered in thin-film techniques. First, an underlying substrate dictates how nuclei are formed, grow laterally, and merge into a coalesced film. Single-crystal substrates such as sapphire are widely utilized for wafer-scale synthesis of atomically thin epitaxial TMD films [13]. The epitaxial relationship of the single-crystal substrate with the growing films is essential to minimize rotationally misaligned grain boundaries that severely deteriorate the electronic transport properties [14]. However, if the substrate is polycrystalline or amorphous, the large area homogeneity and uniformity are lost, limiting the scalable synthesis of single-crystal 2D layers to single-crystal substrates. It is important to note that if the substrate is not the same as that used for device fabrication, the films will have to be transferred to other substrates. However, current state-of-the-art transfer processes yield undesired residues on the surface and require significant improvement. Second, accurate thickness control remains a big challenge. Even though fully coalesced, monolayer films are achieved, secondary nucleation on top of the first layer cannot be fully eliminated. Furthermore, while monolayer films are of great interest for photovoltaics due to the transition from indirect to direct bandgap [15], thicker films (3–4 layers) are needed for other applications, and large-area uniformly thick films are needed and are still very challenging to achieve. Developing new engineering approaches such as self-limiting growth or selective area etching is necessary to achieve reliable scalable synthesis with atomic precision thickness control. Lastly, the growth temperature is equally important as is the substrate. High-growth temperatures ($> 800\,°C$) are necessary to obtain high-quality crystalline films due to the kinetic nature of the growth process. Since sticking coefficients of metal and chalcogen adatoms are different and a direct function of the substrate temperature [13], elevated growth temperatures increase the probability of chalcogen desorption, leading to a high density of chalcogen vacancies [11]. Moreover, high temperatures limit possible substrate choices since not all surfaces (e.g., low dielectric constant oxide materials coated on Si) are thermally stable above about 800 °C. There is an opportunity to develop possible methods for reducing growth temperature including ALD [16], *in situ* plasma sources [17], various growth promoters [18], and even room-temperature solution processes [19].

6.2 Current synthesis techniques

6.2.1 Reactor design

Vapor-phase synthesis reactors are composed of three main components (*i*) source delivery systems (*ii*) deposition chamber, and (*iii*) exhaust mechanism. The purpose of this section is to present the most common source delivery system and deposition chamber designs used to synthesize 2D transition metal dichalcogenides and discuss their impact on resulting film quality. For a review on exhaust systems, the reader is encouraged to read sections in thin-film deposition books that discuss vacuum technology like *"Materials Science of Thin Films"* by Milton Ohring, *"Thin-Film Deposition: Principles and Practice"* by Donald Smith, or *"A User's Guide to Vacuum Technology"* by John O'Hanlon.

The purpose of source delivery systems is to transport precise amounts of reactants to the growth substrate in a reliable fashion. In solid-source chemical vapor deposition (SS-CVD) systems, carrier gases are flowed over heated high-purity powders held in crucibles. Carrier gases commonly used in SS-CVD are Ar, N_2, and H_2; where H_2 has been used to enhance precursor reduction [20–22]. Carrier gases are held in pressure regulated gas cylinders connected to the chamber by stainless-steel tubing. Before reaching the chamber, the gas lines typically go through gas filters and mass flow controllers (MFCs) [23]. MFCs are a fundamental component of any vapor deposition system for precisely controlling the amount of carrier gas flowing into the chamber and, ultimately, the precursor flux to the substrate. It is important for the researcher to select the appropriately rated MFC based on the target gas flow range and gas type. A good design practice is to place various valves between the various components of the gas delivery system prior to the intake manifold to facilitate leak-detection [23]. In SS-CVD systems the source powders are generally located upstream of the substrate either inside the growth chamber or, outside of it, in an independently controlled zone, where the temperature can be regulated by using PID (proportional integral derivative) (Fig. 6.1a). In the ubiquitous SS-CVD growth scheme for MoS_2 that employs MoO_3 and S powders, MoO_3 is held in the growth chamber heated to temperatures typically within 700–900 °C while S is located outside of the reactor chamber further away from the growth substrate where it is heated at 130–200 °C by heating belts [14,18,24–29]. The difference in temperature between the two powder sources is a consequence of the vapor pressure of MoO_3 is much lower than that of S [14]. Other than the source's vapor pressure, careful consideration must also be given to the positioning of the source powders with respect to the source substrates as nucleation density and final film morphology are highly dependent on pressure [18]. During growth, both unreacted sublimed source powders and particles generated by gas phase reaction will redeposit on the inner wall surfaces at the furnace downstream where the temperature is near room temperature. The advantage of gas-source CVD (GS-CVD) over SS-CVD is its ability to introduce precursors in the chamber independently and its higher control over precursor flux [30].

Metal organic chemical vapor deposition (MOCVD) has been widely explored for synthesizing uniform and large-area TMD [11,30–33]. In MOCVD, metal-organics (MO) precursors are employed due to their relatively high vapor pressures with respect to other metal-containing compounds [34]. They are stored in bubblers outside of the growth chamber such that their vapor pressure can be regulated by external temperature and/or background gas. The carrier gasses (Ar N_2, and H_2) are flowed (or bubbled) through the bubbler by MFCs and deliver the vapor source into the growth chamber. Early reports utilized MO precursors such as diethyl sulfide [30] and dimethyl selenium [32] for chalcogen sources. The use of hydrate-based sources including H_2S and H_2Se is also common and might be more preferrable due to a higher than a factor of 10 reduction in carbon contamination, and larger demonstrated grain sizes [35]. However, H_2S and H_2Se are highly toxic and need to be handled with great caution. Various levels of engineering controls such as gas alarms and automatic shut-off valves must be installed for safe operation. Background gasses are also introduced into MOCVD systems in addition to the source carrier gas to build up a background pressure and establish controlled laminar flows. Hybrid SS- and MO-CVD systems that have both solid- and gas-precursor sources are also used to grow TMD crystals [36,37].

Vapor-phase synthesis reactors can be divided into categories based on their geometry: vertical *vs.* horizontal and their heating mechanism and hot-wall *vs.* cold-wall [38–40]. In the vertical reactor design, substrates are placed in a susceptor typically made of graphite and perpendicular to the gas flow. To obtain more uniform films over the substrate surface susceptor rotation (clockwise *vs* counterclockwise) as well as showerhead designs are commonly utilized [41,42]. A simple schematic of a vertical, cold-wall MOCVD system used for syntheis of 2D TMD is shown in Fig. 6.1b. In horizontal reactors, the precursor flow is parallel to substrate. Since this geometry inherently results in a precursor gradient over the substrates, in some cases the sample is tilted in the direction of the flow to improve film uniformity [43]. Advantages of the vertical geometry over the horizontal geometry are faster deposition rate, higher precursor efficiency, and better film uniformity due to the uniform boundary layer thickness on top of the substrate and a more homogenous temperature field. Vertical reactor chambers, however, have a more complex flow system in which it is more difficult to maintain laminar flow. Additionally, there is also a possibility for convection currents to carry gas containing spent reaction products back towards the inlet gas flow. Nevertheless, vertical CVD systems used in both semiconductor industries and academic research [42] typically have a pancake-type inductive heater for substrate heating and a showerhead for feeding precursors into the chamber to improve film uniformity. Advantages of horizontal reactor design are its simplicity and a gas flow dynamic that is easier to model. The non-uniformity of horizontal systems originates from the non-homogeneous boundary layer thickness over the substrates which results in anisotropic growth rates with respect to distance to the source [44]. This results in the presence of growth plume with a nonuniform grain size and film coalescence on the growth substrate [18,25]. MOCVD systems, on the other hand, can be either horizontal or vertical. Another approach to categorize vapor-phase synthesis reactors is based on their heating mechanism. As the name suggests, in hot-wall systems the entire reactor tube is heated typically through resistive heating. Contrarily, in cold-wall systems only the susceptor is heated. This is usually accomplished mainly through inductive heating and resistive heating, although other methods such as laser heating [45] could also be used. The advantage of this design choice is that no parasitic reactions occur on the chamber walls if its temperature is kept higher than the precursor sublimation temperature to prevent condensation. While parasitic reactions may occur in hot-wall systems, they are much simpler to set-up and the temperature profile within the reaction chamber is more easily controlled.

Molecular beam epitaxy (MBE) systems operate in ultra-high vacuum (UHV) and use a much lower precursor flux, compared to SS-CVD and MOCVD systems. A schematic of a standard MBE systems is illustrated in Fig. 6.1c. The main components of any MBE chamber are (*i*) molecular beam sources, (*ii*) sample manipulator, and (*iii*) vacuum pumping systems. High-purity source materials are contained in effusion cells for which the temperature can be regulated with high-precision ($\pm 1\,°C$) by thermocouples and proportional-integral-derivative (PID) controllers. Based on the source material, by adjusting the temperature of the source material and the pressure of the cell, a precise amount of precursor can be delivered to the substrate. Quartz crystal monitors are often used to accurately measure the flux from effusion cells. Mechanical shutters are mounted on top of effusion cells to effectively turn off precursor flux instantly. An electron beam is also used in some instances to evaporate the source material especially if high-growth rates are desired or if the source material has a high melting point like refractory metals. Due to the UHV conditions, the evaporated material behaves more like a beam of molecules (hence the name of the technique) than a gas. The beam expands from the effusion cell orifice to the growth substrate where condensation occurs. Substrates are mounted on sample manipulators which are heated to the growth temperature to aid in crystallization. Some systems have rotating sample manipulators to improve film uniformity over large substrate areas. Pumping systems in MBE tools are often combinations of cryogenic adsorption and turbomolecular pumps as well as cryogenic panels to further increase pumping speed. The stainless-steel growth chamber is often also isolated from other portions of the system by gate valves and load-lock modules to minimize base pressure and avoid contamination. An advantage that MBE systems have over SS-CVD and MOCVD chambers is *in situ* reflection high energy electron diffraction (RHEED) that can be readily incorporated into the UHV chamber.

6.2.2 Solid-source chemical vapor deposition (SS-CVD)

A widely used and straightforward method for the controllable synthesis of transition metal dichalcogenides is SS-CVD, where solid precursors (a metal source and a chalcogen source) are placed within the hot zones of a tube furnace and vaporized for deposition/reaction on a desired substrate. The process of SS-CVD is based on five main steps: evaporation, transport, diffusion, nucleation, and growth [48]. Metal and chalcogen precursors are typically evaporated/volatilized in corresponding hot zones (e.g. 700–900 °C for metal oxides, 130–450 °C for chalcogens [29,49]), and a carrier gas such as argon is flowed through the reaction chamber, thus transporting the evaporated/volatilized precursors to the growth substrate (Fig. 6.2a) [29,50]. This method can also be referred to as solid source chemical vapor

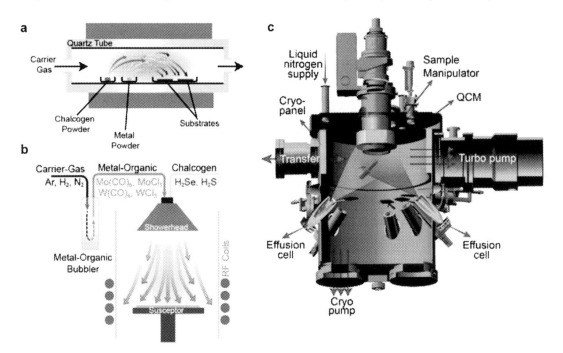

FIGURE 6.1 Summary of growth techniques used for large-area synthesis of semiconducting TMD: (a) Horizontal, hot-wall SS-CVD system. Reproduced with permission [46]. Copyright 2019, IOP Publishing. (b) Vertical, cold-wall MOCVD system. Reproduced with permission [46]. Copyright 2019, IOP Publishing, and (c) MBE system used for synthesis of 2D TMD. Reproduced with permission [47]. Copyright 2017, Wiley-VCH.

deposition. As the precursor atoms adsorb onto the substrate surface, they undergo surface diffusion as adatoms, eventually reaching a saturation concentration and segregating to form nuclei. Further migration of adatoms to the edges of these nuclei allows for grain growth, leading to the micron-scale domains that are the hallmark of SS-CVD. If these domains are allowed to grow until they coalesce, a continuous film can be formed, and secondary nuclei will usually begin to form [48]. The first notable use of this method for use in TMD growth is in Lee's seminal work on the synthesis of MoS_2 in 2012 [50]. Since then, dozens of TMD have been grown *via* this method, including sulfides, selenides, and tellurides of Mo, W, Ti, Hf, Zr, V, Nb, Ta, Pt, Re and Pd among other materials [49]. In SS-CVD of TMD, the defining factor has been the direct heating of solid sources within the furnace to form vapors, which are transported to the substrate surface to redeposit and form the crystalline 2D materials.

Solid-source CVD has several advantages over other TMD synthesis techniques. The domains made using solid sources repeatedly reach tens or hundreds of microns in lateral dimensions [29,49,51]. An additional advantage is the simplicity and straightforward nature and ease of implementation of this technique in a laboratory environment. A tube furnace, a carrier gas, and high purity solid sources are the basic components for a rudimentary setup, although fine pressure and gas flow control are often required as well. A separate upstream heating zone, accomplished via heat tape or a multi-zone furnace, can be used to account for the difference in vaporization temperatures between the metal and chalcogen precursors (Fig. 6.1a–b) [29]. Such simplicity is coupled with a correspondingly low cost, making these types of systems far more accessible than a state-of-the-art MOCVD or MBE systems. Also, solid-source precursors are typically safer to work with than, for example, metalorganic/pyrophoric precursors, often due to nearly negligible vapor pressures at room temperature. Thus, simple ventilation systems and appropriate personal protective equipment (PPE) can be used instead of costly and complicated advanced safety infrastructure, such as scrubbers/waste mitigation measures and toxic gas monitoring systems.

Conversely, there are some notable shortcomings for this technique. Due to the narrow temperature and precursor reaction regime range, the region where optimal, large domains are found is limited. In fact a "plume" is usually observed where the precursor concentration is at its highest (Fig. 6.2c) [29]. Within the plume, larger domains could be found at particular spots (Fig. 6.2d–e). This limit to the uniformity across a sample usually precludes SS-CVD from truly wafer-scale synthesis. For this reason, it might be more fit for fundamental studies on film characteristics and for proof-of-the-concept experiments than for a scalable synthesis and mass production required for practical industrial applications. Additionally, because the partial pressure of each precursor in the reaction chamber is dependent on the temperature of its respective hot zone (versus a nearly instantaneous introduction/cessation of flow in MOCVD and MBE), SS-CVD lacks the fine control over gas-phase reactor chemistry present in some other synthesis techniques like MBE or MOCVD [52]. This most commonly manifests as a gradual decrease in vapor pressure of one precursor relative to the other, and it is impossible to control the metal-to-chalcogen ratio during growth, or to keep it constant during ramping and cooling.

6.2.3 Metal-organic chemical vapor deposition (MOCVD)

Metal-organic chemical vapor deposition utilizes chemical compounds with low to moderate vapor pressure as precursors. [13,53] The precursors that are used for the synthesis of two-dimensional (2D) layered TMD (with the general formula MX_2, where M=W, Mo, and Nb, and X=S, Se, and Te) include transition metal carbonyls ($Mo(CO)_6$ and $W(CO)_6$), halides ($MoCl_5$, WCl_6, and $NbCl_5$), organo-chalcogen compounds ((C_2H_5)S, (C_4H_9)S, (CH_3)Se, (C_2H_5)Se, (C_4H_9)Se, (C_2H_5)$_2$Te, and (C_3H_7)$_2$Te), and hydrides (H_2S and H_2Se) [13,46]. Similarly, for layered topological insulators Bi_2Se_3 and Bi_2Te_3, (CH_3)$_3$Bi is an ideal precursor for providing Bi. Several compounds in the liquid form have also been used for the synthesis of TMD monolayers [54,55], including (t-BuN=)$_2$W(NMe$_2$)$_2$ and (t-BuN=)$_2$Mo(NMe$_2$)$_2$ for W and Mo, and (t-C$_4$H$_9$)$_2$S$_2$ and (C_2H_5)$_2$Se$_2$ for S and Se. There are also reports that grow *hexagonal* boron nitride (h-BN) layers by MOCVD using NH_3 and (C_2H_5)$_3$B [56]. Organometallics' vapor pressures are generally higher by many orders of magnitude compared with their oxide counterparts. This relatively high vapor pressure enables one to control the flux ratios precisely during growth by placing the organometallic compounds in bubbler manifolds controlled by temperature and carrier gases. Therefore, MOCVD is arguably more reliable than SS-CVD for synthesizing TMD layers in terms of controlling nucleation density, growth rate, spatial homogeneity, and domain geometry. TMD atomically thin films made by this technique exhibit excellent spatial homogeneity in terms of surface continuity and thicknesses. For example, Kalanyan *et al.* used pulsed MOCVD to grow MoS_2 films with controllable

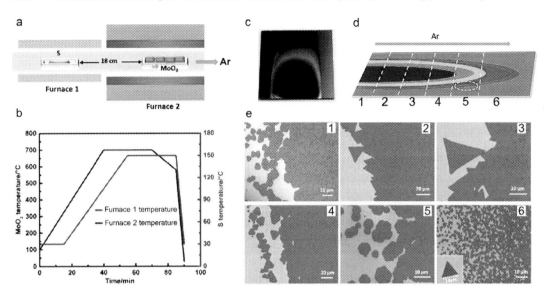

FIGURE 6.2 (a) An illustration of powder-based chemical vapor deposition furnace used for synthesis of MoS$_2$. (b) The temperature profiles of Furnace 1 and Furnace 2 for S and MoO$_3$ evaporation, respectively, in (a). (c) A camera image of the pattern of deposited MoS$_2$ on a thermally oxidized Si wafer. (d) A schematic diagram of the "plume" along with a photograph of the same, depicting the lack of deposition uniformity across the substrate. (e) Scanning electron micrographs of various points on the substrate, depicting differences in domain morphology, but also the large domains and film coalescence possible with SS-CVD. Reproduced with permission. [29] Copyright 2014 American Chemical Society.

thicknesses on 2″ quartz substrates (Fig. 6.3a–b) [55]. Additionally, vapor-phase epitaxial WS$_2$ and MoS$_2$ with good electronic performance are grown by MOCVD at high temperature (e.g., 1000 °C), in a chalcogen-rich environment on 2″ single-crystal sapphire (Fig. 6.3c–d) [11,57], as demonstrated by Chubarov *et al.* [58]. On the other hand, at low temperatures (≤ 400 °C), polycrystalline TMD few-layers with small domains dominate the morphology when grown on non-crystalline substrates, including SiO$_2$/Si [33] at 470 °C, pure glass substrates [59] at 400 °C, and a hybrid SiO$_2$/polymer surface [60] at 250 °C. One advantage of MOCVD is the ability to deliver multiple precursors and/or switch them intermittently during growth. This allows one to fabricate lateral single-crystal 2D epitaxial heterojunctions with atomically sharp interfaces. For example, Kobayashi *et al.* demonstrated MOCVD grown MoS$_2$/WSe$_2$ and WS$_2$/MoSe$_2$ lateral heterojunctions using liquid metal-organic precursors (Fig. 6.3e–f) [54]. Similarly, Xie *et al.* fabricated 2D (WSe$_2$/WS$_2$)$_n$ epitaxial superlattices on SiO$_2$ by MOCVD and engineered the interface strain by controlling the width of each grown constituent during synthesis to tailor the optical properties of WSe$_2$ and WS$_2$ (Fig. 6.3g–h) [61].

6.2.4 Molecular beam epitaxy (MBE)

Molecular beam epitaxy is another industrially accepted deposition method that has been actively utilized to produce electronic-grade III-V and II-VI semiconductors. MBE can be

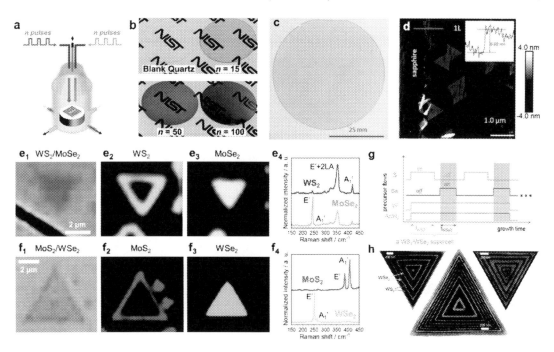

FIGURE 6.3 (a) A vertical MOVCVD system with pulsed precursor delivery. Reproduced with permission [55]. Copyright 2017, American Chemical Society (b) 2″ fused quartz demonstrates wafer-scale growth of MoS$_2$ deposited with increasing numbers of precursor pulses. Reproduced with permission [55]. Copyright 2017, American Chemical Society (c) Photograph of a WS$_2$ film deposited on a 6″ c-plane sapphire wafer using multistep process with varying temperature in MOCVD. Reproduced with permission [58]. Copyright 2021, American Chemical Society. (d) Atomic force micrograph of a WS$_2$ film showing exposed sapphire surface at the left, monolayer coverage with some bilayer lands. The inset shows the height profile along red line in the top left corner corresponding to monolayer thickness. Reproduced with permission [58]. Copyright 2021, American Chemical Society. (e–f) MOCVD grown monolayer heterojunctions of two constituents: (e$_1$) Optical image, (e$_2$–e$_3$) PL intensity maps (1.94 to 2.07 eV for WS$_2$ and 1.46 to 1.65 eV for MoSe$_2$) and (e$_4$) Raman spectra of a WS$_2$-MoSe$_2$ heterostructure. (f$_1$) Optical image, (f$_2$–f$_3$) PL intensity maps (1.77 to 1.94 eV for MoS$_2$ and 1.55 to 1.77 eV for WSe$_2$) and (f$_4$) Raman spectra of a MoS$_2$-WSe$_2$ heterostructure. Reproduced with permission [54]. Copyright 2019, American Chemical Society. (g) Time sequence of the modulated superlattice growth where the growth time for a WS$_2$-WSe$_2$ supercell is t_{WS2} and t_{WSe2}. (h) Scanning electron micrograph of three monolayers WS$_2$-WSe$_2$ superlattices comprised of various widths of WSe$_2$ and WS$_2$ controlled by the modulated program shown in (g). Scale bars: 200 nm. Adapted from Xie *et al.* [62].

simply considered as an advanced form of vacuum evaporation (physical vapor deposition) process with excellent control over component partial pressures, layer thickness, and doping concentration. Ultra-high vacuum environment is an indispensable part of the process and allows in-situ monitoring of thickness, crystallinity, the chemical composition of thin films via reflective high-energy electron diffraction (RHEED), X-ray photoelectron spectroscopy (XPS), low-energy electron diffraction (LEED), and scanning tunneling microscopy (STM). Pioneering work on epitaxial growth of 2D TMDs, led by Koma *et al.*, showed homoepitaxial and heteroepitaxial growth of NbS$_2$, NbSe$_2$, MoSe$_2$ on van der Waals (SnS$_2$, MoSe$_2$) and non-van der Waals surfaces (mica, CaF$_2$, GaAs). [63–66] This was the first demonstration of van der

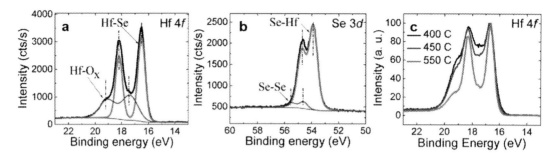

FIGURE 6.4 Chemical compositional analyses of 2D HfSe$_2$ on HOPG substrates. XPS spectra of (a) Hf 4f and (b) Se 3d grown at 550 °C after post-growth air exposure. (c) Core level XPS spectra of HfSe$_2$ synthesized at different temperatures. Chemically stable and more oxidation resistant films are achieved upon increasing the substrate temperature. Reproduced with the permission [67]. Copyright 2014, American Chemical Society.

Waals epitaxy on lattice-mismatched substrates and has been actively re-investigated over the last decade. Yue et al., successfully synthesized 2D HfSe$_2$ on highly oriented pyrolytic graphite (HOPG) and MoS$_2$ substrates [67]. Even though a lattice mismatch as large as >40% is expected, films do not show increased misfit dislocation density, further confirming the benefits of van der Waals (vdW) epitaxial growth of 2D HfSe$_2$ on lattice mismatched 2D surfaces. Furthermore, the chemical stability under ambient conditions of 2D HfSe$_2$ on HOPG is investigated as a function of growth temperature. As the substrate temperature is raised from 400 °C to 550 °C, HfO$_x$ is significantly decreased indicating that films obtained at higher temperatures are more resistant to oxidation and chemically stable due to improved crystal quality (Fig. 6.4a–c) [67]. Similarly, successful realization of other members of the TMD family such as MoSe$_2$ [68,69], WSe$_2$ [70,71], WTe$_2$ [71,72], SnSe$_2$ [73], HfTe$_2$ [74], PtSe$_2$ [75], etc. has been reported on various substrates at a temperature range from 200 °C to 900 °C. While the aforementioned reports may be a first demonstration of 2D thin film growth by MBE, the crystallinity and electrical properties of the films are still far from being considered as acceptable for the semiconductor industry requiring further process optimization to achieve desired level of maturity.

The main challenge encountered by all TMD crystals synthesized by MBE is the presence of various types of point, line, and planar defects. Among all, chalcogen vacancies are the most prevalent due to the difference in sticking coefficients of metal and chalcogen sources at a given substrate temperature [76]. Hence, to achieve near stoichiometric films it is a common practice to utilize very high chalcogen to metal ratios [76]. However, since elemental precursors are used in Knudsen cells, the evaporated flux of chalcogen (X) atoms consists of a range of clusters dominated by X$_2$ dimers which have the largest dissociation energy among all. As a result, the efficiency of obtaining atomic chalcogen species is reduced precluding the formation of stoichiometric films. Recent work by Walsh et al., reported the so-called "beam-interruption" method to synthesize WTe$_2$ films where W flux is sequentially introduced during the growth while Te flux is kept constant. [77] WTe$_2$ synthesized via conventional MBE on HOPG substrates is polycrystalline and Te deficient (Fig. 6.5a–b), whereas films were grown via beam interrupted MBE exhibit improved properties. Interrupting the chalcogen allows more time for Te atoms to react with W and as a result, the obtained films

6.2. Current synthesis techniques 169

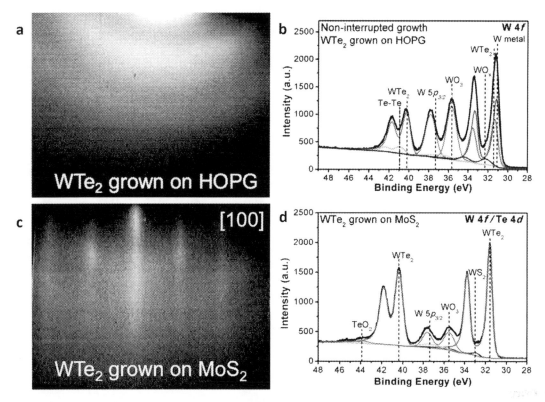

FIGURE 6.5 Structural and chemical compositional analyses of WTe$_2$ films grown on HOPG and MoS$_2$. (a) RHEED pattern of WTe$_2$ on HOPG synthesized without beam interruption demonstrating diffused Debye rings confirming polycrystalline nature of the films. (b) Corresponding XPS spectra of the sample where Te 4d, W 4f, and W 5$p_{3/2}$ core levels show sub-stoichiometric films with reduced Te content. (c) RHEED pattern of WTe$_2$ on MoS$_2$ along [100] direction synthesized with beam interruption in which observed streaks indicate a presence of rotationally aligned crystalline film. (d) The core-level XPS spectra of WTe$_2$/MoS$_2$ confirms a stoichiometric WTe$_2$ crystal can be achieved with the beam interruption method. Reproduced with permission [78]. Copyright 2017, IOP Publishing.

are rotationally aligned with the underlying substrates and nearly stoichiometric with Te: W ratio of 2:1 (Fig. 6.5c–d) [77].

Gas-phase reactions are minimized in MBE where the incoming flux of metal and chalcogen atoms allowed to react on the substrate surface and form a nucleus. Nuclei will grow and form individual islands of MX$_2$ (M=Mo, W, Sn, and X= S, Se, Te) and coalesce into a complete film if the fluxes are kept constant during the entire growth window. Upon the coalescence of domains, anti-phase grain boundaries are observed if the merging islands are misaligned with respect to the substrate other than 0°. Anti-phase grain boundaries severely degrade transport properties and, due to their metallic nature, might globally deteriorate the semiconducting properties of the films. The anti-phase grain boundaries are one of the major challenges that must be resolved for wafer-scale, electronic grade 2D TMD. Until recently, the subject had remained unexplored and a recent report by Fu *et al.* [79], shows that using a "two-step" growth method, wafer-scale MoS_2 can be grown on h-BN/sapphire substrates where anti-phase grain boundaries can be suppressed by 98% (Fig. 6.6a–e) [79]. In the "two-step" process, first Mo and S atoms react and nucleate on h-BN surface at 750 °C for 3–4 hours (Fig. 6.6 b). Mo flux is reduced in the second step and the substrate temperature is increased up to 900 °C to promote the lateral growth of domains (Fig. 6.6c). The authors found that the triangular MoS_2 grains' alignment on h-BN can be regulated by Mo flux and observed the overall grain alignment towards zero degree under the low Mo flux condition (Fig. 6.6d). Fully coalesced MoS_2 on h-BN is obtained in a total of 10 hours. Devices fabricated using these films have shown improved transport properties with an ON/OFF ratio $\sim 10^6$ and average carrier mobility of 21 cm^2/V-s (Fig. 6.6e); these values are still lower than exfoliated flakes but comparable to CVD grown films [30].

Significant progress has been made towards wafer-scale epitaxial 2D TMDs with a low defect density *via* MBE. It also offers an advantage in thickness control down to Angstrom level and *in-situ* synthesis of vertical heterostructures with pristine, intact vdW interfaces. Looking forward, systematic studies on surface engineering of substrates and *in-situ* introduction of growth promoters would be beneficial to increase the growth rate of 2D TMDs without sacrificing the crystal quality.

6.3 Controlling nucleation and crystal growth

6.3.1 Substrate engineering

The 2D nature of TMDs makes their growth mechanisms and properties especially susceptible to the substrate type. The substrate selection will have a significant effect on the nucleation density, grain orientation, adatom surface diffusion, and growth mechanism (layer by layer vs. 3D growth). Epitaxial 2D TMDs has been demonstrated on a variety of substrates most notably c-plane sapphire [11,31,52], Au [80], and van der Waals (vdW) substrates like graphene [24,32,37], h-BN [81–83], graphite [84], and mica [85,86]. Growth on amorphous substrates like glass [28,59] and SiO_2 [33,87,88] has also been investigated. Fig. 6.7a summarizes common substrates used for 2D TMDs synthesis ranking them based on chemical activity and roughness of the surface. Smooth vdW substrates with low surface diffusion barriers could help achieve well-aligned, compact grains in vdW epitaxial growth. [73,77,89,90] It is clear that using different substrates will yield films with vastly different grain sizes and growth morphology which will then determine the possible applications for the growth materials [91]. When growing MoS_2 *via* CVD on $SrTiO_3$ (STO), Zhang *et al.* illustrated that, due to the highly anisotropic surface diffusion energy of the growth substrate, dendritic growth would occur. The dendritic morphology was desirable for hydrogen evolution reaction (HER) applications [92]. Substrates, however, do not just influence growth morphology; rather substrate-2D layer interactions have substantial consequences on the photoluminescence [93–95], carrier lifetime [96], carrier density [97], and mobility [98] of semiconducting TMDs. Particularly, TMD layers are susceptible to dielectric screening, charge transfer, and strain induced by the growth substrate. Fig. 6.7b shows the difference in PL response of ML MoS_2 grown via SS-CVD on sapphire, and transferred onto mica, h-BN, Teflon, and

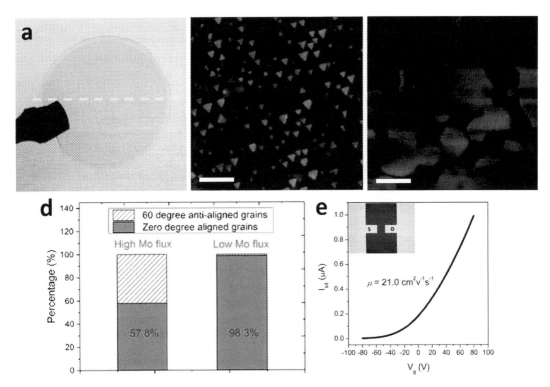

FIGURE 6.6 Wafer-scale growth of electronic-grade MoS$_2$ on h-BN/sapphire substrates via the "two-step" growth method. (a) Photograph of as-grown MoS$_2$ on 2-inch h-BN/sapphire substrates. AFM scans (scale bars 200 nm) (b) at the nucleation (first) step and (c) lateral growth (second) steps where fully coalesced MoS$_2$ is achieved. (d) Statistical analyses of domains on the substrate surface demonstrating superior alignment of domains using the "two-step" growth method. (e) Transport properties of MoS$_2$/h-BN transferred onto 285 nm SiO$_2$/Si substrate for back-gated field-effect transistor measurements demonstrate n-type conduction with an extracted carrier mobility of ~20 cm^2/V-s. Reproduced with permission [79]. Copyright 2017, American Chemical Society.

polystyrene when compared to the suspended layer displaying a large different in both intensity and energy between the different substrates. Substrate engineering efforts exploit this high sensitivity to tailor their properties. Ling *et al.* [18] deposited a variety of planar aromatic molecules on growth substrates to serve as seeding promoters during SS-CVD leading to large, monolayer, MoS$_2$ at 650 °C on Si/SiO$_2$, Au, h-BN, and graphene substrates (Fig. 6.7c). The increase in layer-by-layer growth mode over island growth in the presence of a seeding promoter is hypothesized to originate from the enhancement in surface adhesion force.

The planar aromatic molecules also function as heterogeneous nucleation sites which facilitate synthesis at lower temperatures (Fig. 6.7 d). Surface energy can also be modulated via substrate thermal treatments. Annealing sapphire substrates in air, for example, has enabled the epitaxial growth of MoS$_2$ [51,96,99] and WSe$_2$ [52]. Zhang *et al.* [83] demonstrated that the density of single-atom defects on the h-BN substrate surface alters both the nucleation density and orientation of MOCVD-grown WSe$_2$. With appropriate plasma treatments and NH$_3$

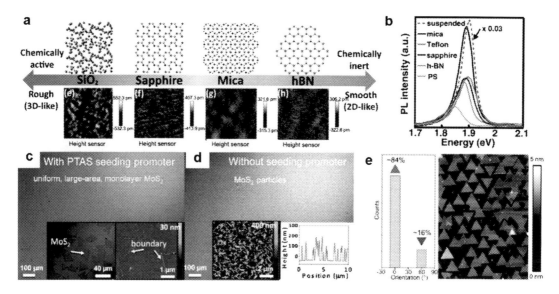

FIGURE 6.7 (a) Common substrates used for TMD synthesis categorized based on chemical activity and surface roughness. Reprinted from [Appl. Phys. Lett. 111, 143106 (2017)] [97], with the permission of AIP Publishing. (b) Comparison of photoluminescence response of CVD-grown MoS$_2$ grown on sapphire and transferred onto a variety of substrates. Reused from [arXiv:1311.3869v1] [94]. (c–d) Optical images and AFM scans of CVD-grown MoS$_2$ grown on Si/SiO$_2$ (c) with and (d) without planar aromatic molecule seeding promoter Reproduced with permission [18]. Copyright 2014, American Chemical Society. (e) AFM scan of MOCVD-grown WSe$_2$ on hBN with modulated single atom defects via plasma treatments and annealing. Reproduced with permission [83]. Copyright 2019, American Chemical Society.

annealing performed on h-BN, the density of single-atom defects was modulated to achieve grains with a high degree of epitaxy and unifies their orientation (Fig. 6.7e).

6.3.2 Precursor chemistry

Vapor phase synthesis techniques rely heavily on the choice of precursor, so much so that methods are often classified by precursor type (solid, liquid, or gas), Volatility and reactivity are two of the main factors in choosing a precursor for CVD. For example, oxides such as WO$_3$ or MoO$_3$ are usually used in SS-CVD instead of metallic W or Mo, due to the high melting point and correspondingly low vapor pressure of these elemental metals compared to their oxides [50]. Furthermore, MoO$_3$ for SS-CVD of MoS$_2$ is facilitated by the reducing nature of sulfur, which forms a more volatile suboxide [50]. Because of this, excess sulfur is used in SS-CVD of sulfides, which also serves to minimize the sulfur vacancies in the film. Reports have shown that hydrogen is key to providing the reducing environment necessary for converting WO$_3$ to WSe$_2$ during SS-CVD growth [100]. Similar to the oxides, ammonium salts (e.g. ammonium molybdate tetrahydrate) can be included as liquid precursors for the metal; these undergo thermal decomposition in elevated temperatures [101]. Typically, elemental chalcogens (except Te) are frequently used for SS-CVD, as their melting temperatures are relatively low (e.g. selenium melts at 221 °C) [29,49,102].

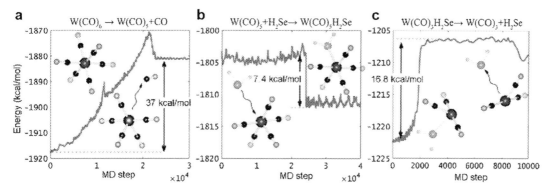

FIGURE 6.8 Simulated energy vs. reaction coordinate plots for (a) abstraction of CO from W(CO)$_6$, the first step in its thermal decomposition, (b) insertion of H$_2$Se into the resulting W(CO)$_5$ molecule, which is a barrierless addition, and (c) the release of bonded H$_2$Se from a similar complex, W(CO)$_3$H$_2$Se, showing the barrier to such a release. Reprinted from Journal of Crystal Growth, 527, Xuan et al., Multi-scale modeling of gas-phase reactions in metalorganic chemical vapor deposition growth of WSe$_2$, 125247, Copyright (2019), with permission from Elsevier. [103].

There are more choices of metalorganic precursors than elemental powders and oxides due to the variety of ligands that can be attached to the metal or chalcogen. Due to their environmental sensitivity, metalorganic precursors are handled in an inert atmosphere and kept inside a stainless-steel bubbler. Carrier gases (H$_2$, Ar and etc.) are passed through a bubbler to deliver volatilized species into a reaction chamber [13,52,53]. Some common precursors of MOCVD for TMDs include carbonyls of transition metals (e.g. W(CO)$_6$, see Fig. 6.8a) [103] and metalorganic chalcogen precursors with simple hydrocarbon ligands (e.g. dimethyl selenide, diethyl sulfide, di-tert-butyl sulfide) [11,13,32,52,103,104]. During thermal decomposition of these precursors, the ligands typically adsorb to the substrate in addition to the intended metal and chalcogen precursors, leading to some degree of carbon contamination in the resulting films [32,105]. This has been shown to have a negative impact on film properties, and has led to the use of highly reactive, carbon-free hydrides such as hydrogen sulfide and hydrogen selenide (visible in Fig. 6.8b–c) [32,104,106]. Another family of metal precursors useful for CVD are the chlorides and oxychlorides, such as MoCl$_5$, WCl$_6$, MoCl$_4$, and WOCl$_5$ can be included in the chamber or separately in bubblers [32,104,106].

Controlling the gas-phase metal-to-chalcogen ratio has a significant impact on nucleation density, which in turn affects domain size and film coalescence [52]. In MOCVD, these ratios can be finely controlled. In a case study of WSe$_2$ MOCVD, it was shown that the cracking efficiency of precursors such as dimethyl selenium is much lower than that of tungsten hexacarbonyl. This is due to a much higher bond strength in the selenium source, and necessitates selenium source to tungsten source ratios of 14 000–20 000 to achieve stoichiometric films [52]. Residence times of the gas-phase precursors, particularly for MOCVD, affect decomposition as well, with longer residence times (typically from lower carrier gas flux) leading to more precursor decomposition [48,104].

6.3.3 Impact of growth temperature

The growth temperature controls the kinetic processes of adatoms on growth substrates during MOCVD of 2D materials. These processes, including the balance between the adsorption and desorption of adatoms and the diffusion and reorientation of adatoms and clusters, are governed by Arrhenius relationships ($\sim e^{\frac{-E}{k_B T}}$) in terms of temperature (T) and activation energy (E), and Boltzmann constant (k_B) [53,107]. Fig. 6.9a illustrates kinetic-limited processes that adatoms go through before they are incorporated into a nucleation site [76]. Thermodynamically, the growth temperature is a first-order parameter that determines the final crystal phase and heterogeneity of a grown layer. For example, a theoretically predicted temperature-composition phase diagram for MoS_2 suggests that the growth window is located between 734 K to 2004 K in the S-rich region at one atmosphere pressure (1 atm) of gaseous species, including Mo, MoS, MoS_2, Mo_2, and S_n $_{(n = 1-8)}$ (Fig. 6.9b) [108]. While commonly used transition metal precursors, $Mo(CO)_6$, $W(CO)_6$, and WCl_6, and the chalcogen, H_2S, H_2Se, and $(CH_3)_2Se$, constituents of TMDs generally dissociate at T < 400 °C [60,109,110], the growth temperature needs to be much higher, i.e., > 700 °C [11], to overcome the kinetic and thermodynamic limitations for near electronic-grade MOCVD grown layers. Because the adsorption energy barrier is one order larger than the surface diffusion barrier [53], a low growth temperature leads to high adsorption and very little desorption for transition metal adatoms and clusters. Even at $\sim$ 2300 °C, the vapor pressure of W is only on the order of 10^{-7} Torr due to its high melting point temperature ($\sim$ 3414 °C) [13,111]. The large difference between adsorption and desorption rate leads to high nucleation density and minimal mobility of adatoms for domain growth. At low growth temperature ($\leq$ 450 °C), MOCVD grown WSe_2 and MoS_2 layers are polycrystalline comprised of sub-micrometer size domains. [33,60,112] The defect density and grain boundary line density are also high in the low-temperature grown layers [11,33]. To obtain large single-crystal domains of TMDs (e.g., films with reduced grain boundaries), ideally, one should minimize the nucleation density and promote the lateral growth around the edges of nucleation sites.

Increasing the growth temperature can increase the adatom desorption rate (exponential relationship) and improve the ripening of adatoms and clusters on the surface effectively reduces the nucleation density. This is shown in a nucleation and ripening study for WSe_2 growth between 600 °C to 900 °C (Fig. 6.9c). To obtain high-quality epitaxial TMD on single-crystal substrates, a high growth temperature ($\geq$ 800 °C) is essential for adatoms and clusters to migrate and reorient to minimize the energy associated with orbital overlap and electrostatic interactions on the substrate [11,13]. Fig. 6.9d shows a continuous lateral growth of epitaxial domains at 800 °C on sapphire under constant pressure and a Se/W flux ratio that eventually leads to a coalesced epitaxial monolayer film (Fig. 6.9d). In addition to sapphire, other common substrates that are suitable for high-temperature MOCVD of TMD include SiC [113], GaN [114], h-BN [83], and graphite/graphene [115]. They are chosen because they are chemically inert to the precursors used for TMDs at high-temperature and their ability to form commensurability with semiconducting TMD layers. Although low-temperature MOCVD grows polycrystalline films and sub-mm size domains, they are beneficial for thin-film transistor production on low thermal budget substrates and back-end-of-line silicon device processes. [116] While the high density of defects and grain boundaries seen in the low temperature grown layers make them unsuitable for electro-optical applications, they can still be used for hydrogen evolution [117] (as also discussed in Chapter 10) and mem-resistive switch devices [118] that rely on the point defects and grain boundaries to function, respectively.

6.3.4 Impact of growth pressure

Empirical studies demonstrate that the growth pressure during the MOCVD of TMDs can impact domain size, domain density, and the number of carbon atoms that incorporate into grown layers [30,32,119–121]. The change in growth pressure also changes the sticking efficiency of adatoms on the surface and the impurity concentration (i.e., water vapors [30] or hydrocarbons dissociated from metal-organic precursors [52,119]). Therefore, optimal pressures for the MOCVD of TMD in the literature could vary widely, depending on types of precursors, gaseous sources, furnace geometry, and substrates used in each study. For example, MOCVD processes for MoS_2 that use $Mo(CO)_6$, H_2S, [121] diethyl sulfide, [120] and other types of metal-organic compounds [54,119] for sulfur are carried out at partial pressure between 10 to 30 Torr. The processes for WSe_2 using $W(CO)_6$, DMSe, and H_2Se are typically carried out at 700 Torr. Eichfeld *et al.* studied the pressure effect for MOCVD grown WSe_2 on graphene at 750 °C in a vertical cold-wall reactor. [32] They revealed that domain density reduces, and domain size increases when the pressure increases from 500 to 700 Torr, as shown in Fig. 6.10, a1–a3. At lower pressure, W and Se adatoms may have a lower sticking coefficient on a high-temperature substrate, leading to desorption before they can ripen into bigger nuclei. The relationships between pressure, temperature, and Se/W ratio were established (Fig. 6.10b). Although the growth pressure can control the nucleation density and growth rate, the growth temperature and Se/W ratio seem to play a more decisive role in growing larger domain sizes.

6.4 Materials engineering

6.4.1 Defect engineering

All vapor-phase synthesis techniques discussed in this chapter inherently lead to a variety of defects in 2D TMD layers [122]. It is imperative to demonstrate control over both intrinsic and extrinsic defects in synthetic 2D TMDs before their integration in next-generation devices to reduce issues such as mid-gap states. [5] Defects can be classified based on their dimensionality: (*i*) zero-dimensional like vacancies and antisites, (*ii*) one-dimensional such as grain boundaries, and (*iii*) two-dimensional like pinholes [123]. All of these defects have significant implications on the electronic and optical properties and subsequent device performance, and significant research effort has gone into analyzing and engineering zero-dimensional defects [124–133]. The low formation energy of chalcogen vacancies [122,134] makes them particularly prominent in synthetic 2D TMDs and have been observed in films grown by SS-CVD [135], MOCVD [11], and MBE [136]. The high-temperature processes and non-equilibrium growth environments typically lead to more defective layers as compared to CVT-grown crystals. Chemical vapor deposition, however, is a much more versatile method than CVT enabling possibilities for *in-situ* defect engineering. Lin *et al.* [11], for instance, added a

176 6. Realization of electronic-grade two-dimensional transition metal dichalcogenides by thin-film deposition techniques

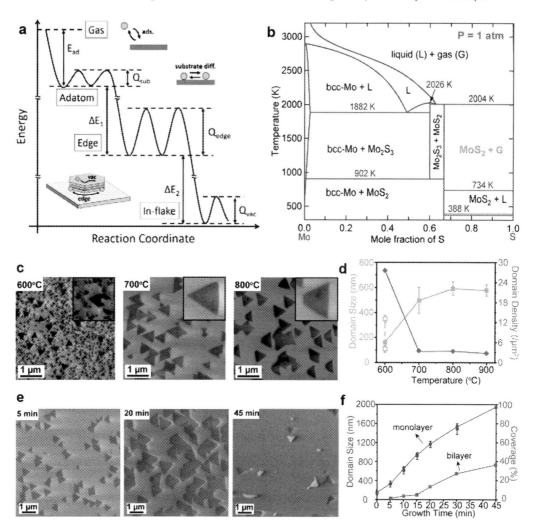

FIGURE 6.9 (a) Reaction energy diagram of thin film growth processes for TMD monolayer. Reproduced under the terms of the CC-BY 3.0 license [76]. Copyright 2017, The Authors, published by IOP. (b) modeled temperature–composition Mo–S phase diagram under pressure P = 1 atm. The ideal growth window for MoS_2 from a thermodynamic perspective is highlighted in green as "MoS_2 + gas (G)", which is in the S-rich region and at between 734 K to 2004 K. Reproduced with permission [108]. Copyright 2016, American Chemical Society. (c) Scanning electron micrograph of WSe_2 grown on sapphire under 30 s of nucleation, 15 min of ripening, and 10 min of lateral growth in increased temperature. During the 10 min lateral growth stage, H_2Se flow rate was held constant at 7 sccm and Se/W ratio was kept at 26000. Inset highlights the shape of the WSe_2 domains. Reproduced with permission. [31] Copyright 2018, American Chemical Society. (d) WSe_2 domain size and domain density as a function of substrate temperature. At low temperature, domain density is high, and size is small due to low desorption rate and low diffuse mobility. Reproduced with permission [31]. Copyright 2018, American Chemical Society. (e) Scanning electron micrograph of WSe_2 grown on sapphire substrate the same nucleation and ripening time, and varied lateral growth time for 5, 20, and 45 min at 800 °C. Reproduced with permission [31]. Copyright 2018, American Chemical Society. (f) Epitaxial monolayer and bilayer domain size and surface coverage as a function of lateral growth time. Reproduced with permission [31]. Copyright 2018, American Chemical Society.

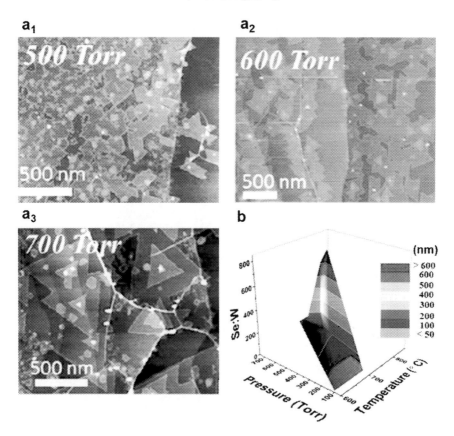

FIGURE 6.10 (a) A pressure study for MOCVD of WSe$_2$ in a vertical reactor. AFM images of WSe$_2$ deposited on graphene at 500 Torr (a$_1$), 600 Torr (a$_2$), and 700 Torr (a$_3$) show the domain size increases and domain density decreases in increased pressure. Reproduced with permission [32]. Copyright 2015, American Chemical Society. (b) The plot of temperature, pressure, and Se/W ratio indicates the pressure has less impact on the domain size, compared to temperature and Se/W ratio. Reproduced with permission [32]. Copyright 2015, American Chemical Society.

post-growth annealing in H$_2$Se following WSe$_2$ MOCVD synthesis to reduce the number of chalcogen vacancies in the films. As demonstrated by STM images of monolayer WSe$_2$ grown on epitaxial graphene (Fig. 6.11a–b), this post-growth annealing process not only lowers the density of V$_{Se}$, but it also reduces the number of metal-rich particles left-over on the surface following the synthesis process. Engineering vacancies, however, can also occur simply by manipulating the chalcogen flow during CVD synthesis [25,135,137,138]. Kim et al. [138] changed both the temperature and exposure time of sulfur powder precursor in the SS-CVD of MoS$_2$ to produce sub-stoichiometric films (from MoS$_2$ to MoS$_{2-x}$) with improved electronic properties as compared to the stoichiometric layers (Fig. 6.11c–d). The versatility of MOCVD was demonstrated by Zhang et al. [31] to control nucleation density and grain boundary density by implementing a multi-stage growth process with nucleation, ripening, and lateral growth steps [31]. In MBE, the poor sticking coefficient of chalcogen atoms under

178   6. Realization of electronic-grade two-dimensional transition metal dichalcogenides by thin-film deposition techniques

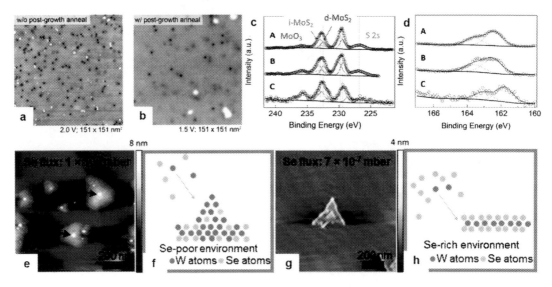

FIGURE 6.11 Scanning tunneling microscopy (STM) image of monolayer WSe$_2$ grown via MOCVD on epitaxial graphene (a) without and (b) with post-growth anneal in H$_2$Se. The samples show both Se vacancies and metal particles (circled in red) which are severely reduced following the annealing protocol. Reproduced with permission [11]. Copyright 2018, American Chemical Society. (c–d) High-resolution XPS spectra of (c) Mo 3d and (d) S 2p of SS-CVD-grown MoS$_2$ at different sulfur vapor exposure with A having the most S content and C having the least. As the sulfur content is reduced the contribution from substoichiometric, defective MoS$_2$ (d-MoS$_2$) intensifies with respect to intrinsic MoS$_2$ (i-MoS$_2$). Reproduced with permission [138]. Copyright 2014, American Chemical Society. (e–h) AFM scans and schematic of MBE-grown WSe$_2$ in (e,f) a Se-poor environment and (g,h) a Se-rich environment. Reproduced under the terms of the CC-BY 3.0 license [76]. Copyright 2017, The Authors, published by IOP.

UHV conditions and the higher growth temperatures required for high metal adatom surface diffusion yields chalcogen-deficient films. Yue et al. [76] employed a low metal flux, high growth temperatures for MBE synthesis of stoichiometric WSe$_2$ large-grain films on vdW substrates (Fig. 6.11e–h). As discussed in the substrate engineering section, 2D TMD quality strongly depends on substrate surface roughness, lattice constant, and chemical activity [139]. Other than *in-situ* defect engineering strategies and substrate choice, defect concentrations in 2D TMDs can be modulated by a broad range of *ex-situ* methods, including gaseous (O$_2$ [140] and N$_2$ [141]) plasma treatment, low-energy ion plasma [142,143], energetic ion bombardment [144], laser irradiation [145,146], and thermal annealing [127].

6.4.2 Heterostructures

Research on heterostructures of two-dimensional materials is an exciting field due to the unique electronic, optoelectronic, and chemical phenomena that occur when stacking or stitching dissimilar 2D materials (Fig. 6.12a) [104,106,147–149]. The initial work on mechanical stacking of graphene and hexagonal boron nitride in 2010 [150] has since expanded to include numerous other 2D materials, more scalable assembly by direct synthesis, and even lateral heterostructures of varying types [104,151]. Mechanical assembly of nanostructures, most often by stacking films grown separately onto one another via transfer processes, has proven invaluable for proof-of-concept devices and testing properties of novel heterostructures. However, mechanical assembly may not be a scalable process; it leads to trapping of contaminants between layers, and lacks the uniformity of deposited films [147,152–154]. Thus, advancements in direct synthesis of vertical heterostructures are essential to realize electronic-grade, scalable material. Vertical heterostructures combining different TMD layers [153,154], graphene with h-BN [150], graphene with TMDs, or h-BN with TMDs [106,148,155,156] have all been demonstrated via direct synthesis. For example, synthesizing a graphene/TMD bilayer or h-BN/TMD heterostructure is readily achievable using epitaxial graphene or h-BN as a growth substrate for TMD synthesis (Fig. 6.12b) [157].

Lateral (or in-plane) heterostructures have seen an increase in interest as well. In studies on vertical and lateral heterostructures of two distinct TMDs, synthesis of one TMD on a substrate where another TMD is already present can lead to either vertical stacking or lateral stitching together of the two, depending on growth temperature [153]. This was demonstrated with WS_2 and MoS_2 in 2014, where high temperatures (850 °C) yielded vertical heterostructures and lower temperatures (650 °C) yielded lateral heterostructures (Fig. 6.12c) [153]. Tellurium was added to the reaction chamber to decrease the melting temperature of metal oxides [158], and the difference in growth rate prevented alloying of the material, thereby preserving discrete MoS_2 and WS_2 regions. Another example of lateral heterostructure growth relied on the use sodium chloride, a growth promoter that drastically increases growth rate [159], to preferentially grow WS_2 from the edge of MoS_2 crystals rather than alloy the two [154]. A quasi-lateral heterostructure formed by slightly overlapping MoS_2 on graphene also show promise for utilizing graphene as a contact to TMD, shown by studies on directly grown MoS_2-on-graphene [156].

6.4.3 Doping and alloying

Doping and alloying are the most common approaches for tuning the electronic and optical properties of any semiconducting material including TMD crystals. Dopants can also add novel functionalities such as magnetism and spintronic properties. Currently, it is difficult to use ion implantation techniques for doping monolayer crystals since the ultrathin body makes it easy to damage, to form unintentional defects by high energy implantation and to control the doping at for atomically thin films [113,143,160]. The methods include substitutional doping commonly used in bulk crystal growth as well as other approaches enabled by the layered and ultrathin body of 2D materials, including charge transfer, ion intercalation, and electrostatic gating. A broad range of choices for doping and alloying 2D materials allows us to control their physical properties and chemical reactivity precisely with respect to 3D materials.

Charge transfer doping is an emerging doping approach made possible by the vdW surfaces of 2D materials and their large surface area. This kind of doping utilizes the interactions between the 2D material and adsorbed dopants on the surface, including adatoms, ions, and molecules. Such an approach avoids lattice distortion in 2D materials and allows the selection of many different types of dopants and efficient doping to avoid degrading the physical properties of 2D materials. However, charge transfer doping is not scalable and does not provide stable recipes for batch fabrication in the electronic industry.

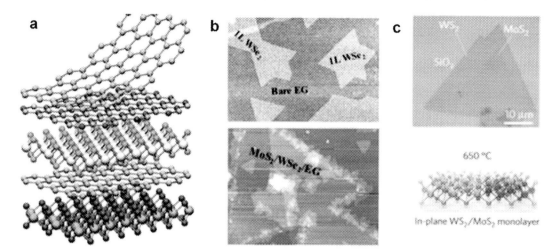

FIGURE 6.12 (a) A schematic depicting a possible vertical heterostructure, depicting a stacking mechanism akin to stacking LEGO blocks. Reprinted from Applied Materials Today, 16, Van der Waals heterostructures for optoelectronics, Progress and prospects, 435, Copyright (2019), with permission from Elsevier [147]. (b) AFM images of two heterostructures grown via sequential CVD: WSe$_2$ on epitaxial graphene and MoS$_2$ on WSe$_2$ on epitaxial graphene. Reproduced under the terms of the CC-BY 3.0 license [157]. Copyright 2015, The Authors, published by Springer Nature. (c) Schematic diagrams and optical microscopy images of WS$_2$/MoS$_2$ lateral heterostructures, grown by Te-assisted synthesis at different temperatures. Reprinted from Applied Materials Today, 16, Van der Waals heterostructures for optoelectronics, Progress and prospects, 435, Copyright (2019), with permission from Elsevier. [147].

Substitutional doping is replacing the cationic or anionic constituents in the lattice with dopants during either synthesis or post-synthesis treatments that either by implantation or by additive doping followed by a diffusion and activation process. The substituted atoms with the same number of valence electrons are isoelectronic dopants, which can be used to tune defect density and concentration. On the other hand, atoms with different valence electron numbers are donors and acceptors for enhancing n-type or p-type conductivity, respectively. However, since substitutional doping requires the replacement of an atomic constituent in the lattice, it may induce lattice distortions and perturbation [161]. As a result, the type of atoms that can be applied to a doped material is limited and may depend on the degree and type of non-stoichiometry. For example, although high-temperature sulfurization [161] or top-down deposition of transition metal oxides [140] are simple post-synthesis approaches for doping foreign atoms in 2D TMD, they can also create defects or distort the structure unavoidably, thereby reducing the material quality or even inducing a new structure to tailor the electronic and catalytic properties.

The following are a few considerations for doping of 2D materials: *i*) what is the initial stoichiometry of the 2D material (M/X ratio); *ii*) can they be incorporated into the 2D material lattice without destabilizing the crystal structure? *iii*) would they induce a new phase? *iv*) how do they dictate the electronic properties and functionality? These considerations should guide one to determine the most suitable dopants and process used to introduce them. It is known that the elements adjacent to Group-VI transition metal (TM) and Group-XVI chalcogen constituents of 1H-TMDs are easier to incorporate into the lattice due to similar atomic size and electronegativity. For example, semiconducting TMD monolayers made of Group-VI elements (Mo and W) and Group-XVI elements (S, Se, and Te) are stable after being substitutionally doped with Group-V (e.g., Nb) and Group-VII (e.g., Re) elements for type and n-type doping, respectively. Similarly, the S, Se, and Te constituents could be substituted with Group-XV (e.g., N and P) and Group-XVII (e.g., F and Cl) elements for p- and n-doping, respectively. A computational analysis using Density Functional Theory (DFT) or the cluster expansion method can also help narrow down achievable doped/alloyed 2D materials. For instance, Kutana *et al.* [162] theoretically suggested that in order to form stable doped 2D semiconductors, the lattice mismatch between the two mixed compounds (e.g., V–Mo dichalcogenides) should be $\leq 3\%$ and also the difference of the metal-chalcogen bond should be ≤ 0.1 Å. Similarly, Komsa and Krasheninnikov's work predicted the stability and electronic structure of chalcogen-based ternary 2D alloys (i.e., $MoS_{2-x}Se_{2x}$) [163] that provide an insight into the stable alloys. Besides electronic dopants, Karthikeyan *et al.* theoretically suggested the most promising metal dopants add magnetism in 2D $MoTe_2$ based on the values of transition metal atomic radii and the number of $(s + d)$ electrons available for bonding [164].

On the experimental side, Zhang *et al.* have demonstrated Re-doped MoS_2 and Nb-doped MoS_2 by CVD (Fig. 6.13a–c) or MOCVD processes (Fig. 6.13d–e) [165,166]. In these two cases, Re and Nb dopants from vapor sources of ReO_3 and $NbCl_5$ were used during CVD and MOCVD for MoS_2 to change the valence electron numbers to modulate n-type and p-type conductivity. CVD processes typically provide individual domains of doped 2D crystals, while MOCVD processes grow doped continuous films. With 1% Re doped into MoS_2 lattice (Fig. 6.13b), a nearly degenerated n-type doping was confirmed by x-ray photoemission spectroscopy, XPS, (Fig. 6.13c) and field-effect transistor measurement [165].

Additionally, *in-situ* Nb doping was performed during MOCVD for MoS_2. The XPS measurements confirmed the presence of Nb-S bonding and continued p-doping with increasing Nb concentration (Fig. 6.13e) [166]. It is also worth noting that dissociated carbon atoms from gaseous methane (CH_4) can also be used as a p-dopant for 2D crystals [167]. Both Nb- and Re-doped MoS_2 and other TMD exhibit potentials not only for transistor devices but also for the diffuse barriers and sensing purposes [166,168]. Isoelectronic substitution doping is another efficient technique to dope 2D semiconductors for modulating the carrier type and reducing dislocations and defects due to the high tolerance of these dopants with the matrix compound [169]. Isoelectronic substitutions do not provide additional electrons to tune the valence band or alter the band structure only utilize electronegativity to modulate the carrier density [170]. For instance, Li *et al.* [170] nuanced the isoelectronic substitution of $MoSe_2$ with a series of W atomic percentage and found that increased W addition ($Mo_{1-x}W_xSe_2$, $x = 0$, 0.02, 0.07, and 0.018) can shift the transistors of $Mo_{1-x}W_xSe_2$ from n-type to p-type without changing their bandgap size.

The challenge of achieving a desirable conductivity in semiconducting TMD monolayers with small doping concentrations is the high ionization energy (~ 0.4 eV, which is much larger than the room temperature thermal energy of 26 meV) due to the quantum confinement effects and reduced dielectric screening in 2D layers. Guo *et al.* [171] has shown that the electrical conductance and the mobility of Nb-MoS_2 only match the expected values that assume the full ionization when the Nb atomic percentage is at 19%. Moreover, similar observations have been reported by Kozhakhmetov *et al.* [172], where intentionally introduced

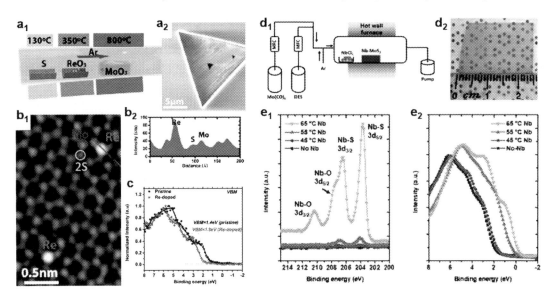

FIGURE 6.13 (a₁) Schematic for the monolayer Re-doped MoS₂ synthesis in a horizontal tube reactor. (a₂) Scanning electron micrograph of Re-doped MoS₂ triangular domain on sapphire. Reproduced with permission [165]. Copyright 2018, Wiley-VCH (b₁) z-contrast scanning transmission electron micrograph of Re-doped MoS₂. Re, Mo, and 2S atoms (brighter atoms, blue circle, and yellow circle, respectively) are identified by the intensity profile (b₂). Reproduced with permission [165]. Copyright 2018, Wiley-VCH (c) Valence band maxima (VBM) comparison between pristine MoS₂ and Re-doped MoS₂. The Fermi level is shifted away from the VBM by 0.5 eV after Re-doping, indicating n-type doping. Reproduced with permission. [165] Copyright 2018, Wiley-VCH (d₁) The hotwall MOCVD set-up for Nb-doped MoS₂. (d₂) Camera image of an MOCVD grown Nb-doped MoS₂. Reproduced with permission [165]. Copyright 2018, Wiley-VCH (e₁-e₂) The X-ray photoemission spectra of Nb 3d and the VBM of MoS₂ and Nb-doped MoS₂ with increasing Nb concentrations, respectively. The Femi level shifts further towards the VBM with increasing Nb, indicating p-type doping. Reproduced with permission [166]. Copyright 2020, Wiley-VCH.

rhenium atoms in monolayer, epitaxial 1.1% Re-WSe₂ remain inactive at room temperature due to the high ionization energy of the dopant at 2D limit. Due to ineffective dopant ionization at the monolayer level, most of so-called "doped 2D semiconductors" are actually 2D alloys from the perspective of Si-technology that considers the standard doping concentration to be on the orders of parts per million (ppm). Therefore, one of the future directions for electronic doping of 2D semiconductors is to alleviate the dopant's high ionization energy either by engineering the dielectric environment of doped monolayers or by growing few-layer thick doped TMDs.

6.5 Summary

Understanding the growth mechanisms and controllability of thin-film deposition is essential for the defect control because defects are mostly incurred during nucleation and growth in a non-equilibrium environment. The nature of each deposition technique, precursors, substrates, as well as kinetic and thermodynamic considerations for crystal growth can all influence the final defect type and density inside the 2D film. At the beginning of this chapter, we provide a basic understanding on some of most popular thin-film deposition systems for 2D TMD materials and representative results of TMD monolayer synthesis. Then we move on to cover many aspects of CVD for TMD such as precursor properties, growth temperatures, growth pressures, and growth substrates, which could be used as tuning knobs to control the crystallinity, impurity density, point defect density, grain boundary density, and resulted heterogeneity in the film. Towards the end, we provide several *ex-situ* methods for engineering defects in the as-grown film and consider forming vertical vdW heterostructures as a means to control their semiconducting properties. Additionally, doping, and alloying TMD not only can modify their electronic properties but can also suppress deep levels of chalcogen vacancies through isoelectronic doping. As with control of defects in any material we might not be able to annihilate defects completely in synthetic TMD crystals, we can only hope to be able to carefully control every aspect of growth to minimize their density and to make the crystals quality acceptable to be considered as electronic grade.

Note

Y.-C.L., R.T., N.A.S., A.K. contributed equally.

Acknowledgments

Y.-C.L., R.T., and J.A.R. acknowledge funding from NEWLIMITS, a center in nCORE as part of the Semiconductor Research Corporation (SRC) program sponsored by NIST through award number 70NANB17H041. A.K. and J.A.R. acknowledge Intel through the Semiconductor Research Corporation (SRC) Task 2746, the Penn State 2D Crystal Consortium (2DCC)-Materials Innovation Platform (2DCC-MIP) under NSF cooperative agreement DMR-1539916, and NSF CAREER Award 1453924 for financial support. N.A.S. and J.A.R. would like to acknowledge Corning Incorporation for support.

References

[1] S. Thompson, N. Anand, M. Armstrong, C. Auth, B. Arcot, M. Alavi, P. Bai, J. Bielefeld, R. Bigwood, J. Brandenburg, M. Buehler, S. Cea, V. Chikarmane, C. Choi, R. Frankovic, T. Ghani, G. Glass, W. Han, T. Hoffmann, M. Hussein, P. Jacob, A. Jain, C. Jan, S. Joshi, C. Kenyon, J. Klaus, S. Klopcic, J. Luce, Z. Ma, B. McIntyre, K. Mistry, A. Murthy, P. Nguyen, H. Pearson, T. Sandford, R. Schweinfurth, R. Shaheed, S. Sivakumar, M. Taylor, B. Tufts, C. Wallace, P. Wang, C. Weber, M. Bohr, A 90 nm logic technology featuring 50 nm strained silicon channel transistors, 7 layers of Cu interconnects, low k ILD, and 1 um2 SRAM cell, in: Technical Digest – International Electron Devices Meeting, 2002, pp. 61–64.

[2] B. Yu, L. Chang, S. Ahmed, H. Wang, S. Bell, C.Y. Yang, C. Tabery, C. Ho, Q. Xiang, T.J. King, J. Bokor, C. Hu, M.R. Lin, D. Kyser, FinFET scaling to 10 nm gate length, Technical Digest – International Electron Devices Meeting (2002) 251–254, https://doi.org/10.1109/IEDM.2002.1175825.

[3] C.A. MacK, Fifty years of Moore's law, in: IEEE Trans. Semicond. Manuf., 2011, pp. 202–207.

[4] D. Akinwande, C. Huyghebaert, C.H. Wang, M.I. Serna, S. Goossens, L.J. Li, H.S.P. Wong, F.H.L. Koppens, Graphene and two-dimensional materials for silicon technology, Nature 573 (2019) 507–518, https://doi.org/10.1038/s41586-019-1573-9.

[5] Y. Liu, X. Duan, H.J. Shin, S. Park, Y. Huang, X. Duan, Promises and prospects of two-dimensional transistors, Nature 591 (2021) 43–53, https://doi.org/10.1038/s41586-021-03339-z.

[6] K. Uchida, H. Watanabe, A. Kinoshita, J. Koga, T. Numata, S.I. Takagi, Experimental study on carrier transport mechanism in ultrathin-body SOI n- and p-MOSFETs with SOI thickness less than 5 nm, in: Technical Digest – International Electron Devices Meeting, 2002, pp. 47–50.

[7] D. Lembke, S. Bertolazzi, A. Kis, Single-layer MoS2 electronics, Accounts of Chemical Research 48 (2015) 100–110, https://doi.org/10.1021/ar500274q.

[8] W. Cao, J. Kang, W. Liu, K. Banerjee, A compact current-voltage model for 2D semiconductor based field-effect transistors considering interface traps, mobility degradation, and inefficient doping effect, I.E.E.E. Transactions on Electron Devices 61 (2014) 4282–4290, https://doi.org/10.1109/TED.2014.2365028.

[9] G. Iannaccone, F. Bonaccorso, L. Colombo, G. Fiori, Quantum engineering of transistors based on 2D materials heterostructures, Nature Nanotechnology 13 (2018) 183–191, https://doi.org/10.1038/s41565-018-0082-6.

[10] D. Rhodes, S.H. Chae, R. Ribeiro-Palau, J. Hone, Disorder in van der Waals heterostructures of 2D materials, Nature Materials 18 (2019) 541–549, https://doi.org/10.1038/s41563-019-0366-8.

[11] Y.-C. Lin, B. Jariwala, B.M. Bersch, K. Xu, Y. Nie, B. Wang, S.M. Eichfeld, X. Zhang, T.H. Choudhury, Y. Pan, R. Addou, C.M. Smyth, J. Li, K. Zhang, M.A. Haque, S. Fölsch, R.M. Feenstra, R.M. Wallace, K. Cho, S.K. Fullerton-Shirey, J.M. Redwing, J.A. Robinson, Realizing large-scale, electronic-grade two-dimensional semiconductors, ACS Nano 12 (2018) 965–975, https://doi.org/10.1021/acsnano.7b07059.

[12] J. Hong, Z. Hu, M. Probert, K. Li, D. Lv, X. Yang, L. Gu, N. Mao, Q. Feng, L. Xie, J. Zhang, D. Wu, Z. Zhang, C. Jin, W. Ji, X. Zhang, J. Yuan, Z. Zhang, Exploring atomic defects in molybdenum disulphide monolayers, Nature Communications 6 (2015) 6293, https://doi.org/10.1038/ncomms7293.

[13] T.H. Choudhury, X. Zhang, Z.Y. Al Balushi, M. Chubarov, J.M. Redwing, Epitaxial growth of two-dimensional layered transition metal dichalcogenides, Annual Review of Materials Research 50 (2020) 155–177, https://doi.org/10.1146/annurev-matsci-090519-113456.

[14] S. Najmaei, Z. Liu, W. Zhou, X. Zou, G. Shi, S. Lei, B.I. Yakobson, J.-C. Idrobo, P.M. Ajayan, J. Lou, Vapour phase growth and grain boundary structure of molybdenum disulphide atomic layers, Nature Materials 12 (2013) 754–759, https://doi.org/10.1038/nmat3673.

[15] A. Splendiani, L. Sun, Y. Zhang, T. Li, J. Kim, C.-Y. Chim, G. Galli, F. Wang, Emerging photoluminescence in monolayer MoS$_2$, Nano Letters 10 (2010) 1271–1275, https://doi.org/10.1021/nl903868w.

[16] W. Hao, C. Marichy, C. Journet, Atomic layer deposition of stable 2D materials, 2D Materials 6 (2019) 012001, https://doi.org/10.1088/2053-1583/aad94f.

[17] P.M. Campbell, C.J. Perini, J. Chiu, A. Gupta, H.S. Ray, H. Chen, K. Wenzel, E. Snyder, B.K. Wagner, J. Ready, E.M. Vogel, Plasma-assisted synthesis of MoS$_2$, 2D Materials 5 (2018) 015005, https://doi.org/10.1088/2053-1583/aa8c96.

[18] X. Ling, Y.-H. Lee, Y. Lin, W. Fang, L. Yu, M.S. Dresselhaus, J. Kong, Role of the seeding promoter in MoS$_2$ growth by chemical vapor deposition, Nano Letters 14 (2014) 464–472, https://doi.org/10.1021/nl4033704.

[19] Y. Sun, K. Fujisawa, Z. Lin, Y. Lei, J.S. Mondschein, M. Terrones, R.E. Schaak, Low-temperature solution synthesis of transition metal dichalcogenide alloys with tunable optical properties, Journal of the American Chemical Society 139 (2017) 11096–11105, https://doi.org/10.1021/jacs.7b04443.

[20] Y. Zhang, Y. Zhang, Q. Ji, J. Ju, H. Yuan, J. Shi, T. Gao, D. Ma, M. Liu, Y. Chen, X. Song, H.Y. Hwang, Y. Cui, Z. Liu, Controlled growth of high-quality monolayer WS$_2$ layers on sapphire and imaging its grain boundary, ACS Nano 7 (2013) 8963–8971, https://doi.org/10.1021/NN403454E.

[21] J.C. Shaw, H. Zhou, Y. Chen, N.O. Weiss, Y. Liu, Y. Huang, X. Duan, Chemical vapor deposition growth of monolayer MoSe$_2$ nanosheets, Nano Research 7 (2015) 511–517, https://doi.org/10.1007/S12274-014-0417-Z.

[22] C. Jung, S.M. Kim, H. Moon, G. Han, J. Kwon, Y.K. Hong, I. Omkaram, Y. Yoon, S. Kim, J. Park, Highly crystalline CVD-grown multilayer MoSe$_2$ thin film transistor for fast photodetector, Scientific Reports 5 (2015) 1–9, https://doi.org/10.1038/srep15313.

[23] L.W. Godwin, D. Brown, R. Livingston, T. Webb, L. Karriem, E. Graugnard, D. Estrada, Open-source automated chemical vapor deposition system for the production of two-dimensional nanomaterials, PLoS ONE 14 (2019) e0210817, https://doi.org/10.1371/JOURNAL.PONE.0210817.

[24] Y. Shi, W. Zhou, A.-Y. Lu, W. Fang, Y.-H. Lee, A.L. Hsu, S.M. Kim, K.K. Kim, H.Y. Yang, L.-J. Li, J.-C. Idrobo, J. Kong, van der Waals epitaxy of MoS$_2$ layers using graphene as growth templates, Nano Letters 12 (2012) 2784–2791, https://doi.org/10.1021/nl204562j.

[25] A.M. van der Zande, P.Y. Huang, D. a Chenet, T.C. Berkelbach, Y. You, G.-H. Lee, T.F. Heinz, D.R. Reichman, D. a Muller, J.C. Hone, Grains and grain boundaries in highly crystalline monolayer molybdenum disulphide, Nature Materials 12 (2013) 554–561, https://doi.org/10.1038/nmat3633.

[26] W. Wu, D. De, S.-C. Chang, Y. Wang, H. Peng, J. Bao, S.-S. Pei, High mobility and high on/off ratio field-effect transistors based on chemical vapor deposited single-crystal MoS_2 grains, Applied Physics Letters 102 (2013) 142106, https://doi.org/10.1063/1.4801861.

[27] M. Amani, M.L. Chin, A.G. Birdwell, T.P. O'Regan, S. Najmaei, Z. Liu, P.M. Ajayan, J. Lou, M. Dubey, Electrical performance of monolayer MoS_2 field-effect transistors prepared by chemical vapor deposition, Applied Physics Letters 102 (2013) 193107, https://doi.org/10.1063/1.4804546.

[28] P. Yang, X. Zou, Z. Zhang, M. Hong, J. Shi, S. Chen, J. Shu, L. Zhao, S. Jiang, X. Zhou, Y. Huan, C. Xie, P. Gao, Q. Chen, Q. Zhang, Z. Liu, Y. Zhang, Batch production of 6-inch uniform monolayer molybdenum disulfide catalyzed by sodium in glass, Nature Communications 9 (2018) 979, https://doi.org/10.1038/s41467-018-03388-5.

[29] S. Wang, Y. Rong, Y. Fan, M. Pacios, H. Bhaskaran, K. He, J.H. Warner, Shape evolution of monolayer MoS_2 crystals grown by chemical vapor deposition, Chemistry of Materials 26 (2014) 6371–6379, https://doi.org/10.1021/cm5025662.

[30] K. Kang, S. Xie, L. Huang, Y. Han, P.Y. Huang, K.F. Mak, C.-J. Kim, D. Muller, J. Park, High-mobility three-atom-thick semiconducting films with wafer-scale homogeneity, Nature 520 (2015) 656–660, https://doi.org/10.1038/nature14417.

[31] X. Zhang, T.H. Choudhury, M. Chubarov, Y. Xiang, B. Jariwala, F. Zhang, N. Alem, G.C. Wang, J.A. Robinson, J.M. Redwing, Diffusion-controlled epitaxy of large area coalesced WSe_2 monolayers on sapphire, Nano Letters 18 (2018) 1049–1056, https://doi.org/10.1021/acs.nanolett.7b04521.

[32] S.M. Eichfeld, L. Hossain, Y.-C. Lin, A.F. Piasecki, B. Kupp, A.G. Birdwell, R.A. Burke, N. Lu, X. Peng, J. Li, A. Azcatl, S. McDonnell, R.M. Wallace, M.J. Kim, T.S. Mayer, J.M. Redwing, J.A. Robinson, Highly scalable, atomically thin WSe_2 grown via metal-organic chemical vapor deposition, ACS Nano 9 (2015) 2080–2087, https://doi.org/10.1021/nn5073286.

[33] A. Kozhakhmetov, J.R. Nasr, F. Zhang, K. Xu, N.C. Briggs, R. Addou, R. Wallace, S.K. Fullerton-Shirey, M. Terrones, S. Das, J.A. Robinson, Scalable BEOL compatible 2D tungsten diselenide, 2D Materials 7 (2020) 015029, https://doi.org/10.1088/2053-1583/AB5AD1.

[34] G.B. Stringfellow, Source molecules, in: Organomet. Vap. Ep., 2nd ed., Academic Press, San Diego, CA, 1999, pp. 151–209.

[35] T.H. Choudhury, H. Simchi, R. Boichot, M. Chubarov, S.E. Mohney, J.M. Redwing, Chalcogen precursor effect on cold-wall gas-source chemical vapor deposition growth of WS2, Crystal Growth & Design 18 (2018) 4357–4364, https://doi.org/10.1021/acs.cgd.8b00306.

[36] J.E. Brom, Y. Ke, R. Du, D. Won, X. Weng, J.C. Gagnon, S.E. Mohney, Q. Li, K. Chen, X.X. Xi, J.M. Redwing, Structural and electrical properties of epitaxial Bi_2Se_3 thin films grown by hybrid physical-chemical vapor deposition, Applied Physics Letters 100 (2012) 162110, https://doi.org/10.1063/1.4704680.

[37] G.V. Bianco, M. Losurdo, M.M. Giangregorio, A. Sacchetti, P. Prete, N. Lovergine, P. Capezzuto, G. Bruno, Direct epitaxial CVD synthesis of tungsten disulfide on epitaxial and CVD graphene, https://doi.org/10.1039/c5ra19698a, 2015.

[38] S. Irvine, P. Capper (Eds.), Metalorganic Vapor Phase Epitaxy (MOVPE), Wiley, 2019.

[39] G.B. Stringfellow, Organometallic vapor-phase epitaxy: theory and practice, 2nd ed., Elsevier, 1999.

[40] M.J. Ludowise, Metalorganic chemical vapor deposition of III-V semiconductors, Journal of Applied Physics 58 (1985) 31, https://doi.org/10.1063/1.336296.

[41] J. Mun, Y. Kim, I.-S. Kang, S.K. Lim, S.J. Lee, J.W. Kim, H.M. Park, T. Kim, S.-W. Kang, Low-temperature growth of layered molybdenum disulphide with controlled clusters, Scientific Reports 6 (2016) 21854, https://doi.org/10.1038/srep21854.

[42] M. Seol, M. Lee, H. Kim, K.W. Shin, Y. Cho, I. Jeon, M. Jeong, H. Lee, J. Park, H. Shin, High-throughput growth of wafer-scale monolayer transition metal dichalcogenide via vertical Ostwald ripening, Advanced Materials 32 (2020) 2003542, https://doi.org/10.1002/adma.202003542.

[43] A. Rockett, The Materials Science of Semiconductors, Springer US, 2008.

[44] L. Tang, T. Li, Y. Luo, S. Feng, Z. Cai, H. Zhang, B. Liu, H.-M. Cheng, Vertical chemical vapor deposition growth of highly uniform 2D transition metal dichalcogenides, ACS Nano 14 (2020) 4646–4653, https://doi.org/10.1021/ACSNANO.0C00296.

[45] N. Azam, Z. Ahmadi, B. Yakupoglu, S. Elafandi, M. Tian, A. Boulesbaa, M. Mahjouri-Samani, Accelerated synthesis of atomically-thin 2D quantum materials by a novel laser-assisted synthesis technique, 2D Materials 7 (2020) 015014, https://doi.org/10.1088/2053-1583/AB53F7.

[46] N. Briggs, S. Subramanian, Z. Lin, X. Li, X. Zhang, K. Zhang, K. Xiao, D. Geohegan, R. Wallace, L.-Q. Chen, M. Terrones, A. Ebrahimi, S. Das, J. Redwing, C. Hinkle, K. Momeni, A. van Duin, V. Crespi, S. Kar, J.A. Robinson, A. van Duin, V. Crespi, S. Kar, J.A. Robinson, A roadmap for electronic grade 2D materials, 2D Materials 6 (2019) 022001, https://doi.org/10.1088/2053-1583/AAF836.

[47] M. Brahlek, A. Sen Gupta, J. Lapano, J. Roth, H.-T. Zhang, L. Zhang, R. Haislmaier, R. Engel-Herbert, Frontiers in the growth of complex oxide thin films: past, present, and future of hybrid MBE, Advanced Functional Materials 28 (2018) 1702772, https://doi.org/10.1002/adfm.201702772.

[48] S. Zhao, L. Wang, L. Fu, Precise vapor-phase synthesis of two-dimensional atomic single crystals, IScience 20 (2019) 527–545, https://doi.org/10.1016/j.isci.2019.09.038.

[49] J. Zhou, J. Lin, X. Huang, Y. Zhou, Y. Chen, J. Xia, H. Wang, Y. Xie, H. Yu, J. Lei, D. Wu, F. Liu, Q. Fu, Q. Zeng, C.H. Hsu, C. Yang, L. Lu, T. Yu, Z. Shen, H. Lin, B.I. Yakobson, Q. Liu, K. Suenaga, G. Liu, Z. Liu, A library of atomically thin metal chalcogenides, Nature 556 (2018) 355–359, https://doi.org/10.1038/s41586-018-0008-3.

[50] Y. Lee, X. Zhang, W. Zhang, M. Chang, C. Lin, K. Chang, Y. Yu, J.T. Wang, C. Chang, L. Li, T. Lin, Synthesis of large-area MoS_2 atomic layers with chemical vapor deposition, Advanced Materials 24 (2012) 2320–2325, https://doi.org/10.1002/adma.201104798.

[51] D. Dumcenco, D. Ovchinnikov, K. Marinov, P. Lazić, M. Gibertini, N. Marzari, O.L. Sanchez, Y.C. Kung, D. Krasnozhon, M.W. Chen, S. Bertolazzi, P. Gillet, A. Fontcuberta, I. Morral, A. Radenovic, A. Kis, Large-area epitaxial monolayer MoS_2, ACS Nano 9 (2015) 4611–4620, https://doi.org/10.1021/acsnano.5b01281.

[52] S.M. Eichfeld, V.O. Colon, Y. Nie, K. Cho, J.A. Robinson, Controlling nucleation of monolayer WSe_2 during metal-organic chemical vapor deposition growth, 2D Materials 3 (2016) 025015, https://doi.org/10.1088/2053-1583/3/2/025015.

[53] M. Ohring, Materials Science of Thin Films: Deposition and Structure, Academic Press, 2002.

[54] Y. Kobayashi, S. Yoshida, M. Maruyama, H. Mogi, K. Murase, Y. Maniwa, O. Takeuchi, S. Okada, H. Shigekawa, Y. Miyata, Continuous heteroepitaxy of two-dimensional heterostructures based on layered chalcogenides, ACS Nano 13 (2019) 7527–7535, https://doi.org/10.1021/acsnano.8b07991.

[55] B. Kalanyan, W.A. Kimes, R. Beams, S.J. Stranick, E. Garratt, I. Kalish, A.V. Davydov, R.K. Kanjolia, J.E. Maslar, Rapid wafer-scale growth of polycrystalline 2H-MoS_2 by pulsed metal-organic chemical vapor deposition, Chemistry of Materials 29 (2017) 6279–6288, https://doi.org/10.1021/acs.chemmater.7b01367.

[56] M. Chubarov, H. Pedersen, H. Högberg, V. Darakchieva, J. Jensen, P.O.Å Persson, A. Henry, Epitaxial CVD growth of sp2-hybridized boron nitride using aluminum nitride as buffer layer, Physica Status Solidi (RRL)—Rapid Research Letters 5 (2011) 397–399, https://doi.org/10.1002/pssr.201105410.

[57] A. Sebastian, R. Pendurthi, T.H. Choudhury, J.M. Redwing, S. Das, Benchmarking monolayer MoS_2 and WS_2 field-effect transistors, Nature Communications 12 (2021) 1–12, https://doi.org/10.1038/s41467-020-20732-w.

[58] M. Chubarov, T.H. Choudhury, D.R. Hickey, S. Bachu, T. Zhang, A. Sebastian, A. Bansal, H. Zhu, N. Trainor, S. Das, M. Terrones, N. Alem, J.M. Redwing, Wafer-scale epitaxial growth of unidirectional WS_2 monolayers on sapphire, ACS Nano 15 (2021) 2532–2541, https://doi.org/10.1021/acsnano.0c06750.

[59] N.A. Simonson, J.R. Nasr, S. Subramanian, B. Jariwala, R. Zhao, S. Das, J.A. Robinson, Low-temperature metalorganic chemical vapor deposition of molybdenum disulfide on multicomponent glass substrates, FlatChem 11 (2018) 32–37, https://doi.org/10.1016/j.flatc.2018.11.004.

[60] J. Mun, H. Park, J. Park, D. Joung, S.-K. Lee, J. Leem, J.-M. Myoung, J. Park, S.-H. Jeong, W. Chegal, S. Nam, S.-W. Kang, High-mobility MoS_2 directly grown on polymer substrate with kinetics-controlled metal–organic chemical vapor deposition, ACS Applied Electronic Materials 1 (2019) 608–616, https://doi.org/10.1021/acsaelm.9b00078.

[61] S. Xie, L. Tu, Y. Han, L. Huang, K. Kang, K.U. Lao, P. Poddar, C. Park, D.A. Muller, R.A. DiStasio, J. Park, Coherent, atomically thin transition-metal dichalcogenide superlattices with engineered strain, Science 359 (2018) 1131–1136, https://doi.org/10.1126/science.aao5360.

[62] S. Xie, L. Tu, Y. Han, L. Huang, K. Kang, K.U. Lao, P. Poddar, D.A. Muller, R.A. DiStasio, J. Park, Coherent atomically-thin superlattices with engineered strain, arXiv:1708.09539, 2017. (Accessed 7 May 2021).

[63] A.K. Keiji Ueno, Toshihiro Shimada, Koichiro Saiki, Heteroepitaxial growth of layered transition metal dichalcogenides on sulfur-terminated GaAs {111} surfaces, Applied Physics Letters 56 (1989) 327–329.

[64] K. Ueno, K. Saiki, T. Shimada, A. Koma, Epitaxial growth of transition metal dichalcogenides on cleaved faces of mica, Journal of Vacuum Science & Technology. A. Vacuum, Surfaces, and Films 8 (1990) 68–72, https://doi.org/10.1116/1.576983.

[65] A.K.F.S. Ohuchi, B.A. Parkinson, K. Ueno, Van der Waals epitaxial growth and characterization of $MoSe_2$ thin films on SnS_2, Journal of Applied Physics 68 (1990) 2168–2175.

[66] A. Koma, K. Saiki, Y. Sato, Heteroepitaxy of a two-dimensional material on a three-dimensional material, Applied Surface Science 41–42 (1990) 451–456, https://doi.org/10.1016/0169-4332(89)90102-5.

[67] R. Yue, A.T. Barton, H. Zhu, A. Azcatl, L.F. Pena, J. Wang, X. Peng, N. Lu, L. Cheng, R. Addou, S. McDonnell, L. Colombo, J.W.P. Hsu, J. Kim, M.J. Kim, R.M. Wallace, C.L. Hinkle, $HfSe_2$ thin films: 2D transition metal dichalcogenides grown by molecular beam epitaxy, ACS Nano 9 (2015) 474–480, https://doi.org/10.1021/nn5056496.

[68] S. Vishwanath, S. Rouvimov, T. Orlova, X. Liu, J.K. Furdyna, D. Jena, H. Grace Xing, Atomic structure of thin $MoSe_2$ films grown by molecular beam epitaxy, Microscopy and Microanalysis 20 (2014) 164–165, https://doi.org/10.1017/S1431927614002542.

[69] L. Jiao, H.J. Liu, J.L. Chen, Y. Yi, W.G. Chen, Y. Cai, J.N. Wang, X.Q. Dai, N. Wang, W.K. Ho, M.H. Xie, Molecular-beam epitaxy of monolayer $MoSe_2$: growth characteristics and domain boundary formation, New Journal of Physics 17 (2015) 053023, https://doi.org/10.1088/1367-2630/17/5/053023.

[70] M. Nakano, Y. Wang, Y. Kashiwabara, H. Matsuoka, Y. Iwasa, Layer-by-layer epitaxial growth of scalable WSe_2 on sapphire by molecular beam epitaxy, Nano Letters 17 (2017) 5595–5599, https://doi.org/10.1021/acs.nanolett.7b02420.

[71] W. Mortelmans, S. El Kazzi, A.N. Mehta, D. Vanhaeren, T. Conard, J. Meersschaut, T. Nuytten, S. De Gendt, M. Heyns, C. Merckling, Peculiar alignment and strain of 2D WSe_2 grown by van der Waals epitaxy on reconstructed sapphire surfaces, Nanotechnology 30 (2019) 465601, https://doi.org/10.1088/1361-6528/AB3C9B.

[72] S. Vishwanath, X. Liu, S. Rouvimov, L. Basile, N. Lu, A. Azcatl, K. Magno, R.M. Wallace, M. Kim, J.-C. Idrobo, J.K. Furdyna, D. Jena, H.G. Xing, Controllable growth of layered selenide and telluride heterostructures and superlattices using molecular beam epitaxy, Journal of Materials Research 31 (2016) 900–910, https://doi.org/10.1557/jmr.2015.374.

[73] Y.W. Park, S.K. Jerng, J.H. Jeon, S.B. Roy, K. Akbar, J. Kim, Y. Sim, M.J. Seong, J.H. Kim, Z. Lee, M. Kim, Y. Yi, J. Kim, D.Y. Noh, S.H. Chun, Molecular beam epitaxy of large-area $SnSe_2$ with monolayer thickness fluctuation, 2D Materials 4 (2017) 014006, https://doi.org/10.1088/2053-1583/aa51a2.

[74] S. Aminalragia-Giamini, J. Marquez-Velasco, P. Tsipas, D. Tsoutsou, G. Renaud, A. Dimoulas, Molecular beam epitaxy of thin $HfTe_2$ semimetal films, 2D Materials 4 (2016) 015001, https://doi.org/10.1088/2053-1583/4/1/015001.

[75] M. Hilse, K. Wang, R. Engel-Herbert, Growth of ultrathin Pt layers and selenization into $PtSe_2$ by molecular beam epitaxy, 2D Materials 7 (2020) 045013, https://doi.org/10.1088/2053-1583/ab9f91.

[76] R. Yue, Y. Nie, L.A. Walsh, R. Addou, C. Liang, N. Lu, A.T. Barton, H. Zhu, Z. Che, D. Barrera, L. Cheng, P.-R. Cha, Y.J. Chabal, J.W.P. Hsu, J. Kim, M.J. Kim, L. Colombo, R.M. Wallace, K. Cho, C.L. Hinkle, Nucleation and growth of WSe_2: enabling large grain transition metal dichalcogenides, 2D Materials 4 (2017) 045019, https://doi.org/10.1088/2053-1583/AA8AB5.

[77] L.A. Walsh, R. Yue, Q. Wang, A.T. Barton, R. Addou, C.M. Smyth, H. Zhu, J. Kim, L. Colombo, M.J. Kim, R.M. Wallace, C.L. Hinkle, WTe2 thin films grown by beam-interrupted molecular beam epitaxy, 2D Materials 4 (2017) 025044, https://doi.org/10.1088/2053-1583/aa61e1.

[78] L.A. Walsh, R. Addou, R.M. Wallace, C.L. Hinkle, Molecular beam epitaxy of transition metal dichalcogenides, in: Mol. Beam Ep., Elsevier, 2018, pp. 515–531.

[79] D. Fu, X. Zhao, Y.Y. Zhang, L. Li, H. Xu, A.R. Jang, S.I. Yoon, P. Song, S.M. Poh, T. Ren, Z. Ding, W. Fu, T.J. Shin, H.S. Shin, S.T. Pantelides, W. Zhou, K.P. Loh, Molecular beam epitaxy of highly crystalline monolayer molybdenum disulfide on hexagonal boron nitride, Journal of the American Chemical Society 139 (2017) 9392–9400, https://doi.org/10.1021/jacs.7b05131.

[80] F. Zhang, K. Momeni, M.A. AlSaud, A. Azizi, M.F. Hainey, J.M. Redwing, L.Q. Chen, N. Alem, Controlled synthesis of 2D transition metal dichalcogenides: from vertical to planar MoS_2, 2D Materials 4 (2017) 025029, https://doi.org/10.1088/2053-1583/aa5b01.

[81] A. Yan, J. Jairo Velasco, S. Kahn, K. Watanabe, T. Taniguchi, F. Wang, M.F. Crommie, A. Zettl, Direct growth of single- and few-layer MoS_2 on h-BN with preferred relative rotation angles, Nano Letters 15 (2015) 6324–6331, https://doi.org/10.1021/ACS.NANOLETT.5B01311.

[82] H. Yu, Z. Yang, L. Du, J. Zhang, J. Shi, W. Chen, P. Chen, M. Liao, J. Zhao, J. Meng, G. Wang, J. Zhu, R. Yang, D. Shi, L. Gu, G. Zhang, Precisely aligned monolayer MoS_2 epitaxially grown on h-BN basal plane, Small 13 (2017) 1603005, https://doi.org/10.1002/smll.201603005.

[83] X. Zhang, F. Zhang, Y. Wang, D.S. Schulman, T. Zhang, A. Bansal, N. Alem, S. Das, V.H. Crespi, M. Terrones, J.M. Redwing, Defect-controlled nucleation and orientation of WSe_2 on hBN: a route to single-crystal epitaxial monolayers, ACS Nano 13 (2019) 3341–3352, https://doi.org/10.1021/acsnano.8b09230.

[84] C. Zhang, A. Johnson, C.-L. Hsu, L.-J. Li, C.-K. Shih, Direct imaging of band profile in single layer MoS_2 on graphite: quasiparticle energy gap, metallic edge states, and edge band bending, Nano Letters 14 (2014) 2443–2447, https://doi.org/10.1021/nl501133c.

[85] R. Schlaf, D. Louder, O. Lang, C. Pettenkofer, W. Jaegermann, K.W. Nebesny, P.A. Lee, B.A. Parkinson, N.R. Armstrong, Molecular beam epitaxy growth of thin films of SnS_2 and $SnSe_2$ on cleaved mica and the basal planes of single-crystal layered semiconductors: reflection high-energy electron diffraction, low-energy electron diffraction, photoemission, and scanning tunnelin, Journal of Vacuum Science & Technology. A. Vacuum, Surfaces, and Films 13 (1998) 1761, https://doi.org/10.1116/1.579766.

[86] Q. Ji, Y. Zhang, T. Gao, Y. Zhang, D. Ma, M. Liu, Y. Chen, X. Qiao, P.-H. Tan, M. Kan, J. Feng, Q. Sun, Z. Liu, Epitaxial monolayer MoS_2 on mica with novel photoluminescence, Nano Letters 13 (2013) 3870–3877, https://doi.org/10.1021/NL401938T.

[87] X. Fan, R. Siris, O. Hartwig, G. Duesberg, F. Niklaus, Rapid and large-area visualization of grain boundaries in MoS_2 on SiO_2 using vapor hydrofluoric acid, ACS Applied Materials & Interfaces 12 (2020) 34049–34057, https://doi.org/10.1021/acsami.0c06910.

[88] M. Berg, K. Keyshar, I. Bilgin, F. Liu, H. Yamaguchi, R. Vajtai, C. Chan, G. Gupta, S. Kar, P. Ajayan, T. Ohta, A.D. Mohite, Layer dependence of the electronic band alignment of few-layer MoS_2 on SiO_2 measured using photoemission electron microscopy, Physical Review B 95 (2017) 235406, https://doi.org/10.1103/PhysRevB.95.235406.

[89] H.C. Diaz, R. Chaghi, Y. Ma, M. Batzill, Molecular beam epitaxy of the van der Waals heterostructure $MoTe_2$ on MoS_2: phase, thermal, and chemical stability, 2D Materials 2 (2015) 044010, https://doi.org/10.1088/2053-1583/2/4/044010.

[90] T. Hotta, T. Tokuda, S. Zhao, K. Watanabe, T. Taniguchi, H. Shinohara, R. Kitaura, Molecular beam epitaxy growth of monolayer niobium diselenide flakes, Applied Physics Letters 109 (2016) 133101, https://doi.org/10.1063/1.4963178.

[91] A. Kozhakhmetov, T.H. Choudhury, Z.Y. Al Balushi, M. Chubarov, J.M. Redwing, Effect of substrate on the growth and properties of thin 3R NbS_2 films grown by chemical vapor deposition, Journal of Crystal Growth 486 (2018) 137–141, https://doi.org/10.1016/j.jcrysgro.2018.01.031.

[92] Y. Zhang, Q. Ji, J. Wen, J. Li, C. Li, J. Shi, X. Zhou, K. Shi, H. Chen, Y. Li, S. Deng, N. Xu, Z. Liu, Y. Zhang, Monolayer MoS_2 dendrites on a symmetry-disparate $SrTiO_3$ (001) substrate: formation mechanism and interface interaction, Advanced Functional Materials 26 (2016) 3299–3305, https://doi.org/10.1002/adfm.201505571.

[93] D. Liu, X.-Q. Yan, H.-W. Guo, Z.-B. Liu, W.-Y. Zhou, J.-G. Tian, Substrate effect on the photoluminescence of chemical vapor deposition transferred monolayer WSe_2, Journal of Applied Physics 128 (2020) 43101, https://doi.org/10.1063/5.0008586.

[94] M. Buscema, G.A. Steele, H.S.J. van der Zant, A. Castellanos-Gomez, The effect of the substrate on the Raman and photoluminescence emission of single-layer MoS_2, Nano Research 7 (2014) 1–11, https://doi.org/10.1007/s12274-014-0424-0.

[95] Y. Yu, Y. Yu, C. Xu, Y.-Q. Cai, L. Su, Y. Zhang, Y.-W. Zhang, K. Gundogdu, L. Cao, Engineering substrate interactions for high luminescence efficiency of transition-metal dichalcogenide monolayers, Advanced Functional Materials 26 (2016) 4733–4739, https://doi.org/10.1002/adfm.201600418.

[96] K. Zhang, N.J. Borys, B.M. Bersch, G.R. Bhimanapati, K. Xu, B. Wang, K. Wang, M. Labella, T.A. Williams, M.A. Haque, E.S. Barnard, S. Fullerton-Shirey, P.J. Schuck, J.A. Robinson, Deconvoluting the photonic and electronic response of 2D materials: the case of MoS_2, Scientific Reports 7 (2017) 16938, https://doi.org/10.1038/s41598-017-16970-6.

[97] W.H. Chae, J.D. Cain, E.D. Hanson, A.A. Murthy, V.P. Dravid, Substrate-induced strain and charge doping in CVD-grown monolayer MoS_2, Applied Physics Letters 111 (2017) 143106, https://doi.org/10.1063/1.4998284.

[98] N. Ma, D. Jena, Charge scattering and mobility in atomically thin semiconductors, Physical Review X 4 (2014) 11043, https://doi.org/10.1103/PhysRevX.4.011043.

[99] Q. Ji, M. Kan, Y. Zhang, Y. Guo, D. Ma, J. Shi, Q. Sun, Q. Chen, Y. Zhang, Z. Liu, Unravelling orientation distribution and merging behavior of monolayer MoS_2 domains on sapphire, Nano Letters 15 (2015) 198–205, https://doi.org/10.1021/nl503373x.

References   **189**

[100] B. Liu, M. Fathi, L. Chen, A. Abbas, Y. Ma, C. Zhou, Chemical vapor deposition growth of monolayer WSe_2 with tunable device characteristics and growth mechanism study, ACS Nano 9 (2015) 6119–6127, https://doi.org/10.1021/acsnano.5b01301.

[101] K. Chen, X. Wan, W. Xie, J. Wen, Z. Kang, X. Zeng, H. Chen, J. Xu, Lateral built-in potential of monolayer MoS_2-WS_2 in-plane heterostructures by a shortcut growth strategy, Advanced Materials 27 (2015) 6431–6437, https://doi.org/10.1002/adma.201502375.

[102] K.K. Liu, W. Zhang, Y.H. Lee, Y.C. Lin, M.T. Chang, C.Y. Su, C.S. Chang, H. Li, Y. Shi, H. Zhang, C.S. Lai, L.J. Li, Growth of large-area and highly crystalline MoS_2 thin layers on insulating substrates, Nano Letters 12 (2012) 1538–1544, https://doi.org/10.1021/nl2043612.

[103] Y. Xuan, A. Jain, S. Zafar, R. Lotfi, N. Nayir, Y. Wang, T.H. Choudhury, S. Wright, J. Feraca, L. Rosenbaum, J.M. Redwing, V. Crespi, A.C.T. van Duin, Multi-scale modeling of gas-phase reactions in metal-organic chemical vapor deposition growth of WSe_2, Journal of Crystal Growth 527 (2019) 125247, https://doi.org/10.1016/j.jcrysgro.2019.125247.

[104] D.H. Lee, Y. Sim, J. Wang, S.Y. Kwon, Metal-organic chemical vapor deposition of 2D van der Waals materials – the challenges and the extensive future opportunities, APL Materials 8 (2020), https://doi.org/10.1063/1.5142601.

[105] X. Zhang, Z.Y. Al Balushi, F. Zhang, T.H. Choudhury, S.M. Eichfeld, N. Alem, T.N. Jackson, J.A. Robinson, J.M. Redwing, Influence of carbon in metalorganic chemical vapor deposition of few-layer WSe_2 thin films, Journal of Electronic Materials 45 (2016) 6273–6279, https://doi.org/10.1007/s11664-016-5033-0.

[106] S.Y. Kim, J. Kwak, C.V. Ciobanu, S.Y. Kwon, Recent developments in controlled vapor-phase growth of 2D group 6 transition metal dichalcogenides, Advanced Materials 31 (2019) 1804939, https://doi.org/10.1002/adma.201804939.

[107] Y. Nie, C. Liang, P.-R. Cha, L. Colombo, R.M. Wallace, K. Cho, A kinetic Monte Carlo simulation method of van der Waals epitaxy for atomistic nucleation-growth processes of transition metal dichalcogenides, Scientific Reports 7 (2017) 2977, https://doi.org/10.1038/s41598-017-02919-2.

[108] S.-L. Shang, G. Lindwall, Y. Wang, J.M. Redwing, T. Anderson, Z.-K. Liu, Lateral versus vertical growth of two-dimensional layered transition-metal dichalcogenides: thermodynamic insight into MoS_2, Nano Letters 16 (2016) 5742–5750, https://doi.org/10.1021/acs.nanolett.6b02443.

[109] K.E. Lewis, D.M. Golden, G.P. Smith, Organometallic bond dissociation energies: laser pyrolysis of $Fe(CO)_5$, $Cr(CO)_6$, $Mo(CO)_6$, and $W(CO)_6$, Journal of the American Chemical Society 106 (1984) 3905–3912, https://doi.org/10.1021/ja00326a004.

[110] R.K. Pearson, G.R. Haugen, Kinetics of the thermal decomposition of H_2Se, International Journal of Hydrogen Energy 6 (1981) 509–519, https://doi.org/10.1016/0360-3199(81)90082-3.

[111] E.R. Plante, A.B. Sessoms, Vapor pressure and heat of sublimation of tungsten, Journal of Research of the National Bureau of Standards. Section A. Physics and Chemistry 77A (1973) 237, https://doi.org/10.6028/JRES.077A.015.

[112] Y. Gong, X. Zhang, J.M. Redwing, T.N. Jackson, Thin film transistors using wafer-scale low-temperature MOCVD WSe_2, Journal of Electronic Materials 45 (2016) 6280–6284, https://doi.org/10.1007/s11664-016-4987-2.

[113] T. Shi, R.C. Walker, I. Jovanovic, J.A. Robinson, Effects of energetic ion irradiation on WSe2/SiC heterostructures, Scientific Reports 7 (2017) 4151, https://doi.org/10.1038/s41598-017-04042-8.

[114] K. Zhang, B. Jariwala, J. Li, N.C. Briggs, B. Wang, D. Ruzmetov, R.A. Burke, J.O. Lerach, T.G. Ivanov, M. Haque, R.M. Feenstra, J.A. Robinson, Large scale 2D/3D hybrids based on gallium nitride and transition metal dichalcogenides, Nanoscale 10 (2018) 336–341, https://doi.org/10.1039/c7nr07586c.

[115] Y.-C. Lin, C.-Y.S. Chang, R.K. Ghosh, J. Li, H. Zhu, R. Addou, B. Diaconescu, T. Ohta, X. Peng, N. Lu, M.J. Kim, J.T. Robinson, R.M. Wallace, T.S. Mayer, S. Datta, L.-J. Li, J.A. Robinson, Atomically thin heterostructures based on single-layer tungsten diselenide and graphene, Nano Letters 14 (2014) 6936–6941, https://doi.org/10.1021/nl503144a.

[116] A. Kozhakhmetov, R. Torsi, C.Y. Chen, J.A. Robinson, Scalable low-temperature synthesis of two-dimensional materials beyond graphene, Journal of Physics: Materials 4 (2020) 012001, https://doi.org/10.1088/2515-7639/ABBDB1.

[117] H. Li, C. Tsai, A.L. Koh, L. Cai, A.W. Contryman, A.H. Fragapane, J. Zhao, H.S. Han, H.C. Manoharan, F. Abild-Pedersen, J.K. Nørskov, X. Zheng, Activating and optimizing MoS_2 basal planes for hydrogen evolution through the formation of strained sulphur vacancies, Nature Materials 15 (2016) 48–53, https://doi.org/10.1038/nmat4465.

[118] V.K. Sangwan, D. Jariwala, I.S. Kim, K.-S. Chen, T.J. Marks, L.J. Lauhon, M.C. Hersam, Gate-tunable memristive phenomena mediated by grain boundaries in single-layer MoS_2, Nature Nanotechnology 10 (2015) 403–406, https://doi.org/10.1038/nnano.2015.56.

[119] M. Marx, A. Grundmann, Y.R. Lin, D. Andrzejewski, T. Kümmell, G. Bacher, M. Heuken, H. Kalisch, A. Vescan, Metalorganic vapor-phase epitaxy growth parameters for two-dimensional MoS_2, Journal of Electronic Materials 47 (2018) 910–916, https://doi.org/10.1007/s11664-017-5937-3.

[120] K. Zhang, B.M. Bersch, F. Zhang, N.C. Briggs, S. Subramanian, K. Xu, M. Chubarov, K. Wang, J.O. Lerach, J.M. Redwing, S.K. Fullerton-Shirey, M. Terrones, J.A. Robinson, Considerations for utilizing sodium chloride in epitaxial molybdenum disulfide, ACS Applied Materials & Interfaces 10 (2018) 40831–40837, https://doi.org/10.1021/acsami.8b16374.

[121] D. Chiappe, J. Ludwig, A. Leonhardt, S. El Kazzi, A.N. Mehta, T. Nuytten, U. Celano, S. Sutar, G. Pourtois, M. Caymax, K. Paredis, W. Vandervorst, D. Lin, S. De Gendt, K. Barla, C. Huyghebaert, I. Asselberghs, I. Radu, Layer-controlled epitaxy of 2D semiconductors: bridging nanoscale phenomena to wafer-scale uniformity, Nanotechnology 29 (2018) 425602, https://doi.org/10.1088/1361-6528/AAD798.

[122] W. Zhou, X. Zou, S. Najmaei, Z. Liu, Y. Shi, J. Kong, J. Lou, P.M. Ajayan, B.I. Yakobson, J.C. Idrobo, Intrinsic structural defects in monolayer molybdenum disulfide, Nano Letters 13 (2013) 2615–2622, https://doi.org/10.1021/nl4007479.

[123] Z. Lin, B.R. Carvalho, E. Kahn, R. Lv, R. Rao, H. Terrones, M.A. Pimenta, M. Terrones, Defect engineering of two-dimensional transition metal dichalcogenides, 2D Materials 3 (2016) 022002, https://doi.org/10.1088/2053-1583/3/2/022002.

[124] J. Jiang, T. Xu, J. Lu, L. Sun, Z. Ni, Defect engineering in 2D materials: precise manipulation and improved functionalities, Research 2019 (2019) 4641739, https://doi.org/10.34133/2019/4641739.

[125] H.-P. Komsa, J. Kotakoski, S. Kurasch, O. Lehtinen, U. Kaiser, A.V. Krasheninnikov, Two-dimensional transition metal dichalcogenides under electron irradiation: defect production and doping, Physical Review Letters 109 (2012) 035503, https://doi.org/10.1103/PhysRevLett.109.035503.

[126] M. Tosun, L. Chan, M. Amani, T. Roy, G.H. Ahn, P. Taheri, C. Carraro, J.W. Ager, R. Maboudian, A. Javey, Air-stable n-doping of WSe_2 by anion vacancy formation with mild plasma treatment, ACS Nano 10 (2016) 6853–6860, https://doi.org/10.1021/acsnano.6b02521.

[127] S. Tongay, J. Suh, C. Ataca, W. Fan, A. Luce, J.S. Kang, J. Liu, C. Ko, R. Raghunathanan, J. Zhou, F. Ogletree, J. Li, J.C. Grossman, J. Wu, Defects activated photoluminescence in two-dimensional semiconductors: interplay between bound, charged, and free excitons, Scientific Reports 3 (2013) 2657, https://doi.org/10.1038/srep02657.

[128] B. Schuler, D.Y. Qiu, S. Refaely-Abramson, C. Kastl, C.T. Chen, S. Barja, R.J. Koch, D.F. Ogletree, S. Aloni, A.M. Schwartzberg, J.B. Neaton, S.G. Louie, A. Weber-Bargioni, Large spin-orbit splitting of deep in-gap defect states of engineered sulfur vacancies in monolayer WS2, Physical Review Letters 123 (2019) 076801, https://doi.org/10.1103/PhysRevLett.123.076801.

[129] S. Barja, S. Refaely-Abramson, B. Schuler, D.Y. Qiu, A. Pulkin, S. Wickenburg, H. Ryu, M.M. Ugeda, C. Kastl, C. Chen, C. Hwang, A. Schwartzberg, S. Aloni, S.K. Mo, D. Frank Ogletree, M.F. Crommie, O.V. Yazyev, S.G. Louie, J.B. Neaton, A. Weber-Bargioni, Identifying substitutional oxygen as a prolific point defect in monolayer transition metal dichalcogenides, Nature Communications 10 (2019), https://doi.org/10.1038/s41467-019-11342-2.

[130] P.K. Chow, R.B. Jacobs-Gedrim, J. Gao, T.-M. Lu, B. Yu, H. Terrones, N. Koratkar, Defect-induced photoluminescence in monolayer semiconducting transition metal dichalcogenides, ACS Nano 9 (2015) 1520–1527, https://doi.org/10.1021/nn5073495.

[131] K. Greben, S. Arora, M.G. Harats, K.I. Bolotin, Intrinsic and extrinsic defect-related excitons in TMDCs, Nano Letters 20 (2020) 2544–2550, https://doi.org/10.1021/acs.nanolett.9b05323.

[132] J. Lu, A. Carvalho, X.K. Chan, H. Liu, B. Liu, E.S. Tok, K.P. Loh, A.H. Castro Neto, C.H. Sow, Atomic healing of defects in transition metal dichalcogenides, Nano Letters 15 (2015) 3524–3532, https://doi.org/10.1021/acs.nanolett.5b00952.

[133] Y. Guo, D. Liu, J. Robertson, Chalcogen vacancies in monolayer transition metal dichalcogenides and Fermi level pinning at contacts, Applied Physics Letters 106 (2015) 173106, https://doi.org/10.1063/1.4919524.

[134] H.-P. Komsa, A.V. Krasheninnikov, Native defects in bulk and monolayer MoS_2 from first principles, Physical Review B 91 (2015) 125304, https://doi.org/10.1103/PhysRevB.91.125304.

[135] N. Peimyoo, J. Shang, C. Cong, X. Shen, X. Wu, E.K.L. Yeow, T. Yu, Nonblinking, intense two-dimensional light emitter: monolayer WS2 triangles, ACS Nano 7 (2013) 10985–10994, https://doi.org/10.1021/nn4046002.

[136] S.M. Poh, X. Zhao, S.J.R. Tan, D. Fu, W. Fei, L. Chu, D. Jiadong, W. Zhou, S.J. Pennycook, A.H.C. Neto, K.P. Loh, Molecular beam epitaxy of highly crystalline MoSe2 on hexagonal boron nitride, ACS Nano 12 (2018) 7562–7570, https://doi.org/10.1021/ACSNANO.8B04037.

[137] H.R. Gutiérrez, N. Perea-López, A.L. Elías, A. Berkdemir, B. Wang, R. Lv, F. López-Urías, V.H. Crespi, H. Terrones, M. Terrones, Extraordinary room-temperature photoluminescence in triangular WS2 monolayers, Nano Letters 13 (2013) 3447–3454, https://doi.org/10.1021/nl3026357.

[138] I.S. Kim, V.K. Sangwan, D. Jariwala, J.D. Wood, S. Park, K.S. Chen, F. Shi, F. Ruiz-Zepeda, A. Ponce, M. Jose-Yacaman, V.P. Dravid, T.J. Marks, M.C. Hersam, L.J. Lauhon, Influence of stoichiometry on the optical and electrical properties of chemical vapor deposition derived MoS_2, ACS Nano 8 (2014) 10551–10558, https://doi.org/10.1021/nn503988x.

[139] Q. Ji, Y. Zheng, Y. Zhang, Z. Liu, Chemical vapour deposition of group-VIB metal dichalcogenide monolayers: engineered substrates from amorphous to single crystalline, Chemical Society Reviews 44 (2015) 2587–2602, https://doi.org/10.1039/c4cs00258j.

[140] Y. Lin, B.M. Bersch, R. Addou, K. Xu, Q. Wang, C.M. Smyth, B. Jariwala, R.C. Walker, S.K. Fullerton-Shirey, M.J. Kim, R.M. Wallace, J.A. Robinson, Modification of the electronic transport in atomically thin WSe_2 by oxidation, Advanced Materials Interfaces 7 (2020) 2000422, https://doi.org/10.1002/admi.202000422.

[141] A. Khosravi, R. Addou, C.M. Smyth, R. Yue, C.R. Cormier, J. Kim, C.L. Hinkle, R.M. Wallace, Covalent nitrogen doping in molecular beam epitaxy-grown and bulk WSe_2, APL Materials 6 (2018) 026603, https://doi.org/10.1063/1.5002132.

[142] B. Huang, F. Tian, Y. Shen, M. Zheng, Y. Zhao, J. Wu, Y. Liu, S.J. Pennycook, J.T.L. Thong, Selective engineering of chalcogen defects in MoS_2 by low-energy helium plasma, ACS Applied Materials & Interfaces 11 (2019) 24404–24411, https://doi.org/10.1021/acsami.9b05507.

[143] Y.-C. Lin, C. Liu, Y. Yu, E. Zarkadoula, M. Yoon, A.A. Puretzky, L. Liang, X. Kong, Y. Gu, A. Strasser, H.M. Meyer, M. Lorenz, M.F. Chisholm, I.N. Ivanov, C.M. Rouleau, G. Duscher, K. Xiao, D.B. Geohegan, Low energy implantation into transition metal dichalcogenide monolayers to form Janus structures, ACS Nano 14 (2020) 3896–3906, https://doi.org/10.1021/acsnano.9b10196.

[144] K. Fujisawa, B.R. Carvalho, T. Zhang, N. Perea-López, Z. Lin, V. Carozo, S.L.L.M. Ramos, E. Kahn, A. Bolotsky, H. Liu, A.L. Elías, M. Terrones, Quantification and healing of defects in atomically thin molybdenum disulfide: beyond the controlled creation of atomic defects, ACS Nano (2021), https://doi.org/10.1021/acsnano.0c10897.

[145] G. Danda, P. Masih Das, M. Drndić, Laser-induced fabrication of nanoporous monolayer WS_2 membranes, 2D Materials 5 (2018) 035011, https://doi.org/10.1088/2053-1583/aabb73.

[146] T. Afaneh, P.K. Sahoo, I.A.P. Nobrega, Y. Xin, H.R. Gutiérrez, Laser-assisted chemical modification of monolayer transition metal dichalcogenides, Advanced Functional Materials 28 (2018) 1802949, https://doi.org/10.1002/adfm.201802949.

[147] A.K. Geim, I.V. Grigorieva, Van der Waals heterostructures, Nature 499 (2013) 419–425, https://doi.org/10.1038/nature12385.

[148] K.S. Novoselov, A. Mishchenko, A. Carvalho, A.H. Castro Neto, 2D materials and van der Waals heterostructures, Science 353 (2016) aac9439, http://science.sciencemag.org/content/353/6298/aac9439. (Accessed 11 June 2017).

[149] S. Manzeli, D. Ovchinnikov, D. Pasquier, O.V. Yazyev, A. Kis, 2D transition metal dichalcogenides, Nature Reviews Materials 2 (2017) 1–15, https://doi.org/10.1038/natrevmats.2017.33.

[150] C.R. Dean, A.F. Young, I. Meric, C. Lee, L. Wang, S. Sorgenfrei, K. Watanabe, T. Taniguchi, P. Kim, K.L. Shepard, J. Hone, Boron nitride substrates for high-quality graphene electronics, Nature Nanotechnology 5 (2010) 722–726, https://doi.org/10.1038/nnano.2010.172.

[151] P.K. Kannan, D.J. Late, H. Morgan, C.S. Rout, Recent developments in 2D layered inorganic nanomaterials for sensing, Nanoscale 7 (2015) 13293–13312, https://doi.org/10.1039/c5nr03633j.

[152] S.J. Haigh, A. Gholinia, R. Jalil, S. Romani, L. Britnell, D.C. Elias, K.S. Novoselov, L.A. Ponomarenko, A.K. Geim, R. Gorbachev, Cross-sectional imaging of individual layers and buried interfaces of graphene-based heterostructures and superlattices, Nature Materials 11 (2012) 764–767, https://doi.org/10.1038/nmat3386.

[153] Y. Gong, J. Lin, X. Wang, G. Shi, S. Lei, Z. Lin, X. Zou, G. Ye, R. Vajtai, B.I. Yakobson, H. Terrones, M. Terrones, B.K. Tay, J. Lou, S.T. Pantelides, Z. Liu, W. Zhou, P.M. Ajayan, Vertical and in-plane heterostructures from WS_2/MoS_2 monolayers, Nature Materials 13 (2014) 1135–1142, https://doi.org/10.1038/nmat4091.

[154] W. Zhan, X. Yong, W. Haolin, W. Ruixue, N. Tang, Z. Yongjie, S. Jing, J. Teng, Z. Ying, L. Yimin, Y. Mei, W. Weidong, Z. Qing, M. Xiaohua, H. Yue, NaCl-assisted one-step growth of MoS_2–WS_2 in-plane heterostructures, Nanotechnology 28 (2017) 325602.

[155] J.A. Miwa, M. Dendzik, S.S. Grønborg, M. Bianchi, J.V. Lauritsen, P. Hofmann, S. Ulstrup, Van der Waals epitaxy of two-dimensional MoS_2-graphene heterostructures in ultrahigh vacuum, ACS Nano 9 (2015) 6502–6510, https://doi.org/10.1021/acsnano.5b02345.

[156] S. Subramanian, D.D. Deng, K. Xu, N. Simonson, K. Wang, K. Zhang, J. Li, R. Feenstra, S.K. Fullerton-Shirey, J.A. Robinson, Properties of synthetic epitaxial graphene/molybdenum disulfide lateral heterostructures, Carbon N.Y. 125 (2017) 551–556, https://doi.org/10.1016/J.CARBON.2017.09.058.

[157] Y.-C. Lin, R.K. Ghosh, R. Addou, N. Lu, S.M. Eichfeld, H. Zhu, M.-Y. Li, X. Peng, M.J. Kim, L.-J. Li, R.M. Wallace, S. Datta, J.A. Robinson, Atomically thin resonant tunnel diodes built from synthetic van der Waals heterostructures, Nature Communications 6 (2015) 7311, https://doi.org/10.1038/ncomms8311.

[158] Y. Gong, Z. Lin, G. Ye, G. Shi, S. Feng, Y. Lei, A.L. Elías, N. Perea, R. Vajtai, H. Terrones, Z. Liu, M. Terrones, P.M. Ajayan, Te-assisted low-temperature synthesis of MoS_2 and WS_2 monolayers, ACS Nano 9 (2015) 11658–11666, https://doi.org/10.1021/acsnano.5b05594.

[159] H. Kim, D. Ovchinnikov, D. Deiana, D. Unuchek, A. Kis, Suppressing nucleation in metalorganic chemical vapor deposition of MoS_2 monolayers by alkali metal halides, Nano Letters 17 (2017), https://doi.org/10.1021/acs.nanolett.7b02311.

[160] Y. Lin, R. Torsi, D.B. Geohegan, J.A. Robinson, K. Xiao, Controllable thin-film approaches for doping and alloying transition metal dichalcogenides monolayers, Advancement of Science 8 (2021) 2004249, https://doi.org/10.1002/advs.202004249.

[161] H. Taghinejad, A.A. Eftekhar, P.M. Campbell, B. Beatty, M. Taghinejad, Y. Zhou, C.J. Perini, H. Moradinejad, W.E. Henderson, E.V. Woods, X. Zhang, P. Ajayan, E.J. Reed, E.M. Vogel, A. Adibi, Strain relaxation via formation of cracks in compositionally modulated two-dimensional semiconductor alloys, npj 2D Materials and Applications 2 (2018) 10, https://doi.org/10.1038/s41699-018-0056-4.

[162] A. Kutana, E.S. Penev, B.I. Yakobson, Engineering electronic properties of layered transition-metal dichalcogenide compounds through alloying, Nanoscale 6 (2014) 5820–5825, https://doi.org/10.1039/c4nr00177j.

[163] H.-P. Komsa, A.V. Krasheninnikov, Two-dimensional transition metal dichalcogenide alloys: stability and electronic properties, Journal of Physical Chemistry Letters 3 (2012) 3652–3656, https://doi.org/10.1021/jz301673x.

[164] J. Karthikeyan, H.P. Komsa, M. Batzill, A.V. Krasheninnikov, Which transition metal atoms can be embedded into two-dimensional molybdenum dichalcogenides and add magnetism?, Nano Letters 19 (2019) 4581–4587, https://doi.org/10.1021/acs.nanolett.9b01555.

[165] K. Zhang, B.M. Bersch, J. Joshi, R. Addou, C.R. Cormier, C. Zhang, K. Xu, N.C. Briggs, K. Wang, S. Subramanian, K. Cho, S. Fullerton-Shirey, R.M. Wallace, P.M. Vora, J.A. Robinson, Tuning the electronic and photonic properties of monolayer MoS_2 via in situ rhenium substitutional doping, Advanced Functional Materials 28 (2018) 1706950, https://doi.org/10.1002/adfm.201706950.

[166] K. Zhang, D.D. Deng, B. Zheng, Y. Wang, F.K. Perkins, N.C. Briggs, V.H. Crespi, J.A. Robinson, Tuning transport and chemical sensitivity via niobium doping of synthetic MoS_2, Advanced Materials Interfaces 7 (2020) 2000856, https://doi.org/10.1002/admi.202000856.

[167] K.A. Cochrane, T. Zhang, A. Kozhakhmetov, J.H. Lee, F. Zhang, C. Dong, J.B. Neaton, J.A. Robinson, M. Terrones, A.W. Bargioni, B. Schuler, Intentional carbon doping reveals CH as an abundant charged impurity in nominally undoped synthetic WS_2 and WSe_2, 2D Materials 7 (2020) 31003, https://doi.org/10.1088/2053-1583/ab8543.

[168] R. Zhao, C. Lo, F. Zhang, R.K. Ghosh, T. Knobloch, M. Terrones, Z. Chen, J. Robinson, Incorporating niobium in MoS_2 at BEOL-compatible temperatures and its impact on copper diffusion barrier performance, Advanced Materials Interfaces 6 (2019) 1901055, https://doi.org/10.1002/admi.201901055.

[169] B. Huang, M. Yoon, B.G. Sumpter, S.H. Wei, F. Liu, Alloy engineering of defect properties in semiconductors: suppression of deep levels in transition-metal dichalcogenides, Physical Review Letters 115 (2015) 126806, https://doi.org/10.1103/PhysRevLett.115.126806.

[170] X. Li, M.-W. Lin, L. Basile, S.M. Hus, A.A. Puretzky, J. Lee, Y.-C. Kuo, L.-Y. Chang, K. Wang, J.C. Idrobo, A.-P. Li, C.-H. Chen, C.M. Rouleau, D.B. Geohegan, K. Xiao, Isoelectronic tungsten doping in monolayer $MoSe_2$ for carrier type modulation, Advanced Materials 28 (2016) 8240–8247, https://doi.org/10.1002/adma.201601991.

[171] H. Gao, J. Suh, M.C. Cao, A.Y. Joe, F. Mujid, K.H. Lee, S. Xie, P. Poddar, J.U. Lee, K. Kang, P. Kim, D.A. Muller, J. Park, Tuning electrical conductance of MoS2Monolayers through substitutional doping, Nano Letters 20 (2020) 4095–4101, https://doi.org/10.1021/acs.nanolett.9b05247.

[172] A. Kozhakhmetov, B. Schuler, A.M.Z. Tan, K.A. Cochrane, J.R. Nasr, H. El-Sherif, A. Bansal, A. Vera, V. Bojan, J.M. Redwing, N. Bassim, S. Das, R.G. Hennig, A. Weber-Bargioni, J.A. Robinson, Scalable substitutional re-doping and its impact on the optical and electronic properties of tungsten diselenide, Advanced Materials 32 (2020) 2005159, https://doi.org/10.1002/adma.202005159.

Materials engineering – defect healing & passivation

Yu Li Huang, Rebekah Chua, and Andrew Thye Shen Wee
Department of Physics, National University of Singapore, Singapore, Singapore

7.1 Introduction

Two-dimensional (2D) materials have great potential for future device applications due to their novel electronic and optical properties. These properties are often strongly modified by defects, which are inevitable during the growth and device processing procedures [1–5]. The common defects observed in 2D materials include intrinsic disorders arising from crystalline imperfections such as vacancy, anti-site, substitution, edge and grain boundary (GB), and extrinsic disorders arising from environments such as strain, adsorbates, surface roughness, charged impurities and oxidations (Fig. 7.1) [1]. These defects can modify the material properties significantly by varying local electrostatic potential [6], modifying electronic band structures [7], scattering charge carrier [8], and acting as recombination centres for excitons [9], which consequently result in much lower carrier mobilities and optical responses compared to theoretical predictions. To realize their full potential, it is crucial to reduce the defects in the 2D crystals as well as minimize their impacts in the device performance.

Compared to graphene and boron nitride (BN), 2D transition metal dichalcogenides (TMDs) are more susceptible to chemical interactions with the environment, and the formation energy for similar defects in TMDs is also much lower. For instance, the formation energy for a carbon vacancy is about 7–8 eV in graphene [10], versus that of only 2.1 eV for a sulfur vacancy in MoS_2 [7,11]. High quality graphene and BN films with low defect densities are achievable by mechanical exfoliation or growth due to their outstanding chemical stability. 2D TMDs exhibit high tunable properties with respect to the defects, particularly the semiconducting bandgap, and carrier density and type. For example, deep gap states can be induced by the presence of defects such as vacancies, lattice antisites, substitutions, dislocations and so on [11–13]. The bandgaps are tunable with strain, adsorbates, and impurities [14–17], and semiconducting-to-metallic transitions have been observed at GBs and edges [6,18,19]. It is of great interest to understand how to control the formation, healing, and passivation of the defects, and thus engineer the material properties.

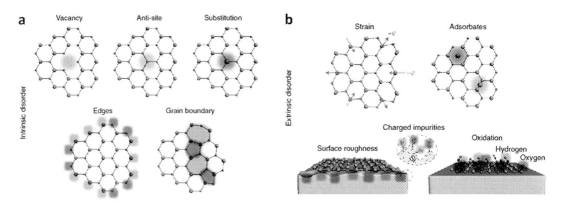

FIGURE 7.1 Types of defects in 2D TMDs. (a) Intrinsic defects: vacancy, anti-site, substitution, edges, and grain boundary (GB). (b) Extrinsic defects: strain, adsorbates, surface roughness, charge impurities and oxidations. Figures adapted with permission from Ref. [1].

In this chapter, we will focus on the recent progress achieved in defect healing and passivation, particularly in 2D TMDs. Many prototypical intrinsic point and line defects will first be discussed, as well as their self-passivated property with oxygen. Next, the underlying mechanism of defect engineering via chemical treatments, e.g., decoration with organic layers, and the improved properties for device applications are summarized. The uses of various external sources, including thermal heating, electron beams, and plasma treatment, also provide promising methods for defect control. In particular, the protection of the 2D crystal surface with encapsulation, e.g., oxide and BN layers, to eliminate extrinsic disorders and prevent oxidation are presented. Finally, we conclude with a perspective on the future challenges for 2D TMDs in harnessing their intrinsic properties.

7.2 Defect formation and healing in 2D TMDs

7.2.1 Point defects

As a prototypical TMD (MX_2), intrinsic defects in MoS_2 monolayer have been intensively studied, and various different types of point and line defects have been observed in samples prepared by physical vapor deposition (PVD), chemical vapor deposition (CVD) as well as mechanical exfoliation (ME) [7,11,17,18]. Fig. 7.2a shows three point defects commonly observed in single-layered (SL) MoS_2 grown by CVD method imaged by scanning transmission electron microscopy (STEM), including the monosulfur vacancy (V_S), disulfur vacancy (V_{S2}), and anti-site defects formed by a S_2 column substituting a Mo atom ($S2_{Mo}$) [11]. Theoretical calculations predict that these defects can introduce localized gap states into the bandgap of MoS_2 (Fig. 7.2b), thus modifying its intrinsic properties. Similar defects have also been observed in other 2D TMD crystals such as $MoSe_2$, WS_2, WSe_2 and so on [5].

The stability of the point defects can be inferred from their formation energy obtained from density functional theory (DFT) calculations. Fig. 7.2c and 7.2d show the formation energy

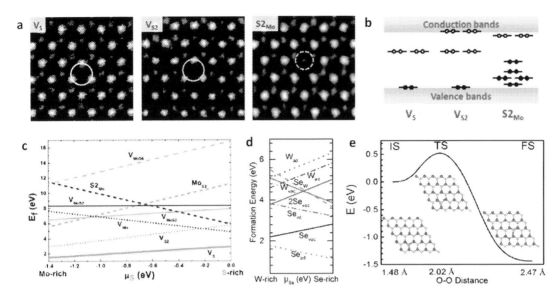

FIGURE 7.2 Common point defects in 2D TMDs. (a) STEM images of various point defects observed in CVD-grown MoS$_2$ monolayer: V$_S$, V$_{S2}$, and S2$_{Mo}$. (b) A schematic depicts the defect levels in MoS$_2$ bandgap. (c) Formation energies of different point defects as functions of sulfur chemical potential (μ_S) in SL-MoS$_2$. (d) Formation energies of various intrinsic defects in SL-WSe$_2$ on graphite. (e) Energy barrier for O$_2$ dissociative adsorption at the Se vacancy site. Figures (a–c) adapted with permission from Ref. [11]; (d–e) from Ref. [20].

of various intrinsic defects in SL-MoS$_2$ and SL-WSe$_2$, respectively [11,20]. Various types of common defects, including X and M vacancies, X adatoms, X and M interstitials, and XX and MX double vacancies in MX$_2$ monolayers, have been considered. Here, M is a group IV, V or VI transition metal (typically Mo and W), and X is a chalcogen (i.e., S, Se or Te). Clearly, the formation of the intrinsic defects greatly depends on the growth processes, and the predominant type and density of the defects can be controlled via process engineering. Among these defects, the X vacancy usually possesses the lowest formation energies in both X-rich and M-rich conditions, apart from the X adatom (X$_{ad}$) [11,20,21]. Although X$_{ad}$ may have the lowest formation energy, it is not the most abundant defect observed in experimental because it can be removed easily by thermal annealing, probing tips or electron beams. The theoretical predictions are consistent with experimental observations that the chalcogen (X) vacancies are the predominant defect found in 2D TMDs, e.g., S$_{vac}$ in SL-MoS$_2$ [7] and Se$_{vac}$ in SL-WSe$_2$ [20].

The chalcogen vacancies (X$_{vac}$) are predominantly understood to be the origin of catalytic activity [22–24], optical responses (e.g., single-photon emission) [25], and transport characteristics [26] of 2D TMDs, but it is susceptible to chemical interactions in ambient. Theoretical calculations have predicted that the chalcogen vacancy defects can be substituted by many non-metal atoms such as O, C, and N, which originate from O$_2$, CO, NO and NO$_2$ molecules used as precursors [20,27]. Among them, only O substitution can remove the gap states induced by the vacancy defect, while N and C substitutions result in new gap states in the 2D TMDs [27,28]. A study by Zheng et al. has revealed that O$_2$ can dissociate easily at the

Se_{vac} site in SL-WSe$_2$, and the dissociation barrier is only 0.52 eV (Fig. 7.2e) [20]. This indicates that oxygen substitution can be activated thermally, and take place at room temperature (a $\sim$0.7 eV barrier estimated by $\sim$30 $k_B T$ at RT). Thus, the chalcogen vacancy is easily passivated spontaneously in ambient conditions.

The spontaneous passivation of chalcogenide vacancies with O$_2$ dissociation has been observed experimentally. The three most commonly observed point defects in SL-WSe$_2$ are O-related defects, namely O bound to Se vacancy (O_{Se}), O interstitial (O_{ins}) in the hexagonal hollow site within the W layer, and O adsorbed atop Se (O_{ad}), as shown in Fig. 7.3a–c [20]. In contrast to the intrinsic defects, all these O-related defects have no gap states. Since the CVD grown SL-WSe$_2$ sample was exposed to ambient and subsequently annealed in vacuum to $\sim$300 °C, it is reasonable to observe the passivation of Se_{vac} by O to form O_{Se} in scanning tunneling microscopy (STM) images, thus removing the gap states. In another study, Barja et al. also found that the chalcogen-site point defects commonly observed in SL-MoSe$_2$ and SL-WS$_2$ grown by MBE and CVD are indeed substitutional oxygen atoms [29]. As shown in Fig. 7.3d–g, using a combination of STM, non-contact atomic force microscopy (nc-AFM) and state-of-the-art theory calculations, the most abundant types of point defects in the MoSe$_2$ sample are identified to be O_{Se} defects, without the presence of gap states. These studies suggest that the chalcogen vacancies in the 2D-TMD semiconductors can be easily passivated by O-related defects after gentle annealing, thereby removing the defect gap states.

The vacancies in TMDs can result in n- or p-doing, and act as nonradiative recombination centres for excitons which are crucial in determining their electronic and optical properties. Strong enhancement of photoluminescence (PL) can be achieved by passivating the chalcogenide vacancies with O substitution, via thermal annealing, oxygen plasma irradiation [30] or laser beam treatment [31]. With the suppression of defect-related mid-gap states, the photoelectric properties of the 2D TMDs can be significantly improved. For instance, the conductivity of monolayer WSe$_2$ can be enhanced by $\sim$400 times, and its photoconductivity by $\sim$150 times after laser beam treatment in air due to the healing of the electronic structure of the materials [31]. The catalytic activity also changes with O substitution, e.g., they act as single-atom reaction centres for the improved efficiency in electrochemical H$_2$ evolution reactions [23,32]. Such substitutional oxygen defects can be healed to recover defect-free 2D crystals via thermally annealing under chalcogen rich conditions [23].

7.2.2 Line defects

As one of the most common defects, grain boundaries (GBs) have been widely observed in monolayer TMDs synthesized by bottom-up CVD and MBE. GBs formed by irregular or regular lines containing 4-, 5-, 7- and 8-membered dislocation cores have been reported in MoS$_2$, MoSe$_2$, WS$_2$, WSe$_2$ and so on, which usually results in deep gap states and/or bandgap tunability [5,11,18,33]. It is therefore important to understand the atomic structures of the GB defects, which could play key roles in determining the material properties and hence potential device applications.

Various prototypical GB defects are illustrated in Fig. 7.4. The STEM image in Fig. 7.4a shows a GB in a CVD-grown MoS$_2$ monolayer forming with a recurring periodic 8-4-4 ring motif, and DFT calculations in Fig. 7.4b reveal that localized mid-gap states develop in the GB [18]. These 8–4 defects cause strong PL enhancement but slightly decrease the electronic con-

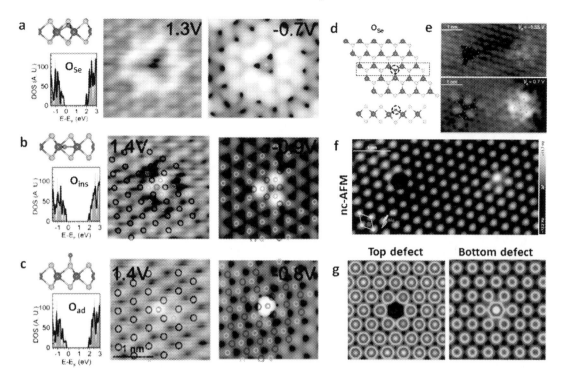

FIGURE 7.3 Chalcogenide vacancies passivated with oxygen. (a–c) Three dominant defects observed in SL-WSe$_2$: (a) O$_{Se}$, (b) O$_{ins}$ and (c) O$_{ad}$. Left panels: Atomic structures and density of state (DOS) of the O-related defects; middle panels: experimental STM images; right panels: simulated STM images. (d–g) O$_{Se}$ defects observed in 2D MoSe$_2$: (d) atomic models, (e) STM images, (f) nc-AFM image and (g) simulated STM images. O$_{Se}$ top and bottom defects in the top and bottom Se layers demonstrate distinct features in STM and nc-AFM images. Figures (a–c) adapted with permission from Ref. [20]; figures (d–g) from Ref. [29].

ductivity. In contrast, mirror twin boundaries (MTBs), which were also observed in the same sample, cause PL quenching and slight electrical conductivity increase [18]. In Fig. 7.4c–d, STM and scanning tunneling spectroscopy (STS) were employed to investigate the atomic and electronic structures of low-angle GBs in SL-WSe$_2$, where the misorientation angles between the neighbor grains are typically in the range of 3–6° [13]. The WSe$_2$ monolayer film was grown by CVD method and transferred onto a graphite substrate for STM/STS measurements. Butterfly features are observed along the GBs, and the interspacing between the butterflies is found to depend on the misorientation angle, in accordance with the Burges model [13]. Distinguishing electronic states arising from the dislocation cores are observed in the SL-WSe$_2$ bandgap (Fig. 7.4d). Theoretical calculations (Fig. 7.4e–g) reveal that these butterfly features correspond to the localized gap states that arise in tetragonal dislocation cores and extend to distorted six-membered rings around the dislocation core.

Fig. 7.4g–i shows the atomic structures and electronic properties of a dense MTB network formed in MBE-grown MoSe$_2$ on graphite surfaces [6]. MTBs in SL-MoSe$_2$ and other 2D TMDs have attracted increasing attention recently due to the direct observation of

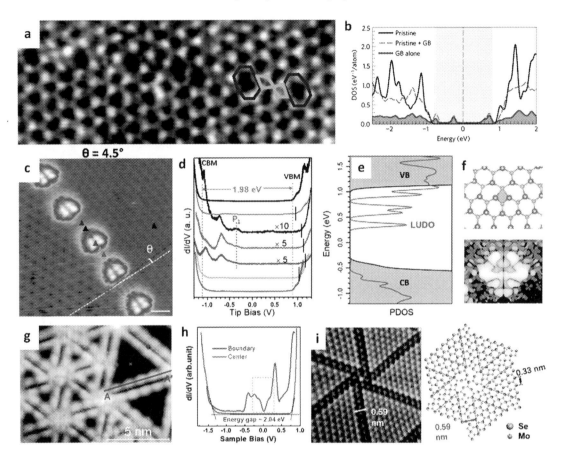

FIGURE 7.4 Grain boundary (GB) defects in 2D TMDs. (a) A GB in monolayer MoS$_2$ with a periodic line of 8-4-4 ring defects, and (b) the total density of states (DOS) of pristine MoS$_2$ and the GB. (c–f) Small-angle GB in SL-WSe$_2$. (c) The dislocation cores appear as butterfly features in a periodic GB of 4.5° misorientation. (d) STS spectra taken on the marked position in panel (c) reveal that deep gap states arise at the dislocation cores. (e) Calculated PDOS of the pristine SL-WSe$_2$ (black) and the dislocation core (red). (f) The atomic model of the tetragonal dislocation core (top) and the simulated STM image (bottom). (g–i) Mirror twin boundaries (MTB) in SL-MoSe$_2$. (g) Dense MTB networks are observed in MBE-grown MoSe$_2$ surface. (h) dI/dV spectra recorded at the domain center (red line) and MTB (blue line) reveal their semiconducting and metallic nature, respectively. (i) A high-resolution nc-AFM image shows a typical wagon-wheel pattern (left) and the corresponding atomistic model (right). Figures (a–b) adopted with permission from Ref. [18]; figures (c–f) from Ref. [13]; and figures (g–i) from Ref. [6].

one-dimensional (1D) charge density waves (CDW) along the MTBs at low temperatures [19,34,35]. In Fig. 7.4h, typical STS spectra taken on the pristine MoSe$_2$ domain (red curve) and the MTBs (blue curve), clearly reveal their semiconducting and metallic characteristics, respectively. Such metallic line defects are usually unfavorable for carrier/energy transport and optical applications. However, they are often catalytically active due to their large DOS, and could be potentially used as templates for self-assembly of molecules [36]. He et al. have reported site-selective adsorption behaviors in the self-assembly of 2,3-diaminophenazine

(DAP) molecules on this dense MTB network, with the formation of a porous supermolecular structure that map onto the wagonwheel patterns of the underlying MoSe$_2$ [6].

Experimentally, it has been observed that GBs in SL-MoS$_2$ heal themselves during sample processing, resulting in an absence of deep gap states but unexpected tunability of the bandgap size [17]. Fig. 7.5a shows a large-scale STM image (75×75 nm^2) of a typical GB with brighter contrast in a SL-MoS$_2$ film, which was directly grown on a graphite substrate by CVD. The atomically resolved images in Fig. 7.5b–d reveal an 18°misorientation between grain I and II. Based on STS measurements, a significant variation of the bandgap respect to the distance from the GB is depicted in the schematic diagram of Fig. 7.5e. The bandgap decreases to ~0.85 eV at the 18° GB, and both the valence band maximum (VBM) and conduction band minimum (CBM) bend upwards at the GB. The change in bandgap also depends on the misorientation angle, where a decrease of only ~0.17 eV is obtained when the misorientation is 3°. In another study, it was observed that the bandgaps at the GB can be either increased or decreased relative to intrinsic SL-MoS$_2$, with the misorientation angle ranging from 0° to 25° [37]. It is known that the bandgap of 2D TMDs is very sensitive to the strain field (Fig. 7.5f) [38]. The probable mechanism contributing to the decrease or increase of the electronic bandgap is the strain induced by lattice distortions around the GB region, which can extend several nanometers from the dislocation cores at the GBs [39]. The upward bending observed both in VBM and CBM is also attributed to charge transfer effects, either from the substrate or the Thomas-Fermi screening effect [40]. Therefore, the bandgap tunability observed at the GBs is due to strain combined with charge-transfer effects (Fig. 7.5g).

The absence of deep gap states in the GB region is in sharp contrast to theoretical predictions. Since deep gap states usually arise from non-six-membered rings that form the line dislocation defects [11,18], such energetically unfavorable bonding configurations are not present in the GBs. As shown in Fig. 7.5b–d, the GB region has a wide width of ~1.3 nm, or four lattice constants, in contrast to previous models with the GB width extending over one lattice constant only (Fig. 7.4). In fact, due to the growth and annealing processes, the GB defects might be healed with extrinsic adsorbates, e.g., hydrogen or oxygen, which passivate the dangling bonds and remove the defect states. Therefore, the GB defects are mainly accommodated through lattice distortions that result in a strain field and hence the bandgap variation.

7.3 Defect engineering by chemical treatment and applications

7.3.1 Vacancy healing

The vacancies in 2D TMDs can be filled by chalcogen atoms with the assistance of chemical treatment, and thus restore the intrinsic properties of the material. For example, S$_{vac}$ defects in MoS$_2$ surface can be repaired by extrinsic S atoms produced from the dissociation of organic molecules containing thiol groups [41–44]. Fig. 7.6a shows an elegant example of (3-mercaptopropyl) trimethoxysilane (MPS) molecules repairing S vacancies in SL-MoS$_2$ via two-step reactions [42]. In the first step, the MPS molecules selectively adsorb onto the S vacancies with S–H bond dissociations to form thiolate surface intermediates. Subsequently, the S–C bond of the thiolate intermediate breaks leaving the lone S atom at the vacancy site. This

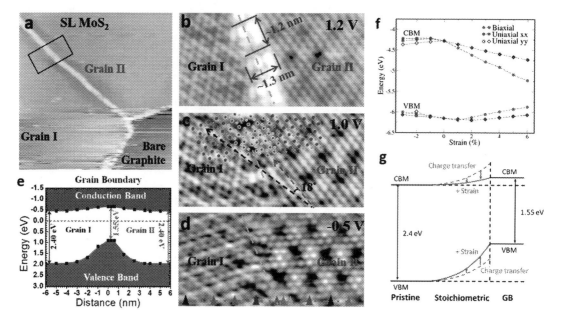

FIGURE 7.5 Unexpected bandgap tunability at the GB in SL-MoS$_2$. (a) Large-scale STM image shows a GB appearing as bright protrusions. (b–d) Bias-dependent images recorded at the boundary region highlighted by a black rectangle in panel a. (e) Schematic diagram shows the bandgap change across the GB. (f) Strain-induced CBM and VBM changes of monolayer MoS$_2$, where the vacuum level is aligned at 0 eV. (g) Schematic illustration demonstrates that the band modulation across the GB is induced by strain and charge transfer. Figures adopted with permission from Ref. [17].

process can be driven by thermal annealing [42] or an electric field [43]. The high-resolution STEM images reveal that the density of S$_{vac}$ reduces dramatically after MPS treatment, from $\sim 6.5 \times 10^{13}$ cm^{-2} for the as-exfoliated samples (Fig. 7.6b) to $\sim 1.6 \times 10^{13}$ cm^{-2} for the top-side treated sample (Fig. 7.6c). Improvements in electrical conductivity and carrier mobility (μ) were also obtained (Fig. 7.6d and 7.6e). A mobility (field effect) as high as 81 cm^{2}V^{-1}S^{-1} was achieved in the double-side treated MoS$_2$ sample.

Fig. 7.6f demonstrates a S vacancy self-healing (SVSH) strategy with poly(4-styrene-sulfonate) (PSS) treatment [45]. Here, the S adatoms pre-existing on the MoS$_2$ surface were guided by the PSS-induced hydrogenation process to repair the vacancies. As shown in Fig. 7.6g–h, a current decrease can be observed in both the output and transfer characteristic curves, and the threshold voltage shifts toward zero in the self-healed sample. It leads a dramatical decrease (by 643 times) of the electron concentration as well as a work function increase of $\sim$150 meV via the SVSH method. Furthermore, this strategy can be employed to fabricate a lateral SL-MoS$_2$ homojunction photodiode with high performance as well as outstanding air-stability [45].

2D TMDs have interesting optical properties, and are being explored for applications in many optoelectronic devices [46,47]. Compared to the bulk, monolayer TMDs have a direct bandgap and strong PL intensities [48,49]. However, the PL quantum yield (QY), defined as the ratio of the population of radiated photons over the population of the total generated

7.3. Defect engineering by chemical treatment and applications

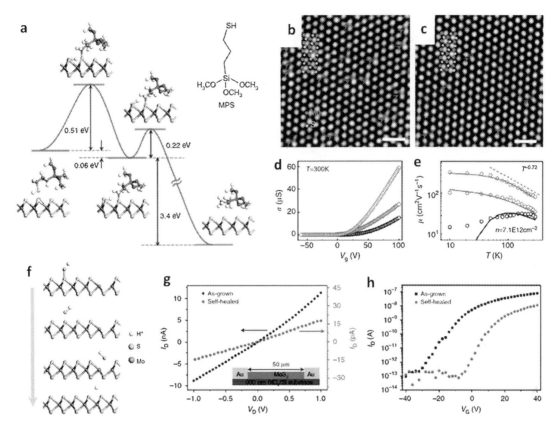

FIGURE 7.6 Sulfur vacancy healing. (a) Atomic plots of the initial, transient and final states involved in the reactions between an S vacancy and an MPS molecule. (b,c) High-resolution TEM images of monolayer MoS$_2$ samples before and after MPS treatment, where a significant reduction of S vacancy density is observed. (d) The conductivity σ as a function of back gate voltage Vg for three devices fabricated by as-exfoliated (black), top-side treated (blue) and double-side treated (red) MoS$_2$ monolayers at room temperature. (e) The corresponding μ–T curves for the three devices at $n = 7.1 \times 10^{12}$ cm^{-2}, where the solid lines are theoretical fittings. (f) Atomic structural model shows the PSS-induced SVSH effect. (g,h) Output characteristics and transfer characteristics of a monolayer MoS$_2$ transistor before and after PSS treatment. Figures (a–e) adopted with permission from Ref. [42]; and figures (f–h) from Ref. [45].

electron-hole pairs, is typically below < 1% due to the large number of defects in the 2D TMDs [9,48,50]. That is, instead of the desired radiative recombination, the defect-mediated nonradiative and biexcitonic recombination are dominant, and electrons and holes are mainly captured by defects via Auger scattering [9]. The healing of chalcogen vacancies via chemical treatments largely eliminates the defect-mediated recombination and hence enhances the optical responses.

Fig. 7.7a–c shows the PL intensity of a MoS$_2$ monolayer has a 190-fold increase after treating with an organic superacid, bis(trifluoromethane) (TFSI) [50]. A PL QY >95% was also obtained from the pump-power dependence of the calibrated PL intensity at a low pump

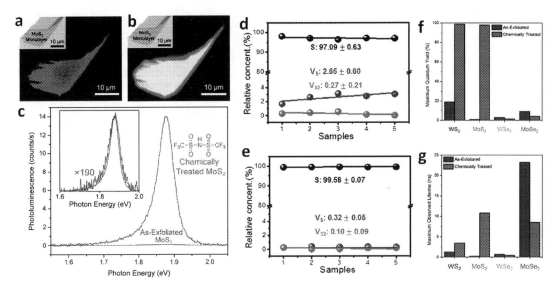

FIGURE 7.7 Significant PL enhancement after vacancy healing. (a,b) PL images of a MoS$_2$ monolayer before and after superacid (TFSI) treatment. Insets show optical images of the sample. (c) The corresponding PL spectra for the as-exfoliated (blue) and TFSI-treated (red) MoS$_2$ monolayers, and the inset shows a 190-fold increase. (d,e) Measured concentrations of the total V$_S$, V$_{S2}$, and S in the five MoS$_2$ samples before and after TFSI-treatment. (f,g) Summaries of the peak QY and lifetimes for various TMDs obtained before and after chemical treatments. Figures (a–c) adopted with permission from Ref. [50]; figures (d,e) from Ref. [43]; and figures (f,g) from Ref. [51].

intensity (10^{-2} W cm^{-2}), and a much longer lifetime of ~10 ns was extracted (compared to that of several tenths ns in as-exfoliated MoS$_2$). The near-unity QY suggests a significant improvement in the sample quality. A follow-up study revealed that most of the S vacancies in SL-MoS$_2$ were directly healed by the extrinsic S atoms dissociated from the TFSI molecules, resulting in the significant PL enhancement [43]. As summarized in Fig. 7.7d and 7.7e, V$_S$ is the dominant defect in the CVD-grown MoS$_2$ monolayer with a concentration of ~2.65% obtained from STEM measurements, which is reduced to 0.32% after TFSI treatment [43]. Similar PL enhancement can be also observed in SL-WS$_2$ with chemical treatment, and a QY over 95% was achieved with the same strategy (Fig. 7.7f) [51]. However, this strategy is far less effective in Se based systems such as MoSe$_2$ and WSe$_2$. TFSI treatment results in a small drop in the PL intensity and radiative lifetime from as-exfoliated MoSe$_2$ and WSe$_2$ samples (Fig. 7.7f and 7.7g). This is because the defects in MoSe$_2$ and WSe$_2$ include not only chalcogen vacancies, but also missing metal atoms [26].

All these studies suggest that simple chemical treatment can be used to heal the vacancy defects in 2D TMDs, thereby improving their carrier mobility and PL properties, and advancing practical optoelectronics applications.

7.3.2 Covalent functionalization

In addition to vacancy healing, organic molecules containing thiol (-SH) groups can be used to covalently functionalize 2D TMDs by selective adsorption on defect sites.[16,52–54]

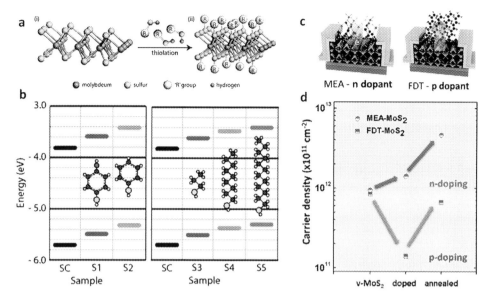

FIGURE 7.8 Surface covalent functionalization with vacancies. (a) Schematic of thiol molecules with "R" group bounded to S vacancy sites to functionalize MoS$_2$ monolayer. (b) Energy diagram shows the change of the MoS$_2$ energy levels with different thiols with various substituent groups and chain lengths. (c) Schematic illustration presents n-type ad p-type doping of a back-gated MoS$_2$ FET with MEA and FDT molecules, respectively. (d) Summary of the changes in carrier density before and after doping process (measured at $V_{GS} = 0$ V). Figures (a,b) adopted with permission from Ref. [16]; and figures (c,d) from Ref. [52].

As demonstrated in the schematic of Fig. 7.8a, a thiol molecule containing a carbon-bonded thiol is depicted by R–SH formula, where R represents an alkyl or other organic functional groups [16]. After thiolation, the molecules graft to the sulfur vacancy sites in the MoS$_2$ flakes via the S atoms, with the dissociation of the S–H bond [16,53]. Depending on the selection of the functional groups, organic thiols are able to tune the energy level alignment, carrier type and density, optical response, as well as the catalytic properties of the 2D system [5,16,52,53,55]. For instance, the energy diagrams in Fig. 7.8b show that VBM and CBM positions of MoS$_2$ vary significantly with the tail groups, while the optical bandgap (~1.9 eV) is maintained. Both VBM and CBM positions shift upward relative to pristine MoS$_2$, induced by the weak Lewis base character and the dipole moments of the functional polar thiolate molecules [16]. The magnitude of the shift can be tuned by the substituent group as well as the chain length, to as much as 0.5 eV with the utilizations of S2 and S5 molecules. The capability of tuning the band position is important for many applications requiring specific band edge matching between TMDs and other device components (e.g., metal electrodes) for optimal device operation.

Other electrical properties of MoS$_2$ can be also effectively engineered by functional molecules chemically adsorbed on the vacancy sites. Cho et al. reported a decrease of the carrier concentration in a MoS$_2$ FET device after passivating sulfur vacancies with thiol molecules, e.g., alkanethiol (HS(CH$_2$)$_{n-1}$CH$_2$) [53]. Simultaneously, the source-drain current (I_{DS}) decreases by 45% and the threshold voltage shifts to the positive gate voltage direc-

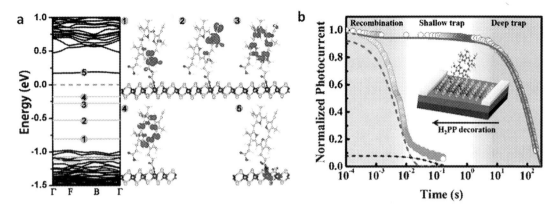

FIGURE 7.9 Charge transfer at the organic-TMD interfaces. (a) Band structures of ReS$_2$ with a protoporphyrin (H$_2$PP) molecule adsorbed on the S vacancy and the partial charge density distributions of the bands near the Fermi level. (b) Transient response of as-prepared and H$_2$PP decorated ReS$_2$, where the deep traps are removed in the later. Inset shows the schematic of the device. Figures adopted with permission from Ref. [56].

tion, similar to the effects from the vacancy healing [45]. This method can also be used to efficiently dope 2D TMDs via charge transfer as well as passivation effect. Sim et al. reported that a NH$_2$-containing thiol molecule (mercaptoethylamine, MEA) can serve as an n-dopant to donate electrons into the MoS$_2$ layers, while a fluorine-rich molecule (1H,1H,2H,2H-perfluorodecanethiol, FDT) can be used as a p-dopant to withdraw electrons from the material due to the high electronegativity value of F (Fig. 7.8c and 7.8d) [52]. Thus, an increase and a decrease of the carrier density were achieved in the MoS$_2$ samples treated with MEA and FDT respectively. Upon further annealing, the carrier density increased in both cases due to the detrapping of electrons from adsorbed H$_2$O and O$_2$ molecules in the MoS$_2$ channel layers.

7.3.3 Interfacial charge transfer

The defect-induced trap states in 2D TMDs can also be passivated by organic molecules without thiol groups, via charge transfer between the organic-TMD interfaces [56,57]. Jiang et al. reported that the localized defect states in ReS$_2$ monolayer disappear upon decoration with protoporphyrin (H$_2$PP) molecules [58]. As shown in Fig. 7.9a, although several new energy states (levels 1–5) appear due to the adsorption of H$_2$PP molecule on the S vacancy, only level 5 resides in ReS$_2$ while levels 1–4 belong to the molecule. A charge transfer of 0.65 e between the ReS$_2$ and molecule was estimated by theoretical calculations. The deep trap centres, which greatly limit the decay time of ReS$_2$ photoconductor, can be successfully passivated by the organic molecules selectively adsorbed on the vacancy sites. As a result, the decay time of the ReS$_2$ device is shortened by several orders after the removal of deep traps via H$_2$PP decoration (Fig. 7.9b), and the specific detectivity of the phototransistor is also enhanced because of the reduced dark current. In another study, a strong negative charge transfer between SL-MoS$_2$ and a monolayer titanyl phthalocyanine (TiOPc) to remove the defect states was observed [57]. As a result of the formation of a van der Waals (vdW) interface, the I_{ON}/I_{OFF} in the backgated MoS_2 transistors increases by more than two orders in magnitude, and the PL signal is greatly enhanced.

Surface adsorption of strong organic acceptors or donors on 2D materials with and without defects is an effective approach for tuning the material properties via interfacial charge transfer. Both theoretical calculations and experimental measurements have revealed that the electronic structures of 2D TMDs can be modified by molecular dopants without the formation of covalent bonding [58,59,61]. For example, benzyl viologen (BV) is a strong electron donor, and is able to donate about 0.25 e per molecule to the semiconducting MoS_2 [60]; while fluorinated fullerene ($C_{60}F_{48}$) is a strong electron acceptor, which can withdraw about 0.29 e per molecule from a WSe_2 monolayer [40]. Such large charge transfer induces significant changes to the energy level alignments at the vdW organic-2D TMD interfaces as well as the device transport properties, and been reviewed in detail elsewhere [2,62–64].

7.4 Defect control by external sources

7.4.1 Thermal annealing

In monolayer TMDs synthesised by bottom-up approaches such as CVD and MBE, large-scale growth is yet to be realized due to the inevitable multiple imperfections, limiting their potential in industrial applications. To realize the fabrication of 2D crystals with high quality, e.g., low defect density and large grain size, it is important to understand the mechanisms and kinetics that control defect formation and annihlation during growth. Zhao et al. recently reported that GBs in multilayered $MoSe_2$ can migrate and subsequently annihilate during thermal annealing, thus healing the stacking faults and rotational disorders in the 2D crystal (Fig. 7.10a and 7.10b) [65]. The atomic resolution STEM images in Fig. 7.10c and 7.10d were recorded at the same region of a bilayer $MoSe_2$ before and after *in situ* annealing at 700 °C for 30 minutes. It can be observed that all 3R (interlayers stacking in AB sequence, Fig. 7.10e) regions transform to 2H (in AA′ sequence, Fig. 7.10f) configurations via a consecutive migration of the 1D MTB (Fig. 7.10g), leading to a phase homogenization [65]. GB sliding can also occur in a vertical $MoSe_2$/HOPG heterostructure driven by interfacial vdW coupling with the incommensurate substrate. In contrast, GBs randomly hop in all directions without obvious annihilation in a freestanding $MoSe_2$ monolayer. This suggests that the correction of the stacking faults and disorders in bilayer and multilayer $MoSe_2$ is facilitated by interlayer vdW coupling to enhance the interlayer binding energy. This offers a promising approach to obtain large-scale defect-free 2D materials via appropriate postgrowth annealing.

Annealing a crystal at elevated temperature can also be utilized to control the formation of defect patterns in 2D TMDs. As shown in Fig. 7.11a–c, distinct line features are observed in an MBE VSe_2 monolayer epitaxially grown on MoS_2 upon *in situ* annealing above 400 °C [66]. The density of the line features is controllable with the annealing temperature as well as duration. For example, after annealing the VSe_2 samples at 400 °C, 430 °C and 470 °C for 30 minutes, the densities of the line patterns are statistically estimated to be ~1%, ~3%, and ~6%, respectively. This formation process is reversible. By depositing Se on the defective sample and annealing to 240 °C, the pristine VSe_2 crystal without patterned features can be recovered (Fig. 7.11d). The 1D defective patterns are found to be composed of 8-member

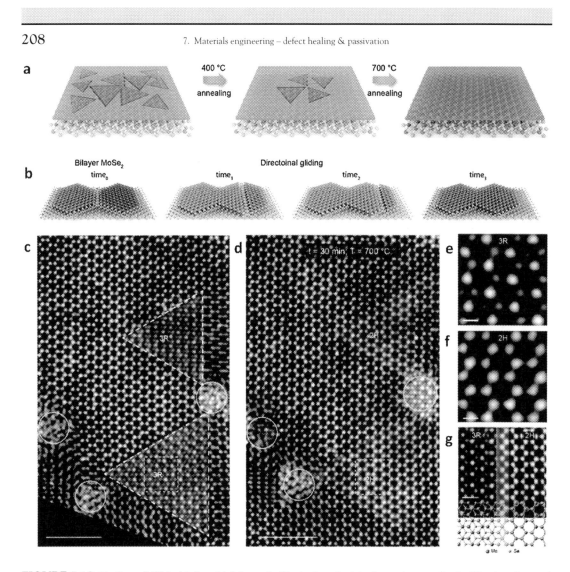

FIGURE 7.10 Healing of GB in MoSe$_2$. (a) Schematic illustration depicts the grain growth via GB migration and annihilation upon thermal annealing. (b) GB glides directionally in bilayer MoSe$_2$. (c,d) STEM images before and after *in situ* annealing at 700 °C for 30 min, showing the healing of stacking faults via MTB migration. Enlarged STEM images of (e) 3R- and (f) 2H-stacked bilayer MoSe$_2$ films derived from the yellow box regions in (c) and (d), respectively. (g) A 2H|3R domain boundary with overlaid atomic model. Figures adopted with permission from Ref. [65].

rings (Fig. 7.11e), which are formed via a Se deficient reconstruction process [66]. More interestingly, the defective VSe$_2$ monolayer displays detectable ferromagnetism signals under X-ray magnetic circular dichroism (XMCD) at room-temperature (Fig. 7.11g), while the pristine sample shows almost negligible signals (Fig. 7.11f). Therefore, this suggests the frustrated intrinsic magnetism in 2D VSe$_2$ [34] can be lifted by the introduction of such Se-deficient defects. Similarly, a formation of well-ordered Se-deficient line defects by annealing has also

7.4. Defect control by external sources 209

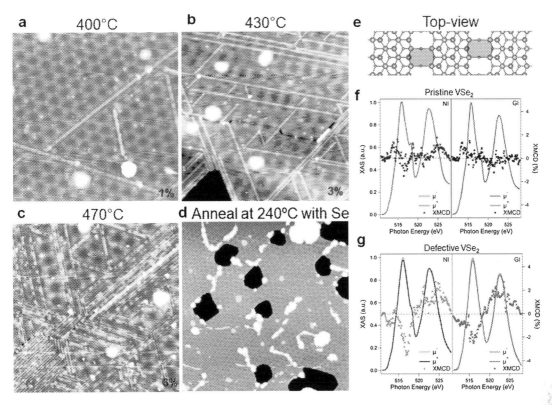

FIGURE 7.11 Formation and annihilation of line defects in VSe$_2$ monolayer. (a–c) The densities of the 1D line defects increase with the thermal annealing temperature: (a) 1% after annealing at 400 °C, (b) 3% at 430 °C, and (c) 6% at 470 °C. (d) The line defects annihilate after depositing Se and annealing at 240 °C. (e) Top-view of the atomic model of the 1D line defect with 4a$_0$ periodicity. (f,g) XAS spectra of V L$_{2,3}$ edge and XMCD spectra at room temperature (μH = ±1 T): pristine and defective VSe$_2$ samples measured at normal incidence (NI) and grazing incidence (GI). Figures adopted with permission from Ref. [66].

been observed in SL-VSe$_2$ grown on HOPG, which demonstrates promising catalytic activity for hydrogen evolution reaction (HER) comparable to Pt surfaces [67]. Therefore, thermal annealing is an efficient method to engineer the formation and annihilation of surface defects which could enrich their pristine properties with potential applications in devices.

7.4.2 Electron beam irradiation

STEM is a leading characterization tool in the field of 2D materials, due to its capability of achieving atomic resolution structural and elemental analysis [3,5,68–70]. However, 2D crystals with atomic thickness are susceptible to electron beams, where structural distortions or atomic defects can be created by knock-on damage even at low accelerating voltages below 80 kV. The energy transferred from the electron beam can be used to modify 2D materials via defect engineering, by precisely controlling the accelerating voltages as well as the irradiation time and region [3,69]. Fig. 7.12a and 7.12b illustrate the use of electron beams to generate patterned hole arrays in SL-MoS$_2$ with pore sizes down to 0.6 nm [69]. A loss of a single Mo atom with its neighboring six S atoms can be generated by shorter irradiation times, while nanopores form with longer irradiation times. During the irradiation, the loss of S atoms could induce the destabilization of the next-nearest Mo atoms, thus facilitating the formation of extended nanopores. In addition, bond rotations [71], defect migrations [39] and phase transition [72] can also be driven by electron beam irradiation. Fig. 7.12c–e demonstrate the formation process of trefoil defects via a 60° rotation of three W–Se bonds around the central W atom in ML-WS$_2$. During prolonged STEM imaging, the induced chalcogenide vacancy defects could be healed by extra chalcogenide adatoms or passivated with impurities (e.g., oxygen) [73]. Fig. 7.12f–k show a series of STEM images recorded at the same region in a GaSe monolayer. Beam-induced defects are observed to form, migrate to adjacent region, and then heal with relevant adatoms to recover the perfect lattice structure within approximately 10 seconds of consecutive STEM imaging. Similar phenomena have also been observed in a InSe monolayer [73]. The electron beams provide new opportunities to precisely engineer the atomic structures of 2D materials and their properties.

7.4.3 Plasma treatment

Plasma treatment with partially ionized gas (e.g., O$_2$, H$_2$, N$_2$, He, Ar, and NH$_3$) is another feasible and powerful method to modulate the properties of atomically thin 2D materials [74]. It can also be used to improve the stability, mobility and response of TMD-based devices via etching, surface modification, doping and so on. Many studies have focused on controlling the carrier types and concentrations by creating vacancies or introducing heteroatoms through plasma treatment in TMDs. For example, mild H$_2$, He or Ar plasma bombardment can induce chalcogenide vacancies in MoS$_2$ and WSe$_2$ flakes, thereby decreasing the PL intensity due to n-type doping [75,76]; while N$_2$ plasma can effectively induce N atoms doping in the chalcogenide vacancies and thus p-doping single- and few-layer MoS$_2$ and WS$_2$ [28,77].

Plasma techniques have been extensively used for thinning 2D crystals even down to monolayer via layer-by-layer etching, e.g., monolayer MoS$_2$ obtained by Ar plasma treatments [78] and monolayer black phosphorene by O$_2$ plasma treatments (Fig. 7.13) [79]. Furthermore, oxide layers can form on the surface during the O$_2$ etching process, thus improving the stability of the materials by preventing further degradation [79–81]. As shown in Fig. 7.13a–d, P$_x$O$_y$ layers and MoO$_x$ layers form on black phosphorene and MoS$_2$ surface after O$_2$ plasma treatment respectively, where the samples were further coated with protective Al$_2$O$_3$ layers. Liu et al. compared the transfer curves of MoS$_2$-based devices with and without O$_2$ plasma treatment [80]. A large hysteresis due to charge trapping is observed in pristine MoS$_2$ without any encapsulation, as a result of the adsorption of O$_2$ and H$_2$O in the ambient atmosphere on the surface. With O$_2$ plasma treatment and Al$_2$O$_3$ encapsulation, the hysteresis was almost completely eliminated. Mild O$_2$ plasma irradiation can also passivate the chalcogenide vacancies with the fills of O atoms to modulate the optical properties of the 2D semiconductors [82]. Plasma bombardment is an effective approach for diverse modifications of 2D TMD materials, due to its high controllability and energy efficiency.

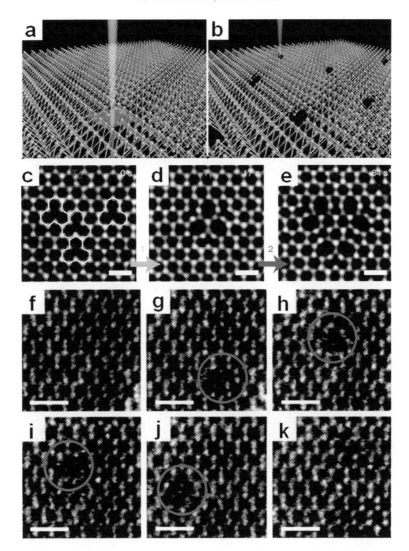

FIGURE 7.12 Engineering defects by electron beams. (a,b) Schematic illustration of the electron beam drilling sub-nanometer holes in monolayer MoS$_2$. (c–e) A series of STEM images showing the formation of a trefoil defects in WSe$_2$. (f–k) A series of consecutive STEM images showing formation, motion, and healing of a small defect cluster in monolayer GaSe. Time between frames is ~1s with an electron fluence per frame of ~1.3×10^5 e$^-$Å^{-2}. Figures (a,b) adopted with permission from Ref. [69]; figure (c–e) from Ref. [71]; figure (f–k) from Ref. [73].

7.4.4 Encapsulation

Surface protection and substrate flatness are critical for obtaining high-quality, high performance, stable TMD-based devices. As has been demonstrated in Fig. 7.1b, extrinsic disorders in 2D materials, such as adsorbates and oxidation under ambient conditions, and charge impurities and roughness from the supporting substrates (e.g., SiO$_2$), are obstacles in the

212　　　　　　　　7. Materials engineering – defect healing & passivation

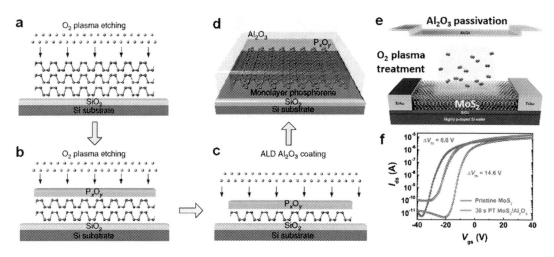

FIGURE 7.13 Plasma induced surface passivation. (a–d) Schematic illustration shows the fabrication of air-stable mono- and few-layer black phosphorene via the formation of the oxide layer. Figures adopted with permission from Ref. [79].

realization of practical applications. Therefore, it is important to develop effective strategies to create devices with protection for both top and bottom surfaces to minimize such extrinsic effects. To realize this aim, two main approaches have been utilized. The first approach is the deposition of high-k dielectric materials such as HfO_2 (Fig. 7.14a) and Al_2O_3 to screen and reduce charged impurities formed by surface oxidation [83]. Another approach is encapsulation with hexagonal boron nitride (hBN) (Fig. 7.14b), which provides efficient passivation due to its chemical stability, absence of dangling bond and a wide bandgap ($\sim$6 eV). hBN can also be used as dielectric layers to replace conventional dielectrics such as SiO_2.

Improvements of the electronic transport and optical properties can be achieved by reducing the extrinsic disorders in 2D TMDs with the encapsulation layers. Fig. 7.14c shows the temperature-dependent carrier mobilities of MoS_2 monolayers obtained under three different environments: on bare SiO_2 surface without encapsulation (black) [84] and with top HfO_2 encapsulation (blue) [85], and encapsulated between two hBN layers (red) [86]. For the SiO_2-supported device without encapsulation, the mobility of MoS_2 is $\sim$10 cm^2 V^{-1} S^{-1} at room temperature (RT) and limited to $\sim$170 cm^2 V^{-1} S^{-1} at low temperature (LT), far below the theoretically predicted phonon-limited value [8]. Obviously, the mobility of MoS_2 is significantly improved with both the HfO_2 and hBN encapsulations at RT, and with hBN encapsulation at LT. The encapsulation can also efficiently reduce and screen the inhomogeneous charge disorders that contribute to the exciton linewidth in PL measurements. Fig. 7.14d–f demonstrates that substantial narrowing of the optical transition linewidths is observed in MoS_2, WS_2, $MoSe_2$, and WSe_2 monolayers after encapsulation between hBN layers [87]. In particular, a full width at half maximum (FWHM) down to 2 meV is obtained at LT ($\sim$ 4 K) for the encapsulated MoS_2, much smaller than that obtained on bare SiO_2 (Fig. 7.14f) [87]. Furthermore, the improvements in TMD-based devices with hBN encapsulation even enable the observation of quantum transport characteristics in high magnetic fields, e.g., Shubnikov–de

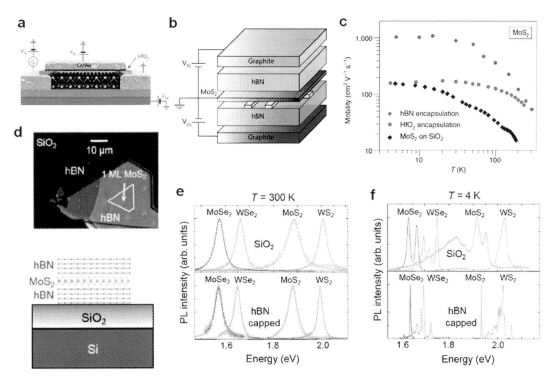

FIGURE 7.14 2D TMDs encapsulated with oxide and hBN layers. (a) A schematic shows a MoS$_2$-based device with HfO$_2$ encapsulation. (b) Device structure of a SL-MoS$_2$ encapsulated between two BN flakes and graphite flakes used as bottom and top gate. (c) The temperature-dependent mobility of SL-MoS$_2$ on SiO$_2$ without and with HfO$_2$ encapsulation, and with BN encapsulation. (d) Optical image (top) of a hBN/SL-MoS$_2$/hBN heterostructure and the corresponding schematic of the sample (bottom). (e,f) PL spectra for different TMD monolayers on SiO$_2$ without and with hBN encapsulation measured at (e) 300 K and (f) 4 K. Figure (a) adopted with permission from Ref. [85]; figure (b) from Ref. [86]; figure (c) from Ref. [1]; figures (d–f) from Ref. [87].

Haas (SdH) resistance oscillations in MoS$_2$ [88] and WSe$_2$ [89] originated from valley Zeeman splitting.

This encapsulation method has been widely applied to many air-sensitive TMDs (e.g., NbSe$_2$ [90] and HfS$_2$ [91]) as well as other layered crystals (e.g., BP [92] and CrI$_3$ [93]), and provides excellent protection from degradation due to oxidation. Thus, the novel intrinsic properties, e.g., superconductivity in NbSe$_2$ [90] and ferromagnetism in CrI$_3$ [93], are well preserved by encapsulating these 2D crystals with hBN layers in an inert atmosphere.

7.5 Future perspectives

Reducing defects in 2D materials is essential in order to fully realize their novel intrinsic properties and potential applications. In this chapter, we provide an overview of the current research effort devoted to defect healing and passivation, particularly in 2D TMDs. Towards this aim, the types and sources of the defects must be identified, and more work is needed to develop methods for rapid characterization of material quality. The common defects observed in 2D TMDs can be categorized into intrinsic defects arising from crystalline imperfects, and extrinsic defects arising from exposure to the environment. The formation of the intrinsic defects greatly depends on the growth environments, and self-healing can occur spontaneously in ambient conditions at RT or post-annealing. Ultimately, large-scale growth methods are needed to increase grain sizes and reduce defects in 2D crystals. Significant progress has been achieved in the synthesis of wafer-scale single-crystal graphene [94], hBN [95], and MoS_2 [96], as well as clean transfer techniques [97,98]. Scale-up encapsulation approaches are also needed for surface protection to reduce extrinsic defects, which is facilitated by the synthesis of high quality layered metal oxides as well as hBN films with large grain size. Further progress in this field requires rapid development in material synthesis and processing techniques to reduce defect density on a large scale.

Various strategies have been used to minimize the impact of defects on the material properties, including chemical treatment, electron beam irradiation, plasma treatment, and so on. Among them, chemical treatment and plasma treatment provide promising approaches for flexible modification of 2D materials over a large area and scale, allowing effective healing of vacancies, passivation of defective states, formation of surface protection layers, as well as doping of the materials. The utilization of STEM electron beams, however, allows precise atomic-scale control of the defects but with low efficiency. Significant improvement of material properties, e.g., carrier mobility and optical response, have been achieved for high performance electronic and optoelectronic devices. Intensive efforts are nevertheless still needed to develop material engineering techniques that enable precise control with high efficiency and stability.

References

[1] D. Rhodes, S.H. Chae, R. Ribeiro-Palau, J. Hone, Disorder in van der Waals heterostructures of 2D materials, Nature Materials 18 (2019) 541–549.

[2] S. Bertolazzi, M. Gobbi, Y. Zhao, C. Backes, P. Samori, Molecular chemistry approaches for tuning the properties of two-dimensional transition metal dichalcogenides, Chemical Society Reviews 47 (2018) 6845–6888.

[3] S. Wang, A. Robertson, J.H. Warner, Atomic structure of defects and dopants in 2D layered transition metal dichalcogenides, Chemical Society Reviews 47 (2018) 6764–6794.

[4] F. Banhart, J. Kotakoski, A.V. Krasheninnikov, Structural defects in graphene, ACS Nano 5 (2011) 26–41.

[5] Z. Lin, B.R. Carvalho, E. Kahn, R. Lv, R. Rao, H. Terrones, M.A. Pimenta, M. Terrones, Defect engineering of two-dimensional transition metal dichalcogenides, 2D Materials (3) (2016) 022002.

[6] X. He, L. Zhang, R. Chua, P.K.J. Wong, A. Arramel, Y.P. Feng, S.J. Wang, D. Chi, M. Yang, Y.L. Huang, A.T.S. Wee, Selective self-assembly of 2, 3-diaminophenazine molecules on MoSe2 mirror twin boundaries, Nature Communications 10 (2019) 2847.

[7] J. Hong, Z. Hu, M. Probert, K. Li, D. Lv, X. Yang, L. Gu, N. Mao, Q. Feng, L. Xie, J. Zhang, D. Wu, Z. Zhang, C. Jin, W. Ji, X. Zhang, J. Yuan, Z. Zhang, Exploring atomic defects in molybdenum disulphide monolayers, Nature Communications (6) (2015) 6293.

[8] N. Ma, D. Jena, Charge scattering and mobility in atomically thin semiconductors, Physical Review X (4) (2014) 011043.

[9] H. Wang, C. Zhang, F. Rana, Ultrafast dynamics of defect-assisted electron-hole recombination in monolayer MoS_2, Nano Letters 15 (2015) 339–345.

[10] S.T. Skowron, I.V. Lebedeva, A.M. Popov, E. Bichoutskaia, Energetics of atomic scale structure changes in graphene, Chemical Society Reviews 44 (2015) 3143–3176.

References

[11] W. Zhou, X. Zou, S. Najmaei, Z. Liu, Y. Shi, J. Kong, J. Lou, P.M. Ajayan, B.I. Yakobson, J.C. Idrobo, Intrinsic structural defects in monolayer molybdenum disulfide, Nano Letters 13 (2013) 2615–2622.

[12] X. Zou, Y. Liu, B.I. Yakobson, Predicting dislocations and grain boundaries in two-dimensional metal-disulfides from the first principles, Nano Letters 13 (2013) 253–258.

[13] Y.L. Huang, Z. Ding, W. Zhang, Y.H. Chang, Y. Shi, L.J. Li, Z. Song, Y.J. Zheng, D. Chi, S.Y. Quek, A.T. Wee, Gap states at low-angle grain boundaries in monolayer tungsten diselenide, Nano Letters 16 (2016) 3682–3688.

[14] J. Feng, X. Qian, C.-W. Huang, J. Li, Strain-engineered artificial atom as a broad-spectrum solar energy funnel, Nature Photonics 6 (2012) 866–872.

[15] H.J. Conley, B. Wang, J.I. Ziegler, R.F. Haglund Jr., S.T. Pantelides, K.I. Bolotin, Bandgap engineering of strained monolayer and bilayer MoS_2, Nano Letters 13 (2013) 3626–3630.

[16] E.P. Nguyen, B.J. Carey, J.Z. Ou, J. van Embden, E.D. Gaspera, A.F. Chrimes, M.J. Spencer, S. Zhuiykov, K. Kalantar-Zadeh, T. Daeneke, Electronic tuning of 2D MoS_2 through surface functionalization, Advanced Materials 27 (2015) 6225–6229.

[17] Y.L. Huang, Y. Chen, W. Zhang, S.Y. Quek, C.H. Chen, L.J. Li, W.T. Hsu, W.H. Chang, Y.J. Zheng, W. Chen, A.T. Wee, Bandgap tunability at single-layer molybdenum disulphide grain boundaries, Nature Communications (6) (2015) 6298.

[18] A.M. van der Zande, P.Y. Huang, D.A. Chenet, T.C. Berkelbach, Y. You, G.H. Lee, T.F. Heinz, D.R. Reichman, D.A. Muller, J.C. Hone, Grains and grain boundaries in highly crystalline monolayer molybdenum disulphide, Nature Materials 12 (2013) 554–561.

[19] S. Barja, S. Wickenburg, Z.-F. Liu, Y. Zhang, H. Ryu, Miguel M. Ugeda, Z. Hussain, Z.-X. Shen, S.-K. Mo, E. Wong, Miquel B. Salmeron, F. Wang, M.F. Crommie, D.F. Ogletree, Jeffrey B. Neaton, A. Weber-Bargioni, Charge density wave order in 1D mirror twin boundaries of single-layer $MoSe_2$, Nature Physics 12 (2016) 751–756.

[20] Y.J. Zheng, Y. Chen, Y.L. Huang, P.K. Gogoi, M.Y. Li, L.J. Li, P.E. Trevisanutto, Q. Wang, S.J. Pennycook, A.T.S. Wee, S.Y. Quek, Point defects and localized excitons in 2D WSe2, ACS Nano 13 (2019) 6050–6059.

[21] S. Haldar, H. Vovusha, M.K. Yadav, O. Eriksson, B. Sanyal, Systematic study of structural, electronic, and optical properties of atomic-scale defects in the two-dimensional transition metal dichalcogenides MX_2 (M = Mo, W, X = S, Se, Te), Physical Review B 92 (2015) 235408.

[22] H. Li, C. Tsai, A.L. Koh, L. Cai, A.W. Contryman, A.H. Fragapane, J. Zhao, H.S. Han, H.C. Manoharan, F. Abild-Pedersen, J.K. Norskov, X. Zheng, Corrigendum: activating and optimizing MoS_2 basal planes for hydrogen evolution through the formation of strained sulphur vacancies, Nature Materials (15) (2016) 364.

[23] J. Peto, T. Ollar, P. Vancso, Z.I. Popov, G.Z. Magda, G. Dobrik, C. Hwang, P.B. Sorokin, L. Tapaszto, Spontaneous doping of the basal plane of MoS_2 single layers through oxygen substitution under ambient conditions, Nat. Chem. 10 (2018) 1246–1251.

[24] D. Voiry, R. Fullon, J. Yang, E.S.C. de Carvalho Castro, R. Kappera, I. Bozkurt, D. Kaplan, M.J. Lagos, P.E. Batson, G. Gupta, A.D. Mohite, L. Dong, D. Er, V.B. Shenoy, T. Asefa, M. Chhowalla, The role of electronic coupling between substrate and 2D MoS_2 nanosheets in electrocatalytic production of hydrogen, Nature Materials 15 (2016) 1003–1009.

[25] S. Zhang, C.G. Wang, M.Y. Li, D. Huang, L.J. Li, W. Ji, S. Wu, Defect structure of localized excitons in a WSe_2 monolayer, Physical Review Letters 119 (2017) 046101.

[26] H. Qiu, T. Xu, Z. Wang, W. Ren, H. Nan, Z. Ni, Q. Chen, S. Yuan, F. Miao, F. Song, G. Long, Y. Shi, L. Sun, J. Wang, X. Wang, Hopping transport through defect-induced localized states in molybdenum disulphide, Nature Communications 4 (2013) 2642.

[27] D. Ma, Q. Wang, T. Li, C. He, B. Ma, Y. Tang, Z. Lu, Z. Yang, Repairing sulfur vacancies in the MoS2 monolayer by using CO, NO and NO_2 molecules, Journal Materials Chemistry C 4 (2016) 7093–7101.

[28] J. Jiang, Q. Zhang, A. Wang, Y. Zhang, F. Meng, C. Zhang, X. Feng, Y. Feng, L. Gu, H. Liu, L. Han, A facile and effective method for patching sulfur vacancies of WS_2 via nitrogen plasma treatment, Small 15 (2019) e1901791.

[29] S. Barja, S. Refaely-Abramson, B. Schuler, D.Y. Qiu, A. Pulkin, S. Wickenburg, H. Ryu, M.M. Ugeda, C. Kastl, C. Chen, C. Hwang, A. Schwartzberg, S. Aloni, S.K. Mo, D. Frank Ogletree, M.F. Crommie, O.V. Yazyev, S.G. Louie, J.B. Neaton, A. Weber-Bargioni, Identifying substitutional oxygen as a prolific point defect in monolayer transition metal dichalcogenides, Nature Communications (10) (2019) 3382.

[30] K. Cho, M. Min, T.-Y. Kim, H. Jeong, J. Pak, J.-K. Kim, J. Jang, S.J. Yun, Y.H. Lee, W.-K. Hong, T. Lee, Electrical and optical characterization of MoS_2 with sulfur vacancy passivation by treatment with alkanethiol molecules, ACS Nano 9 (2015) 8044–8053.

[31] J. Lu, A. Carvalho, X.K. Chan, H. Liu, B. Liu, E.S. Tok, K.P. Loh, A.H. Castro Neto, C.H. Sow, Atomic healing of defects in transition metal dichalcogenides, Nano Letters 15 (2015) 3524–3532.

[32] Q. Liang, Q. Zhang, J. Gou, T. Song, Chen H. Arramel, M. Yang, S.X. Lim, Q. Wang, R. Zhu, N. Yakovlev, S.C. Tan, W. Zhang, K.S. Novoselov, A.T.S. Wee, Performance improvement by ozone treatment of 2D $PdSe_2$, ACS Nano (2020), https://doi.org/10.1021/acsnano.0c00180, online.

[33] X. Zhao, D. Fu, Z. Ding, Y.Y. Zhang, D. Wan, S.J.R. Tan, Z. Chen, K. Leng, J. Dan, W. Fu, D. Geng, P. Song, Y. Du, T. Venkatesan, S.T. Pantelides, S.J. Pennycook, W. Zhou, K.P. Loh, Mo-terminated edge reconstructions in nanoporous molybdenum disulfide film, Nano Letters 18 (2018) 482–490.

[34] P.K.J. Wong, W. Zhang, F. Bussolotti, X. Yin, T.S. Herng, L. Zhang, Y.L. Huang, G. Vinai, S. Krishnamurthi, D.W. Bukhvalov, Y.J. Zheng, R. Chua, A.T. N'Diaye, S.A. Morton, C.Y. Yang, K.H. Ou Yang, P. Torelli, W. Chen, K.E.J. Goh, J. Ding, M.T. Lin, G. Brocks, M.P. de Jong, A.H. Castro Neto, A.T.S. Wee, Evidence of spin frustration in a vanadium diselenide monolayer magnet, Advanced Materials 31 (2019) e1901185.

[35] H.-P. Komsa, A.V. Krasheninnikov, Engineering the electronic properties of two-dimensional transition metal dichalcogenides by introducing mirror twin boundaries, Advanced Electronic Materials 3 (2017) 1600468.

[36] M. Chhowalla, H.S. Shin, G. Eda, L.J. Li, K.P. Loh, H. Zhang, The chemistry of two-dimensional layered transition metal dichalcogenide nanosheets, Nat. Chem. 5 (2013) 263–275.

[37] D. Wang, H. Yu, L. Tao, W. Xiao, P. Fan, T. Zhang, M. Liao, W. Guo, D. Shi, S. Du, G. Zhang, H. Gao, Bandgap broadening at grain boundaries in single-layer MoS_2, Nano Research 11 (2018) 6102–6109.

[38] K. He, C. Poole, K.F. Mak, J. Shan, Experimental demonstration of continuous electronic structure tuning via strain in atomically thin MoS_2, Nano Letters 13 (2013) 2931–2936.

[39] A. Azizi, X. Zou, P. Ercius, Z. Zhang, A.L. Elias, N. Perea-Lopez, G. Stone, M. Terrones, B.I. Yakobson, N. Alem, Dislocation motion and grain boundary migration in two-dimensional tungsten disulphide, Nature Communications 5 (2014) 4867.

[40] Z. Song, T. Schultz, Z. Ding, B. Lei, C. Han, P. Amsalem, T. Lin, D. Chi, S.L. Wong, Y.J. Zheng, M.Y. Li, L.J. Li, W. Chen, N. Koch, Y.L. Huang, A.T.S. Wee, Electronic properties of a 1D intrinsic/p-Doped heterojunction in a 2D transition metal dichalcogenide semiconductor, ACS Nano 11 (2017) 9128–9135.

[41] M. Makarova, Y. Okawa, M. Aono, Selective adsorption of thiol molecules at sulfur vacancies on $MoS_2(0001)$, followed by vacancy repair via S–C dissociation, Journal of Physical Chemistry C 116 (2012) 22411–22416.

[42] Z. Yu, Y. Pan, Y. Shen, Z. Wang, Z.Y. Ong, T. Xu, R. Xin, L. Pan, B. Wang, L. Sun, J. Wang, G. Zhang, Y.W. Zhang, Y. Shi, X. Wang, Towards intrinsic charge transport in monolayer molybdenum disulfide by defect and interface engineering, Nature Communications 5 (2014) 5290.

[43] S. Roy, W. Choi, S. Jeon, D.H. Kim, H. Kim, S.J. Yun, Y. Lee, J. Lee, Y.M. Kim, J. Kim, Atomic observation of filling vacancies in monolayer transition metal sulfides by chemically sourced sulfur atoms, Nano Letters 18 (2018) 4523–4530.

[44] A. Forster, S. Gemming, G. Seifert, D. Tomanek, Chemical and electronic repair mechanism of defects in MoS_2 monolayers, ACS Nano 11 (2017) 9989–9996.

[45] X. Zhang, Q. Liao, S. Liu, Z. Kang, Z. Zhang, J. Du, F. Li, S. Zhang, J. Xiao, B. Liu, Y. Ou, X. Liu, L. Gu, Y. Zhang, Poly(4-styrenesulfonate)-induced sulfur vacancy self-healing strategy for monolayer MoS_2 homojunction photodiode, Nature Communications 8 (2017) 15881.

[46] O. Lopez-Sanchez, D. Lembke, M. Kayci, A. Radenovic, A. Kis, Ultrasensitive photodetectors based on monolayer MoS_2, Nature Nanotechnology 8 (2013) 497–501.

[47] K.F. Mak, J. Shan, Photonics and optoelectronics of 2D semiconductor transition metal dichalcogenides, Nature Photonics 10 (2016) 216–226.

[48] K.F. Mak, C. Lee, J. Hone, J. Shan, T.F. Heinz, Atomically thin MoS_2: a new direct-gap semiconductor, Physical Review Letters 105 (2010) 136805.

[49] A. Splendiani, L. Sun, Y. Zhang, T. Li, J. Kim, C.Y. Chim, G. Galli, F. Wang, Emerging photoluminescence in monolayer MoS_2, Nano Letters 10 (2010) 1271–1275.

[50] M. Amani, D.-H. Lien, D. Kiriya, J. Xiao, A. Azcatl, J. Noh, S.R. Madhvapathy, R. Addou, S. KC, M. Dubey, K. Cho, R.M. Wallace, S.-C. Lee, J.-H. He, J.W.A. III, X. Zhang, E. Yablonovitch, A. Javey, Near-unity photoluminescence quantum yield in MoS_2, Science 350 (2015) 1065.

[51] M. Amani, P. Taheri, R. Addou, G.H. Ahn, D. Kiriya, D.H. Lien, J.W. Ager 3rd, R.M. Wallace, A. Javey, Recombination kinetics and effects of superacid treatment in sulfur- and selenium-based transition metal dichalcogenides, Nano Letters 16 (2016) 2786–2791.

[52] D.M. Sim, M. Kim, S. Yim, M.-J. Choi, J. Choi, S. Yoo, Y.S. Jung, Controlled doping of vacancy-containing few-layer MoS_2 via highly stable thiol-based molecular chemisorption, ACS Nano 9 (2015) 12115–12123.

[53] K. Cho, M. Min, T.-Y. Kim, H. Jeong, J. Pak, J.-K. Kim, J. Jang, S.J. Yun, Y.H. Lee, W.-K. Hong, T. Lee, Electrical and optical characterization of MoS_2 with sulfur vacancy passivation by treatment with alkanethiol molecules, ACS Nano 9 (2015) 8044–8053.

[54] Q. Ding, K.J. Czech, Y. Zhao, J. Zhai, R.J. Hamers, J.C. Wright, S. Jin, Basal-plane ligand functionalization on semiconducting 2H-MoS2 monolayers, ACS Applied Materials & Interfaces 9 (2017) 12734–12742.

[55] Y. Zhao, S. Bertolazzi, M.S. Maglione, C. Rovira, M. Mas-Torrent, P. Samori, Molecular approach to electrochemically switchable monolayer MoS_2 transistors, Advanced Materials (2020) e2000740.

[56] J. Jiang, C. Ling, T. Xu, W. Wang, X. Niu, A. Zafar, Z. Yan, X. Wang, Y. You, L. Sun, J. Lu, J. Wang, Z. Ni, Defect engineering for modulating the trap states in 2D photoconductors, Advanced Materials (2018) e1804332.

[57] J.H. Park, A. Sanne, Y. Guo, M. Amani, K. Zhang, H.C.P. Movva, J.A. Robinson, A. Javey, J. Robertson, S.K. Banerjee, A.C. Kummel, Defect passivation of transition metal dichalcogenides via a charge transfer van der Waals interface, Science Advances 3 (2017) e1701661.

[58] W. Chen, S. Chen, D.C. Qi, X.Y. Gao, A.T.S. Wee, Surface transfer p-type doping of epitaxial graphene, Journal of the American Chemical Society 129 (2007) 10418–10422.

[59] H.Y. Mao, Y.H. Lu, J.D. Lin, S. Zhong, A.T.S. Wee, W. Chen, Manipulating the electronic and chemical properties of graphene via molecular functionalization, Progress in Surface Science 88 (2013) 132–159.

[60] Y. Jing, X. Tan, Z. Zhou, P.W. Shen, Tuning electronic and optical properties of MoS_2 monolayer via molecular charge transfer, Journal of Materials Chemistry A 2 (2014) 16892–16897.

[61] D. Kiriya, M. Tosun, P.D. Zhao, J.S. Kang, A. Javey, Air-stable surface charge transfer doping of MoS_2 by benzyl viologen, Journal of the American Chemical Society 136 (2014) 7853–7856.

[62] K. Cho, J. Pak, S. Chung, T. Lee, Recent advances in interface engineering of transition-metal dichalcogenides with organic molecules and polymers, ACS Nano 13 (2019) 9713–9734.

[63] Y.L. Huang, Y.J. Zheng, Z. Song, D. Chi, A.T.S. Wee, S.Y. Quek, The organic-2D transition metal dichalcogenide heterointerface, Chemical Society Reviews 47 (2018) 3241–3264.

[64] Z. Hu, Z. Wu, C. Han, J. He, Z. Ni, W. Chen, Two-dimensional transition metal dichalcogenides: interface and defect engineering, Chemical Society Reviews 47 (2018) 3100–3128.

[65] X. Zhao, Y. Ji, J. Chen, W. Fu, J. Dan, Y. Liu, S.J. Pennycook, W. Zhou, K.P. Loh, Healing of planar defects in 2D materials via grain boundary sliding, Advanced Materials 31 (2019) e1900237.

[66] R. Chua, J. Yang, X. He, X. Yu, W. Yu, F. Bussolotti, P.K.J. Wong, K.P. Loh, M.B.H. Breese, K.E.J. Goh, Y.L. Huang, A.T.S. Wee, Can reconstructed Se-deficient line defects in monolayer VSe_2 induce magnetism?, Advanced Materials 32 (2020) 2000693.

[67] Z.L. Liu, B. Lei, Z.L. Zhu, L. Tao, J. Qi, D.L. Bao, X. Wu, L. Huang, Y.Y. Zhang, X. Lin, Y.L. Wang, S. Du, S.T. Pantelides, H.J. Gao, Spontaneous formation of 1D pattern in monolayer VSe_2 with dispersive adsorption of Pt atoms for HER catalysis, Nano Letters 19 (2019) 4897–4903.

[68] X. Zhao, J. Kotakoski, J.C. Meyer, E. Sutter, P. Sutter, A.V. Krasheninnikov, U. Kaiser, W. Zhou, Engineering and modifying two-dimensional materials by electron beams, MRS Bulletin 42 (2017) 667–676.

[69] S. Wang, H. Li, H. Sawada, C.S. Allen, A.I. Kirkland, J.C. Grossman, J.H. Warner, Atomic structure and formation mechanism of sub-nanometer pores in 2D monolayer MoS_2, Nanoscale 9 (2017) 6417–6426.

[70] J. Jiang, T. Xu, J. Lu, L. Sun, Z. Ni, Defect engineering in 2D materials: precise manipulation and improved functionalities, Research (Wash D C) 2019 (2019) 4641739.

[71] Y.C. Lin, T. Bjorkman, H.P. Komsa, P.Y. Teng, C.H. Yeh, F.S. Huang, K.H. Lin, J. Jadczak, Y.S. Huang, P.W. Chiu, A.V. Krasheninnikov, K. Suenaga, Three-fold rotational defects in two-dimensional transition metal dichalcogenides, Nature Communications 6 (2015) 6736.

[72] Y.C. Lin, D.O. Dumcenco, Y.S. Huang, K. Suenaga, Atomic mechanism of the semiconducting-to-metallic phase transition in single-layered MoS_2, Nature Nanotechnology 9 (2014) 391–396.

[73] D.G. Hopkinson, V. Zolyomi, A.P. Rooney, N. Clark, D.J. Terry, M. Hamer, D.J. Lewis, C.S. Allen, A.I. Kirkland, Y. Andreev, Z. Kudrynskyi, Z. Kovalyuk, A. Patane, V.I. Fal'ko, R. Gorbachev, S.J. Haigh, Formation and healing of defects in atomically thin GaSe and InSe, ACS Nano 13 (2019) 5112–5123.

[74] H. Nan, R. Zhou, X. Gu, S. Xiao, K. Ken Ostrikov, Recent advances in plasma modification of 2D transition metal dichalcogenides, Nanoscale 11 (2019) 19202–19213.

[75] M. Tosun, L. Chan, M. Amani, T. Roy, G.H. Ahn, P. Taheri, C. Carraro, J.W. Ager, R. Maboudian, A. Javey, Air-stable n-Doping of WSe2 by anion vacancy formation with mild plasma treatment, ACS Nano 10 (2016) 6853–6860.

218   7. Materials engineering – defect healing & passivation

[76] S. Bertolazzi, S. Bonacchi, G. Nan, A. Pershin, D. Beljonne, P. Samori, Engineering chemically active defects in monolayer MoS_2 transistors via ion-beam irradiation and their healing via vapor deposition of alkanethiols, Advanced Materials 29 (2017) 1606760.

[77] B. Tang, Z.G. Yu, L. Huang, J. Chai, S.L. Wong, J. Deng, W. Yang, H. Gong, S. Wang, K.W. Ang, Y.W. Zhang, D. Chi, Direct n- to p-type channel conversion in monolayer/few-layer WS_2 field-effect transistors by atomic nitrogen treatment, ACS Nano 12 (2018) 2506–2513.

[78] Y. Liu, H. Nan, X. Wu, W. Pan, W. Wang, J. Bai, W. Zhao, L. Sun, X. Wang, Z. Ni, Layer-by-layer thinning of MoS_2 by plasma, ACS Nano 7 (2012) 4202–4209.

[79] J. Pei, X. Gai, J. Yang, X. Wang, Z. Yu, D.Y. Choi, B. Luther-Davies, Y. Lu, Producing air-stable monolayers of phosphorene and their defect engineering, Nature Communications 7 (2016) 10450.

[80] N. Liu, J. Baek, S.M. Kim, S. Hong, Y.K. Hong, Y.S. Kim, H.S. Kim, S. Kim, J. Park, Improving the stability of high-performance multilayer MoS_2 field-effect transistors, ACS Applied Materials & Interfaces 9 (2017) 42943–42950.

[81] H. Nan, S. Guo, S. Cai, Z. Chen, A. Zafar, X. Zhang, X. Gu, S. Xiao, Z. Ni, Producing air-stable InSe nanosheet through mild oxygen plasma treatment, Semiconductor Science and Technology 33 (2018) 074002.

[82] H. Nan, Z. Wang, W. Wang, Z. Liang, Y. Lu, Q. Chen, D. He, P. Tan, F. Miao, X. Wang, Jinlan Wang , Z. Ni, Strong photoluminescence enhancement of MoS_2 through defect engineering and oxygen bonding, ACS Nano 8 (2014) 5738–5745.

[83] Z. Yu, Z.Y. Ong, Y. Pan, Y. Cui, R. Xin, Y. Shi, B. Wang, Y. Wu, T. Chen, Y.W. Zhang, G. Zhang, X. Wang, Realization of room-temperature phonon-limited carrier transport in monolayer MoS_2 by dielectric and carrier screening, Advanced Materials 28 (2016) 547–552.

[84] B.W. Baugher, H.O. Churchill, Y. Yang, P. Jarillo-Herrero, Intrinsic electronic transport properties of high-quality monolayer and bilayer MoS_2, Nano Letters 13 (2013) 4212–4216.

[85] B. Radisavljevic, A. Kis, Mobility engineering and a metal-insulator transition in monolayer MoS_2, Nature Materials 12 (2013) 815–820.

[86] R. Pisoni, A. Kormanyos, M. Brooks, Z. Lei, P. Back, M. Eich, H. Overweg, Y. Lee, P. Rickhaus, K. Watanabe, T. Taniguchi, A. Imamoglu, G. Burkard, T. Ihn, K. Ensslin, Interactions and magnetotransport through spin-valley coupled Landau levels in monolayer MoS_2, Physical Review Letters 121 (2018) 247701.

[87] F. Cadiz, E. Courtade, C. Robert, G. Wang, Y. Shen, H. Cai, T. Taniguchi, K. Watanabe, H. Carrere, D. Lagarde, M. Manca, T. Amand, P. Renucci, S. Tongay, X. Marie, B. Urbaszek, Excitonic linewidth approaching the homogeneous limit in MoS_2-based van der Waals heterostructures, Physical Review X 7 (2017) 021026.

[88] X. Cui, G.H. Lee, Y.D. Kim, G. Arefe, P.Y. Huang, C.H. Lee, D.A. Chenet, X. Zhang, L. Wang, F. Ye, F. Pizzocchero, B.S. Jessen, K. Watanabe, T. Taniguchi, D.A. Muller, T. Low, P. Kim, J. Hone, Multi-terminal transport measurements of MoS_2 using a van der Waals heterostructure device platform, Nature Nanotechnology 10 (2015) 534–540.

[89] B. Fallahazad, H.C. Movva, K. Kim, S. Larentis, T. Taniguchi, K. Watanabe, S.K. Banerjee, E. Tutuc, Shubnikov-de Haas oscillations of high-mobility holes in monolayer and bilayer WSe_2: Landau level degeneracy, effective mass, and negative compressibility, Physical Review Letters 116 (2016) 086601.

[90] N.E. Staley, J. Wu, P. Eklund, Y. Liu, L. Li, Z. Xu, Electric field effect on superconductivity in atomically thin flakes of $NbSe_2$, Physical Review B 80 (2009) 184505.

[91] S.H. Chae, Y. Jin, T.S. Kim, D.S. Chung, H. Na, H. Nam, H. Kim, D.J. Perello, H.Y. Jeong, T.H. Ly, Y.H. Lee, Oxidation effect in octahedral hafnium disulfide thin film, ACS Nano 10 (2016) 1309–1316.

[92] L. Li, F. Yang, G.J. Ye, Z. Zhang, Z. Zhu, W. Lou, X. Zhou, L. Li, K. Watanabe, T. Taniguchi, K. Chang, Y. Wang, X.H. Chen, Y. Zhang, Quantum Hall effect in black phosphorus two-dimensional electron system, Nature Nanotechnology 11 (2016) 593–597.

[93] B. Huang, G. Clark, E. Navarro-Moratalla, D.R. Klein, R. Cheng, K.L. Seyler, D. Zhong, E. Schmidgall, M.A. McGuire, D.H. Cobden, W. Yao, D. Xiao, P. Jarillo-Herrero, X. Xu, Layer-dependent ferromagnetism in a van der Waals crystal down to the monolayer limit, Nature 546 (2017) 270–273.

[94] J.-H. Lee, E.K. Lee, W.-J. Joo, Y. Jang, B.-S. Kim, J.Y. Lim, S.-H. Choi, S.J. Ahn, J.R. Ahn, M.-H. Park, C.-W. Yang, B.L. Choi, S.-W. Hwang, D. Whang, Wafer-scale growth of single-crystal monolayer graphene on reusable hydrogen-terminated germanium, Science 344 (2014) 286–289.

[95] T.A. Chen, C.P. Chuu, C.C. Tseng, C.K. Wen, H.P. Wong, S. Pan, R. Li, T.A. Chao, W.C. Chueh, Y. Zhang, Q. Fu, B.I. Yakobson, W.H. Chang, L.J. Li, Wafer-scale single-crystal hexagonal boron nitride monolayers on Cu (111), Nature 579 (2020) 219–223.

[96] K. Kang, S. Xie, L. Huang, Y. Han, P.Y. Huang, K.F. Mak, C.J. Kim, D. Muller, J. Park, High-mobility three-atom-thick semiconducting films with wafer-scale homogeneity, Nature 520 (2015) 656–660.

[97] L. Gao, G.X. Ni, Y. Liu, B. Liu, A.H. Castro Neto, K.P. Loh, Face-to-face transfer of wafer-scale graphene films, Nature 505 (2014) 190–194.

[98] K. Kang, K.H. Lee, Y. Han, H. Gao, S. Xie, D.A. Muller, J. Park, Layer-by-layer assembly of two-dimensional materials into wafer-scale heterostructures, Nature 550 (2017) 229–233.

CHAPTER 8

Nonequilibrium synthesis and processing approaches to tailor heterogeneity in 2D materials☆

David B. Geohegan, Kai Xiao, Alex A. Puretzky, Yu-Chuan Lin, Yiling Yu, and Chenze Liu

Functional Hybrid Nanomaterials Group, Center for Nanophase Materials Sciences, Oak Ridge National Laboratory, Oak Ridge, TN, United States

8.1 Introduction

Defects in two-dimensional (2D) materials impose some of the greatest limitations for the practical application of these materials in optoelectronics. Key properties such as carrier mobility and photoresponsivity are greatly affected by scattering or recombination of carriers at a variety of different heterogeneities, such as vacancies and other point defects, dopants, and grain boundaries. Monolayer or few-layer 2D crystals synthesized from the bottom-up physical vapor deposition (PVD) processes are inherently more defective than crystalline layers exfoliated from bulk crystals because of the highly nonequilibrium nature of nucleation and growth processes on growth substrates. Such heterogeneities incurred during bottom-up synthesis currently limit the scaled production of high quality, low defect density, 2D materials at the wafer-scale that is necessary for optoelectronic applications.

Nucleation and growth involve a variety of processes that each can be modeled with activation energy barriers in the Arrhenius equation. Such processes include the sticking and

---

☆ Notice: This manuscript has been authored by UT-Battelle, LLC, under Contract No. DE-AC05-00OR22725 with the U.S. Department of Energy. The United States Government retains and the publisher, by accepting the article for publication, acknowledges that the United States Government retains a non-exclusive, paid-up, irrevocable, world-wide license to publish or reproduce the published form of this manuscript, or allow others to do so, for United States Government purposes. The Department of Energy will provide public access to these results of federally sponsored research in accordance with the DOE Public Access Plan (http://energy.gov/downloads/doe-public-access-plan).

chemical decomposition of precursor species on substrates, the diffusion of secondary intermediates to crystal edges, their incorporation and crystallization into the crystal, and defect formation energies that arise due to chemical reactions at open edges or due to such factors as strain that is present at grain boundaries between crystalline domains as they coalesce [1–23]. As growth proceeds and substrate coverage changes, the dynamic equilibrium and competition between these processes also changes as second layers may begin to nucleate and grow. In addition, etching may become favored where growth just occurred. In short, all growth processes that are typically employed for bottom-up 2D crystal growth are highly dynamic and sensitive to the growth environment, and inherently non-equilibrium.

Unfortunately, the vast number of 2D crystalline materials that are being synthesized, and the variety of physical and chemical vapor deposition approaches that are being utilized for their bottom-up synthesis both contribute to the complexity of modeling and understanding of their synthesis. Moreover, very few growth studies employ *in situ* diagnostics of the growth environment in typical growth reactors. Such diagnostics are desperately needed to provide information on the role of temperature, the nature of the precursor flux and its kinetic energy upon arrival at the substrate, the deposition rate, and evolving strain. In this regard, non-equilibrium growth processes that can controllably vary these parameters have distinct advantages in understanding elements of the growth process. Pulsed processes allow time-resolved *in-situ* diagnostics measurements of the growth dynamics and the evolving crystalline structure, including identification of metastable phases and growth pathways.

With such understanding, the overarching goal of such processes is to *control heterogeneities* during synthesis – not only to enable methods for their reliable scaled production, but to enable new materials properties. For example, control over the edge structure, phase (stacking), or dipole moment of the 2D transition metal dichalcogenides (TMDs) has been shown to govern their catalytic activity [24,25]. Point defects and strain in h-BN or TMDs can enable single photon emission for quantum information applications [26,27]. Defects can induce magnetism in non-magnetic 2D materials [28]. Controlling vacancies and substitutional dopants can enable the tunability of the band gap and carrier type of semiconducting TMDs. A combination of bottom-up synthesis (precursor type, temperature, gas composition, kinetic energy, substrate, patterned substrates, etc.) and top-down processing (etching, plasma exposure, thermal annealing, etc.) approaches have been used to control heterogeneities for these applications.

However, the fundamental understanding linking synthesis and processing to materials properties remains a grand challenge to enable transformative manufacturing of materials, especially 2D materials for quantum technology [29,30]. One advantage of 2D materials is that they are computationally tractable materials for which a wide variety of quantum phases and properties have been predicted, however the synthetic pathways to enable these metastable, kinetically stabilized, and thermodynamic phases must first be determined, understood, and eventually controlled.

In this chapter, we present a few examples of nonequilibrium growth methods and *in situ* diagnostic techniques that are being developed to understand and control materials synthesis pathways toward these goals. The general strategy for such studies is shown in Fig. 8.1. In the plot, according to Ostwald's law of stages [17], during the crystallization of a material, the reaction proceeds through a variety of metastable polymorphs, separated by small changes in free energy, as it proceeds toward the most energetically stable crystalline form.

8.1. Introduction

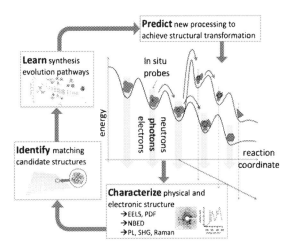

FIGURE 8.1 General strategy to understand and control the synthesis of 2D materials, utilizing non-equilibrium processing to overcome energy barriers and *in situ* diagnostic probes to follow the evolution of material structure. Atomistic characterization and computational modeling are utilized to identify structures and defects in 2D materials, and learn synthesis pathways. This data should enable computational predictions of processing changes to achieve new structural transformations and the roles of defects in synthesis and processing.

With *in situ* probes (including spectroscopic diagnostics utilizing electrons, photons, X-rays, neutrons), one can characterize both the physical and electronic structure of a material during its transition through each of these metastable states. Nonequilibrium processing (especially pulsed processing) can explore the energy barriers to reveal these metastable states. For 2D materials, the *atomistic* structure can be directly obtained because the atomically-thin materials allow direct imaging by techniques such as scanning transmission electron microscopy (STEM) or scanning tunneling microscopy (STM). This information must then be compared with computational models of such structures to understand and identify the evolution of the growth processes. Through repeated exploration of different pathways, a predictive and mechanistic control of the synthesis process can be developed. *In situ* diagnostics that are developed can serve to fulfill the grand challenge to achieve a mechanistic control over synthesis to achieve directed synthesis with real-time adaptive control [29,30].

We will primarily focus on the semiconducting 2D TMD materials because their optical properties enable remote optical spectroscopic diagnostics that can be used to correlate the evolution of their atomic structure and various signatures of heterogeneities during synthesis with theory and modeling. We first focus on the effects of changing the chemical potential on the evolution of their domain shape and their edge structure. Then we examine how strain during growth can introduce heterogeneities to alter their morphology and growth rate. We then move to explore pulsed laser processes to understand how amorphous cluster and nanoparticle precursors and hyperthermal kinetic energies are utilized and controlled to deposit and convert 2D TMD materials. In each case, we describe the development of *in situ* spectroscopic diagnostics that can enable the general strategy in Fig. 8.1.

8.2 Non-equilibrium synthesis – effects of chemical potential on the heterogeneity of 2D materials

Heterogeneities can both negatively influence the structure and properties of 2D materials but can also bring new functionalities [31,32]. Therefore, it is necessary to precisely control heterogeneity in 2D materials through controlled synthesis and processing to tailor their properties. According to the Gibbs phase rule, the compositional variety of TMDs provides new degrees of freedom to modulate their growth behavior [33]. In terms of the chemical potential, the driving force for growth is $\Delta\mu_{MX} = (\mu_M + \mu_X) - \mu_{MX}$, where μ_M or μ_X are the chemical potentials of the constituent atoms, and μ_{MX} is the energy of an MX pair in a MX crystal that allows variation of the equilibrium composition and morphology depending on thermodynamic conditions. Coupled with their atomically thin nature, this thermodynamic range of freedom brings about new phenomena for 2D materials, an ideal platform where different heterogeneities not only can be found but also can be reversibly tuned by controlling the chemical potentials of the constituent elements during growth [34]. Therefore, this interplay between the formation energy of each heterogeneity in 2D materials and the freedom to vary the chemical potential during synthesis offers a new way to control the structure and morphology of 2D materials to tune their properties. Various types of heterogeneities, such as vacancy levels, substitutional dopants, and edges, can be manipulated by controlling the chemical potential of constituent elements during synthesis [35]. Compared to the ground state energy of bulk-phase 2D crystals, heterogeneous 2D materials have higher energies, which can require nonequilibrium growth and processing methods for their formation. Recent developments in theory and computational modeling enable the calculation of such total energies and formation energies for complex, heterogeneous 2D material systems, providing an impetus for the rational design of specific heterogeneities by controlling the growth conditions [33,36]. In addition, atomic resolution high angle annular dark field scanning transmission electron microscopy (HAADF-STEM) imaging techniques, electron energy loss spectroscopy (EELS) and data analytics enables the unambiguous identification of the precise structure of heterogeneities in 2D materials by quantitatively analyzing image intensity [37–39]. Therefore, such atomistic characterization techniques must be combined with a full theoretical understanding of the thermodynamic targets in order to design synthetic pathways to induce and control heterogeneity in 2D materials by bottom-up synthesis.

In this section, we present some recent examples that illustrate this strategy to control heterogeneities including defects, edges, and shape morphologies in 2D materials through a synergistic approach that combines non-equilibrium synthesis and processing, direct atomic resolution imaging, and first principles calculations. We begin with controlling specific point defects by tuning the chemical potentials of constituent elements during synthesis. Then we introduce *in situ* ADF-STEM analyzes of edge reconstructions in TMDs under the influence of different chemical potentials, and how such knowledge can impact the edge-determined shape evolution of 2D materials during CVD growth.

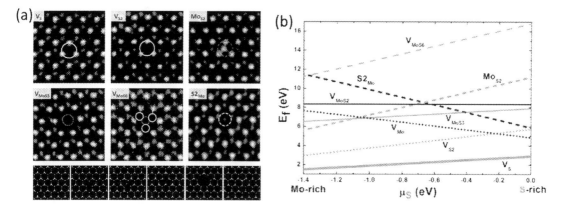

FIGURE 8.2 Point defects in monolayer MoS$_2$. (a) Atomic resolution ADF-STEM images and the relaxed structures of various point defects in monolayer MoS$_2$, including V$_S$, V$_{S2}$, Mo$_{S2}$, V$_{MoS3}$, V$_{MoS6}$, and S2$_{Mo}$. From left to right of the relaxed structures of the defects: V$_S$, V$_{S2}$, Mo$_{S2}$, V$_{MoS3}$, V$_{MoS6}$, and S2$_{Mo}$. Purple, yellow, and white balls represent Mo, top layer S, and bottom layer S, respectively. (b) The formation energies of various defects in MoS$_2$ at different chemical potentials of S. Note that this figure provides a simple criterion in terms of defect formation energy: V$_S$ possess the lowest formation energies in both the Mo-rich and S-rich conditions, thus it is commonly observed in MoS$_2$ samples [4]. Reprinted with permission from Ref. [4], Copyright © 2015, American Chemical Society.

8.2.1 Point defects control by nonequilibrium laser-based synthesis and Au-assisted CVD growth

Point defects, including vacancies, substitutional dopants, antisite occupancy, and interstitials, are the most common defects in 2D TMDs. As shown in Fig. 8.2a, the structures of various point defects in monolayer MoS$_2$ can be clearly characterized by atomic resolution HAADF-STEM imaging. Their formation during growth depends on the formation energy and the chemical potentials of the constituent atoms as previously noted, which is largely controlled by the growth environment and by the growth kinetics [4,40,41]. In addition, the diversity of chemical precursors utilized for the synthesis of 2D TMDs grown by bottom-up approaches (e.g., CVD, MOCVD, PVD, MBE), can give rise to different densities and types of defects, as each introduces additional chemical pathways. For instance, STEM imaging of MoS$_2$ reveals that CVD-grown films are dominated by S vacancies, whereas MoS$_2$ films grown by PVD contain predominantly antisite defects [40]. As shown in Fig. 8.2b, the S vacancy in MoS$_2$ has the lowest formation energy among all the defects, which is the most commonly observed defect in MoS$_2$, especially in CVD-grown monolayer TMDs. In thermal equilibrium, for example, the formation energy (E$_f$) of monolayer MoS$_2$ with different defects is defined as $E_f = E(\text{defect}) - E(\text{pristine}) + N \cdot \mu(\text{removed atom})$, where E(defect) and E(pristine) refer to the total energy of the defective and pristine monolayer MoS$_2$, respectively, where N is the number of the atoms removed from the system and μ(removed atom) (i.e., μ_{Mo} and μ_S) is the chemical potential of the removed atoms [4,13,42]. Note that the formation energy of Mo vacancies as well as the antisite defect (S2$_{Mo}$) is low in film grown in S-rich environment as compared to the Mo-rich environment. Since the formation energy of a S vacancy is much lower in a S-deficient environment than under S-rich condition, when

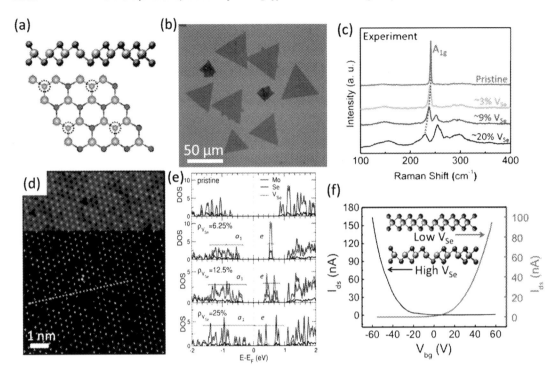

FIGURE 8.3 Synthesis of highly Se-deficient MoSe$_{2-x}$ by a laser evaporation method. (a) Illustrations showing the side and top view of a Se-deficient MoSe$_2$ crystal. (b) An optical image of as-grown monolayer triangular MoSe$_2$ crystals on SiO$_2$ substrate. (c) Raman spectra of pristine and Se-deficient crystals clearly show that the A$_{1g}$ Raman peak at 230 cm^{-1} shifted toward the pristine peak at 240 cm^{-1}, and the new defect peak at 250 cm^{-1} decreased as the vacancies were annihilated by the post selenization process. (d) Z-STEM image of monolayer MoSe$_2$ crystal with ~20% Se vacancies based on Z-contrast STEM intensity analysis. Part of the image was colored using ImageJ for better visibility. Note that the bright dots in the image are dual Se columns and if a pair of selenium atoms is missing from the top and the bottom of a sheet, hence leaving a hole in the hexagonal lattice. Otherwise, they are single Se vacancies in Se position of lattice. The concentration of single vacancies was quantified in each micrograph by counting the total number of columns that were determined to be missing a single Se atom from the Z-contrast STEM images divided by the total number of columns. (e) Calculated local electronic density of states of V$_{Se}$, Mo, and Se in pristine and Se-deficient material with various V$_{Se}$ concentrations. (f) Electrical transport properties transitioning from n- to p-type characteristics for pristine and highly deficient MoSe$_2$. Reprinted with permission from Reference [11], Copyright © 2016, American Chemical Society.

MoS$_2$ is grown in a Mo-rich condition a high concentration of sulfur vacancies, V$_S$ is expected. A similar scenario applies to MoSe$_2$ grown under Se-deficient conditions [4,43].

Recently, we developed a laser-based non-equilibrium growth approach for highly Se-deficient MoSe$_{2-x}$ monolayers [11]. Under continuous laser-evaporation of MoSe$_2$ powders, MoSe$_2$ preferentially loses Se first, so nucleation is in a Se-rich environment, however for most of the other growth processes the chemical potential is skewed to Se-poor growth conditions. Based on DFT calculations, the formation energy of Se vacancies is much lower under this Se-poor environment, giving rise to a high concentration of Se vacancies far beyond intrinsic levels – up to 20% in ML MoSe$_2$ grown by this non-equilibrium process. Remarkably, these highly deficient $MoSe_{2-x}$ crystals are still thermodynamically stable and retain their single-crystal structure and triangular morphology (Fig. 8.3a, b). Interestingly, these Se vacancies can be repaired by a post-selenization process using low-KE pulsed laser evaporated Se atoms combined with thermal annealing (see Fig. 8.14). Atomic-scale Z-contrast HAADF-STEM imaging was used to identify the single and double Se vacancies and accurately determine the concentration of Se vacancies (Fig. 8.3c). High concentrations of Se vacancies can also induce new defect-activated Raman peaks and a high-intensity defect-related bound exciton states revealed by low temperature photoluminescence spectroscopy (Fig. 8.3d). DFT calculations showed that a high density of Se vacancies also gives rise to strong hybridization of localized defect states and leads to the band edge of acceptor-like defect state moving closer to the valence band maximum (Fig. 8.3e), thus facilitating the production of holes in highly Se-deficient MLs to induce a p-type transport behavior. Interestingly, the post-selenization of highly defective $MoSe_2$ effectively leads to the restoration of the original n-type transport property (Fig. 8.3f), so the system can be potentially used in rewritable electronics. Therefore, with a full understanding of the thermodynamic parameters at play, in principle, we can rationally design and produce specific defects in 2D materials through a nonequilibrium synthesis process.

Antisite defects in 2D materials can induce many interesting properties. For example, the antisite defects on the W sites, Se_W, in WSe_2 have been shown to induce spatially-localized excitons for long-lifetime single photon emitters [26,44]. The antisite defects on the S sites, Mo_S, have been found as the predominant defects in PVD-MoS_2 monolayers and possess a local magnetic moment [40]. The W_{S2} antisite defects undergo additional atomic displacement from the breaking of the local threefold symmetry. Therefore, antisite defects tend to have much higher formation energies compared to the vacancy defects, which are thermodynamically unfavorable under equilibrium conditions [41]. However, the formation energy of antisite defects can be significantly reduced by changing the thermodynamic parameters through non-equilibrium synthesis and processing method. For example, previous reports showed that antisite defects, where Mo atoms replace S atoms, are the dominant point defects in MoS_2 MLs grown by physical vapor deposition due to Mo-deficient growth conditions, whereas defects in CVD-grown ML MoS_2 are predominantly sulfur vacancies [40]. As shown in Fig. 8.4, the antisite defect (S_W), where S occupies the W vacancy positions, in monolayer WS_2 has much lower formation energy in the S-rich environment than in the W-rich environment. Therefore, in order to create this type of defect in WS_2, the W-poor growth condition should be achieved to reduce the formation energy of S_W during the synthesis. We recently developed a Au-assisted CVD method to grow WS_2 monolayers on Au-coated W foils [42]. Our study showed that S_W antisite defects are the predominant defect in WS_2 monolayers on Au-coated W foils since solubility of W in Au is very low, thus limiting the W diffusion through the Au to the Au surface, inducing the W-poor growth conditions. Atomic-resolution STEM characterizations clearly verify the structure of S_W antisite defects. The strategy of lowering the formation energy of antisite defects by tuning the chemical potential of reactants during synthesis should be applicable for achieving specific defects in 2D TMDs that are hard to form under thermodynamic equilibrium conditions.

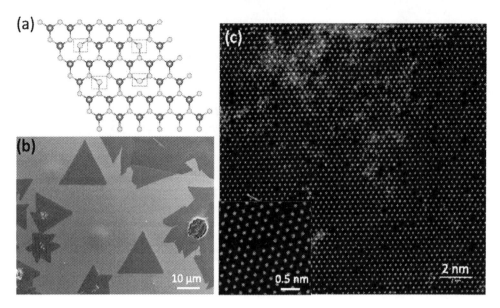

FIGURE 8.4 Nonequilibrium synthesis of S_W antisite defects in monolayer WS_2 grown on Au-coated W. (a) Top view of the atomic structures of S_W defects in WS_2. Gray and yellow balls represent the W and S atoms, respectively. (b) A SEM image of monolayer WS_2 crystals grown on the Au surface. (c) A Z-STEM image of monolayer WS_2 with predominately S_W defects. Inset is an enlarged STEM image which clearly shows one sulfur atom in the W vacancy position (unpublished).

8.2.2 Forming line defects, edges, and morphologies of 2D materials through controlled kinetics

When the chalcogen vacancy concentration in TMDs increases, they can aggregate to form extended line defects [45,46], thereby resulting in a local electronic structure change, which can potentially be useful for electronic and catalytic applications. For example, Se vacancies in $MoSe_2$ can agglomerate and form mirror twin boundaries that have particular atomic structures between regions with 60° relative orientation (Fig. 8.5a). This occurs because the formation energy of twin boundaries is significantly lower than isolated Se vacancies [47]. Based on DFT calculations as shown in Fig. 8.5b, these twin boundaries have low formation energies under metal-rich conditions [13]. Therefore, the formation of these line defects requires either addition of metal atoms or removal of chalcogen atoms in/from the lattice. In addition, it is also possible to control the types of line defects by controlling the growth condition, specifically, the chemical potentials of metal and chalcogen atoms [12]. For examples, by depositing extra Mo atoms on MBE-grown $MoSe_2$ monolayers to create a metal-rich condition, a large density of mirror twin boundaries was observed in $MoSe_2$ following thermal annealing at 500°C [48]. With a similar strategy to create Se-poor conditions, electron-beam irradiation [13] and thermal annealing [46] have been used to produce different twin boundaries in CVD grown-MoS_2 and MBE-grown $MoSe_2$. Under the electron beam irradiation, the chalcogen atoms in TMDs are preferentially knocked out of their lattice position where the chalcogen vacancies are created first then diffuse to form line defects with the aid of ki-

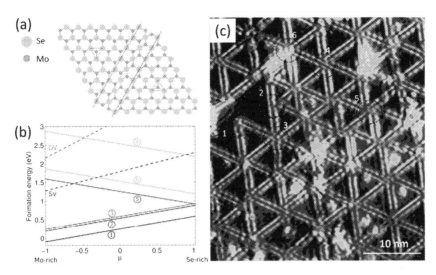

FIGURE 8.5 Formation of mirror twin boundaries under metal rich growth conditions. (a) Atomic structure of a mirror twin grain boundary in monolayer MoSe$_2$ [12]. (b) The calculated formation energies of different mirror twin boundaries as well as single (SV) and double vacancy (DV) lines as a function of the Se chemical potential [13]. (c) STM image at 15 K of mirror twin boundary network. The grain boundaries appear as two parallel lines, which correspond to the atomic position next to the grain boundary. Also note that the weak corrugation along the grain boundaries is due to the charge density wave in the low-temperature image. The periodicity of the charge density wave is 3 times the lattice constant. At least six different intersection points for the grain boundaries are identified in the STM image. Reprinted with permission from References [12,13], Copyright © 2017, American Chemical Society [12,13].

netic energy absorbed from the energetic electrons. Interestingly, unlike the localized states in the bandgap of TMDs observed for point defects, the mirror twin boundaries usually show large dispersing bands covering almost the whole band gap range of the TMDs [47] which show the one-dimensional metallic behavior, such as charge density waves [49] and the Tomonaga-Luttinger Liquid (TLL) [50] that have been verified following STM/STS measurements (Fig. 8.5c).

In addition to line defects, edges are also important extended defects in 2D TMDs. 2D TMDs have diverse edge structures due to the low lattice symmetry of binary or ternary compositions. Many novel properties are associated with the terminated edges and corresponding atomic structures in 2D TMDs. For example, the presence of unpaired electrons induced by the structural discontinuity at the edges of 2D material can increase chemical reactivity for enhanced catalytic performance [51,52]. Also the novel edge structures can introduce additional local magnetism in nonmagnetic 2D TMDs [22,53]. However, only limited types of edge structures have been observed experimentally in 2D TMD materials and controlling these edge structures is still a great challenge. For example, the TMDs grown by CVD often show triangular shapes featuring Mo edges, but by changing the growth conditions, hexagonally-shaped flakes with both S- and Mo-terminated edges can be prepared [1]. Therefore, understanding edge evolution in TMDs is critical to the design and growth of new edge structures for novel functionality.

Recently, distinct edge reconstructions in 2D materials were captured, including transition states, at the atomic scale on suspended 2D crystals by in-situ heating in a STEM [22]. In addition to the low-energy Mo- and Se-terminated zigzag (ZZ) edges, many new types of edges were observed, presumably formed under a Mo-rich conditions. In this work, CVD-grown $Mo_{1-x}W_xSe_2$ monolayers were transferred onto an *in situ* heating microchip platform for exploring the thermodynamics and kinetics of chemistry-dependent edge reconstruction. The few, interspersed W-atoms are the brightest atoms in the HAADF STEM images as shown in Fig. 8.6 and served as high-contrast fiducial markers to trace the diffusion of Se vacancies and Mo atoms during the edge reconstruction. The Mo chemical potential μ_{Mo}, was controlled by two key parameters: (1) the local carbon residue concentration and (2) the heating time. Under *in situ* heating at 500 °C in the STEM, Se atoms are desorbed from the zigzag ZZSe edge, increasing the local chemical potential of Mo, μ_{Mo}. With the μ_{Mo} increasing, the edge structures change from random edges to Se-terminated edges, then Mo_6Se_6 nanowire (NW)-terminated edges (Fig. 8.6a). Using DFT calculations, the formation energies of 66 different edges were calculated at different chemical potentials, revealing several Se- and Mo-oriented edges with low formation energies within their respective μ_{Mo} window as shown in Fig. 8.6b. At low μ_{Mo}, the formation energies of nanowire (NW)-terminated edges are seen to be higher than that of both ZZSe and ZZMo edges of $MoSe_2$, so both edges are hard to transform into NW-terminated edges. However, with μ_{Mo} increasing, the formation energies of NW-terminated edges are significantly reduced and even lower than that of ZZSe and ZZMo edges. Therefore, when the environment becomes Mo-rich (high μ_{Mo}), the ZZSe edges of $MoSe_2$ will first transform to GB4-reconstructed edges, then to Mo-rich NW-terminated edges. In agreement with in-situ experimental observations, NW-terminated edges were areas with high μ_{Mo}. In addition, as μ_{Mo} increased with heating time, the captured dynamic edge evolution was mediated by μ_{Mo} from a media carbon residue area to confirm the edge transformation pathways from ZZSe edges to ZZ-Se-GB4-Se edges, then to ZZSe-Mo-NW30 edges.

HAADF-STEM images in Fig. 8.6a show the atomic structure of the four most frequently observed edges in $Mo_{1-x}W_xSe_2$ during in situ heating: ZZSe edges terminated with -NW and -Se, ZZMo edges terminated with -NW, and -Se because they have the lowest formation energies suppressed in the carbon residue-rich areas due to a low μ_{Mo} but formed readily in carbon-free within their respective μ_{Mo} window. Note, here the carbon residues result from the sample transfer process. In order to accommodate lattice mismatch strain between the hexagonal lattice and the NW edge, a pentagon-heptagon (5|7) ring defect was reconstructed at the interface in ZZSe-Mo-NW edge. In addition, the mirror twin boundary was also presented in ZZSe-GB4-Se edges and turned the ZZSe edge into a ZZMo edge passivated by Se atoms. These unique defect structures significantly influence the edge electronic and magnetic properties. For example, our DFT calculations showed that the spin splitting of Mo d_{z2} and d_{x2-y2} orbitals in the 5|7 pair at the interface (Fig. 8.6a) could induce the local magnetic moment, therefore giving rise to ferromagnetism of the ZZSe-Mo-NW edge. The intrinsic 1D metallic GB4 mirror twin boundary in ZZSe-GB4-Se edges could potentially enable the edge exhibiting charge density waves and TLL behaviors [50]; these however need the experimental verification. In addition, freestanding Mo_6Se_6 nanowires have been demonstrated as 1D metallic system [54], enabling to confine electron-electron interactions. Therefore, as-formed NW-terminated edges including ZZSe-Mo-NW30 and ZZMo-NW30 could inherit such 1D metallic behavior. Interestingly, Xia's [18] following work successfully proved that these

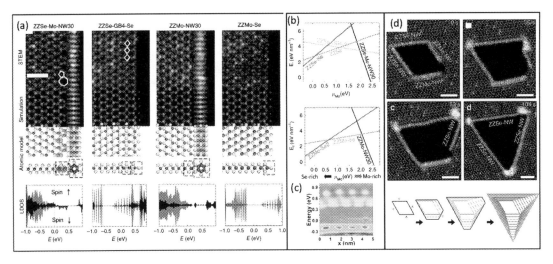

FIGURE 8.6 Edge reconstruction in $Mo_{1-x}W_xSe_2$ monolayer during in situ heating in STEM, where ZZ and NW refer to zigzag and nanowire, respectively. (a) Experimental (the 1st row) and simulated (the 2nd row) HAADF-STEM images, as well as corresponding atomic models (the 3rd row) and local density of states (LDOS) with arbitrary unit (the 4th row) of ZZSe-Mo-NW30, ZZSe-GB4-Se, ZZMo-NW30, and ZZMo-Se edges, respectively. Red and black LDOS plots are calculated from atoms framed by red and black dashed boxes, respectively. Spin-magnetization density is overlaid on the atomic structure of ZZSe-Mo-NW30. Scale bar corresponds to 1 nm. (b) DFT-calculated formation energy per unit length (E_f) as a function of the Mo chemical potential, μ_{Mo}, for four ZZSe edges and three ZZMo edges. The two vertical dashed lines denote the range $0\,eV < \mu_{Mo} < 2.04\,eV$. (c) Spatially resolved STS maps of ZZSe-Mo-NW30 edge revealing the quantum well states at the edge [18]. (d) Atomic resolution HAADF-STEM images showing the shape evolution of an initial NW-terminated parallelogram pore (top). (bottom) Simulated etching progress of the NW-terminated parallelogram pore shown in top. The ZZSe-NW edges (top and left edges, in brown etch more slowly than the ZZMo-NW edge (blue). Scale bars are 2 nm. Reprinted with permission from Springer Nature, Copyright © 2018, The Authors [22].

NW-terminated edges exhibited quantum confined TLL behavior, as evidenced by scanning tunneling microscopy and spectroscopy (Fig. 8.6c). The rich edge structures and enhanced interaction of edges with other defects offer exciting possibilities of controlling the precise edge structures and thus the local properties of edges through fine-tuning the chemical environment during growth.

The edge structure also influences the growth and/or etching kinetics of 2D materials and determines their morphologies and shapes. In equilibrium, edge energy defines the shape of 2D materials as predicted by traditional Wulff construction theory, whereas the kinetics of edge defect formation and subsequent growth control the shape of crystals during their nonequilibrium growth. When the chemical potential of precursor atoms on the substrate is less than that in bulk 2D materials, the atoms in the edge detach easily, thereby making 2D materials etching possible. As shown in Fig. 8.6d, during extended *in situ* heating, at high μ_{Mo}, the initial parallelogram-shaped etched pores with two ZZSe-NW edges and two ZZMo-NW edges transformed to a triangular shape with ZZSe-NW edges on all sides [22]. This can be explained by the kinetic Wulff construction (KWC) theory that simply states that low energy edges will be preferred during etching [55]. In other words, during the etching process, edges with fast etching rate disappear and the morphology of the etched pores is determined by the slow etching edges [36]. Therefore, the observed ZZSe-NW dominated edge evolution can be interpreted from the estimated etching rates of the edges, resulting from their different formation energies with varying chemical potentials of μ_{Mo}. Due to the fast etching rate of ZZMo-NW edges (low formation energy), these edges finally disappear and the etched pores change to a triangle with ZZSe-NW edges which have a much lower etching rate.

In order to further understand how edge structures determine the morphology evolution of 2D materials, a growth-etching-regrowth process was developed by changing the chemical potentials of reactants during the vapor phase deposition of monolayer GaSe on the substrate [7]. By turning the Ar gas flow on and off during growth, the growth of monolayer triangular GaSe domain could be switched to the etching mode, forming pores with different morphologies (hexagon, truncated triangle, and triangle) (Fig. 8.7a) in the single crystal domains under different chemical environments. The etching of monolayers occurs not only from the edges, but also from the surface, creating pores in the crystals (Fig. 8.7b). The etched pores can also be refilled to form the monolayer GaSe again by re-exposure to the precursors. As in previous reports, the morphologies were determined by either energies of the edges at equilibrium or by nonequilibrium growth kinetics at the edges [36]. In order to clearly understand this morphological evolution pathway, DFT calculations with thermodynamic Wulff construction theory were performed. As shown in Fig. 8.7c, GaSe monolayers have two types of ZZ 19.1° edges (Ga- and Se-terminated), as well as one type of armchair (AC) edge. Increasing the chemical potential difference $\Delta\mu$ between Ga and Se atoms, the equilibrium shape of monolayer GaSe crystals during the growth should change from a truncated triangular shape to a nonagonal shapefollowed by a hexagonal shape. However, almost all as-grown GaSe crystals are triangular shapes, which is not consistent with the predicted shapes by equilibrium edge energy calculations. In addition, the GaSe ML crystals still retain their crystal structure after etching, indicating that the etching is a steady state process controlled by the kinetics of atom removal at the edges. Therefore, the nonequilibrium growth kinetics at the edge should be considered. The growth/etching kinetics of GaSe flakes is determined by the atom attachment/detachment to/from ZZ, AC, and tilted edges. Based on the KWC theory, the growth morphology should be determined by the slowest-growing edges, resulting from their high formation energies [33,36]. As shown in Fig. 8.7d, the formation energy for initiating the growth of different edges at different $\Delta\mu$ was calculated. Note that the ZZGa and AC edges have constant formation energy while the formation energy of ZZSe edge decreases as $\Delta\mu$ increases. Therefore, by changing the growth conditions, the morphology of GaSe with ZZ or AC edges can be tuned through the change of relative growth rates between ZZ and AC edges induced by different $\Delta\mu$. For example, the ZZSe edge grows at a slower rate than that of the ZZGa edge at low $\Delta\mu$, so the growth morphology of GaSe will form a truncated triangle with ZZSe and ZZGa edges, then transform to a triangle with only ZZSe edges. In contrast, at high $\Delta\mu$, the ZZGa edge has a lower growth rate compared to the ZZSe edgeleading to the growth of a triangle with only ZZGa edges. This is consistent with our experimental results that the as-grown monolayer GaSe generally has triangular shapes with ZZSe edges.

Compared to the growth, the etching process is only governed by atomic detachment from edges, therefore the edges with the highest etching rates disappear, and the edges with the lowest etching rates determine the shape of the GaSe domains. As shown in the DFT cal-

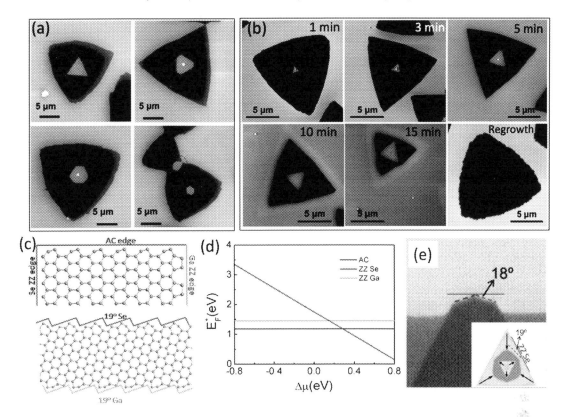

FIGURE 8.7 Edge-determined etching and regrowth of monolayer GaSe. (a) SEM images of GaSe monolayers on SiO$_2$ substrate after being etched 5 min, showing different shapes of the etched pores. (b) SEM images of typical triangular GaSe monolayers after being etched for different time and regrown GaSe monolayers. (c) Different edge structures of hexagonal GaSe, including ZZGa, ZZSe, AC 19° tilted edges. (d) Change of the formation energy costs for initiating the growth of ZZ and AC edges as a function of $\Delta\mu$. (e) SEM image of a bilayer GaSe flake with a nearly 19° tilt edge at the initial etching stage. Inset shows the simulated etching process of a triangular GaSe domain with ZZ Se edges. The etching starts from both the domain center and the domain perimeter. Reprinted with permission from Reference [7], Copyright © 2017, American Chemical Society [7].

culations (Fig. 8.7e), initially, 19° edges appeared at the vertices of the domain, and the ZZ Se edges were shortened. With further etching, ZZ Se edges disappeared and the shape of the GaSe domain was changed into a dodecagon with only 19° edges. However, only 9° tilt edges were observed in monolayer GaSe due to effects of the substrate during the etching process which caused deviations from the theoretical prediction. Tilt edges close to the 19° edges were indeed observed during the etching of the top layer of the bilayer GaSe domain because of the weak interaction between GaSe layers. In addition to etchof the outside edges, the center of flakes of GaSe is also etched and develops into an inverse triangular pore with straight ZZ edges, that is the domain shape of the etched pores are usually opposite to that of the outer circumference of the 2D crystal [55]. Based on theoretical calculations, the pore morphology changes from a hexagonal shape to a truncated triangle with ZZ Se and ZZ Ga edges, and then to a triangle with only ZZ Se edges. Despite the hexagonal and truncated triangular shape of the etched pores at the early stage of growth, the pores finally evolve into the most stable triangles as modeled by KWC. Note that the edge and shape evolution of the etched pores during etching is the same as that of the flakes during growth, indicating the edge propagation of etched holes during etching equals the edge evolution of morphology of 2D materials during growth. Such experimental results and theoretical modeling indicate that morphology and edge control of GaSe on monolayer flakes is possible by adjusting $\Delta\mu$ through tuning the local fluctuation of precursor concentration in the nonequilibrium synthesis.

In summary, various heterogeneities can be formed by controlling the chemical potential of the constituent TMD atoms during nonequilibrium synthesis and processing approaches. Further, understanding of the heterogeneity formation mechanism in 2D materials by advanced theoretical calculations and modeling, atomic-resolution STEM imaging, and *in situ* diagnostics at multiple length scales, in addition to control of the type and density of defects, edge morphologies in 2D TMDs can be achieved by tuning the thermodynamic and kinetic growth processes. This basic understanding can pave the way towards engineering the electronic structure and properties for designed functionality.

8.3 Strain induced phenomena in 2D materials

One of the outstanding properties of 2D materials is their ability to reversibly withstand huge strains that may exceed those in their bulk counterparts by orders of magnitude. This unique ultra-strength of 2D materials has attracted much attention as a controlled way to introduce heterogeneity to modify and tailor their properties. For example, precise and spatially selective tuning of the electronic band structures, transport properties, and excitonic energy levels is possible by strain manipulations. As predicted theoretically, strain induced tuning could result in a direct to indirect band gap transition in MoS_2 and semiconducting to metal phase transitions at larger strains [56]. The attractive feature of 2D materials is that strained regions can be highly localized to specific locations using for example, nanoindentation or other approaches to produce deformation at the nanoscale. Such deformation induces strain gradients that funnel excitons toward the indentation center at the highest strain [57] which can then locally enhance photoluminescence, that may result in single photon emission [58–60]. The possibility of a continuous band gap tuning using strain opens new ways to design solar cells [57] and photodetectors [61,62] with enhanced characteristics, and potentially to other optoelectronic devices. Different methods have been employed to generate strain in 2D materials including bending, uniaxial strain through substrate stretching, piezo-electric stretching, mismatch of thermal expansions between a substrate and a 2D material, and wrinkle formation in 2D crystals that have been reviewed in Ref. [63]. Many publications have been devoted to different aspects of strain in 2D materials (see review [63]). The goal of this section is not to provide a comprehensive review of this subject, but to highlight some interesting new aspects of strain related effects. Here, we mainly focus on strain generated during growth on patterned substrates and strains introducing during merging of monolayer crystals to form multidomain films. To demonstrate strain effects developed during CVD growth, we select donut patterns of different heights that are etched into SiO_2/Si substrate because of their nonzero, positive and negative Gaussian curvatures (Figs. 8.8a, b) [5]. Another reason to consider this type of pattern is that the biaxial strain distribution of a 2D crystal, which conforms to a donut can be predicted theoretically using well established continuum elasticity models [5].

First, we consider the most common optical methods to assess strain in 2D materials that include (1) photoluminescence (PL) or absorption, (2) Raman scattering, and (3) second harmonic generation (SHG). The first approach is based on strain induced changes in electronic properties of a 2D crystal and is directly related to changes in the band gap. The second method monitors structural changes of 2D materials under strain through analysis of the lattice vibrations and the third one is based on changes of the nonlinear susceptibility tensor due to strain via a photoelastic effect.

8.3.1 Strain estimates from PL/absorption spectra

To measure strain, shifts in PL or absorption bands of A, B, or I excitons (where I stand for indirect exciton) induced by strain are usually employed. This approach is attractive since it allows direct estimates of band gap changes under strain, which is a key parameter for using strain to modify the optoelectronic properties of 2D materials. Two main outcomes are desired from PL/absorption measurements, i.e., magnitude of strain including its sign at the specific locations and strain spatial distribution. Although, the latter can be achieved relatively easily by PL mapping, getting the magnitude of strain is more challenging since it requires calibration of the PL shifts, or in other words, determining gauge factors defined as the exciton peak energy shifts per 1% of strain. For example, Fig. 8.8c shows different shifts of the A exciton peak of a WS_2 monolayer grown over a donut of a height of 20 nm (Figs. 8.8a, b) exhibiting uniform tensile strain distribution in the center region of the donut as shown by the PL map in the inset [5].

Large numbers of publications have been devoted to the determination of the excitonic band gauge factors in monolayers and few layers of MoS_2 [57,61,62,64–75] using different approaches to induce strain. However, only a limited number of studies have been devoted to other 2D materials: WS_2 [73,76,77], WSe_2 [58–60,78,79], $ReSe_2$ [80], and heterostructures of WS_2/MoS_2 [81,82]. When analyzing strain to determine the gauge factors from the measured shifts in the PL/absorption spectra, it is important to distinguish between different types of strain, i.e., compressive $vs.$ tensile, uniaxial $vs.$ biaxial, and homogeneous $vs.$ inhomogeneous. The gauge factors can also depend on the number of layers in a 2D crystal. The common problem in measuring the gauge factors is incomplete transfer of strain from a substrate to a 2D material to explain differences in the measured results [63]. Therefore, special attention should be devoted to substrate selection for these measurements [73]. For example, the careful selection of a polymer substrate to generate biaxial tensile strain based on thermal expansion followed by measuring the strain induced shifts in the differential reflection spectra of TMD monolayers allowed the authors of Ref. [73] to deduce the following order of the gauge factors for the A exciton in the following TMDs:

$$MoSe_2(|-33|\ meV/\%) < MoS_2(|-51|\ meV/\%) < WSe_2(|-63|\ meV/\%) < WS_2(|-94|\ meV/\%)$$

Note that the negative shifts reflect a red shift in the A exciton PL/absorption bands that is a characteristic feature of tensile strain in these 2D crystals. The same trend in the absolute values of the gauge factors for this set of TMDs has been predicted theoretically. However, the values of the calculated strains are systematically larger than those obtained experimentally, which can be explained by incomplete transfer of strain from the substrate to the 2D crystal [73,83].

8.3.2 Strain estimates from Raman spectra

Raman spectroscopy is another common approach to estimate strain in 2D materials by measuring shifts of the Raman peaks caused by changes in the lattice vibrations due to strain. Similar to the PL approach described above the types of strain should be distinguished since they result in different lattice symmetries under the applied strain. Usually, symmetric out-of-plain (A_1') and the doubly degenerate in-plain (E') modes are used to characterize strain in TMD monolayers with D_{3h} symmetry. Note that due to the symmetry change from odd (D_{3h}) to even (D_{3d}) number of layers and to bulk (D_{6h}) (the symbols D_{3h}, D_{3d}, and D_{6h} are the Schoenflies notations for the corresponding crystallographic symmetry point groups) the notations for these modes also change from A_1', E' to A_{1g}, E_g and to A_{1g}, E_{2g}^1, respectively [84]. In the literature, often the bulk notations are used for all the cases which may introduce some confusion. As an example, Fig. 8.8d shows changes in the Raman spectra of a WS_2 monolayer grown over a 20 nm height donut at three different locations where bulk notations are used for the corresponding modes. Like strain induced shifts in the PL spectra, the Raman shifts can be used to characterize strain in 2D materials. The majority of studies described in the literature have been devoted to monolayers and few layers of MoS_2 [61,64–66,71–74,85–87] and only a few papers considered other TMDs, e.g., WS_2 [76,88] and WSe_2 [78,88]. The common feature of these studies is that for TMD monolayers the degenerate E' mode is very sensitive to strain and in the most studied case of uniaxial tensile strain this mode exhibits a red shift and splitting to E'^+ and E'^- components at the specific magnitudes of strain, e.g., $\sim 1\%$ for MoS_2 [66,86] and $\sim 1.4\%$ for WS_2 [76] monolayers. After the split both components show a linear decrease of the Raman shifts with strain and the corresponding slopes can be used as gauge factors for strain assessment. Unfortunately, as in the case of PL gauge factors a large variation is found among the various studies. For example, in the case of MoS_2 monolayers, *uniaxial tensile* strains of -4.5 ± 0.3 cm^{-1}/% (E'^-) and -1.0 ± 1 cm^{-1}/% (E'^+) were reported in Ref. [66] and [-2.5 ± 0.3 cm^{-1}/% (E'^-) and -0.8 ± 0.1 cm^{-1}/% (E'^+)] were measured in Ref. [85] (see also discussion in Ref. [74]). A better agreement between different studies exists for the case of *biaxial tensile* strain in MoS_2 monolayers, e.g., (-4.48 cm^{-1}/% (E'); -1.02 cm^{-1}/% (A_1') Ref. [71]) and (-5.2 cm^{-1}/% (E'); -1.7 cm^{-1}/% (A_1') Ref. [72]). Nevertheless, shifts and splitting of these modes can be used as a useful tool for strain assessment, and mapping of strain spatial distribution, thus distinguishing its type based on red (tensile) or blue (compressive) shifts of the Raman lines. Another interesting possibility to assess strain is based on analysis of the intensity ratio of A_1' and E' modes in TMDs [89].

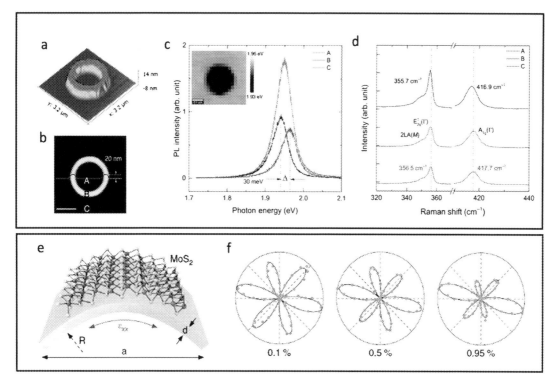

FIGURE 8.8 Different approaches to generate and estimate strain in 2D materials. (a)–(d) *Generation of strain during growth, PL and Raman strain assessment.* Direct growth of 2D materials over patterns with nonzero Gaussian curvatures, e.g., over a donut structure (2-μm outer diameter, 20 nm height) as shown by an AFM image in (a) and (b). (c)–(d) Strain can be estimated based on shifts in PL and Raman scattering spectra as shown by examples of measuring strain at different points (A-C) of a WS$_2$ monolayer crystal grown over a donut. The measured PL shift of 30 meV and the corresponding Raman shifts of both A'_1 and E' modes of 0.8 cm^{-1} from A to C indicate appearance of strain in the A and B regions. Reprinted from Reference [5], Copyright 2019 © The Authors. (e)–(f) *Two-point bending method to apply variable tensile strain with its assessment by an SHG* in a monolayer MoS$_2$ crystal. (e) Schematic of the bending method. (f) SHG polarization diagrams for different values of applied strains (0.1, 0.5, and 0.95%). Reprinted from Ref. [16]. Copyright 2018 © The Authors.

8.3.3 Second harmonic generation (SHG) for strain estimates

Recently, second harmonic generation (SHG) has emerged as a powerful technique for strain assessment in 2D crystals [16,90,91]. Since strain results in reduction of the crystal symmetry, it significantly affects its second order nonlinear polarization, and changes can be used to determine the strain tensor. The mathematical treatment and assessment of this problem for TMD monolayers with D$_{3h}$ symmetry is given in Refs. [16,90,91].

Figs. 8.8e and 8.8f show a schematic of two-point bending method used to apply variable tensile strain for the measurements described in Ref. [16] with the strain assessment by an SHG based on changes in its polarization diagrams, $I_{2\omega}(\theta)$. For an unstrained MoS$_2$ monolayer crystal and the incident laser polarization collinear with that of an analyzer, the SHG polarization diagram is determined by $I_{2\omega}(\theta) \propto \cos^2(3\theta + \theta_0)$ (where θ_0 is the starting rotation angle with respect to the armchair direction of the crystal) with all six petals identical to each other. However, as one can see from Fig. 8.8f even a small strain of $\sim 0.1\%$ causes visual changes to the polarization diagram that can be used as a sensitive tool for strain assessment.

Mennel et al. [16,91], used two different approaches to estimate the localized strain in 2D TMD crystals with D_{3h} symmetry based on the polarization SHG measurements. The first approach uses the minimum amount of measured data, i.e., only three points of a polarization diagram measured along the armchair directions at $0°$, $60°$, and $120°$ to characterize its asymmetry [16]. The limited number of measured points to determine the local strain with submicron spatial resolution allows acquiring spatial strain maps. The second approach requires acquisition of full SHG polarization diagrams at each point, which is time consuming and is not convenient for detail mapping of strain distribution [91].

8.3.4 Extended compressive strain at grain boundaries of merged monolayer crystals

Coalescence of monolayer 2D crystal domains is a common phenomenon that occurs during 2D film growth for different technique including CVD. This process can result in the formation of grain boundaries (GBs) between the merging crystal domains. Recently, we discovered that an extended (over a few microns) compressive strain develops around the grain boundaries during this merging process. Grain boundaries in merged monolayer crystal domains can be predicted based on a kinetic Monte Carlo model developed in Ref. [92]. According to this model the growth of a crystal edge starts with nucleation kink and proceeds with its propagation along the crystal's edge with velocity, $\overrightarrow{v}$. In the case of two merging edges growing with kink propagation velocities $\overrightarrow{v}_1$ and $\overrightarrow{v}_2$, the grain boundary direction will be determined by the sum of these vectors as shown in Fig. 8.9a where an optical image of two merging MoS_2 monolayer crystals with misorientation angle of $11°$ is presented and the predicted grain boundaries are marked by black dashed lines. When $|\overrightarrow{v}_1| = |\overrightarrow{v}_2|$ the grain boundary will be simply determined by bisections of the angles between the merging sides. The yellow circle marks the merge point of the crystals and their shape at this point is outlined by the yellow dashed lines. Optical images do not show GBs, but they can be visualized using SHG mapping as shown in Fig. 8.9b. The comparison between the GBs revealed by the SHG mapping and those predicted based on the model (white dashed line, same as in Fig. 8.9a) confirms that this simple model well describes the shape of the GBs. Now, we can employ Raman mapping to understand strain generation and distribution along the grain boundaries of merged monolayer crystals. Fig. 8.9c shows an example of a map of the E' mode shift for two merged monolayer MoS_2 crystals with the misorientation angle of $25°$. Fig. 8.9c clearly shows that compressive strain ($\sim -0.4\%$) is developed along the grain boundary outlined by the white dashed line predicted by the vector model. Detailed analysis of many other merged monolayer MoS_2 crystals showed that the development of strain along GBs is a common phenomenon.

To understand how strain develops and evolves during the coalescence of growing monolayer MoS_2 crystals, we conducted reactive atomistic molecular dynamic (MD) simulations of the growth process [3]. The growth was simulated by supplying and removing of Mo and S atoms at each iteration step that formed MoS_2 precursors followed by structural relaxation

8.3. Strain induced phenomena in 2D materials 239

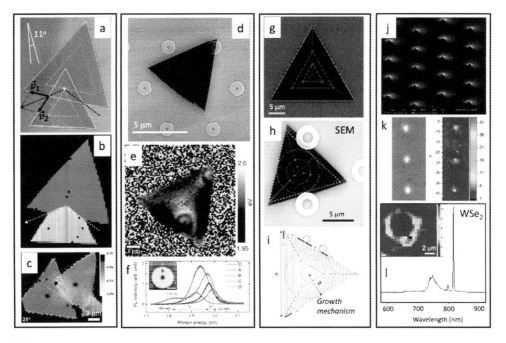

FIGURE 8.9 Strain related phenomena in 2D materials. (a)–(c) *Development of extended compressive strain at grain boundaries of merging monolayer crystals during CVD growth.* (a) An optical image of two merged MoS_2 monolayer crystals with misorientation angle of 11° forming stitched GBs outlined by black dashed lines based on the prediction of a kink nucleation and propagation model in Ref. [92]. The yellow circle marks the merge point of the crystals with their shape at the onset of merging outlined by yellow dashed lines. (b) SHG intensity map of the crystals in (a) revealing the GBs with white dashed line indicated the predicted grain boundaries same as in (a). (c) Raman map of E' peak position showing strain distribution around the GBs of two merged monolayer crystals (marked by white dashed lines) with misorientation angle of 25°. Figure Reprinted with permission from Ref. [3], Copyright © 2021, American Chemical Society. (d)–(f) *Strain generation by growth on curved surfaces.* (d) SEM images of WS_2 monolayer crystals grown by CVD over donut-like structures with the height of 40 nm. (e) Map of PL peak positions, showing dark (tensile) and bright (compressive) regions of strain. (f) Examples of PL spectra measured at the center and edge of a donut in (e), the neighboring flat substrate, and strained area as labeled by A-D in the inset AFM image with the reference PL spectrum measured on a flat region of the same substrate. Bandgap shifts of −150 meV due to tensile strain (spectra O vs. C) and 22 meV due to compressive strain (spectra O versus A) are obtained. The white arrow in the inset marks the outward propagation direction of the crystal edge during growth. (g)–(i) *Strain induced growth acceleration.* (g)–(h) SEM images of monolayer WS_2 crystals showing centrosymmetric growth on a flat substrate (g) and non-centrosymmetric growth on a substrate with obstacles (180 nm-height donuts) exhibiting faster growth of the facets that interfere with the obstacles (h). Inner solid and dashed equilateral triangles are guides to assess relative facet propagation. (i) Illustration of step nucleation (blue dots) and propagation (red arrows) model for growth of a monolayer crystal. The corner nucleation sites are initially responsible for a centrosymmetric growth mode. Each triangle boundary corresponds to a subsequent time increment. The yellow circles indicate obstacles that generate additional nucleation steps (blue dots) that accelerate the growth of the affected facets. Reprinted with permission Reprinted from Ref. [5], Copyright © The Authors. (j)–(l) *Strain induced exciton funneling.* (j) SEM image of WS_2 monolayer grown on patterned Si cone structures. The top view is shown in (k) (left panel). (k) (right panel) PL intensity map showing strong increase in the PL intensity and the bandgap changes on cone tips due to pattern induced strain. [Unpublished] (l) PL spectrum of a WSe_2 monolayer at 4K (the corresponding PL map is shown in the inset) taken from a hot spot of the strained region around the circumference of a 5 μm TEM grid hole, showing appearance of a narrow peak that indicates possible single photon emission. [Unpublished].

through the MD runs. As the two crystals begin to merge to form a GB, different types of initial defect configurations appeared after the relaxation of crystal lattices that can be qualitatively described as unsaturated defects with a large open space and higher energy and saturated defects with a smaller open space e.g., a five-member ring. We found that unsaturated defects absorb the precursors incorporating them into the crystal lattices, however, saturated defects do not absorb the MoS_2 precursors. In the latter case, the precursors either stabilize on the top of the monolayer or desorb. Our simulations show that insertion of the atomic precursors into the dislocations results in their stabilization and generates strain that transfers through the nearby regions of the crystals and extends by a few microns from the GBs. Eventually this process leads to saturation of the GB defects and to termination of further precursor absorption. The magnitude of the generated strain depends on the misorientation angle of the merging crystals with the minimum strain observed near $0°$ and $60°$. The detailed description of this phenomenon along with theoretical interpretation is given in Ref. [3].

8.3.5 Strain generation by growth on curved surfaces: strain tolerant growth

Another interesting effect of strain occurs during a 2D crystal growth on patterned substrates when a growing crystal encounters an obstacle. Depending on the height of the obstacles, a TMD crystal can simply grow over the obstacles thus maintaining its crystallinity or completely change the growth mode from a single crystal to a multidomain crystal at a donut height ≥ 40 nm [5]. For example, a monolayer WS_2 crystal can grow over the 20 nm-height donuts shown in Figs. 8.8a,b developing only small tensile strain in the center of the donut as shown in Fig. 8.8c. Fig. 8.9d shows an SEM image of a monolayer WS_2 crystal grown over 40 nm-high donuts that is on the onset of branching due to accumulated strain. The corresponding map of the PL peak positions presented in Fig. 8.9f displays regions of compressive (bright) and tensile (dark) strain with the magnitudes at points A and C of $\sim 0.2\%$ and $\sim 1.1\%$, respectively derived based on the shifts in the PL spectra Fig. 8.9f [5]. This example gives an estimate of the threshold for strain-tolerant growth regime as $\leq 1.1\%$ and provides an interesting method for strain engineering.

8.3.6 Strain induced 2D crystal growth acceleration

Now, we consider how strain in the strain-tolerant growth regime affects monolayer crystal growth based on analysis of its shape. As shown by the SEM image in Fig. 8.9g, on a flat substrate all facets of a monolayer WS_2 crystal grow with the same rate forming a centrosymmetric 2D crystal. This can be explained simply by a kink nucleation and propagation model assuming that kink nucleation occurs preferentially at the triangle corners (marked by the blue circles) followed by growth propagation along the crystal facets [5] (see also Ref. [92]). The nucleation at the corners is supported by the appearance of three dark stripes in the corresponding PL map (see Ref. [5]) that come from defect-induced quenching of the PL and traced by the blue circles in the SEM image (Fig. 8.9g).

However, the situation changes when the growing crystal encounters obstacles as those shown in an SEM image in Fig. 8.9h. As one can see from the SEM image, the two crystal edges that interfere with the obstacles grow faster than the free edge. Similar analysis of many other monolayer WS_2 crystals described in Ref. [5] confirmed this observation. The accelerated growth induced by strain generated by the interaction with obstacles has been explained based on the kink nucleation and propagation model as outlined in Fig. 8.9i. According to this model the strain developed during crystal growth creates new nucleation sites near the obstacles in addition to the ones at the triangle corners that result in the faster growth of the corresponding facets and non-centrosymmetric growth mode. A detailed model based on this assumption is described in Ref. [5]. This phenomenon opens up opportunities in the use of strain to engineer the growth directions and other properties of 2D crystals.

8.3.7 Strain induced exciton funneling: single photon emitters

Another interesting application of strain in 2D materials is related to exciton funneling for deterministic fabrication of arrays of single photon emitters (SPEs). SPEs are non-classical photon sources that produce a stream of identical photons with fixed time spacing. SPEs are considered as promising candidates for quantum information technologies including quantum cryptography, quantum computing, and quantum teleportation [93,94].

For example, 2D materials such as hexagonal boron nitride (hBN) or WSe_2 that exhibit single photon emission at room temperature or low temperatures, respectively, can be transferred to or grown directly on a nanopillar array (Fig. 8.9j) which can result in the enhancement of PL intensity at the locations with the maximum strain as shown in Fig. 8.9k due to an exciton funneling effect. This approach is very attractive and allows not only deterministic positioning of SPEs, but also creation of arrays of SPEs [58–60]. Another interesting example from our studies is related to SPEs developing in 2D materials (e.g., WSe_2) transferred to a TEM grid with circular holes. This approach allows one to investigate defects responsible for the single photon emission directly using atomic resolution STEM. The PL map of WSe_2 at low temperature (4K) shown in the insert in Fig. 8.9l displays bright spots localized along the circumference of the grid opening ($\sim 5\ \mu m$ in diameter) that exhibit very narrow and intense lines in their PL spectra as that shown in Fig. 8.9l indicating a possible presence of SPEs.

In summary, we gave a short overview of strain assessment in 2D materials using common optical approaches such as PL, Raman scattering, and SHG. Based on this optical strain assessments, we described three interesting strain related phenomena that include development of extended strain regions near grain boundaries of growing and merging monolayer crystals; strain creation during monolayer crystals growth over obstacles on a patterned substrate that resulted in accelerated growth of the specific crystal facets; and exciton funneling due to localized tensile strain, which can produce single photon emission.

8.4 Heterogeneity introduced by the self-assembly of nanoscale 'building blocks'

As shown in the previous two sections, chemical potential and strain play major roles in the heterogeneity of 2D crystals as they grow. However, it is equally important to understand and control the size, stoichiometry, and crystallinity of the precursor species that serve as the 'building blocks' for the nucleation and growth of 2D crystals, and the mechanisms and kinetics of their assembly. In this section alternate, emerging nonequilibrium approaches are

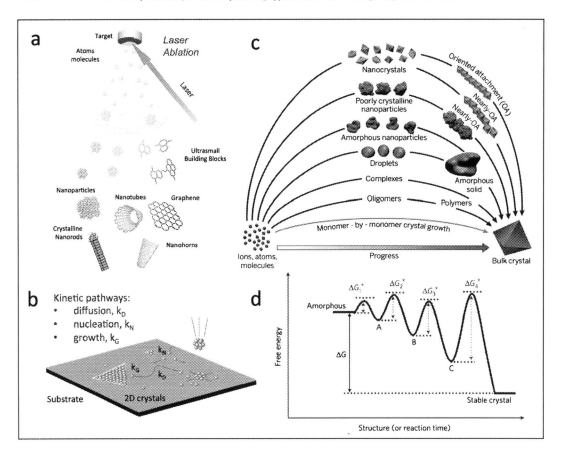

FIGURE 8.10 Concepts of non-equilibrium synthesis and assembly from amorphous precursors. (a) Schematic illustration of the energetic expansion of laser-vaporized atoms, ions, and molecules from a solid target, the subsequent formation of ultrasmall amorphous cluster and nanoparticle 'building blocks' formed by condensation and reactions during collisions with background gas atoms in the gas phase, and the assembly of various crystalline nanomaterials that have been observed by altering the conditions for their self-assembly at high temperatures in the gas-phase. (b) Illustration of the kinetic pathways for assembly of 2D crystals or thin films on a substrate, including rates for adsorption/desorption, diffusion, nucleation, and growth, upon arrival of different ultrasmall 'building blocks'. (c) Concepts of crystallization by particle attachment and the many pathways and types of intermediates, including amorphous or crystalline nanoparticles, that can assemble to form a bulk crystal. [Reprinted with permission from Reference [6] Copyright © 2015, American Association for the Advancement of Science.] (d) Concept of Ostwald's Law of Stages depicting the evolution of an amorphous phase to a stable crystal through a series of metastable phases, each separated by barriers in free energy ΔG_n^*. Reprinted with permission from Reference [17].

described to provide an understanding on the origin of defects in 2D crystals as they grow using bottom-up growth techniques. The kinetic energy of the arriving species is also important in such nonequilibrium processes, with relevance to the formation, conversion, and doping of novel quantum materials by top-down processing. This will be discussed in the final section of the chapter with relevance to the formation of mechanisms of defect formation and rearrangement of existing 2D crystals, enabling materials such as Janus monolayers.

Both concepts are involved in pulsed laser deposition (PLD) as illustrated in Fig. 8.10a. PLD is a highly non-equilibrium process typically employed for the stoichiometric deposition of thin films where the size and crystallinity of the material that is laser-vaporized from a solid target and arrives at a substrate (in Fig. 8.10b), as well as its kinetic energy, can be varied to explore the kinetic pathways for the assembly of different types of precursor 'building blocks', (that can be adjusted from atoms, ions, and molecules to amorphous clusters and nanoparticles, to crystalline nanostructures of different types) and the effects on the nature of defects in 2D materials [95–98]. In PLD, pulsed lasers with nanoseconds pulse widths are typically used to focus light with < 100 mJ/pulse to energy densities of $1–5$ J/cm^2, which exceeds the ablation threshold of nearly all materials. This typically results in the ablation of $\sim 10^{15}$ atoms/pulse and the formation of a weakly-ionized laser plasma with electron temperatures of ≤ 1 eV. Driven by pressure gradients and Coulombic forces, this material begins to freely expand in vacuum while it is very dense. During the nanoseconds-long pulses, the expanding plasmas can absorb energy from the latter part of the laser pulse, accelerating the plasma plume. Collisions during this rapid expansion results in a propagating plume with a characteristic shifted Maxwell-Boltzmann velocity distribution with different materials traveling with a common center-of-mass velocity that is typically $v \sim 1$ cm/µs. Estimating the translational kinetic energies (in eV) of atoms with atomic mass Z (in amu) and velocity v (cm/ms) using $0.52\, Z\, v^2$, we see that Mo atoms ($Z = 96$) have 50 eV/atom while sulfur atoms ($Z = 32$) carry ~ 17 eV/atom – although the fastest material in the plume easily can reach double these velocities (4X kinetic energies). Such energies can propagate across a vacuum and drive the formation of metastable phases, as in the deposition of amorphous diamond from laser vaporized carbon targets, where plasma plumes traveling with velocities of about 4 cm/µs provide about 100 eV/C-atom kinetic energies [96].

However, with spatial confinement provided by a background gas, the plasma plume is thermalized and slowed, which can result in self-assembly of novel nanomaterials *in the gas phase, without a substrate* [98,99]. For the ablation of carbon, this can be tailored for a remarkably high yield of carbon fullerenes [100], single-wall carbon nanotubes [101–103] and nanohorns [104,105] or graphene [106,107]. With the addition of small quantities of a catalyst, the efficient production of vast numbers of nanowires can also be formed in the gas phase by VLS (vapor liquid solid) growth [108,109], where the vapor often depicted as just atoms and molecules, can clearly involve larger clusters and even nanoparticles. As shown in Fig. 8.10c, there are many routes for ions, atoms, and molecules to form crystals; from the typically depicted monomer-addition route, to interactions between amorphous or crystalline nanoparticles (CPA, crystallization by particle attachment) [6]. CPA is typically observed in liquids by *in situ* TEM analysis. However, growth of nanomaterials by CPA can be inferred from gas-phase *in situ* spectroscopic diagnostics (in the same way as efficient catalyst-assisted synthesis of SWNTs) [102,104,110] or from the intentional deposition of nanoparticulates for the growth of nanostructured films [111–114]. In general, amorphous "precursors" are favored to explore the synthesis of materials because, as shown in Fig. 8.10d, they have a higher free energy than the lowest energy phase of a bulk crystal to provide many possible crystallization pathways. In addition, these pathways may include intermediate metastable phases that are separated by free energy barriers, and might be accessed in a stepwise fashion as crystallization proceeds by the supply of external energy [17]. Since clusters and nanoparticles can form in the gas-phase or on the substrate during growth, understanding the mechanisms for their assembly is crucial to enable control of heterogeneities in 2D crystals.

As described earlier, the growth of non-stoichiometric 2D crystals via non-congruent laser vaporization can be used to control the chemical potential during growth for the synthesis of highly chalcogen-deficient single crystals (Fig. 8.3). Stoichiometry control, in fact, is one of the key problems encountered in PLD, which is often compensated using chalcogen rich targets or atmospheres [115]. However, the single-step growth of stoichiometric 2D crystals can also be accomplished by the deposition of ultrasmall nanoparticles (UNPs) of the proper stoichiometry, as shown in Figs. 8.11a,b [14]. Amorphous UNPs can be selectively formed and deposited on substrates by thermalizing laser plumes in background gases to condense nanoparticles and taking advantage of their longer range compared to atoms and smaller clusters [14,111–114,116], although these can assemble to form continuous films, the high nucleation density can lead to highly inhomogeneous alignment, with grain sizes in the ~20 nm range as shown in Fig. 8.11b. These same PLD-delivered nanoparticles shown in Fig. 8.11a, when deposited with different thicknesses on a substrate at room temperature, can serve as sources for the growth of large (~100 μm or larger) 2D crystals of variable thickness by the 'digital transfer growth' process shown in Fig. 8.11c [15]. In this process, a second 'receiver' substrate is brought into near-contact with the first, and the nanoparticles on the 'source' substrate are heated, introducing their re-evaporation. Within this highly-confined region, at moderate inert gas pressures of ~10–20 Torr, these precursors reassemble to grow large 2D crystals of high quality as shown in Figs. 8.11d,e. This process has recently been perfected into a single step as shown in Fig. 8.11f by eliminating the first deposition step and by decoupling the source temperature from the substrate temperature [19]. In this method, the source is a TMD in a boat heated by a laser and placed in the proximity of the substrate inside a tube furnace maintained at high background gas pressure. Laser heating of the boat with the source allows for rapid evaporation of the stoichiometric source material which then condenses onto a substrate held at lower-temperature, allowing the rapid growth of large, high-quality crystals as shown in Figs. 8.11g,h,i. These approaches show that pulsed deposition of stoichiometric particles, either directly on substrates at high temperatures or predeposited on substrates at room temperature as precursors for subsequent thermal processing, are viable methods for the synthesis of 2D materials.

Tuning the PLD process for selective deposition of nanoparticles requires some *in situ* gas phase diagnostics such as intensified charge coupled device (ICCD)-imaging or ion probes to record the thermalization of the laser plasma [95,111], which will be illustrated in the next section. However, very few simple *in situ* diagnostics have been developed to understand the nucleation and growth of 2D crystalline films by PLD or CVD, especially when the growth of monolayer films involves sub-monolayer sensitivity. One simple technique that is well suited for *in situ* diagnostics of 2D crystal growth is shown in Fig. 8.12: *in situ* time-resolved reflectivity (TRR). TRR takes advantage of the high optical contrast that allows the imaging of atomically-thin 2D crystals in optical microscopes. In practice, a single wavelength CW-laser simply reflects from a SiO_2/Si substrate during PLD (Fig. 8.12a) [10]. Utilizing a standard model that calculates the Fresnel transmission and reflection coefficients for a uniform, semitransparent TMD crystal on a SiO_2/Si substrate (Fig. 8.12b) the optical contrast can be predicted for a given choice of TMD and SiO_2 thicknesses. By adjusting the thickness of the SiO_2 layer for a given TMD crystal, the predicted optical contrast can be maximized (*e.g.*,

8.4. Heterogeneity introduced by the self-assembly of nanoscale 'building blocks'   245

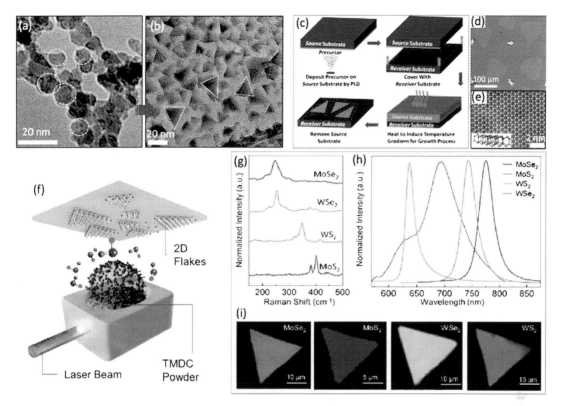

FIGURE 8.11 (a) Ultrasmall amorphous nanoparticles (UNPs) of GaSe collected by pulsed laser deposition at room temperature form (b) few-layer crystals of GaSe when deposited at 600 °C (unpublished) following the procedure described in Ref. [14]. (c) Digital transfer growth process utilizes UNPs deposited at room temperature that serve as a source, encapsulated by a receiver substrate.[Reprinted with permission from Ref. [15], Copyright © 2014, American Chemical Society] Heating the source substrate grows large crystalline domains shown in (d) 100 mm in size with (e) excellent crystal quality as shown in AR-HAADF STEM. (f) Rapid growth of 2D TMDs by the LAST technique within a tube furnace maintains the substrate temperature, with rapid heating of a crucible by a laser utilizing stoichiometric powder. (g) Raman spectra, (h) Photoluminescence, and (i) PL mapping show the excellent quality of the crystals grown by LAST. [Reprinted from Ref. [19] Copyright © 2020 IOP Publishing.

green curve in Fig. 8.12c, which assumes a uniform deposition rate). The reflectivity contrast measured during TMD crystal growth during PLD (red curve) reveals the incremental deposition from each laser pulse (at 1 Hz) in Fig. 8.12c as a monolayer is constructed over ~20 deposition pulses and continues. Figs. 8.12d,e,f,g show the actual evolution of films deposited on nearby TEM grids under the same conditions, stopped after the indicated changes in reflectivity, in excellent agreement with the predicted 'average' layer number by TRR. However, as shown in Fig. 8.12e for example (the predicted 1L thickness) although 76% of the film is monolayer, ~10% of the film is already 2L while 14% of the first layer is still not formed. The valuable *in situ* data shows that growth is not uniform, indicating a nucleation stage then accelerated growth as islands form and grow. Using this simple diagnostic to observe nucleation and growth dynamics as the deposition flux and pulse rate are controlled at different

246   8. Nonequilibrium synthesis and processing approaches to tailor heterogeneity in 2D materials

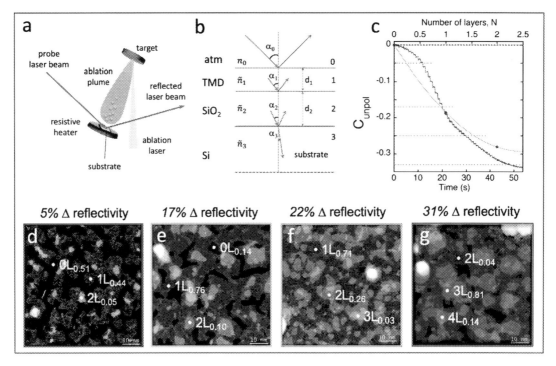

FIGURE 8.12 (a) Schematic of *in situ* time-resolved reflectivity arrangement to monitor the nucleation and growth of 2D layer growth during PLD. A CW-laser reflects from the substrate and the intensity is monitored by a photodiode. (b) Standard model used to calculate the Fresnel transmission and reflection coefficients as a function of thicknesses: d_1 of TMD, for a given d_2 of SiO_2 on Si. The SiO_2 layer thickness d_2 is chosen to maximize the contrast ratio for unpolarized light C_{unpol} between the reflection of the uncoated substrate and that of a monolayer. (c) theoretically-predicted contrast ratio (in green) for uniform deposition rate shows predicted contrast of 19% and 28% for ML and 2L deposition, respectively, of $MoSe_2$ on 80 nm-thick-SiO_2/Si. Measured reflectivity curve (in red) shows incremental deposition every PLD pulse at 1 Hz at 600 °C in vacuum on SiO_2/Si, reaching the predicted monolayer thickness after 20s, and revealing that nucleation is not uniform. (d)–(f) ADF-STEM images of 2D layers of $MoSe_2$ grown on graphene TEM grids by PLD at 600 °C in vacuum under identical conditions as the red curve in (c), after stopping growth at the indicated reflectivity changes. The fractional coverage of bare graphene grid (0L), monolayer (1L), bilayer (2L), etc. are shown, and show excellent agreement with the theoretically predicted values on SiO_2/Si substrates. Reprinted from Ref. [10], Copyright © 2020 IOP Publishing.

substrate temperatures, nucleation and growth can be separately and 'digitally' controlled by pulsing the laser, allowing feedback-controlled growth kinetics [10].

To begin to understand how continuous 2D crystals transform at high temperatures or interact with molecules, clusters, or films that are deposited by various approaches, *in situ* electron microscopy approaches have been used with continuous heating, as shown earlier in Fig. 8.6 [117,118]. However, to understand the non-classical crystallization pathways illustrated in Fig. 8.10c by which amorphous cluster or nanoparticle precursors may assemble into 2D crystals under the highly non-equilibrium conditions inherent during PLD at high temperatures, or by rapid thermal processing after deposition at room temperature, laser heating within a TEM has recently been developed as schematically shown in Fig. 8.13a,b

[2,9]. Laser pulses ranging from a few nanoseconds to longer times can explore how amorphous precursors (from smooth films to nanoparticulate aggregates) crystallize and coalesce through metastable phases, assembling into quasi-continuous 2D crystals and van der Waals heterostructures, allowing details of crystallization mechanisms by particle attachment to be observed by *in situ* HRTEM and HAADF imaging. Electron energy loss spectroscopy (EELS) can be used to simultaneously assess the temperature of the sample (via known shifts of excitonic peaks in low-loss EELS) as well as the evolution in stoichiometry (via core-loss EELS). Selected area electron diffraction is used to assess the orientation and crystallinity of the various domains as they evolve.

Such studies are relevant not only to infer how amorphous precursors form continuous films during PLD or other physical vapor deposition approaches, but also for laser crystallization approaches that are being developed for the direct-write growth of 2D crystals at low-temperature as shown in Figs. 8.13c,d. Using sputter-deposited amorphous films on a variety of substrates, including highly flexible PDMS as shown, laser heating can produce highly crystalline 2D films under proper conditions [20]. By mapping the Raman A_{1g} peak intensity for different laser pulse lengths and intensities in Fig. 8.13d, optimized conditions for assembly into 2D crystals can be established, including scale-up to lamp-based annealing [21]. Coupling such lab-scale conditions with parallel nanoscale observations within the TEM are important to further optimize the heterogeneity of 2D crystals.

8.5 The effects of kinetic energy on defects and doping: hyperthermal implantation for the formation of Janus monolayers

A key variable in the nonequilibrium synthesis of 2D crystals, and the evolution of defects during growth, is the kinetic energy (KE) of the precursor species arriving at the substrate surface. The use of this energy can overcome the free energy barriers to enable crystallization or metastable phase formation (Fig. 8.10c) by processes such as subplantation, when ions create 'thermal spikes' by phonon or electron excitations [119,120]. However, electrons [121] and ions [122] arriving at the substrate with excessive KE can damage growing films as discussed in other chapters within this book. The ability to tune the KE is generally considered an advantage of PLD [95,97], however detailed understanding of the effects of KE on the heterogeneity of crystalline films as they grow has been difficult to assess experimentally. 2D materials present an excellent platform for such studies [123–125] and theory can predict formation energies ($\sim$ few eV) for the insertion of dopant atoms [126], however experimental studies are limited by the availability of ion sources for different materials in this energy range [127].

Recently, the thresholds for chalcogen substitution into 2D monolayers were experimentally investigated by tuning PLD to these low (< 10 eV/atom) energies for the successful formation of Janus 2D monolayers by hyperthermal implantation of WS_2 and MoS_2 monolayers by Se [8], a technique with broad relevance to doping of 2D crystals for novel functionality [126]. Other methods to induce the replacement of the top layer of chalcogen atoms in a TMD monolayer to form Janus monolayers typically have involved high-temperature treatments to phases with higher cohesive energy, for example sulfurization of $MoSe_2$ to form MoS_2 [128].

248   8. Nonequilibrium synthesis and processing approaches to tailor heterogeneity in 2D materials

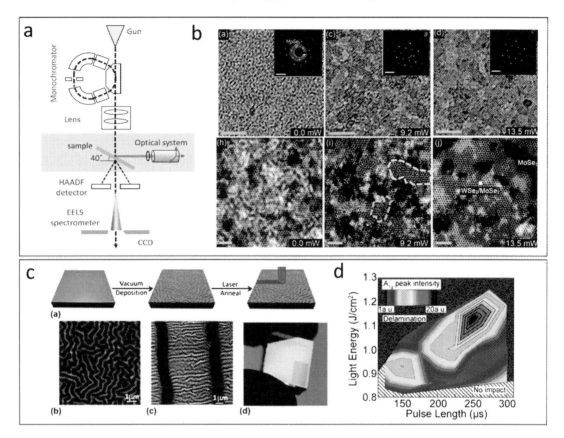

FIGURE 8.13 Laser crystallization approaches for 2D TMD synthesis from amorphous cluster precursors. (a) Schematic of laser heating within a TEM for *in situ* studies of laser-crystallization of amorphous precursors on TEM grids. A fiber-coupled laser is introduced into a TEM and focused to a spot (~3.7 micron) on a tilted TEM-grid, where a monochromated e-beam allows TEM imaging, electron diffraction, and EELS in real time. Reprinted with permission from Ref. [2] Copyright © 2018, The Authors. (b) Example *in situ* HRTEM images and SAED patterns (top row) of amorphous tungsten selenide precursors on ML $MoSe_2$ crystal show how crystallization proceeds via a series of metastable phases as the laser power is raised to the indicated powers (10 ms pulses) into a van der Waals epitaxial Heterobilayer. *Ex situ* UltraSTEM HAADF images (bottom row, scale bar 2 nm) corresponding to the HRTEM images above reveal the aggregate chains of amorphous tungsten selenide on crystalline $MoSe_2$, and the formation of 2D flakes that assemble by particle attachment processes. Reprinted with permission from Ref. [9] Copyright © 2021, American Chemical Society. (c) Sputtered amorphous MoS_2 onto PDMS induces wrinkles. Laser crystallization along different raster lines is shown, producing a flexible polycrystalline 2D film. Reprinted from Ref. [20] Copyright © 2016, The Materials Research Society. (d) The annealing process is gauged as a function of laser pulse length and energy by the Raman spectra A_{1g} peak intensity. Reprinted from Ref. [21] © 2019 The Royal Society of Chemistry. (d) The annealing process is gauged as a function of laser pulse length and energy by the Raman spectra A_{1g} peak intensity. Reprinted with permission from Ref. [21]. ©The Royal Society of Chemistry.

Selenizing MoS_2, on the other hand, has been attempted by more aggressive two-step approaches, such as 50-ev Ar^+-sputtering to try and remove the top layer of S atoms, followed by thermal Se treatments [129]. Both of these methods induce significant defects and cracking

8.5. The effects of kinetic energy on defects and doping: hyperthermal implantation for the formation of Janus monolayers 249

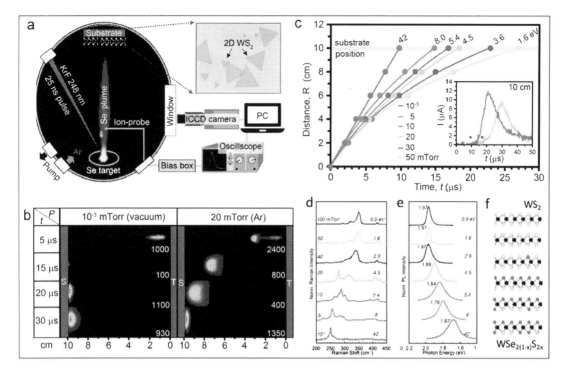

FIGURE 8.14 (a) Experimental arrangement of laser vaporization for generation of hyperthermal species. Pulsed KrF-laser ablation of solid Se target in vacuum generates a forward-directed plume of Se species to impinge upon 2D WS$_2$ monolayers. (b) ICCD-photography of the weakly-luminescent plasma plume trace the free expansion in vacuum and slowing in background Ar (relative intensities shown at the indicated times). (c) R-t plots track the progression of the plume leading edge from ICCD imaging, or ion probe (inset) with curve-fits from a modified drag model to yield the maximum velocity at each point, and maximum kinetic energies at the 10 cm substrate position. (d) Raman spectra and (e) photoluminescence spectra of the WS$_2$ after 400 plume pulses show the transition in properties from unaffected WS$_2$ at < 2 eV/Se atom, to Janus WSSe at 4.5 eV/Se, to WSe$_2$ at > 8 eV/Se (f) Schematic illustrating corresponding structures for the spectra in (d) and (e) as sulfur atoms become replaced on top and bottom layers of a WS$_2$ monolayer. Reprinted with permission from Ref. [8] Copyright © 2020, American Chemical Society.

of the films. Alternatively, the single-step, low-temperature PLD method involves implanting atoms at < 5 eV/atom.

The experimental arrangement and results are shown in Fig. 8.14, where a weakly ionized beam of Se clusters is directed toward monolayer crystals of WS$_2$ either supported on substrates or suspended over TEM grid holes as shown in Fig. 8.14a. The propagation of the plume is captured by intensified CCD-array photography and ion probes to reveal a leading-edge velocity of ∼ 1.0 cm/μs in vacuum (42 eV/Se atom). As shown in Figs. 8.14b,c the kinetic energies of the fastest-moving species can be controllably slowed to 1.6 eV/Se atom upon arrival at the substrate through the introduction of <50 mTorr background Ar gas. The impingement of these Se clusters at these energies begins to replace the S atoms on the crystal surface at different rates, when the KE exceeds a specific threshold. This is evident in the Raman spectra (Fig. 8.14d) and photoluminescence spectra (Fig. 8.14e) that show a stepwise progression, after hundreds of such pulses, from little to no change in properties for energies below 2.9 eV/Se atom, to full conversion into WSe_2 above 8 eV/Se atom as schematically illustrated in Fig. 8.14f. Calculations of Raman spectra and band gap energy of the Janus WSSe matched the experimental spectra for the 3–5 eV/Se atom window [8].

The process of Janus WSSe crystal formation by hyperthermal implantation is schematically shown in Fig. 8.15a. Tilted HAADF STEM images and image simulations that confirmed that Se atoms only implant to replace S atoms within the topmost layer of S atoms in the WS_2 monolayer are also shown. The experimental results, which were confirmed by force-field and DFT molecular dynamics simulations, are shown in Fig. 8.15b for PLD-deposition of Se at 300 °C (a temperature that is sufficient to re-evaporate Se that is not implanted). The results indicate a very narrow window of $\sim$ 3–5 eV/Se atom where the range of Se clusters ($Se_2, \ldots, Se_9$) implant into the topmost layer only. Above these energies, simulations reveal that the larger Se_9 clusters begin to penetrate into the lower layer of S atoms, despite an 8 eV migration barrier for a single Se adatom to diffuse to the lower layer. Figs. 8.15c,d,e show color-coded HAADF-STEM images to indicate the fraction of columns containing 2S, Se-S pairs, and 2Se atoms, W-vacancies, and Se-substituted antisite defects for comparison with XPS spectra. Insets show snapshots of force-field simulations (50-ps after implantation) for each kinetic energy. The simulation results are consistent with the HAADF-STEM observations, revealing that implantation of Se atoms occurs interstitially at 4.5 eV/atom for Janus monolayer formation, followed by displacement of W atoms at 8 eV/atom, or nanopores at 20 eV/atom where the simulations show the irreparable loss of W atoms due to sputtering. The healing process by which the nanophase implanted regions of the crystals expel excess S and Se atoms to recrystallize was simulated by force-field MD for 2 ns following implantation.

Using the energy regime determined for top layer selenization of WS_2 to form the Janus monolayers by PLD, MoS_2 was similarly converted to MoSSe, indicating that PLD could be universally applicable for the formation of other Janus structures. Previously it was shown that sulfurization of $MoSe_2$ crystals could be stepwise converted to MoS_2 by PLD of S species, and that this process could be combined with lithographic patterning to form sharp $MoS_2/MoSe_2$ heterojunctions in arbitrary patterns in a single step [130].

8.6 Summary and outlook

In summary, in this chapter a few examples were presented to illustrate why atomically thin 2D crystals are promising platforms to explore the effects of non-equilibrium synthesis and processing techniques that affect, and can be used to control, heterogeneity in these materials. This chapter has reinforced the synergistic strategy that (a) *in situ* process diagnostics of the non-equilibrium growth environment *and* evolving structure, (b) atomistic characterization of the resulting structure to correlate with measured properties, and (c) computational simulations of the dynamics of synthesis provide unique structural information on 2D crystals. Currently, the vast number of 2D materials systems and heterostructures, coupled with the great variety in synthesis approaches, is providing an impetus for artificial intelligence and machine learning approaches to explore materials science and 'optimize' synthesis.

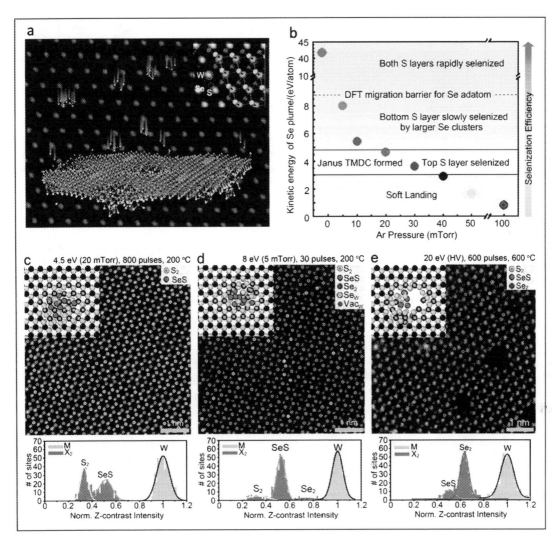

FIGURE 8.15 (a) Snapshot from DFT-MD simulation of Se$_n$ clusters (orange, $n = 2$ to 9) arriving with hyperthermal kinetic energies to implant monolayer WS$_2$ and form Janus WSeS (Mo:blue, S:yellow) within a narrow energy range. Inset shows tilted ADF-STEM image confirming that Se atoms (dimmer than W atoms) and S atoms (very dim) are on different sides of the monolayer. (b) Summary of results of observed selenization vs. maximal kinetic energies observed at different Ar background pressures. Below 3 eV/Se atom no conversion is observed in a "soft landing" regime. From 3–5 eV/Se atom the clusters selenize only the topmost S atoms in a WS$_2$ monolayer. Above 5 eV/Se atom, the bottom S layer also selenizes. (c)–(e) show ADF-STEM images (not tilted) with chalcogen columns color-coded as shown according to intensity, revealing that S$_2$ pairs are replaced by Se-S pairs (but no Se$_2$ pairs) at 4.5 eV, rapid conversion to Se-S pairs and threshold formation of Se$_2$ pairs at 8 eV/Se, and rapid conversion to Se$_2$ pairs (with irreparable loss of atom) at 20 eV/Se and above. Insets in each frame show MD simulations (top views at 50 ps after Se$_9$ arrival to a WS$_2$ monolayer) showing implanted Se atoms taking interstitial positions at 4.5 eV, the displacement of W atoms at 8 eV, and to formation of holes at 20 eV/atom, in agreement with the experimental results. Reprinted with permission from Ref. [8] Copyright © 2020, American Chemical Society.

While such methods can also be applied to 2D materials, understanding of single layer 2D materials assembly mechanisms and simple heterostructures require coordinated characterization and computational simulation. Currently such dynamic, first-principles computational simulations of the simplest multicomponent 2D monolayers and bilayer heterostructures push the limits of atomic resolution electron microscopy analysis and computational methods. These current limitation in supercomputing power and atomistic characterization reinforces why atomically thin 2D materials are unique testbeds for explorations of nonequilibrium processing to induce and control heterogeneity. Hopefully, through the examples in this chapter, the ability to 'see' and control the knobs of synthesis, such as chemical potential, strain, and energetic 'building blocks', has illustrated the novelty and opportunity of atomically thin 2D materials as a platform for heterogeneous quantum materials.

Acknowledgments

The authors gratefully acknowledge funding support from the U.S. Department of Energy, Office of Science, Basic Energy Sciences: (1) Materials Science and Engineering Division for the fundamental studies of nanomaterial growth mechanisms and *in situ* diagnostics development, (2) Scientific User Facilities Division for the development of the characterization techniques represented in this chapter. This work was performed at the Center for Nanophase Materials Sciences, which is a DOE Office of Science User Facility.

References

[1] A.M. van der Zande, P.Y. Huang, D.A. Chenet, T.C. Berkelbach, Y.M. You, G.H. Lee, T.F. Heinz, D.R. Reichman, D.A. Muller, J.C. Hone, Grains and grain boundaries in highly crystalline monolayer molybdenum disulphide, Nature Materials 12 (6) (2013) 554–561.

[2] Y.Y. Wu, C.Z. Liu, T.M. Moore, G.A. Magel, D.A. Garfinkel, J.P. Camden, M.G. Stanford, G. Duscher, P.D. Rack, Exploring photothermal pathways via in situ laser heating in the transmission electron microscope: recrystallization, grain growth, phase separation, and dewetting in $Ag_{0.5}Ni_{0.5}$ thin films, Microscopy and Microanalysis 24 (6) (2018) 647–656.

[3] Y.L. Yu, G.S. Jung, C.Z. Liu, Y.C. Lin, C.M. Rouleau, M. Yoon, G. Eres, G. Duscher, K. Xiao, S. Irle, A.A. Puretzky, D.B. Geohegan, Strain-induced growth of twisted bilayers during the coalescence of monolayer MoS_2 crystals, Acs Nano 15 (3) (2021) 4504–4517.

[4] W. Zhou, X.L. Zou, S. Najmaei, Z. Liu, Y.M. Shi, J. Kong, J. Lou, P.M. Ajayan, B.I. Yakobson, J.C. Idrobo, Intrinsic structural defects in monolayer molybdenum disulfide, Nano Letters 13 (6) (2013) 2615–2622.

[5] K. Wang, A.A. Puretzky, Z.L. Hu, B.R. Srijanto, X.F. Li, N. Gupta, H. Yu, M.K. Tian, M. Mahjouri-Samani, X. Gao, A. Oyedele, C.M. Rouleau, G. Eres, B.I. Yakobson, M. Yoon, K. Xiao, D.B. Geohegan, Strain Tolerance of two-dimensional crystal growth on curved surfaces, Science Advances 5 (5) (2019) aav4028.

[6] J.J. De Yoreo, P.U.P.A. Gilbert, N.A.J.M. Sommerdijk, R.L. Penn, S. Whitelam, D. Joester, H.Z. Zhang, J.D. Rimer, A. Navrotsky, J.F. Banfield, A.F. Wallace, F.M. Michel, F.C. Meldrum, H. Colfen, P.M. Dove, Crystallization by particle attachment in synthetic, biogenic, and geologic environments, Science 349 (6247) (2015) aaa6760.

[7] X.F. Li, J.C. Dong, J.C. Idrobo, A.A. Puretzky, C.M. Rouleau, D.B. Geohegan, F. Ding, K. Xiao, Edge-controlled growth and etching of two-dimensional GaSe monolayers, Journal of the American Chemical Society 139 (1) (2017) 482–491.

[8] Y.C. Lin, C.Z. Liu, Y.L. Yu, E. Zarkadoula, M. Yoon, A.A. Puretzky, L.B. Liang, X.R. Kong, Y.Y. Gu, A. Strasser, H.M. Meyer, M. Lorenz, M.F. Chisholm, I.N. Ivanov, C.M. Rouleau, G. Duscher, K. Xiao, D.B. Geohegan, Low energy implantation into transition-metal dichalcogenide monolayers to form Janus structures, ACS Nano 14 (4) (2020) 3896–3906.

[9] C. Liu, Y.-C. Lin, M. Yoon, Y. Yu, A.A. Puretzky, C.M. Rouleau, M.F. Chisholm, K. Xiao, G. Eres, G. Duscher, D.B. Geohegan, Understanding substrate-guided assembly in van der Waals epitaxy by in situ laser crystallization within a transmission electron microscope, ACS Nano 15 (5) (2021) 8638–8652.

[10] A.A. Puretzky, Y.C. Lin, C.Z. Liu, A.M. Strasser, Y.L. Yu, S. Canulescu, C.M. Rouleau, K. Xiao, G. Duscher, D.B. Geohegan, In situ laser reflectivity to monitor and control the nucleation and growth of atomically thin 2D materials, 2D Materials 7 (2) (2020) 025048.

[11] M. Mahjouri-Samani, L.B. Liang, A. Oyedele, Y.S. Kim, M.K. Tian, N. Cross, K. Wang, M.W. Lin, A. Boulesbaa, C.M. Rouleau, A.A. Puretzky, K. Xiao, M. Yoon, G. Eres, G. Duscher, B.G. Sumpter, D.B. Geohegan, Tailoring vacancies far beyond intrinsic levels changes the carrier type and optical response in monolayer $MoSe_{2-x}$ crystals, Nano Letters 16 (8) (2016) 5213–5220.

[12] Y.J. Ma, S. Kolekar, H.C. Diaz, J. Aprojanz, I. Miccoli, C. Tegenkamp, M. Batzill, Metallic twin grain boundaries embedded in $MoSe_2$ monolayers grown by molecular beam epitaxy, ACS Nano 11 (5) (2017) 5130–5139.

[13] O. Lehtinen, H.P. Komsa, A. Pulkin, M.B. Whitwick, M.W. Chen, T. Lehnert, M.J. Mohn, O.V. Yazyev, A. Kis, U. Kaiser, A.V. Krasheninnikov, Atomic scale microstructure and properties of Se-deficient two-dimensional $MoSe_2$, ACS Nano 9 (3) (2015) 3274–3283.

[14] M. Mahjouri-Samani, R. Gresback, M.K. Tian, K. Wang, A.A. Puretzky, C.M. Rouleau, G. Eres, I.N. Ivanov, K. Xiao, M.A. McGuire, G. Duscher, D.B. Geohegan, Pulsed laser deposition of photoresponsive two-dimensional gase nanosheet networks, Advanced Functional Materials 24 (40) (2014) 6365–6371.

[15] M. Mahjouri-Samani, M. Tian, K. Wang, A. Boulesbaa, C.M. Rouleau, A.A. Puretzky, M.A. McGuire, B.R. Srijanto, K. Xiao, G. Eres, G. Duscher, D.B. Geohegan, Digital transfer growth of patterned 2d metal chalcogenides by confined nanoparticle evaporation, ACS Nano 8 (11) (2014) 11567–11575.

[16] L. Mennel, M.M. Furchi, S. Wachter, M. Paur, D.K. Polyushkin, T. Mueller, Optical imaging of strain in two-dimensional crystals, Nature Communications 9 (2018) 516.

[17] S.Y. Chung, Y.M. Kim, J.G. Kim, Y.J. Kim, Multiphase transformation and Ostwald's rule of stages during crystallization of a metal phosphate, Nature Physics 5 (1) (2009) 68–73.

[18] Y.P. Xia, B. Wang, J.Q. Zhang, Y.J. Jin, H. Tian, W.K. Ho, H. Xu, C.H. Jin, M.H. Xie, Quantum confined tomonaga-Luttinger liquid in Mo_6Se_6 nanowires converted from an epitaxial $MoSe_2$ monolayer, Nano Letters 20 (3) (2020) 2094–2099.

[19] N. Azam, Z. Ahmadi, B. Yakupoglu, S. Elafandi, M.K. Tian, A. Boulesbaa, M. Mahjouri-Samani, Accelerated synthesis of atomically-thin 2d quantum materials by a novel laser-assisted synthesis technique, 2D Materials 7 (1) (2020) 015014.

[20] M.E. McConney, N.R. Glavin, A.T. Juhl, M.H. Check, M.F. Durstock, A.A. Voevodin, T.E. Shelton, J.E. Bultman, J. Hu, M.L. Jespersen, M.K. Gupta, R.D. Naguy, J.G. Colborn, A. Haque, P.T. Hagerty, R.E. Stevenson, C. Muratore, Direct synthesis of ultra-thin large area transition metal dichalcogenides and their heterostructures on stretchable polymer surfaces, Journal of Materials Research 31 (7) (2016) 975.

[21] R.H. Kim, J. Leem, C. Muratore, S. Nam, R. Rao, A. Jawaid, M. Durstock, M. McConney, L. Drummy, R. Rai, A. Voevodin, N. Glavin, Photonic crystallization of two-dimensional MoS_2 for stretchable photodetectors, Nanoscale 11 (28) (2019) 13260–13268.

[22] X.H. Sang, X.F. Li, W. Zhao, J.C. Dong, C.M. Rouleau, D.B. Geohegan, F. Ding, K. Xiao, R.R. Unocic, In situ edge engineering in two-dimensional transition metal dichalcogenides, Nature Communications 9 (2018) 2051.

[23] S.S. Wang, A. Robertson, J.H. Warner, Atomic structure of defects and dopants in 2d layered transition metal dichalcogenides, Chemical Society Reviews 47 (17) (2018) 6764–6794.

[24] D. Voiry, J. Yang, M. Chhowalla, Recent strategies for improving the catalytic activity of 2d TMD nanosheets toward the hydrogen evolution reaction, Advanced Materials 28 (29) (2016) 6197–6206.

[25] D.Q. Er, H. Ye, N.C. Frey, H. Kumar, J. Lou, V.B. Shenoy, Prediction of enhanced catalytic activity for hydrogen evolution reaction in janus transition metal dichalcogenides, Nano Letters 18 (6) (2018) 3943–3949.

[26] G. Grosso, H. Moon, B. Lienhard, S. Ali, D.K. Efetov, M.M. Furchi, P. Jarillo-Herrero, M.J. Ford, I. Aharonovich, D. Englund, Tunable and high-purity room temperature single-photon emission from atomic defects in hexagonal boron nitride, Nature Communications 8 (2017) 705.

[27] Y.M. He, G. Clark, J.R. Schaibley, H. He, M.C. Chen, Y.J. Wei, X. Ding, Q. Zhang, W. Yao, X.D. Xu, C.Y. Lu, J.W. Pan, Single quantum emitters in monolayer semiconductors, Nature Nanotechnology 10 (6) (2015) 497–502.

[28] A. Avsar, A. Ciarrocchi, M. Pizzochero, D. Unuchek, O.V. Yazyev, A. Kis, Defect induced, layer-modulated magnetism in ultrathin metallic $PtSe_2$, Nature Nanotechnology 4 (7) (2019) 674–678.

[29] J. De Yoreo, D. Mandrus, L. Soderholm, T. Forbes, M. Kanatzidis, J. Erlebacher, J. Laskin, U. Wiesner, T. Xu, S. Billinge, S. Tolbert, M. Zaworotko, G. Galli, J. Chan, J. Mitchell, L. Horton, A. Kini, B. Gersten, G. Maracas, R. Miranda, M. Pechan, K. Runkles, Basic Research Needs Workshop on Synthesis Science for Energy Relevant Technology, USDOE Office of Science (SC), United States, 2016, p Medium: ED, Size, 178 p.

[30] C. Broholm, I. Fisher, J. Moore, M. Murnane, A. Moreo, J. Tranquada, D. Basov, J. Freericks, M. Aronson, A. MacDonald, E. Fradkin, A. Yacoby, N. Samarth, S. Stemmer, L. Horton, J. Horwitz, J. Davenport, M. Graf, J. Krause, M. Pechan, K. Perry, J. Rhyne, A. Schwartz, T. Thiyagarajan, L. Yarris, K. Runkles, Basic Research Needs Workshop on Quantum Materials for Energy Relevant Technology, USDOE Office of Science (SC), United States, 2016, p Medium: ED, Size, 170 p.

[31] X.F. Li, A.A. Puretzky, X.H. Sang, K.C. Santosh, M.K. Tian, F. Ceballos, M. Mahjouri-Samani, K. Wang, R.R. Unocic, H. Zhao, G. Duscher, V.R. Cooper, C.M. Rouleau, D.B. Geohegan, K. Xiao, Suppression of defects and deep levels using isoelectronic tungsten substitution in monolayer $MoSe_2$, Advanced Functional Materials 27 (19) (2017) 1603850.

[32] X.F. Li, J.J. Zhang, A.A. Puretzky, A. Yoshimura, X.H. Sang, Q.N. Cui, Y.Y. Li, L.B. Liang, A.W. Ghosh, H. Zhao, R.R. Unocic, V. Meunier, C.M. Rouleau, B.G. Sumpter, D.B. Geohegan, K. Xiao, Isotope-engineering the thermal conductivity of two-dimensional MoS_2, ACS Nano 13 (2) (2019) 2481–2489.

[33] X.L. Zou, B.I. Yakobson, An open canvas-2d materials with defects, disorder, and functionality, Accounts of Chemical Research 48 (1) (2015) 73–80.

[34] X.F. Li, M.W. Lin, L. Basile, S.M. Hus, A.A. Puretzky, J. Lee, Y.C. Kuo, L.Y. Chang, K. Wang, J.C. Idrobo, A.P. Li, C.H. Chen, C.M. Rouleau, D.B. Geohegan, K. Xiao, Isoelectronic tungsten doping in monolayer $MoSe_2$ for carrier type modulation, Advanced Materials 28 (37) (2016) 8240–8247.

[35] Z. Lin, A. McCreary, N. Briggs, S. Subramanian, K.H. Zhang, Y.F. Sun, X.F. Li, N.J. Borys, H.T. Yuan, S.K. Fullerton-Shirey, A. Chernikov, H. Zhao, S. McDonnell, A.M. Lindenberg, K. Xiao, B.J. LeRoy, M. Drndic, J.C.M. Hwang, J. Park, M. Chhowalla, R.E. Schaak, A. Javey, M.C. Hersam, J. Robinson, M. Terrones, 2d materials advances: from large scale synthesis and controlled heterostructures to improved characterization techniques, defects and applications, 2D Materials 3 (4) (2016) 1801583.

[36] J. Dong, L. Zhang, F. Ding, Kinetics of graphene and 2d materials growth, Advanced Materials 31 (9) (2019).

[37] X.H. Sang, Y. Xie, M.W. Lin, M. Alhabeb, K.L. Van Aken, Y. Gogotsi, P.R.C. Kent, K. Xiao, R.R. Unocic, Atomic defects in monolayer titanium carbide ($Ti_3C_2T_x$) Mxene, ACS Nano 10 (10) (2016) 9193–9200.

[38] G.D. Nguyen, L.B. Liang, Q. Zou, M.M. Fu, A.D. Oyedele, B.G. Sumpter, Z. Liu, Z. Gai, K. Xiao, A.P. Li, 3d imaging and manipulation of subsurface selenium vacancies in $PdSe_2$, Physical Review Letters 121 (8) (2018) 086101.

[39] A. Maksov, O. Dyck, K. Wang, K. Xiao, D.B. Geohegan, B.G. Sumpter, R.K. Vasudevan, S. Jesse, S.V. Kalinin, M. Ziatdinov, Deep learning analysis of defect and phase evolution during electron beam-induced transformations in WS_2, npj Computational Materials 5 (2019) 12.

[40] J.H. Hong, Z.X. Hu, M. Probert, K. Li, D.H. Lv, X.N. Yang, L. Gu, N.N. Mao, Q.L. Feng, L.M. Xie, J. Zhang, D.Z. Wu, Z.Y. Zhang, C.H. Jin, W. Ji, X.X. Zhang, J. Yuan, Z. Zhang, Exploring atomic defects in molybdenum disulphide monolayers, Nature Communications 6 (2015) 6293.

[41] W.F. Li, C.M. Fang, M.A. van Huis, Strong spin-orbit splitting and magnetism of point defect states in monolayer WS_2, Physical Review B 94 (19) (2016) 195425.

[42] K. Wang, L. Zhang, G.D. Nguyen, X. Sang, C. Liu, Y. Yu, W. Ko, R.R. Unocic, A. Puretzky, C.M. Rouleau, D. Geohegan, L. Fu, G. Duscher, A.-P. Li, M. Yoon, K. Xiao, Rational synthesis of WS_2 monolayers with preferred antisite defects, Advanced Materials (2021), submitted.

[43] Z. Lin, H. Yan, J.W. Liu, Y.K. An, Defects engineering monolayer $MoSe_2$ magnetic states for 2d spintronic device, Journal of Alloys and Compounds 774 (2019) 160–167.

[44] Chandriker Kavir Dass, M.A. Khan, Genevieve Clark, Jeffrey A. Simon, Ricky Gibson, X.X. Shin Mou, Michael N. Leuenberger, Joshua R. Hendrickson, Ultra-long lifetimes of single quantum emitters in monolayer WSe_2/hBN heterostructures, Advanced Quantum Technology 2 (2020) 1900022.

[45] H.P. Komsa, S. Kurasch, O. Lehtinen, U. Kaiser, A.V. Krasheninnikov, From point to extended defects in two-dimensional MoS_2: evolution of atomic structure under electron irradiation, Physical Review B 88 (3) (2013) 035301.

[46] J.H. Lin, S.T. Pantelides, W. Zhou, Vacancy-induced formation and growth of inversion domains in transition-metal dichalcogenide monolayer, ACS Nano 9 (5) (2015) 5189–5197.

[47] H.P. Komsa, A.V. Krasheninnikov, Engineering the electronic properties of two-dimensional transition metal dichalcogenides by introducing mirror twin boundaries, Advanced Electronic Materials 3 (6) (2017) 1600468.

[48] P.M. Coelho, H.P. Komsa, H.C. Diaz, Y.J. Ma, A.V. Krasheninnikov, M. Batzill, Post-synthesis modifications of two-dimensional $MoSe_2$ or $MoTe_2$ by incorporation of excess metal atoms into the crystal structure, ACS Nano 12 (4) (2018) 3975–3984.

References

[49] S. Barja, S. Wickenburg, Z.F. Liu, Y. Zhang, H.J. Ryu, M.M. Ugeda, Z. Hussain, Z.X. Shen, S.K. Mo, E. Wong, M.B. Salmeron, F. Wang, M.F. Crommie, D.F. Ogletree, J.B. Neaton, A. Weber-Bargioni, Charge density wave order in 1d mirror twin boundaries of single-layer $MoSe_2$, Nature Physics 12 (8) (2016) 751–756.

[50] Y.J. Ma, H.C. Diaz, J. Avila, C.Y. Chen, V. Kalappattil, R. Das, M.H. Phan, T. Cadez, J.M.P. Carmelo, M.C. Asensio, M. Batzill, Angle resolved photoemission spectroscopy reveals spin charge separation in metallic $MoSe_2$ grain boundary, Nature Communications 8 (2017) 14231.

[51] T.F. Jaramillo, K.P. Jorgensen, J. Bonde, J.H. Nielsen, S. Horch, I. Chorkendorff, Identification of active edge sites for electrochemical H-2 evolution from MoS_2 nanocatalysts, Science 317 (5834) (2007) 100–102.

[52] S.M. Poh, S.J.R. Tan, X.X. Zhao, Z.X. Chen, I. Abdelwahab, D.Y. Fu, H. Xu, Y. Bao, W. Zhou, K.P. Loh, Large area synthesis of 1d-$MoSe_2$ using molecular beam epitaxy, Advanced Materials 29 (12) (2017) 1605641.

[53] M. Saab, P. Raybaud, Tuning the magnetic properties of MoS_2 single nanolayers by 3d metals edge doping, Journal of Physical Chemistry C 120 (19) (2016) 10691–10697.

[54] L. Venkataraman, Y.S. Hong, P. Kim, Electron transport in a multichannel one-dimensional conductor: molybdenum selenide nanowires, Physical Review Letters 96 (7) (2006) 076601.

[55] T. Ma, W.C. Ren, X.Y. Zhang, Z.B. Liu, Y. Gao, L.C. Yin, X.L. Ma, F. Ding, H.M. Cheng, Edge-controlled growth and kinetics of single-crystal graphene domains by chemical vapor deposition, Proceedings of the National Academy of Sciences of the United States of America 110 (51) (2013) 20386–20391.

[56] M. Ghorbani-Asl, S. Borini, A. Kuc, T. Heine, Strain-dependent modulation of conductivity in single-layer transition-metal dichalcogenides, Physical Review B 87 (23) (2013) 235434.

[57] J. Feng, X.F. Qian, C.W. Huang, J. Li, Strain-engineered artificial atom as a broad-spectrum solar energy funnel, Nature Photonics 6 (12) (2012) 865–871.

[58] A. Branny, S. Kumar, R. Proux, B.D. Gerardot, Deterministic strain-induced arrays of quantum emitters in a two-dimensional semiconductor, Nature Communications 8 (2017) 15053.

[59] H. Moon, E. Bersin, C. Chakraborty, A.Y. Lu, G. Grosso, J. Kong, D. Englund, Strain-correlated localized exciton energy in atomically thin semiconductors, ACS Photonics 7 (5) (2020) 1135–1140.

[60] W. Wu, C.K. Dass, J.R. Hendrickson, R.D. Montano, R.E. Fischer, X.T. Zhang, T.H. Choudhury, J.M. Redwing, Y.Q. Wang, M.T. Pettes, Locally defined quantum emission from epitaxial few-layer tungsten diselenide, Applied Physics Letters 114 (21) (2019) 213102.

[61] S.W. Wang, H. Medina, K.B. Hong, C.C. Wu, Y.D. Qu, A. Manikandan, T.Y. Su, P.T. Lee, Z.Q. Huang, Z.M. Wang, F.C. Chuang, H.C. Kuo, Y.L. Chueh, Thermally strained band gap engineering of transition-metal dichalcogenide bilayers with enhanced light matter interaction toward excellent photodetectors, ACS Nano 11 (9) (2017) 8768–8776.

[62] P. Gant, P. Huang, D.P. de Lara, D. Guo, R. Frisenda, A. Castellanos-Gomez, A strain tunable single-layer MoS_2 photodetector, Materials Today 27 (2019) 8–13.

[63] R. Roldan, A. Castellanos-Gomez, E. Cappelluti, F. Guinea, Strain engineering in semiconducting two-dimensional crystals, Journal of Physics. Condensed Matter 27 (31) (2015) 313201.

[64] A. Castellanos-Gomez, R. Roldan, E. Cappelluti, M. Buscema, F. Guinea, H.S.J. van der Zant, G.A. Steele, Local strain engineering in atomically thin MoS_2, Nano Letters 13 (11) (2013) 5361–5366.

[65] Y.Y. Hui, X.F. Liu, W.J. Jie, N.Y. Chan, J.H. Hao, Y.T. Hsu, L.J. Li, W.L. Guo, S.P. Lau, Exceptional tunability of band energy in a compressively strained trilayer MoS_2 sheet, ACS Nano 7 (8) (2013) 7126–7131.

[66] H.J. Conley, B. Wang, J.I. Ziegler, R.F. Haglund, S.T. Pantelides, K.I. Bolotin, Bandgap engineering of strained monolayer and bilayer MoS_2, Nano Letters 13 (8) (2013) 3626–3630.

[67] K. He, C. Poole, K.F. Mak, J. Shan, Experimental demonstration of continuous electronic structure tuning via strain in atomically thin MoS_2, Nano Letters 13 (6) (2013) 2931–2936.

[68] C.R. Zhu, G. Wang, B.L. Liu, X. Marie, X.F. Qiao, X. Zhang, X.X. Wu, H. Fan, P.H. Tan, T. Amand, B. Urbaszek, Strain tuning of optical emission energy and polarization in monolayer and bilayer MoS_2, Physical Review B 88 (12) (2013) 5246.

[69] Z. Liu, M. Amani, S. Najmaei, Q. Xu, X.L. Zou, W. Zhou, T. Yu, C.Y. Qiu, A.G. Birdwell, F.J. Crowne, R. Vajtai, B.I. Yakobson, Z.H. Xia, M. Dubey, P.M. Ajayan, J. Lou, Strain and structure heterogeneity in MoS_2 atomic layers grown by chemical vapour deposition, Nature Communications 5 (2014).

[70] G. Plechinger, A. Castellanos-Gomez, M. Buscema, H.S.J. van der Zant, G.A. Steele, A. Kuc, T. Heine, C. Schuller, T. Korn, Control of biaxial strain in single-layer molybdenite using local thermal expansion of the substrate, 2D Materials 2 (1) (2015) 015006.

[71] H. Li, A.W. Contryman, X.F. Qian, S.M. Ardakani, Y.J. Gong, X.L. Wang, J.M. Weisse, C.H. Lee, J.H. Zhao, P.M. Ajayan, J. Li, H.C. Manoharan, X.L. Zheng, Optoelectronic crystal of artificial atoms in strain-textured molybdenum disulphide, Nature Communications 6 (2015) 7381.

[72] D. Lloyd, X.H. Liu, J.W. Christopher, L. Cantley, A. Wadehra, B.L. Kim, B.B. Goldberg, A.K. Swan, J.S. Bunch, Band Gap engineering with ultralarge biaxial strains in suspended monolayer MoS_2, Nano Letters 16 (9) (2016) 5836–5841.

[73] R. Frisenda, M. Druppel, R. Schmidt, S.M. de Vasconcellos, D.P. de Lara, R. Bratschitsch, M. Rohlfing, A. Castellanos-Gomez, Biaxial strain tuning of the optical properties of single-layer transition metal dichalcogenides, NPJ 2D Materials Applications 1 (2017) 10.

[74] J.W. Christopher, M. Vutukuru, D. Lloyd, J.S. Bunch, B.B. Goldberg, D.J. Bishop, A.K. Swan, Monolayer MoS_2 strained to 1.3% with a microelectromechanical system, Journal of Microelectromechanical Systems 28 (2) (2019) 254–263.

[75] Y.K. Ryu, F. Carrascoso, R. Lopez-Nebreda, N. Agrait, A. Castellanos-Gomez, Microheater actuators as a versatile platform for strain engineering in 2d materials, Nano Letters 20 (2020) 5339–5345.

[76] Y.L. Wang, C.X. Cong, W.H. Yang, J.Z. Shang, N. Peimyoo, Y. Chen, J.Y. Kang, J.P. Wang, W. Huang, T. Yu, Strain-induced direct-indirect bandgap transition and phonon modulation in monolayer WS_2, Nano Research 8 (8) (2015) 2562–2572.

[77] Q.H. Zhang, Z.Y. Chang, G.Z. Xu, Z.Y. Wang, Y.P. Zhang, Z.Q. Xu, S.J. Chen, Q.L. Bao, J.Z. Liu, Y.W. Mai, W.H. Duan, M.S. Fuhrer, C.X. Zheng, Strain relaxation of monolayer WS_2 on plastic substrate, Advanced Functional Materials 26 (47) (2016) 8707–8714.

[78] S.B. Desai, G. Seol, J.S. Kang, H. Fang, C. Battaglia, R. Kapadia, J.W. Ager, J. Guo, A. Javey, Strain-induced indirect to direct bandgap transition in multi layer Wse2, Nano Letters 14 (8) (2014) 4592–4597.

[79] O.B. Aslan, M.D. Deng, T.F. Heinz, Strain tuning of excitons in monolayer Wse2, Physical Review B 98 (11) (2018) 115308.

[80] S.X. Yang, C. Wang, H. Sahin, H. Chen, Y. Li, S.S. Li, A. Suslu, F.M. Peeters, Q. Liu, J.B. Li, S. Tongay, Tuning the optical, magnetic, and electrical properties of $ReSe_2$ by nanoscale strain engineering, Nano Letters 15 (3) (2015) 1660–1666.

[81] X. He, H. Li, Z.Y. Zhu, Z.Y. Dai, Y. Yang, P. Yang, Q. Zhang, P. Li, U. Schwingenschlogl, X.X. Zhang, Strain engineering in monolayer WS2, MoS2, and the WS2/MoS2 heterostructure, Applied Physics Letters 109 (17) (2016) 173105.

[82] S. Pak, J. Lee, Y.W. Lee, A.R. Jang, S. Ahn, K.Y. Ma, Y. Cho, J. Hong, S. Lee, H.Y. Jeong, H. Im, H.S. Shin, S.M. Morris, S. Cha, J.I. Sohn, J.M. Kim, Strain-mediated interlayer coupling effects on the excitonic behaviors in an epitaxially grown MoS_2/WS_2 van der Waals heterobilayer, Nano Letters 17 (9) (2017) 5634–5640.

[83] K. Zollner, P.E. Faria, J. Fabian, Strain-tunable orbital, spin-orbit, and optical properties of monolayer transition-metal dichalcogenides, Physical Review B 100 (2019) 195126.

[84] X. Zhang, Q.H. Tan, J.B. Wu, W. Shi, P.H. Tan, Review on the Raman spectroscopy of different types of layered materials, Nanoscale 8 (12) (2016) 6435–6450.

[85] C. Rice, R.J. Young, R. Zan, U. Bangert, D. Wolverson, T. Georgiou, R. Jalil, K.S. Novoselov, Raman-scattering measurements and first-principles calculations of strain-induced phonon shifts in monolayer MoS_2, Physical Review B 87 (8) (2013) 081307(R).

[86] Y.L. Wang, C.X. Cong, C.Y. Qiu, T. Yu, Raman spectroscopy study of lattice vibration and crystallographic orientation of monolayer MoS_2 under uniaxial strain, Small 9 (17) (2013) 2857–2861.

[87] J.U. Lee, S. Woo, J. Park, H.C. Park, Y.W. Son, H. Cheong, Strain-shear coupling in bilayer MoS_2, Nature Communications 8 (2017) 1370.

[88] A.M. Dadgar, D. Scullion, K. Kang, D. Esposito, E.H. Yang, I.P. Herman, M.A. Pimenta, E.J.G. Santos, A.N. Pasupathy, Strain engineering and Raman spectroscopy of monolayer transition metal dichalcogenides, Chemistry of Materials 30 (15) (2018) 5148–5155.

[89] Y. Zhang, H.H. Guo, W. Sun, H.Z. Sun, S. Ali, Z.D. Zhang, R. Saito, T. Yang, Scaling law for strain dependence of Raman spectra in transition-metal dichalcogenides, Journal of Raman Spectroscopy 51 (8) (2020) 1353–1361.

[90] J. Liang, J. Zhang, Z.Z. Li, H. Hong, J.H. Wang, Z.H. Zhang, X. Zhou, R.X. Qiao, J.Y. Xu, P. Gao, Z.R. Liu, Z.F. Liu, Z.P. Sun, S. Meng, K.H. Liu, D.P. Yu, Monitoring local strain vector in atomic-layered $MoSe_2$ by second-harmonic generation, Nano Letters 17 (12) (2017) 7539–7543.

[91] L. Mennel, M. Paur, T. Mueller, Second harmonic generation in strained transition metal dichalcogenide monolayers: MoS2, MoSe2, WS2, and WSe2, APL Photonics 4 (3) (2019).

References

[92] S. Chen, J.F. Gao, B.M. Srinivasan, G. Zhang, M. Yang, J.W. Cha, S.J. Wang, D.Z. Chi, Y.W. Zhang, Revealing the grain boundary formation mechanism and kinetics during polycrystalline MoS_2 growth, ACS Applied Materials & Interfaces 11 (49) (2019) 46090–46100.

[93] I. Aharonovich, D. Englund, M. Toth, Solid-state single-photon emitters, Nature Photonics 10 (10) (2016) 631–641.

[94] B. Gil, G. Cassabois, R. Cusco, G. Fugallo, L. Artus, Boron Nitride for excitonics, nano photonics, and quantum technologies, Nanophotonics 9 (11) (2020) 3483–3504.

[95] D.B. Geohegan, Diagnostics and characteristics of pulsed laser deposition laser plasmas, in: D.B. Chrisey, G.K. Hubler (Eds.), Pulsed Laser Deposition of Thin Films, 1st edition, Wiley-Interscience, New York, 1994, pp. 115–165.

[96] D.H. Lowndes, D. Geohegan, A. Puretzky, D. Norton, C. Rouleau, Synthesis of novel thin-film materials by pulsed laser deposition, Science 273 (1996) 898–903.

[97] R. Eason, Pulsed Laser Deposition of Thin Films: Applications-Led Growth of Functional Materials, Wiley, 2006, ISBN: 978-0-471-44709-2.

[98] Laser-Surface Interactions for New Materials Production: Tailoring Structure and Properties, vol. 130, Springer-Verlag, Berlin Heidelberg, 2010.

[99] P.J.F. Harris, New perspectives on the structure of graphitic carbons, Critical Reviews in Solid State and Materials Sciences 30 (4) (2005) 235–253.

[100] T. Guo, P. Nikolaev, A.G. Rinzler, D. Tomanek, D.T. Colbert, R.E. Smalley, Self-assembly of tubular fullerenes, Journal of Physical Chemistry 99 (27) (1995) 10694–10697.

[101] C.D. Scott, S. Arepalli, P. Nikolaev, R.E. Smalley, Growth mechanisms for single-wall carbon nanotubes in a laser-ablation process, Applied Physics. A, Materials 72 (5) (2001) 573–580.

[102] A.A. Gorbunov, R. Friedlein, O. Jost, M.S. Golden, J. Fink, W. Pompe, Gas-dynamic consideration of the laser evaporation synthesis of single-wall carbon nanotubes, Applied Physics. A, Materials (1999) S593–S596.

[103] A.A. Puretzky, D.B. Geohegan, X. Fan, S.J. Pennycook, Dynamics of single-wall carbon nanotube synthesis by laser vaporization, Applied Physics. A, Materials 70 (2) (2000) 153–160.

[104] A.A. Puretzky, D.J. Styers-Barnett, C.M. Rouleau, H. Hu, B. Zhao, I.N. Ivanov, D.B. Geohegan, Cumulative and continuous laser vaporization synthesis of single wall carbon nanotubes and nanohorns, Applied Physics. A, Materials 93 (4) (2008) 849–855.

[105] S. Iijima, M. Yudasaka, R. Yamada, S. Bandow, K. Suenaga, F. Kokai, K. Takahashi, Nano-aggregates of single-walled graphitic carbon nano-horns, Chemical Physics Letters 309 (3-4) (1999) 165–170.

[106] R. Yuge, S. Bandow, M. Yudasaka, K. Toyama, S. Iijima, T. Manako, Boron- and nitrogen-doped single-walled carbon nanohorns with graphite-like thin sheets prepared by CO_2 laser ablation method, Carbon 111 (2017) 675–680.

[107] W.D. Tennyson, M.K. Tian, A.B. Papandrew, C.M. Rouleau, A.A. Puretzky, B.T. Sneed, K.L. More, G.M. Veith, G. Duscher, T.A. Zawodzinski, D.B. Geohegan, Bottom up synthesis of boron-doped graphene for stable intermediate temperature fuel cell electrodes, Carbon 123 (2017) 605–615.

[108] A.M. Morales, C.M. Lieber, A laser ablation method for the synthesis of crystalline semiconductor nanowires, Science 279 (5348) (1998) 208–211.

[109] C.M. Lieber, Semiconductor nanowires: a platform for nanoscience and nanotechnology, MRS Bulletin 36 (12) (2011) 1052–1063.

[110] D.B. Geohegan, H. Schittenhelm, X. Fan, S.J. Pennycook, A.A. Puretzky, M.A. Guillorn, D.A. Blom, D.C. Joy, Condensed phase growth of single-wall carbon nanotubes from laser annealed nanoparticulates, Applied Physics Letters 78 (21) (2001) 3307–3309.

[111] M. Mahjouri-Samani, M.K. Tian, A.A. Puretzky, M.F. Chi, K. Wang, G. Duscher, C.M. Rouleau, G. Eres, M. Yoon, J. Lasseter, K. Xiao, D.B. Geohegan, Nonequilibrium synthesis of Tio2 nanoparticle "building blocks" for crystal growth by sequential attachment in pulsed laser deposition, Nano Letters 17 (8) (2017) 4624–4633.

[112] G. Nava, F. Fumagalli, S. Neutzner, F. Di Fonzo, Large area porous 1d photonic crystals comprising silicon hierarchical nanostructures grown by plasma-assisted, nanoparticle jet deposition, Nanotechnology 29 (46) (2018) 465603.

[113] A. Perego, G. Giuffredi, P. Mazzolini, M. Colombo, R. Brescia, M. Prato, D.C. Sabarirajan, I.V. Zenyuk, F. Bossola, V. Dal Santo, A. Casalegno, F. Di Fonzo, Hierarchical tin nanostructured thin film electrode for highly stable pem fuel cells, ACS Applied Energy Materials 2 (3) (2019) 1911–1922.

[114] L. Passoni, F. Ghods, P. Docampo, A. Abrusci, J. Marti-Rujas, M. Ghidelli, G. Divitini, C. Ducati, M. Binda, S. Guarnera, A.L. Bassi, C.S. Casari, H.J. Snaith, A. Petrozza, F. Di Fonzo, Hyperbranched quasi-1d nanostructures for solid-state dye-sensitized solar cells, ACS Nano 7 (11) (2013) 10023–10031.

[115] M.I. Serna, S.H. Yoo, S. Moreno, Y. Xi, J.P. Oviedo, H. Choi, H.N. Alshareef, M.J. Kim, M. Minary-Jolandan, M.A. Quevedo-Lopez, Large-area deposition of MoS_2 by pulsed laser deposition with in situ thickness control, ACS Nano 10 (6) (2016) 6054–6061.

[116] S. Elafandi, Z. Ahmadi, N. Azam, M. Mahjouri-Samani, Gas-phase formation of highly luminescent 2d gase nanoparticle ensembles in a nonequilibrium laser ablation process, Nanomaterials-Basel 10 (5) (2020) 908.

[117] L.F. Fei, S.J. Lei, W.B. Zhang, W. Lu, Z.Y. Lin, C.H. Lam, Y. Chai, Y. Wang, Direct tem observations of growth mechanisms of two-dimensional MoS_2 flakes, Nature Communications 7 (2016).

[118] X.H. Sang, X.F. Li, A.A. Puretzky, D.B. Geohegan, K. Xiao, R.R. Unocic, Atomic insight into thermolysis-driven growth of 2d MoS_2, Advanced Functional Materials 29 (52) (2019), https://doi.org/10.1002/adfm.201902149.

[119] Y. Lifshitz, S.R. Kasi, J.W. Rabalais, Subplantation model for film growth from hyperthermal species – application to diamond, Physical Review Letters 62 (11) (1989) 1290–1293.

[120] K.J. Boyd, D. Marton, J.W. Rabalais, S. Uhlmann, T. Frauenheim, Semiquantitative subplantation model for low energy ion interactions with surfaces. I. Noble gas ion-surface interactions, Journal of Vacuum Science & Technology 16 (2) (1998) 444–454.

[121] H.P. Komsa, J. Kotakoski, S. Kurasch, O. Lehtinen, U. Kaiser, A.V. Krasheninnikov, Two-dimensional transition metal dichalcogenides under electron irradiation: defect production and doping, Physical Review Letters 109 (3) (2012) 035503.

[122] M. Ghorbani-Asl, S. Kretschmer, D.E. Spearot, A.V. Krasheninnikov, Two-dimensional MoS_2 under ion irradiation: from controlled defect production to electronic structure engineering, 2D Materials 4 (2) (2017) 025078.

[123] Z.H. Cheng, H. Abuzaid, Y.F. Yu, F. Zhang, Y.L. Li, S.G. Noyce, N.X. Williams, Y.C. Lin, J.L. Doherty, C.G. Tao, L.Y. Cao, A.D. Franklin, Convergent ion beam alteration of 2d materials and metal-2d interfaces, 2D Materials 6 (3) (2019).

[124] B.J. Huang, F. Tian, Y.D. Shen, M.R. Zheng, Y.S. Zhao, J. Wu, Y. Liu, S.J. Pennycook, J.T.L. Thong, Selective engineering of chalcogen defects in MoS_2 by low-energy helium plasma, ACS Applied Materials & Interfaces 11 (27) (2019) 24404–24411.

[125] U. Bangert, A. Stewart, E. O'Connell, E. Courtney, Q. Ramasse, D. Kepaptsoglou, H. Hofsass, J. Anioni, J.S. Tu, B. Kardynal, Ion-beam modification of 2-d materials – single implant atom analysis via annular dark-field electron microscopy, Ultramicroscopy 176 (2017) 31–36.

[126] J. Karthikeyan, H.P. Komsa, M. Batzill, A.V. Krasheninnikov, Which transition metal atoms can be embedded into two-dimensional molybdenum dichalcogenides and add magnetism?, Nano Letters 19 (7) (2019) 4581–4587.

[127] C.D. Cress, S.W. Schmucker, A.L. Friedman, P. Dev, J.C. Culbertson, J.W. Lyding, J.T. Robinson, Nitrogen-doped graphene and twisted bilayer graphene via hyperthermal ion implantation with depth control, ACS Nano 10 (3) (2016) 3714–3722.

[128] J. Zhang, S. Jia, I. Kholmanov, L. Dong, D.Q. Er, W.B. Chen, H. Guo, Z.H. Jin, V.B. Shenoy, L. Shi, J. Lou, Janus monolayer transition-metal dichalcogenides, ACS Nano 11 (8) (2017) 8192–8198.

[129] A.Y. Lu, H.Y. Zhu, J. Xiao, C.P. Chuu, Y.M. Han, M.H. Chiu, C.C. Cheng, C.W. Yang, K.H. Wei, Y.M. Yang, Y. Wang, D. Sokaras, D. Nordlund, P.D. Yang, D.A. Muller, M.Y. Chou, X. Zhang, L.J. Li, Janus monolayers of transition metal dichalcogenides, Nature Nanotechnology 12 (8) (2017) 744.

[130] M. Mahjouri-Samani, M.W. Lin, K. Wang, A.R. Lupini, J. Lee, L. Basile, A. Boulesbaa, C.M. Rouleau, A.A. Puretzky, I.N. Ivanov, K. Xiao, M. Yoon, D.B. Geohegan, Patterned arrays of lateral heterojunctions within monolayer two-dimensional semiconductors, Nature Communications 6 (2015) 7749.

CHAPTER

9

Two-dimensional materials under ion irradiation: from defect production to structure and property engineering

Mahdi Ghorbani-Asl[a], Silvan Kretschmer[a], and Arkady V. Krasheninnikov[a,b]

[a]Helmholtz-Zentrum Dresden-Rossendorf, Institute of Ion Beam Physics and Materials Research, Dresden, Germany [b]Department of Applied Physics, Aalto University, Aalto, Finland

9.1 Introduction

The response of materials to the irradiation by energetic particles such as electrons and ions has been intensively investigated since the beginning of the 20th-century [1]. Irradiation gives rise to defects in solids, and ultimately to the deterioration of their properties, so that the initial motivation for studying the influence of irradiation on materials was the requirement to understand their behavior in radiation-hostile environments, e.g., fission reactors or open space. It was also realized that in spite of the damage, beams of energetic particles may have overall beneficial effects on the irradiated system. A good example is the industrially important ion implantation of dopants into semiconductors every integrated circuit production line has an ion implantation system [2]. Ion and electron beams are also used to alter the structure of materials, by welding and cutting them, drilling holes, modifying the surfaces, etching, making nanostructures by top-down methods, etc.

The isolation of a single sheet of graphene [3] in 2004 has attracted the attention of the scientific community to two-dimensional (2D) systems. Since then many other 2D materials, e.g., hexagonal boron nitride (h-BN) and transition metal dichalcogenides (TMDs) have been prepared by mechanical [4] and chemical [5] exfoliation from their layered bulk counterparts, as well as by the techniques based on chemical vapor deposition (CVD) [6] or catalyst-free vapor-solid methods [7]. It was immediately realized that ion beams can be used for the efficient processing of 2D materials, because (i) contrary to the bulk systems, the atomic structure of the whole sample can be modified by ions, as ion ranges already at low ion energies

---

Defects in Two-Dimensional Materials
https://doi.org/10.1016/B978-0-12-820292-0.00015-X

Copyright © 2022 Elsevier Inc. All rights reserved.

are much larger than the target thickness; (ii) due to the very geometry of 2D systems (surface only) a post-irradiation treatment (e.g., exposure to gases) can be used to passivate the irradiation-induced defects and thus further engineer the atomic structure; (iii) focused ion beams can easily be used for cutting and patterning 2D materials.

However, while the response of 3D materials (bulk – cubic, wurtzite, amorphous, layered systems like graphite, etc.) are well understood [2,8], much less is known on the interaction of energetic particles with few-layer Van der Waals materials or single-layer sheets. The same is relevant to bulk materials composed from 2D components, e.g., graphene bucky paper [9]. The concepts developed for assessing radiation effects in bulk solids are not always suitable for nanomaterials and nano-objects, as these systems are quite different from their bulk counterparts. In fact, the irradiation of nanostructures has been shown to give rise to quite unexpected and even counter-intuitive results. For example, annealing of graphene [10] and ordering of fullerene and carbon nanotube thin films [11] under low-fluence high energy ion irradiation has been reported. The response of Si nanowires to ion irradiation was also found to be different from that in bulk Si [12].

It is clear that the reduced dimensionality of 2D materials should affect the behavior of these systems under ion irradiation. Due to the planar geometry of the target, the sputtering yield and the average energy taken away from the system by the sputtered atoms and thus energy deposition should be different, and the development of collisional cascades should be suppressed. As a result, one can expect that although the number of the produced defects in 2D materials consisting of a single or a few layers of atoms will first naturally increase with ion energy, Fig. 9.1, it will drop at higher energies due to a smaller probability for the ion to displace an atom, as well established for various systems within the framework of the binary collision approximation [2,8,13]. For 3D materials, this implies that the ion will penetrate deeper into the material, but overall more damage in the target will be created at higher ion energies. This is not the case for 2D systems, especially for the free-standing sheets (deposited on a grid or suspended over a trench on the substrate); at higher energies ions should go through the 2D system without producing much damage through ballistic collisions, unless defects at higher ion energies start appearing again due to the enhanced energy deposition into electronic excitations followed by the conversion of the excitation energy into defects.

The situation may be different for the supported 2D materials. It is intuitively clear that at low ion energies the substrate should decrease the amount of damage created by the energetic ions. On the other hand, for light ions with medium energies, the defect production in 2D systems can be governed by the backscattered ions and atoms sputtered from the substrate rather than by the direct ion impacts. The evolution of defects in the 2D material, e.g., vacancies, may also be dominated by the interaction of defects with the substrate.

Moreover, 2D materials are very interesting in the context of the fundamentals of ion-solid interaction. For example, by measuring the energy and charge state of the ion before and after passing through a monoatomic membrane, like graphene or h-BN, along with ion scattering angle, one can get valuable information on the outcome of a single collision between the ion and target atoms and juxtapose it to the theoretical results without averaging over many collisions, as in the case of thicker targets. Additional opportunities to get microscopic insights into the effects of ion irradiation on solid targets in various regimes corresponding to the nuclear and electronic stopping, or the interaction of slow but highly-charged ions with the targets come from the possibility to directly 'see' the irradiation-induced defects in the 2D

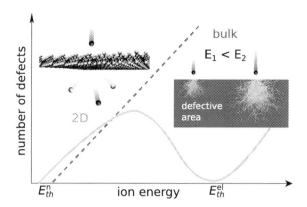

FIGURE 9.1 Schematic illustration of the main differences in defect production in bulk and free-standing 2D materials under impacts of energetic ions. In both systems, defects appear through ballistic displacement of atoms when ion energy exceeds a certain threshold value E_{th}^n, and the number of defects starts growing up with ion energy. However, in 2D materials the number of defects decreases at some point and less ion energy is deposited into the sample due to a decrease in cross-section and the absence of collisional cascades. In contrast, all ion energy is eventually transferred to defects and heat in the bulk system, and more defects are produced, although deeper in the sample. When energy deposited in electronic excitations exceeds also the threshold value E_{th}^{el}, the number of defects in the 2D system can start growing again.

target using transmission electron or scanning tunneling microscopy (TEM/STM), as, e.g., has been demonstrated for graphene [14].

In this chapter, we illustrate the fundamental points presented above by examples taken from the recent theoretical and experimental studies on the response of 2D materials to ion irradiation and discuss general trends in their behavior under irradiation. We address at length the response of 2D materials to ion irradiation in different regimes and compare it to that of 3D materials with the main focus on the role of the reduced dimensionality. We stress that our main goal is not to give a complete overview of the results obtained so far in this field; for this, we refer the interested reader to the recent comprehensive review articles [15–17] and references therein. Finally, we discuss how ion beams can be used to engineer the structure and properties of 2D materials.

9.2 Response of two-dimensional materials to ion irradiation: theoretical aspects

9.2.1 Theoretical background and methods

Various theoretical approaches have been developed to assess the amount of damage in the target produced by ion irradiation [8,13,18]. Among them, atomistic computer simulations, which describe the system as a collection of interacting atoms, have provided lots of insight into the behavior of the materials under the impact of energetic ions. These methods can give precise information on the types of the irradiation-induced defects formed under the impact of energetic ions, their stability and evolution, as well as on ion ranges and energy losses. In most cases multi-scale modeling is employed, where the choice of the simulation methods is dictated by the system size and the required level of sophistication associated with the computational costs. One tries normally to find a suitable compromise between the necessary computational resources and accuracy. In the following subsection, a brief overview of the available computational techniques is presented, starting from the coarser models suitable for the simulations of large systems up to very accurate but computationally demanding first-principles non-adiabatic simulation techniques, which can at the moment be applied for rather small systems composed of tens of atoms.

The most widely used approach to evaluate the amount of damage produced by radiation in a solid target at the atomic level is based on the binary collision approximation, which is implemented in the TRIM/SRIM code [19]. The code employs Monte-Carlo algorithms to calculate how the moving ion loses its energy in the target. It treats the collisions as independent binary collisions regardless of the collision density and amount of energy deposited into the system. The crystal structure of the target is not explicitly accounted for, so that irradiation effects associated with the crystallinity of the target, e.g., channeling [20] cannot be reproduced. Although the code often provides reasonable results for bulk systems, it gives absolutely wrong results for the probability for defect production in graphene, a strictly 2D carbon system [21]. This is due to the extremely small thickness of 2D materials: the collision cascades are essentially absent, and the projectile can lose only a tiny fraction of its energy. The method can be used, however, to account for the effects of the substrate on the production of defects in supported 2D materials [22,23] under ion irradiation, as discussed in detail below.

To study the evolution of the irradiation-induced defects, that is their migration and annihilation, the kinetic Monte Carlo (kMC) approach is used [24]. The kMC simulations are a suitable tool for studying the micro-structure evolution of large systems (millions of atoms) at macroscopic time scales. The approach can be used to describe phenomena such as defect annealing and pattern formation [25–27]. In kMC simulations, the system evolves stochastically along a trajectory, where the transition probabilities between different atomic configurations, e.g. equivalent positions for a diffusing vacancy, are determined from a pre-calculated probability distribution. Sometimes this parameterization can be directly done from the experimental data. Normally more accurate methods such as first-principles calculations are employed to parametrize the set of transition probabilities by accessing the energetics of common configurations and the corresponding transition barriers, e.g. by nudged elastic band (NEB) calculations [28].

Analytical potential molecular dynamics (MD) offer a scheme where atoms move according to the Newton's equation of motion. The forces on the atoms are determined from pre-defined analytical potentials of the form

$$V(r) = \sum_i^N V_{ext}(\mathbf{r}_i) + \sum_{i,j} V_2(\mathbf{r}_i, \mathbf{r}_j) + \sum_{i,j,k} V_3(\mathbf{r}_i, \mathbf{r}_j, \mathbf{r}_k) + ..., \tag{9.1}$$

where the first term is an external potential, followed by two-body and three-body interatomic potentials. For typical interatomic distances, the potentials of Stillinger-Weber [29] or Tersoff [30] types are sufficient to describe the interactions of atoms in the 2D target system. However, the impact of high-energy ions pushes the atoms to small distances for which common analytical potentials are not valid. In this regime suitable purely repulsive potentials which describe the interaction of atoms at small separations, e.g. the universal Ziegler-Littmark-Biersack (ZBL) potential [31] needs to be smoothly joined with the material specific analytical potential at small distances. This modification requires to carefully check the validity and reproducibility of physical quantities for the combined many-body potential. Recently the applicability of machine-learning potentials, which generally do not require analytical expressions as in Eq. (9.1), for radiation damage simulations has also been demonstrated [32,33].

To account for the deposition of energy into electronic excitations due to electronic stopping of the ion, MD simulations can be combined with the two-temperature model [34]. The model assumes that the time evolution of the system after the impact of the ion can be described independently by the ionic (phonon) and electronic subsystems with the corresponding temperatures and an additional electron-phonon coupling term which depends on the temperature difference. The electronic temperature profile after the impact is described by electronic thermal transport equation, while the atomic motion and, correspondingly, ionic temperature, can be accounted for by the MD algorithm coupled to the thermal transport equation. In practice, the energy is deposited into the electronic subsystem of the target at the beginning of the simulation, then the heat conductance equation is solved numerically in suitable, e.g. cylindrical, coordinates, and the energy from the electronic subsystem is transferred at each MD step to the ionic subsystem by assigning random velocities to the atoms, so that the system locally heats up, which may give rise to melting and formation of defects.

An even higher level of sophistication can be achieved, at the price of higher computational complexity, by employing ab-initio molecular dynamics. For ab-initio MD the equations of motion are solved from first-principles without the use of empirical parameters that have to be adjusted. In practice, the full many-body Schrödinger equation is impossible to solve exactly, analytically or numerically, except for very simple systems, like hydrogen atom, due to the complexity involved when considering all the degrees of freedom for ions and electrons. Thus various approximations must be made. The nuclei in the otherwise fully quantum mechanical approach are approximated as classical particles. For that, the Born-Oppenheimer approximation is employed assuming that the electronic subsystem relaxes instantaneously to its ground state for each new configuration of the nuclei. To calculate the energy of the system and forces acting on atoms, it remains to solve the time-independent Schrödinger equation for different atomic configurations. This task is usually accomplished by making use of density functional theory (DFT), which states that all physical properties can be described in terms of the electron density (instead of the far more complex multi-electron wavefunction). The forces on the nuclei are then calculated by virtue of the Hellman-Feynman theorem [35,36]

$$m\ddot{\mathbf{R}} = -\langle \Psi_0 | \nabla_\mathbf{R} \mathcal{H}_e | \Psi_0 \rangle, \tag{9.2}$$

where $\mathbf{R}$ denote the nuclei positions, Ψ_0 the ground-state wavefunction and $\mathcal{H}_e$ the electronic Hamiltonian, where the ground state wave function is given implicitly by the ground state electron density. Nowadays, ab-initio molecular dynamics simulations are feasible up to thousands of atoms and a time scale of several picoseconds. The use of supercomputers makes it possible to describe various irradiation-induced effects such as displacement of tar-

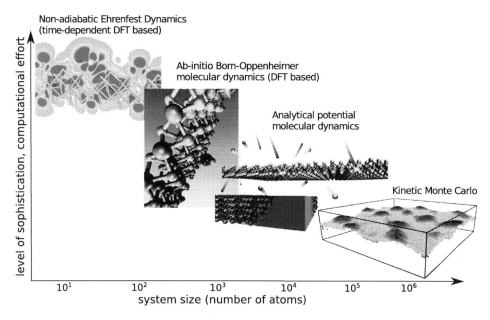

FIGURE 9.2 "Jacob's Ladder" of the approximations and simulation approaches used for modeling of ion irradiation effects in solids.

get atoms [37,38], defect formation and migration [21,39] or implantation of low-energy ions into the 2D target material [40,41].

Reiterating, Born-Oppenheimer MD is capable of describing processes on the electronic ground state configuration landscape (the so-called Born-Oppenheimer energy surface), but it fails, e.g., to describe the transitions between electronic states or the neutralization of highly charged ions in the vicinity of a 2D target. In these situations more sophisticated, nonadiabatic methods need to be employed. Ehrenfest dynamics combined with time-dependent DFT (see Fig. 9.2) is one such method. As a mean-field approach to the ion-electron interaction in the full many-body Schrödinger equation, the ions move in the average potential of the electrons whereas the electrons are evolved according to the time-dependent Schrödinger equation solved within the framework of the time-dependent DFT.

Due to its complexity and the high computational costs, the evolution of systems comprising tens of atoms can be simulated nowadays on a time scale up to picoseconds using massive parallel computers, like CRAY. The method has been successfully applied to electronic stopping power calculation in various (including 2D) targets [39,42–44], excitation mediated diffusion modeling [45], and simulations of the behavior of inorganic 2D materials under electron beam [38,46], but its applicability is limited by the mean-field approximation and the resulting propagation of a mixed state (for example, the sputtered atom may have a fractional charge), so that the results of Ehrenfest dynamics simulations must be interpreted with great care.

As we show below, all these simulation techniques have been used to model the effects of ion irradiation on 2D materials and to understand the behavior of the irradiation-induced defects. The simulations have not only provided lots of insight into the response of 2D systems

9.2.2 Simulations of ion impacts on free-standing 2D materials

Atomistic simulations have extensively been employed to study the ion bombardment of both suspended and supported 2D materials. Depending on ion energy and type, one can expect four different outcomes of an ion impact onto graphene or other 2D materials [47]: the reflection of the ion from the surface, the transmission of the ion through the sheet, adsorption on the surface and the insertion of the ion into the lattice either as a substitutional impurity (the host atom must be displaced) or, in 2D materials composed from several atomic layers, such as, e.g., $MoTe_2$, as an interstitial [48]. Interstitial-type defects are also possible in multi-layer systems, e.g., bi-layer graphene [49]. At low ion energies (~ 10 eV) and low ion charge states, the ions normally neutralize and adsorb on the surface or are reflected back provided that the interaction between the ion and surface atoms is weak. In the medium energy range (tens to several hundreds of eV), the target atoms may be knocked-out from the lattice. The incident ions may also replace the recoil atom as a substitutional impurity or become an interstitial, although the probability for the latter process is usually small. The recoil atoms may also collide with other nearby atoms causing further damage and/or sputtering. At high ion energies, the ions normally pass through the material without displacing any target atoms. Only ballistic charge transfer was assumed in the above discussion, but at high ion energies, when the contribution of the electronic stopping to energy deposition is very large, the formation of defects becomes possible through electronic excitations and ionization, as schematically illustrated in Fig. 9.1.

The irradiation response of graphene, the flagship of 2D materials, has theoretically been investigated more than that for any other member of the 2D materials family. The results can be found in Refs. [21,50–56], just to mention a few. MD simulations [50] of the irradiation process of a free-standing graphene sheet with low-energy carbon ions showed that four types of processes–absorption, reflection, transmission, and defect formation – are indeed possible, depending on ion energy, as discussed above. At energies below 10 eV, the dominant process was reflection; between 10 and 100 eV, absorption; and between 100 eV and 100 keV, the dominant process is transmission. Vacancy-type defects were predicted to appear at energies above 30 eV. Using the MD with analytical potentials, Lehtinen et al. [21] studied the response of the free-standing graphene to single ion impacts of various noble gases. The report addressed the effects of ion energy on defect types and concentrations based on the statistics obtained from a significant number of impact points. In addition to single and double vacancies, some triple vacancies, spatially close Frenkel pairs i.e., adatom-vacancy pairs and Stone-Wales defects were observed to appear due to inplane recoils, Fig. 9.3. It is also evident from the figure that there is a maximum on the curve "number of produced defects vs. ion energy", as discussed in Section 9.1. The position of the maximum depends on ion mass, and, going beyond graphene, naturally on the chemical bonding and masses of the atoms in the 2D target. Similar behavior was found for the free-standing h-BN [57,58] and black phosphorus [59] sheets. The sputtering yield was also calculated for various noble-gas ions in a wide range of energies and angles of incidence, so that the results of simulations could be used for

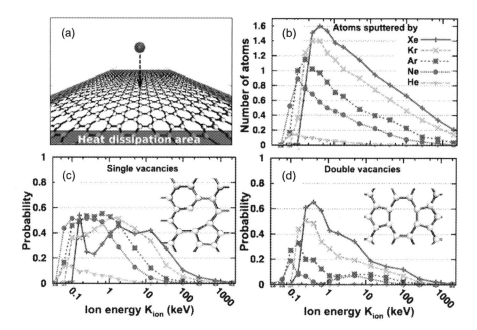

FIGURE 9.3 (a) Simulation setup for ion irradiation of graphene by the analytical potential molecular dynamics. (b) Number of sputtered atoms per ion impact as a function of ion energy. Panel (c) and (d) probability for single and double vacancy formation as a function of ion energy. Reprinted with permission from Ref. [21], Copyright (2010) American Physical Society.

the optimization of ion processing of single-layer and bulk hexagonal boron nitride samples, and for the prediction of material behavior in radiation-hostile environments.

A kMC method was developed to study morphological changes in graphene under multiple ion impacts, from tens of eV to 10 MeV, and angles of incidence between 0 and 88° [25]. Based on the results, cutting and patterning of graphene using focused ion beams could be performed by using optimum energy and angle of incidence. We note that nickel and silver nanoparticles can also be used to etch single-layer graphene to produce sub-10-nm nanoribbons and other graphene nanostructures with edges aligned along a particular crystallographic direction. [60,61] In another study, MD simulations based on the reactive force fields were employed to systematically investigate the bombardment of graphene by iron, gold, and oxygen ions. [62] The effects of ion type, ion energy and impact site on the probability, controllability, and types of defects were studied. Later simulations showed that oxygen ions with an incident angle of 70° gave the highest probability for ion substitution, and the ions at 40–60 eV and 70° yielded the highest efficiency for doping with minimum other defects [63].

The development of damage in graphene under high-fluence noble gas irradiation followed by in-situ annealing, which gave rise to the coalescence of individual vacancies into structures composed of agglomerations of non-hexagonal rings, was studied theoretically as well [52]. The simulated atomic structure of graphene irradiated with Ne ions before and after annealing is shown in Fig. 9.4(a,b). It is evident that the majority of isolated vacancies

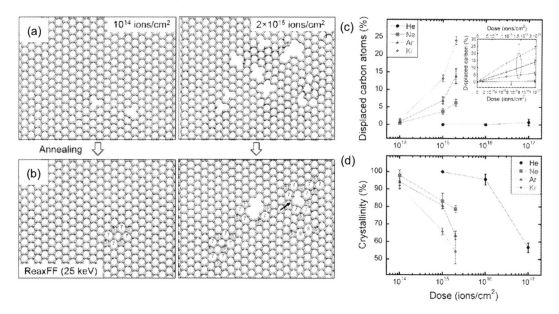

FIGURE 9.4 Simulated atomic structure of graphene irradiated with Ne ions before (a) and after (b) annealing. The majority of isolated vacancies merge into agglomerations of non-hexagonal rings, but small holes also remain. (c) Change in the displaced carbon atoms (%), defined as the number of displaced carbon atoms over the number of total carbon atoms in pristine graphene within the impacted region, with respect to dose. Linear x-scale plot with the same data (inset). Data for He is shown on the bottom x-axis, while those for Ne, Ar, and Kr refer on the top x-axis. (d) Change in the crystallinity (%), defined as the number of six-membered rings over the total number of rings in the impacted region, with respect to a fluence. The images adapted with permission from Ref. [52], Copyright (2016) American Chemical Society.

merge into agglomerates of non-hexagonal rings, but small holes also remain. Changes in the displaced carbon atoms, defined as the number of displaced carbon atoms over the number of total carbon atoms in pristine graphene within the impacted region, with respect to fluence are also presented in Fig. 9.4(c), as well as changes in the crystallinity, defined as the number of six-membered rings over the total number of rings in the impacted region.

Numerous computational studies [64–67] have been carried out in the context of direct ion implantation into graphene and other 2D materials, both free-standing and supported. The simulations provide insights into the optimum implantation energies, as discussed at length in Section 9.3 along with the relevant experiments.

Ghorbani-Asl et al. [68] studied the effects of ion irradiation on free-standing MoS_2 monolayer by using analytical potential MD simulations combined with static DFT calculations, Fig. 9.5(a). Note that contrary to graphene, TMDs consist of three atomic layers. The types and abundance of point defects created by low-fluence irradiation with various noble-gas ions (He, Ar, Ne, Kr, Xe) were investigated for a wide range of ion energies (10 eV to 1 MeV) and incident angles (between 0° and 75°). The authors systematically assessed the ability of various interatomic potentials for describing the atomic structure and energetics of point defects. Among the available potentials, the many-body REBO [69] and Stillinger-Weber (SW) potential [70] provided the best agreement in comparison to the DFT results [71] and they have been used for ion irradiation simulations. The displacement thresholds for the SW potential were found to be $T_d^S = 5.0$ eV and $T_d^{Mo} = 31.7$ eV for molybdenum and sulfur, respectively. Similar to graphene [21] and h-BN [57], sputtering yield in MoS_2 initially grows with ion energy up to a maximum value and then decreases at high energies, Fig. 9.5(b). Very similar results were also obtained by Yin et al. [72], who studied the formation of small holes in the free-standing MoS_2 sheets in the context of gas separation.

In case of light ions, e.g. He, two noticeable peaks were observed in the number of sputtered atoms as a function of ion energy. The first peak is mainly governed by the sputtering of atoms from the upper sulfur layer, while the second peak is triggered by the sputtering from lower sulfur layer. No such features have been found for 2D materials consisting of atoms of same type (graphene) [21] or atoms with roughly the same mass (h-BN) [57]. In addition, the cross-sections for defect production of Mo and S vacancy correlate with the number of sputtered atoms for normal incidence, Fig. 9.5(c). It was found that depending on the angle of incidence, ion type and energy, sulfur atoms can be knocked out mostly from the top or bottom layer, suggesting unique opportunities for patterning free-standing MoS_2 monolayer if focused ion beams combined with the exposure of the system to a precursor gas, Fig. 9.5(d). The first-principles calculations showed that mixed MoSX compound (X are chemical elements from group V or VII) can be stable with quite different electronic structures from MoS_2 e.g. a metallic character for MoSF [68]. Such an irradiation-induced patterning method can provide a unique opportunity for designing lateral metal/semiconductor/metal junctions with minimum contact resistance.

9.2.3 Simulations of ion irradiation of supported 2D materials

The response of free-standing 2D materials to ion bombardment has been considered so far, but in many experiments, the irradiated 2D materials are supported by a substrate.

At the same time, as discussed in Section 9.1, the response of supported 2D materials to ion impacts can be quite different from what the free-standing model gives. In fact, the experiments [23,53,73] indeed clearly indicate that the substrate dramatically influences the defect production in 2D materials under ion irradiation, especially for light ions such as He and Ne. The evolution of the defects can also be affected by the substrate [53], as the atoms sputtered away from a free-standing 2D material may stay in between the 2D sheet and substrate thus increasing the probability for the annealing of vacancies.

Specifically, in addition to the (i) direct ion impacts into the supported 2D material, damage may be induced by (ii) backscattered ions and (iii) sputtered substrate atoms. These additional channels are often referred to as indirect sputtering. By performing MD combined with Monte Carlo/binary collision simulations, the impact of ions into the supported system were modeled [22]. The characteristics (abundance, energy and angle distribution) of the secondary projectiles (backscattered ions and atoms sputtered from the substrate) were calculated using the TRIDYN code [74,75] in order to access each of these channels separately. The central results of this approach, which was used to study the response of 2D MoS_2 and graphene supported by SiO_2 as one of the most common substrate materials, are summarized in Fig. 9.6 for He ion irradiation. The sputtering yield per ion impact was evaluated based on the average number of sputtered S and C atoms, for 2D MoS_2 and graphene, respectively. In

9.2. Response of two-dimensional materials to ion irradiation: theoretical aspects 269

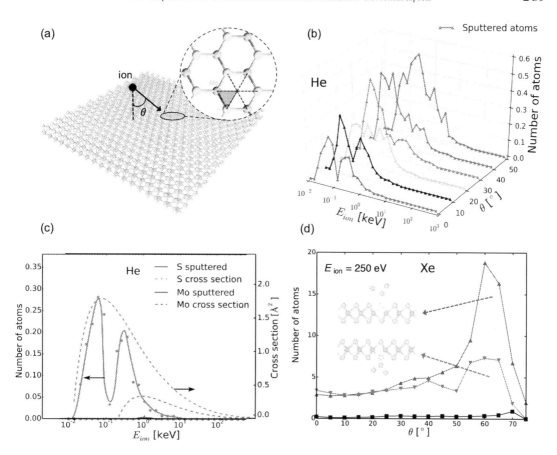

FIGURE 9.5 (a) Schematic of the setup for ion irradiation simulations of free-standing MoS$_2$. The red triangle indicates the minimum irreducible area used for ion impacts. Average number of atoms sputtered per (b) He ion impact. (c) Average numbers of S and Mo vacancies in the upper and lower S layers as functions of ion incident angle. (d) Comparison between the scattering cross section and sputtering yield for Mo and S atoms. Adapted from Ref. [68], under the Creative Commons Attribution 3.0 license.

order to closely match the most common experimental setup, the results are presented for the normal incidence of the primary ion.

The data for the suspended system is shown for comparison. For MoS$_2$, S atoms are predominantly sputtered, as momentum transfer to Mo atoms are much less due to a large mass ratio $m_{Mo}/m_{He} = 24$, while $m_S/m_{He} = 8$. The typical two-peak structure is observed in agreement with the previous results for the free-standing MoS$_2$ monolayer [68]. In the presence of the substrate, the second peak, which corresponds to sputtering from the bottom sulfur layer, is suppressed due to the reduced forward sputtering. For the same reason, the direct defect production below 800 eV is smaller than the case without the supporting material, since the substrate "stops" the recoil atoms and facilitates vacancy-interstitial recombination. At the same time, indirect sputtering is the dominant mechanism for typical energies in the helium

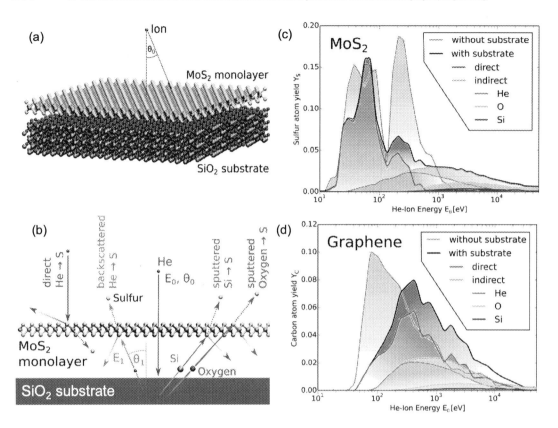

FIGURE 9.6 (a) Atomistic model of MoS$_2$ on a SiO$_2$ substrate. (b) The channels for defect production in a supported 2D material under ion bombardment. Average number of atoms sputtered from MoS$_2$ (c) and graphene (d) per He ion impact. The corresponding numbers for free-standing monolayers (grey) and the direct sputtering (red) are shown for comparison. The images adapted with permission from Ref. [22], Copyright (2018) American Chemical Society.

ion microscope (HIM) in the range of from 10 to 30 keV. For energies below 2 keV backscattered ions generate most of the defects, whereas for higher ion energies sputtered substrate atoms dominate. For graphene, a similar effect is noticeable, which is in agreement with the results of the previous MD study on graphene supported by SiO$_2$ substrate [76]. At low energies, the direct defect production is inhibited by the substrate, and this effect is even more pronounced than for MoS$_2$, since graphene is a true monoatomic layer. Similarly to MoS$_2$ the substrate leads to an increase in the defect production for higher energies.

The results indicate that for light ions, such as He, the substrate causes a five-fold larger defect production in 2D MoS$_2$ and graphene. This dramatic effect was observed experimentally for He irradiated graphene on Si/SiO$_2$ substrate [77]. Although the total sputtering yield increases substantially for heavier ions, indirect sputtering is less pronounced for more massive projectiles. However, for Ne ions the indirect sputtering still constitutes a major contribution that needs to be accounted for in the simulations. The increase in the sputtering yield on supported 2D targets has been observed in several irradiation experiments. The results of the combined MD/MC simulations [22] agree well to the experimental findings, as the simulations for free-standing 2D targets suggest much lower sputtering yield and the simulations for supported systems agree even quantitatively for both graphene [53,73,77–79] and 2D MoS_2 [80,81].

Not only the number of defects but also their spatial distribution is influenced by the substrate. This fact is of paramount importance as the patterning resolution, e.g., when cutting nanoribbons using helium ion microscopy, crucially depends on the presence of the substrate. Fortunately, the spatial distribution can be directly accessed from the combined MD/MC approach. It has been found that for typical energies in the HIM the damaged region increases from 1 nm for suspended to 8–16 nm for supported targets.

The irradiation response of graphene on metal substrates has also been studied both theoretically and experimentally [54,82,83]. Normal and grazing incidences of ions have been considered. By combining STM probing of the irradiated surface with MD simulations, it was shown that under grazing incidence Xe ions are channeled in between graphene and the Ir substrate [54], giving rise to chains of vacancy clusters with their edges bending down towards the substrate, Fig. 9.7(a–c), thus dramatically enhancing defect production with regard to the free-standing graphene. DFT simulations indicated that the bending occurs due to a strong interaction of graphene dangling bonds with the substrate, Fig. 9.7(e). It has also been demonstrated using analytical potential MD [84] that 3 keV Ar and Xe ions can undergo interface channeling between graphene and the first SiC surface layer, and similar to graphene on Ir, channeling gave rise to abundant damage production and sputtering in the graphene layer. Annealing of the irradiated sample at high temperatures was shown to give rise to vacancy agglomeration in specific areas of the Moire supercell [54] giving rise to the formation of vacancy clusters. The clusters can further self-organize to form a graphene nanomesh.

Qualitatively similar results were obtained for single h-BN layers on rhodium substrates [85]. The exposure of h-BN sheets to low energy ions led to the formation of vacancies, which were mobile at elevated temperatures and self-assembled into a nanomesh with 3 nm lattice constant, as observed by STM. In addition, the "can-opener" effect was reported [85]: parts of the h-BN sheet with a diameter of about 2 nm were cut out. DFT simulations showed that this effect also comes from the diffusion of single and double vacancies into specific areas of the Moire pattern, where they strongly interact with the substrate atoms, so that the material was essentially cut so that lids and stable voids in the h-BN layer were formed. These voids repelled each other, which made it possible to form arrays with the nearest neighbor distance down to about 8 nm.

The behavior of graphene on a Cu substrate under a high fluence Ga ion bombardment using focused ion beams has also been investigated through MD simulations [82]. The results indicate that, similar to the SiO_2 substrate, more defects in the supported graphene were created than in the free-standing system. The possibility to reduce the amount of damage in the system outside the cut by tilting ion beam incidence angles was studied. It was demonstrated that tilting the beam results in an asymmetric damage distribution, which can be leveraged to pattern graphene/2D materials by utilizing the lightly-damaged side for device fabrication while discarding the heavily damaged side, Fig. 9.7.

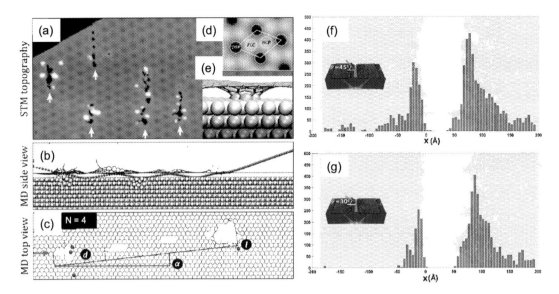

FIGURE 9.7 (a) Experimental STM image of a graphene sheet on the Ir(111) surface after room temperature irradiation with 5 keV Xe ions at 75° off normal. Individual defect patterns are marked by arrows pointing along the direction of the impinging ions. (d) Atomically resolved image of the Moire pattern where HCP, FCC, and TOP denote regions of high symmetry. (b) Snapshot from an MD simulation showing the side view of a single 5 keV Xe ion impinging on graphene/Pt(111) at 75° off normal (Xe is impinging from the left). The size of the simulated layer is 8.7 nm × 22 nm. The time-lapse motion of the ion (red circle) is shown in steps of 2 fs (position of graphene and iridium atoms at a simulation time of 170 fs). (c) Top view of the same simulated impact at 1000 fs. For clarity, the image shows the resulting graphene layer and metal adatoms (blue spheres) only. (e) Top and side view of a tetravacancy in graphene as calculated using DFT. Adapted with permission from Ref. [54], Copyright (2013) American Chemical Society (f,g) Distribution of undercoordinated carbon atoms along the x-direction after tilted Ga ion irradiation of graphene on Cu substrate with an ion fluence of 2×10^{15} ion/cm^2, as obtained using MD simulations. The ion incident angles are depicted schematically in the insets. Note the non-symmetric distribution of defects (blue histograms). Reprinted from Ref. [82], Copyright (2020) Elsevier.

9.2.4 Simulations of the interaction of light or swift ions with two-dimensional materials when electronic stopping dominates

For light or swift heavy ions, when nuclear stopping is small, the energy deposition into the 2D target is governed by the interaction of the ion with the electronic subsystem of the target. The simulations should give answers to two questions: (i) how much energy is deposited into electronic excitations and the associated ionization; (ii) how this energy is converted into defects in the material.

To get insights into energy deposition in these excitations, TD-DFT combined with Ehrenfest dynamics has been used to calculate energy loss of energetic ions in graphene [39,42,86,87] and h-BN [87]. The simulations gave values of energy losses which were in very good agreement with the experimental data for graphitic targets. Moreover, it was possible to estimate the spatial variation of the deposited energy (electronic stopping power), as this approach enables one to simulate the propagation of the ion with a specific impact parameter through the system. Normally, ions with relatively high energies (keV range) were considered, but Ehrenfest dynamics/TD-DFT simulations were also carried out for low energy (460 eV) Cl atoms colliding with a $MoSe_2$ monolayer, and significant energy transfer (14 eV) from the projectile to the electronic excitation in the $MoSe_2$ system were reported. [88] It was found that charged projectiles damage the targets more severely than neutral, but mostly single and double-ionized atoms were considered. Simulations for projectiles with higher charge states (e.g., Ar^{+7}) were also carried out [89], but the simulation cell may not have been large enough to adequately describe the behavior of the system, especially taking into account the necessity to introduce the neutralizing background when periodic boundary conditions are used, and the associated problems related to the spurious interaction of the image charges. Most of the simulations were carried out for protons and He ions, as they do not have core states, while it is computationally very expensive to do all-electron calculations. However, it was also demonstrated [42] that by treating core electrons as valence electrons within the projected augmented wave framework, the role of core electron excitations could be assessed.

With regard to ionization, the emission of electrons from graphene upon impact of 30 keV He ions has been investigated in the context of image formation in the helium ion microscope using Ehrenfest dynamics combined with TD-DFT [90]. It was found that emission depends on the impact point, which suggests the possibility of obtaining a highly accurate image of the honeycomb pattern of the suspended graphene using this technique. The comparison of the results obtained for neutral and ionized He atoms showed that electron emission is governed by the impact ionization instead of Auger processes initiated by the neutralization of He ions.

To answer the second question, analytical potential MD simulations combined with the two-temperature model have been carried out for free-standing single-layer [91] and multi-layer [92] graphene to understand the response of these systems to the impacts of swift heavy ions. The results of simulations (and also experiments carried out by the authors) indicated that swift heavy ions can be used to produce holes in graphene once a threshold value of 1.22–1.48 keV/layer of energy deposition is reached and that the size of the pores can be tuned between 1 and 4 nm depending on the ion energy. The effects of the SiO_2 substrate on defect production in supported graphene under the impact of swift heavy ions have also been investigated [93]. The results indicated that the presence of the substrate decreases the damage threshold of graphene, although the threshold value for the free-standing graphene (8 keV/nm, that is about 2.7 keV/layer) was larger by a factor of two than that reported in Ref. [91]. The damage of supported graphene was revealed to be due to the combined effects of the direct collision and the substrate, and that the rupture of graphene was initiated by the atoms sputtered from the core region of the track formed in the substrate.

9.3 Experiments on ion irradiation of two-dimensional materials

In this Section we present a brief overview of the most important experimental results with the main focus being on the differences between defect production in 2D materials and 3D (bulk) materials under irradiation. Depending on ion mass and energy, we differentiate, as usual, between light/heavy and low (and medium)-energy/swift-heavy ions. We also consider separately rather slow, but highly charged ions.

9.3.1 Low- and medium-energy heavy ion irradiation of two-dimensional materials and direct ion implantation

A considerable number of experimental studies has been carried out to get insights into the behavior of 2D materials under ion irradiation and/or to modify their properties. Graphene has received particular attention: as mentioned previously, graphene is a very interesting target from the viewpoint of fundamental aspects of ion-solid interaction: contrary to bulk solids, every displacement of an atom from a suspended graphene sheet should give rise to the formation of a defect: the atoms are sputtered away and the recombination of vacancy-interstitial pairs is not possible. An additional motivation was to correlate concentration of defects with the changes in the electronic transport. Further, direct implantation of impurities, such as B, N or P beyond the solubility limit of these impurities in sp^2-hybridized carbon was assumed to be a promising way to tailor graphene properties.

The first ion irradiation experiments have been carried out on graphene samples on SiO_2 substrates. Mechanically exfoliated graphene layers were irradiated with 30 keV Ar ions, [94] followed by STM and electronic transport measurements. The defect sites were determined, and new defect associated states close to Fermi energy were revealed. The types of the defects were not identified, however, and the density of the defects obtained during STM character-ization was much higher than what could be expected from the used irradiation fluence. The results indicated that most defects were created not by the ion beam directly but by the atoms sputtered from the substrate, as discussed in Section 9.2. Defects produced by low-energy (140 eV) Ar ions were thoroughly characterized by low-temperature STM in Ref. [95], though. It was found that Ar irradiation frequently gives rise to the formation of divacancies, a very stable defect in graphene, as it does not have any dangling bonds.

Raman spectroscopy is a powerful non-destructive technique which can provide infor-mation not only on the amount of disorder, but also on the types of defects present in the sp^2-bonded carbon [97] and TMDs [98]. The irradiated graphene samples have extensively been characterized using this technique to get the information on defect concentration and amorphization under prolonged irradiation. The effects of low energy (90 eV) Ar^+ ion bom-bardment on single-layer graphene were studied [96] using Raman spectroscopy and STM, Fig. 9.8. A phenomenological model, which correlates the intensity of the defect-activated D-band with defect separation (or, correspondingly, concentration), was developed. The model has been extensively used since then to assess the amount of irradiation-induced damage in graphene.

Low energy nitrogen ion beam irradiation was also used to join two monolayer graphene flakes, which were exposed to 40 eV ions with a fluence of 1.0×10^{14} cm^2 [49]. The joining processes were further studied by classical MD simulations and attributed to two different mechanisms. First, dangling bonds of the defects in the different sheets can be saturated by the formation of the covalent bonds between the sheets as in the irradiated graphite [99]. Second, possibly neutralized N ions are trapped in the space between two layers also making bonds between the layers. The bonds formed by the second mechanism were found to be dominant in the ion energy range from 20 to 100 eV.

A lot of work has been done on direct ion implantation into graphene. Ion implantation is a well-established method of doping bulk semiconductors [100]. However, ion implantation onto 2D materials is a challenging task because ions with energies in a very narrow range,

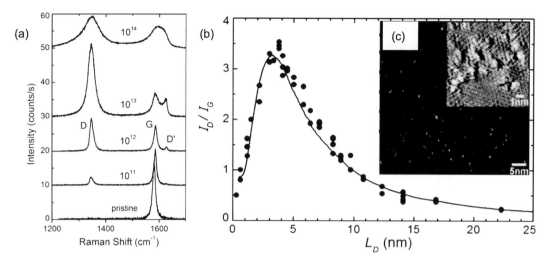

FIGURE 9.8 (a) Evolution of the first-order Raman spectra of a single-layer graphene on an SiO$_2$ substrate after ion bombardment with the ion fluences being indicated next to the respective spectrum in units of ions per cm^2. The spectra are displaced vertically for clarity. (b) The ratio of the D and G peaks intensities I_D/I_G from three different single-layer graphene samples as a function of the average distance L_D between defects, induced by the ion bombardment. The solid line is the result given by the theoretical model. (c) STM images of defects which appear under similar irradiation conditions, albeit in graphite. Reprinted with permission from Ref. [96], Copyright (2010) Elsevier.

and typically low energies (below 100 eV), which technically is not straightforward to implement, should be used to place the atoms in the atomically thin materials. As usual, to avoid significant damage in the lattice, the irradiation fluence must be precisely controlled. Note that contrary to bulk materials, where the fluence is selected based on the desired doping level, it is also governed by the probability that the ion gets implanted into the 2D material.

The possibility of direct doping of N and B atoms in graphene lattice via low-energy ion implantation was first predicted by classical MD combined with DFT calculations [64]. The simulation results showed that the optimum irradiation energy is 50 eV with substitution probabilities of 55% for N and 40% for B. Later on, this prediction was realized in experiments performed by Bangert et al. [65]. In this study, high-resolution TEM combined with electron energy loss spectroscopy showed that N and B particles with sufficiently low energies (below 50 eV) can substitute carbon atoms in the graphene lattice, while a minor fraction residing in defect-related sites was observed, Fig. 9.9(a). In a later experiment the implantation of phosphorus ions (with the energy of 30 eV) into graphene was also confirmed by using atomic-resolution imaging and electron energy loss spectroscopy, Fig. 9.9(b) [101]. The agreement between the measured phosphorous L-edge from energy loss signal with an ab–initio simulation confirmed the presence of the P in a buckled substitutional configuration. Direct implantation of N ions into graphene on Cu and SiO$_2$ substrates was also systematically studied by Cress et al. [66] using STM and Raman spectroscopy experiments combined with MD simulations, and the optimum ion energy was found to be between 30 and 50 eV. Low-energy (25–150 eV) N ion implantation was also carried out for graphene on Ni(111) substrate [102].

A concentration of N impurities of up to 0.05 monolayers was achieved, and the evolution of graphitic (substitutional) and pyridinic (at the edges of multi-vacancies) N impurities upon annealing was investigated using high resolution X-ray photo-electron spectroscopy (XPS). Boron atoms have also been successfully introduced in graphene on top of SiC [103].

Ion implantation in graphene has been also performed using heavier ions. Tripathi et al. [41] reported the implantation of Ge ions into graphene monolayer. Scanning TEM combined with image simulations showed that Ge atoms can either directly substitute for a single carbon atom in a buckled out-of-plane configuration, or occupy an in-plane position in a divacancy, Fig. 9.9(c–f). Such a non-chemical doping process can be an alternative to complex chemical doping routes which often suffer from poor controllability in doping site selectivity, adsorption contamination from different chemical residuals, and secondary impurities.

Alternative ion-beam-mediated approaches can also be used to dope graphene and other 2D materials. Wang et al. [104] demonstrated that low-energy ($\sim 10^2$ eV) irradiation or plasma treatment followed by metal atom deposition can be used for introducing impurities. Different elements (Pt, Co, and In) have been successfully doped in the single-atom form. According to the charge analysis, graphene can be either p-doped with Pt and Co dopants or n-doped with In dopants. Nanda et al. [105] showed that irradiation of graphene encapsulated between h-BN sheets can lead to the substitution of carbon atoms with nitrogen, as manifested in the n-type doping of graphene, so that spatially localized irradiation can be used to realize graphene-based p-n junction devices. Similarly, the possibility of direct doping of Cl atoms in few-layer $MoSe_2$ flakes via low-energy ion implantation was shown [106]. An enhancement of electron concentration with increasing Cl fluence was determined from Raman spectra and the electrical measurements confirmed the n-type doping of Cl-implanted $MoSe_2$.

To compositionally engineer 2D monolayers, a treatment of the sample by hyperthermal species from pulsed laser plasmas was recently demonstrated as a top-down process [107]. Exposing suspended WS_2 monolayers to Se clusters with kinetic energies of less than 10 eV/atom led to selective top layer replacement of sulfur by selenium without causing W atom displacement, enabling the formation of a Janus WSSe structures. MD simulations revealed that implanted Se clusters form disordered metastable alloy regions, which then recrystallize to highly ordered structures.

Xenon ions with higher energies (225 keV) have been used to introduce defects in single-, few-layer, and bulk MoS_2 [108]. It was shown that the stoichiometry, S:Mo atomic ratio, is a function of ion fluence and the number of MoS_2 layers. Also, there was a complex interplay between defect production, crystal thickness and interlayer interactions in this TMD. Overall, the results demonstrated that low- and medium-energy ion irradiation is an effective tool to tailor the electronic, vibrational and structural properties of MoS_2.

9.3.2 High-energy proton irradiation

The response of 2D materials to high-energy proton irradiation has also been studied [9, 109–111]. Note that in this regime energy deposition is fully governed by electronic excitations and ionization.

The damage threshold of graphene, subjected to 2 MeV proton irradiation, was found to increase with the number of layers [109]. More damage was reported for graphene on a SiO_2 substrate, as revealed by Raman probing. A model of intense electronically-stimulated des-

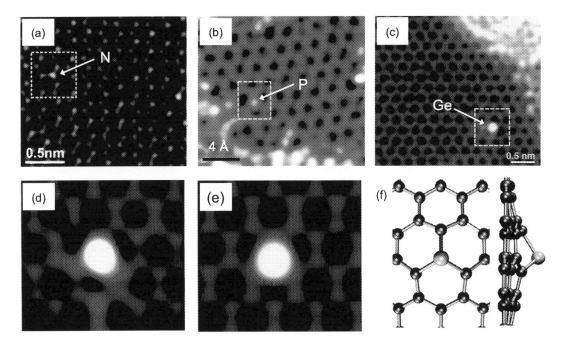

FIGURE 9.9 (a) TEM images of impurities (shown by arrow) in graphene introduced by ion implantation. (a) Nitrogen implanted at 25 eV. Adapted with permission from Ref. [65], Copyright (2013) American Chemical Society. (b) Phosphorus implanted at 30 eV. Adapted from Ref. [101], under the Creative Commons Attribution 3.0 license. (c) Germanium implanted at ion energy of 20 eV. (d) The filtered image and (e) simulated image of germanium implantation in graphene. (f) Atomic structure of three-coordinated germanium atom in graphene lattice. Adapted with permission from Ref. [41], Copyright (2018) American Chemical Society.

orption of the atoms from graphene was proposed as the most likely process for the damage mechanism.

The behavior of graphene laminates under bombardment with 350 keV protons has been studied in the context of defect-mediated magnetism in this system [9,110]. It was found that irradiation gave rise to the development of paramagnetism, which was associated with vacancies and presumably H adatoms on graphene surface; these defects have been shown to possess localized magnetic moments [112,113].

The effects of 10 MeV irradiation of MoS_2 field-effect transistors have also been studied [114]. The electrical characteristics of the devices were measured before and after proton irradiation. For a low irradiation fluence of 10^{12} ion/cm^2, the electrical properties of the devices were nearly unchanged. In contrast, for a fluence of 10^{13} ion/cm^2 and higher the current and conductance of the devices significantly decreased after proton bombardment. This is due to the fact that irradiation of electrons and protons on silicon-based FET devices can ionize atoms and generate electron-hole pairs in the oxide layer. The holes and protons formed in SiO_2 layer will transfer to the SiO_2 layer/MoS_2 interface, forming interface trap states, which act as negatively charged electron trap centers in n-channel transistors. Positive oxide traps in the SiO_2 layer boost the gate electric field, resulting in higher carrier concentrations. Negative interface trap states, on the other hand, serve as electron trap centres, resulting in a decrease in carrier concentration. Thus, the combined effects of positive oxide-charge traps in the SiO_2 and the interface trap state can be due to the effects of proton irradiation on MoS_2 FET devices.

In connection to the magnetism observed in graphene after proton irradiation, it is worth mentioning that ferromagnetism was found [115] in bulk MoS_2 samples after 2 MeV proton irradiation. However, the magnetic properties of irradiated single layer MoS_2 and other TMDs have not yet been studied so far.

9.3.3 Swift heavy ions

Swift heavy ions (SHIs) have extensively been used to modify the properties of 2D materials. The interaction of matter with swift heavy ions is mainly dominated by electronic excitations and ionization processes (electronic stopping), whereas elastic collisions (nuclear stopping) with target atoms have a minor effect. In SHI beams, the kinetic energy of the ion is normally chosen around the maximum of the energy loss vs. energy curve, which is in the range of energies of 1–10 MeV per nucleon. SHIs can cause substantial modifications to the target material such as nano-sized amorphous tracks and surface hillocks. The first irradiation of 2D materials with SHI was performed on graphene using 90 MeV Xe ions in a grazing incidence geometry. The AFM results showed that SHI irradiation creates foldings in graphene [116]. The folding mechanism was described to occur in two steps: in the first step, SHI irradiation presumably leads to extended line defects in graphene via direct atom sputtering followed by the creation of non-hexagonal rings due to the reconstruction of vacancy-type defects; in the second step, the hillocks formed by the interaction of SHI with the substrate surface fold the graphene by unzipping the graphene sheet along the SHI trajectory. It was found later [117] that folding is an intrinsic response of graphene to SHI irradiation and that it can occur even in a free-standing material. Ochedowski et al. [118] systematically investigated the behavior of various suspended and supported 2D materials under the impact of SHIs. The formation of similar foldings was observed in graphene, h-BN, and MoS_2, Fig. 9.10(a,b). However, foldings in MoS_2 were accompanied by rifts and no folding was seen in bilayer MoS_2. The effect can be attributed to the higher bending modulus in MoS_2 (by a factor of seven) than that in graphene due to its finite thickness [119]. In addition, the presence of the substrate and the crystallographic direction of the SHI beam was found to affect the shape of foldings in 2D materials [117]. At larger angles, multiple foldings formed along with low-indexed crystallographic directions of graphene, while under grazing incidence the azimuthal angle determined the direction of the foldings as shown in Fig. 9.10(c).

9.3.4 Highly charged ions

Highly charged ions (HCIs) usually correspond to ions with low to moderate kinetic energies and high charge states. Due to the removal of electrons from a neutral atom, the energy required to create the ion with a specific charge state is accumulated in the ion as the potential energy, which can be in range of tens of eV to tens of keV. The interaction of HCI with the target is mainly taking place in the vicinity of the surface [121] with potential energy transfer through a series of ion neutralization and deexcitation processes, such as Auger electron

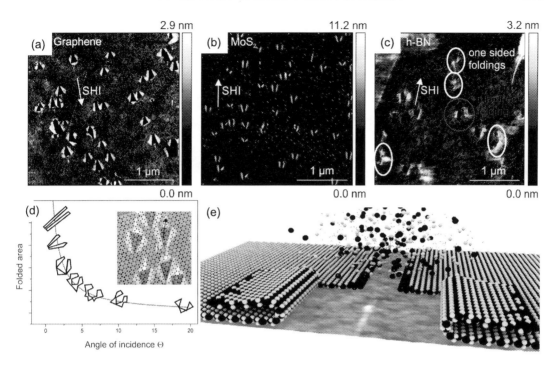

FIGURE 9.10 AFM images showing folding created by SHI irradiation in (a) graphene, (b) MoS$_2$, and (c) h-BN. Reprinted with permission from Ref. [118], Copyright (2014) Elsevier. (d) The shape of the folding pattern depends on the angle of incidence. The image adapted with permission from Ref. [116]. The image adapted from Ref. [16], under the Creative Commons Attribution licence. (e) Schematic illustration of the folding in MoS$_2$ monolayer. The image adapted from Ref. [120], under the Creative Commons Attribution 3.0 licence.

and X-ray emission. It has also been demonstrated [122] that additional channels, referred to as multiple interatomic Coulombic decays, exist in graphene and possibly other 2D materials. The energy deposition (up to tens of keV) by such neutralization processes may even be larger than the energy transfer by ion stopping [123]. In addition, the highly localized excitation in the impact zone and surface selectivity, which are not available with conventional beams, indicates that HCI irradiation is a promising tool for surface modification, and for, e.g., atomic-scale carving of nanopores in van der Waals Heterostructures [124].

Irradiation of 2D materials with HCI can be used for the formation of local defective areas and nanostructures. Since the energy transfer from the HCI to the target is directly related to the charge transfer to the ion, formation of locally charged areas followed by Coulombic explosion is possible, so that 2D materials with semiconductor and insulator character are more susceptible to defect formation upon HCI impacts than metallic systems.

The first experiment on the bombardment of exfoliated MoS$_2$ thin films with HCIs was performed by Hopster et al. [125]. The AFM probing of the irradiated samples demonstrated the appearance of hillocks induced by single ion impacts with their diameter and height being dependent on the charge state of the ion and number of layers of the target material. Later on, Gruber et al. [126] reported that nearly all HCIs passing through suspended single-layer graphene become neutral, but no defects were formed. The unique electronic and thermal properties of graphene provide ultra-fast neutralization within a few femtoseconds timescale. Depending on the initial charge state, the HCIs capture and stabilize between 20 and 30 electrons during their transmission through graphene resulting in the stabilization of the missing electrons in the projectiles, Fig. 9.11(a–b). It is interesting to note that ion impacts do not create extended defects such as nanometre-sized holes in the graphene. At the same time, the HCI irradiation of single-layer h-BN gave rise to nanometer-sized defects under ambient conditions [127]. Pore creation in single layers of TMDs have also been reported [128,129]; the recent experiments by Kozubek et al. [128] demonstrated that the irradiation of free-standing MoS_2 monolayers with HCI produces pores over a broad range of charge states. The size of the pores scaled with the potential energy of the HCI, i.e. changing the charge state of the projectile, ranging from 0.55 to 2.65 nm, Fig. 9.11(c–f). Molecular dynamics simulations including charge-state-dependent nuclear stopping indicated a perforation of the MoS_2 monolayer with pore shapes similar to the experimental results [128]. However, the calculated pore radii were significantly smaller than the experimentally observed ones suggesting that the charge-state-dependent electronic stopping and the potential energy to be the driving force in the defect formation process, Fig, 9.11(g). These effects can hardly be taken into account in the simulations without using adjustable parameters. Besides this, not all of the energy stored in the projectile is deposited in the layer. To improve the computational models, further experimental information about the interaction of the HCIs with 2D materials, such as energy loss, charge transfer, etc. need to be obtained.

9.3.5 Atomic structure engineering by using focused ion beams

Focused ion beams (FIBs) have been routinely used for patterning 2D materials on a nanoscale [77,80,130–134]. This is in part due to the recent progress in helium ion microscopy (HIM) [135], which, in addition to the sputtering of atoms from the target system, can provide at small doses (less than 10^{13} cm^{-2}) nearly non-destructive imaging of 2D target. In HIM, He ions with typical energies in the range of 5–30 keV, can be focused down to sub-nanometer scales. It was shown that using HIM, graphene supported on SiO_2 can be cut into ribbons with a width of less than 40 nm [132], while free-standing graphene can be cut into ribbons with a width of about 10 nm [130,131]. Apart from the applications in lithography and imaging, focused He ion beams were used to produce spatially-localized defects at low fluences or controlled removal of atoms at high fluences in 2D materials [14,77,80,136]. FIBs of heavier atoms, e.g. Ne [137] or Ga can also be used to pattern graphene [138–140] and other 2D materials.

As for 2D materials beyond graphene, Fox et al. [80] studied the possibility of nanometer-scale precision milling of a few-layer MoS_2 sample using the helium-ion beam. They reported the structural and electronic modification of the irradiated samples for a wide range of ion doses and beam profiles. The electrical measurements demonstrated tunable resistivity of the sample by ion dose and intriguing semiconductor-to-insulator-to-metal-to-insulator transitions depending on the irradiation dose, Fig. 9.12. Such changes in the electronic properties are not fully understood at this time. The initial drop in the resistivity may be associated with the sputtering of adatoms from the surface and passivation of dangling bonds in the pre-existing vacancies with hydrogen atoms, followed by the accumulation of vacancy-type

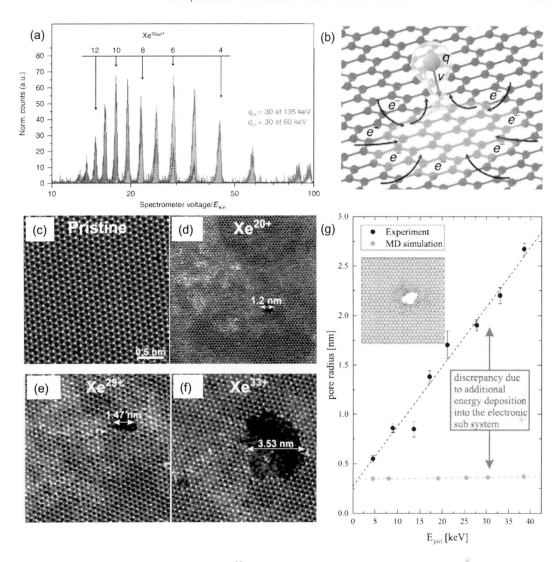

FIGURE 9.11 (a) Measured spectra of a Xe^{+30} beam at kinetic energies of 135 and 60 keV transmitted through a free-standing graphene sheet. The spectra provide information on the distribution of the transmitted ions over the charge states. (b) Schematic of the interaction process between free-standing graphene and an approaching highly charged ion. Adapted from Ref. [126], under the Creative Commons Attribution 4.0 International License. STEM images of MoS$_2$ sheet before (c) and after irradiation with (d) Xe^{+20}, (e) Xe^{+29}, (f) Xe^{+33} ions. Pores of round shapes with a diameter in the nanometer regime are visible. (g) Average pore radius created in MoS$_2$ as a function of the potential energy obtained by STEM measurements and by MD simulations. Reprinted with permission from Ref. [128], Copyright (2019) American Chemical Society.

defects. The metallic-like behavior was attributed to the dominant sputtering of sulfur and the high density of remaining molybdenum in the sample, which formed metallic channels.

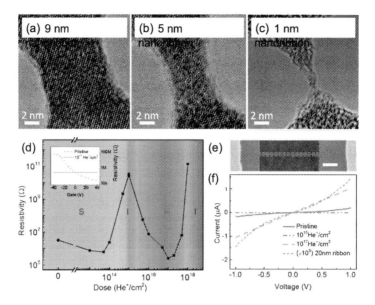

FIGURE 9.12 (a–c) TEM images of MoS$_2$ (amorphous) nanoribbons fabricated using He ion irradiation. AC-STEM image of the pristine and the Ga ion irradiated MoS$_2$ with different ion doses. (d) Electrical characterization of He-ion-beam modified devices. The letters S, I, and M indicate regions with semiconducting, insulating, and metallic-like behavior, respectively. (e–d) The image and current-voltage characteristics of the device with 20 nm wide MoS$_2$ nanoribbon under various ion doses. Reprinted with permission from Ref. [80], Copyright (2015) American Chemical Society.

Based on this report, amorphous nanoribbons with widths down to 1 nm could be produced. Focused helium ion beam was also used [141] to selectively pattern the out-of-plane piezo-electricity via defect engineering in a another TMD: MoTe$_2$. The out-of-plane piezoelectricity generated in the desired area was quantitatively examined using AFM, and the formation of irradiation-induced defects that gave rise to inversion symmetry breaking was confirmed. It is expected that this approach can also be applied to other TMDs and give rise to new applications of TMD-based devices.

In addition to direct nanomaterial processing approaches, beam-mediated strategies may be used by combining irradiation with exposure to precursor gases. In this case, atoms may be sputtered away from a 2D material, accompanied by access to chemical species that fill the vacancies, meaning that impurities with a high spatial resolution can be added into the sample by using focused beams, see Fig. 9.13(a). The precursor gas can also be added during the irradiation process, and the beam will break gas molecules thus providing reactive species. The FIBs based on Ga or Ar ions have been successfully used for the functionalization of various 2D materials. For example, the recent experiments using FIB with 30 keV Ga ion together with the simultaneous introduction of XeF$_2$ displayed selective fluorination of graphene [142]. The fluorine atoms were predominantly localized at the produced defects as indicated by STM and XPS. Irradiation with Ga ions followed by fluorination has also been used to enhance the vertical stiffness of graphene sheets [143]. In addition, the feasibility of

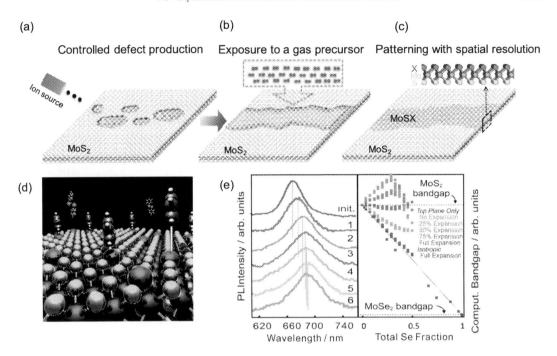

FIGURE 9.13 (a–c) Schematic illustration of ion-beam mediated production of mixed MoSX material starting from MoS$_2$ monolayer. The ion irradiation is used to sputter S atoms from one side of MoS$_2$ monolayer, followed by exposure to a gas precursor. The images are adapted from Ref. [68], under the Creative Commons Attribution 3.0 licence. (d) Schematic illustration of substitution of sulfur atoms by selenium in MoS$_2$ through a process of gentle sputtering followed by exposure to a selenium precursor, and annealing. (e) Normalized room-temperature PL spectra of a single-layer MoS$_2$ after sputtering and Se insertion cycles. (f) The calculated band gap of MoS$_{(1-x)}$Se$_{2x}$ as a function of Se content. Adapted with permission from Ref. [144], Copyright (2014) American Chemical Society.

selective desulfurization of MoS$_2$ monolayer through a low-energy Ar ion irradiation [81] and the post-growth tuning of the band gap by substitution of top-layer sulfur atoms with selenium atoms [144] has been shown, Fig. 9.13(b). Ion irradiation assisted chemical functionalization of MoS$_2$ monolayer FETs was recently reported [145]. The performance of argon-ion irradiated FETs was significantly recovered by exposing the devices to vapors of short linear thiolated molecules. The solvent-free technique resolved the secondary healing impact of oxygen or oxygen-containing compounds, such as ethanol, and could be further employed to functionalize 2D MoS$_2$ sheets with different types of molecules containing thiol groups

The recent study on exposing single-layer MoS$_2$ to focused Ga ions with a dose varied from 6.25×10^{12} ions/cm^2 to 2.50×10^{13} ions/cm^2 reported the production of pores with average and maximum diameters of 0.5 nm and 1.0 nm [140]. While the median vacancy area was almost independent of the dose, the number of vacancies (pores) increased with the ion dose. The measurements of ionic transport through the perforated membrane revealed nonlinear ionic current-voltage properties with a conductivity equivalent to that of ~ 1 nm diameter single MoS$_2$ pores, indicating that the smaller pores in the distribution exhibit negligible conductivity.

9.3.6 Irradiation tolerance

The irradiation tolerance, which can also be referred to as radiation hardness or susceptibility, is a characteristic of a material subjected to beams of energetic particles and γ-rays [146]. There is no strict definition of this quantity, and normally it is discussed with regard to physical damage during radiation exposure and device degradation in a radiation-hostile environment such as open cosmic space or nuclear facilities. Due to the atomically thin nature, low-weight and power consumption, one can expect applications of 2D materials for the outer space. To this end, studies of the 2D materials in the context of radiation tolerance are important.

The radiation hardness of 2D material-based devices has been investigated in several experiments. Ochedowski et al. [147] studied the deterioration of field-effect transistors based on graphene and MoS_2 under 1 GeV uranium beam with three different fluences. The highest applied fluence of 4×10^{11} ions/cm^2 was found to destroy the MoS_2-based transistor, while the device with graphene was still operable, although its performance was degraded.

The behavior of MoS_2- and WS_2-based devices such as field-effect transistors and single-photon sources under combined γ-ray, proton and electron beam has also been studied [148] in the context of their radiation tolerance for space applications. The devices showed negligible changes in performance after the irradiation with the doses equivalent to what one can expect after being for 10^3 years at 500 km above the polar caps. Counter-intuitively, defect densities were found to be lower in WS_2 monolayer under high-dose γ-radiation, as identified by an increase in photo-luminescence, carrier lifetime and a change in doping ratio proportional to the photon flux. The underlying mechanism was attributed to radiation-induced defect healing, possibly due to the passivation of sulfur vacancies by the dissociated oxygen. Similar experiments were also done for WSe_2 and MoS_2 monolayers under gamma-radiation. A slight increase and no change in PL intensity and carrier lifetime were observed for WSe_2 and MoS_2, respectively. It should be noted that the PL emission from MoS_2 monolayer is much weaker than that of other TMDs, making any alteration more difficult to detect.

Kim et al. [114] studied the evolution of the electrical characteristics of a MoS_2-based field-effect transistor under irradiation with 10 MeV proton beams. In the case of low proton fluence of 10^{12} ions/cm^2, the device performance remained almost the same as before irradiation. However, for high proton fluence of 10^{13} or 10^{14} ions/cm^2, a dramatic reduction in the conductance of the devices was observed after the irradiation. The deterioration of the device performance was attributed to the presence of charge traps in the oxide substrate and their subsequent impact on the electrical characteristics of the device. However, the assessment of the radiation hardness of 2D materials using typical electrical measurements is not straightforward due to the fact that the interface traps and oxide charges cause compensating effect, i.e., they give rise to the opposite shift of the I–V curve.

To further investigate how the defects produced in the substrate affect the performance of the irradiated devices based on a 2D material, in a recent study by Arnold et al. [149], a new experimental setup was used to decouple the radiation impact due to the 2D semiconducting channel and the change in the oxide dielectric. It was found that the interface states play a major role in the electrical characteristic following irradiation with a He ion fluence of 10^{15} ions/cm^2, whereas oxide charges have dominant effect for the case exposed to a proton fluence of 1.26×10^{16} ions/cm^2. According to the electrical characteristics, ultrathin MoS_2

nanosheets can withstand proton and helium irradiation with fluences as high as $\sim 10^{16}$ and $\sim 10^{15}$ ions/cm^2, respectively, which was interpreted by the authors as an "extraordinary" radiation hardness. Similar conclusions were made after exposure of multilayer MoS$_2$ FETs to 2 MeV He ions [150].

The high irradiation tolerance of 2D materials (as compared to the operation of bulk devices under similar irradiation conditions) found in the above experiments can be explained by three factors: (i) the intrinsic tolerance of the 2D material [151] related to where the defect-induced electronic states appear (in the gap or in the valence/conduction band) of the irradiated 2D material and ultimately to its chemical content; (ii) Smaller (as compared to 3D materials) probability for defect production at higher ion energies, Fig. 9.1; (iii) In-situ healing of defects during the irradiation due to the interaction with the environment. The last factor may, however, have the opposite effect depending on the material and the environment [152].

9.4 Applications

As already evident from the above, the bombardment with energetic ions can effectively be used to tailor the electronic, optical and catalytic properties of 2D materials [15,153]. In particular, the modification of the electronic properties of 2D semiconducting targets by creating defects and introducing impurities is a promising route for tuning charge-carrier mobility and the threshold voltage in FET devices. Nakaharai et al. [154] reported the operation of a graphene transistor with a He ion irradiated channel in which a defect-induced transport gap was formed. The increase of the on/off current by a factor of two was achieved by increasing the helium ion doses from 2.2×10^{15} to 1.3×10^{16} cm^2. The e-beam lithography combined with He ion beam milling was also used to fabricate single-electron transistors based on graphene quantum dots [155].

The experiment by Valerius et al. [156] demonstrated a crystalline-to-amorphous transformation of the MoS$_2$ monolayer grown on graphene/Ir(111) under 500 eV Xe$^+$ irradiation with grazing trajectories. The STM probing confirmed changes in the electronic properties from semiconductor-to-metal due to the amorphization associated with the extinction of photoluminescence. MD simulations exactly mimicking the experimental irradiation conditions displayed selective sputtering of the top S-layer, whereas the Mo-layer and the bottom S-layer were disordered, but not sputtered. The supportive graphene layer stayed unchanged and no implantation of Xe atoms was found as the Xe projectiles were reflected with 100% probability.

The possibility of changing defect concentrations or inducing local amorphization of a 2D material opens a path for tuning its physical properties via a combination of thermal treatment and a reactive vapor. More recently, lateral memristors from single layer MoS$_2$ were made by utilizing a focused helium ion beam [157]. Site-specific irradiation with the helium ion beam was used for the creation of nanoscale defect-rich regions in the MoS$_2$ lattice. The reversible drift of these defects by an applied electric field controlled the resistance of the channel, enabling versatile memristive functionality, Fig. 9.14 (a,b). In addition, an ion beam was used to alter metal-2D materials interfaces for reducing the contact resistance. As an example, the 60 eV Ar ion beam yields improvement in the metal-MoS$_2$ interfaces by decreasing the contact resistance from 17.5 k$\Omega \cdot$ μm to 6 k$\Omega \cdot$ μm [158]. A similar method has been suggested to enhance the Pd/graphene contact [159].

The creation of defects introduces midgap states in the band structures of specific 2D materials, extending the broadband applications in photonics. A recent report showed that Ga irradiation can enhance the photoresponsivity of a vertical WSe$_2$/graphene-based photodetector [160]. While Se vacancies induced by irradiation quench the in-plane charge transport of WSe$_2$ due to the localization of the electrons with weak dispersion, the out-of-plane charge transport of WSe$_2$ was increased, Fig. 9.14 (c,d). As a result, ion bombardment could improve the photoresponse of the WSe$_2$/graphene heterostructure by a factor of 2 in comparison to the pristine heterostructure. Stanford et al. [161] studied the photoresponse of the WSe$_2$ lateral homo-junction fabricated by the He ion irradiation. The report showed the photovoltaic effect with a considerable open-circuit photovoltage of 220 mV. Apart from that, the ion beam technique exhibited promising results for nonlinear absorption of modified 2D materials. In another experiment, ion irradiation was suggested for modulation of the interlayer coupling in the graphene/WSe$_2$ heterostructure for photoluminescence and ultrafast absorption [162]. The heterostructure with ion beam modified interlayer spacing possesses larger modulation depth and lower saturation intensity than each separate 2D material for pulsed laser applications. Ma et al. [163] reported the sulfur vacancies created by Ar$^+$ ion beam enhance near-infrared (NIR) absorption of WS$_2$. In addition, a Q-switched pulsed laser was produced by using the irradiated WS$_2$ monolayer as a saturable absorber in the waveguide cavity. The efficiency of the Q-switched laser was significantly enhanced as compared to the as-prepared WS$_2$.

Tailoring of optically active defect centers in 2D materials can also open a pathway to host single-photon emitters. The effect of focused helium ion on the valleytronic properties of atomically thin MoS$_2$ was studied [136], Fig. 9.14 (e,f). Robust valley polarization was reported for free excitons up to room temperature even for high irradiation doses. More recently, controlled generation of luminescent centers in h-BN is shown via ion irradiation with a fluence of 1×10^{14} ion/cm^2 to 1×10^{15} ion/cm^2. [164] Both the experiments and MD simulations indicated that the density of created luminescent centers can be adjusted using both irradiation energy and fluence, in which high fluence leads to a significant increase in density [165,166]. It was suggested that luminescent centers are formed by irradiating h-BN and then recrystallizing it through annealing, with the luminescent centers forming in the topmost layers (see Fig. 9.15).

Ion implantation has also been employed to synthesize various 2D materials. For instance, Tsai et al. [167,168] synthesized multilayer silicene and HfSe$_2$ by ion implantation. More recently, it was shown that the single-layer graphene synthesized by carbon ion implantation can be applied to fabricate FET devices with promising transfer and output characteristics [169].

It is well known that defects such as vacancies have a remarkable influence on the catalytic efficiency of 2D materials. Ion beam-assisted treatments can improve the density of chemically-active sites for catalytic reactions [170,171]. Furthermore, electrical conductivity can be modulated by the introduction of defects [172]. Therefore, the catalytic performance of 2D materials can be enhanced by irradiation-mediated defect engineering. It has been shown that sulfur vacancies created via exposure to an argon plasma cause the activation of the basal plane of MoS$_2$ monolayers for hydrogen evolution reactions (HER) [173]. Similarly, O$_2$ plasma

9.4. Applications

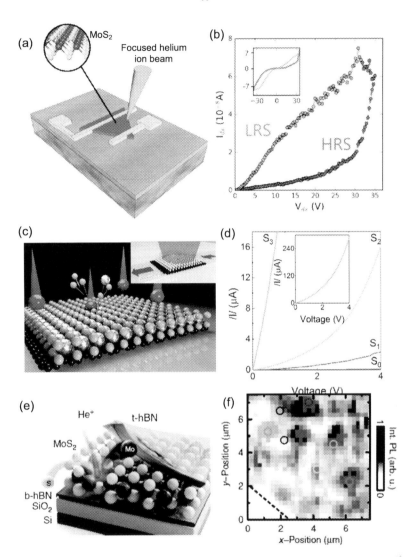

FIGURE 9.14 (a) Illustration of the MoS$_2$ memristors fabricated using He ion beam irradiation (b) Positive-bias sweep region of the memtransistor recorded at Vg = 0 V. Inset shows the full range of the sweep in the same unit. Adapted with permission from Ref. [157], Copyright (2019) American Chemical Society. (c) Schematic illustration of the Ga irradiation in the WSe$_2$/Graphene based photodetector. (d) Transfer characteristics of the WSe$_2$/Graphene photodiode at different irradiation time: S1 = 2 s, S2 = 4 s and S3 = 8 s. Adapted with permission from Ref. [160], Copyright (2018) American Chemical Society. (e) Schematic illustration of the exposed MoS$_2$/h-BN heterostructure by focused helium ions. (f) Spatially resolved and spectrally integrated photoluminescence mapping of the irradiated heterostructure. Adapted from Ref. [136], under the Creative Commons Attribution 4.0 International License.

was used to create more active sites via the formation of Mo-O and S-O bonds, enhancing the HER activity of MoS$_2$ sheets [174].

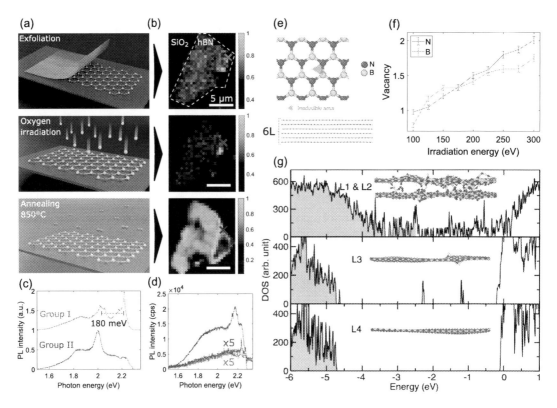

FIGURE 9.15 (a) Schematic representation of the three fabrication stages for generating luminescent centers in h-BN: mechanical exfoliation, irradiation with oxygen atoms, and thermal annealing in nitrogen. (b) Photoluminescence (PL) maps after each fabrication step showing the evolution of one h-BN flake. (c) Typical spectra of luminescent centers of group I and group II. (d) Spectral evolution of an individual luminescent center. (e) Setup for MD simulations of oxygen irradiation. (f) Average numbers of Nitrogen vacancy and Boron vacancy produced at different irradiation energies. L1 and L2 stand for the topmost and second topmost layer, respectively. Adapted from Ref. [164], under a Creative Commons Attribution Non-Commercial License 4.0.

As discussed previously, FIBs can be used to pattern 2D materials and ultimately produced arrays of holes with diameters of a few nanometers, that is nanomeshes, which are interesting in the context of tuning the electronic properties of 2D systems [175,176] and membrane separation technology such as DNA sequencing, gas separation, water desalination and many more [177–182]. As an example, Surwade et al. [183] showed that a nonporous graphene membranes produced by ion irradiation can reach a salt rejection rate of almost 100% and rapid water transport. Theoretical simulations reported that pores with sub-nanometer size in a MoS_2 sheet can be created by cluster-ion bombardment [184]. These produced nanopores demonstrated uniform distribution with the diameter being dependent on cluster size and energy.

Nanomeshes with a sub-10 nm pitch and a 4 nm pore diameter were also manufactured in suspended monolayer graphene by direct helium ion beam milling [134]. Electrical transport measurements revealed an effective energy gap opening of up to ~ 450 meV; this approach may open a path toward nanomesh-based, room-temperature semiconducting applications. Nanoporous graphene-PMMA composite was also produced using SHIs [185].

An interesting application of graphene in the context of the response to irradiation has recently been suggested [186] based on the STM investigation of the behavior of single-layer graphene upon heavy bombardment with various low-energy (200–500 eV) ions at high temperatures. The conclusion was that graphene can be used as an efficient sputter protection for the underlying surfaces for a broad range of metals and alloys. By quantitatively evaluating the sputtered material a drastic decrease in metal sputtering for the graphene protected areas was found. The sputter protection was argued to come from the self-repair of the ion damage in graphene, which takes place at elevated temperatures.

9.5 Summary, challenges, and outlook

In this chapter, we presented an overview of recent advances in our understanding of the response of 2D materials to ion irradiation. As evident from the large body of experimental and theoretical results discussed above, the behavior of 2D materials under ion bombardment is different from the behavior of bulk 3D systems under ion irradiation.

The main difference is that although the number of the produced defects in 2D materials increases with ion energy at small ion energies as in bulk systems, it will reach a maximum and drop at higher energies due to a smaller probability for the ion to displace an atom in the atomically thin target, while in bulk materials the ion with higher energy always creates more damage, but deeper in the sample. This means that there is no need for high-energy ions when irradiating 2D materials, and the whole 2D sample can easily be modified, as ion ranges already at low ion energies are much larger than the thickness of the material. In fact, ultra-low ion energies (below 100 eV) are required for, e.g., direct ion implantation into 2D materials, which is technically a challenge, as it is quite difficult to provide narrow ion distribution over energy in this energy range. Other problems are mass separation and positioning of magnetic lenses, as well as beam focusing and guiding at such low energies.

The second important point is that the behavior of free-standing (suspended) and supported (on a substrate) 2D materials under the ion beam can be quite different, as the backscattered ions or atoms sputtered from the substrate can completely govern defect production. For the supported irradiated 2D materials, the evolution of the defects will also be affected by the substrate. Atoms sputtered away from a free-standing 2D material may stay in between the 2D sheet and the substrate thus increasing the probability for the annealing of vacancies. On the other hand, vacancies in 2D materials, e.g., graphene or h-BN sheets, can strongly interact with the substrates. The atoms with dangling bonds at vacancies can form strong covalent bonds with metal substrates, which completely changes the behavior of vacancies, that is migration and agglomeration.

The geometry of 2D materials also indicates that the environment will strongly affect the behavior of the irradiation-induced defects – additional defects can appear (e.g., due to the interaction with the reactive species like oxygen molecules) or the other way around, disappear – an example is the self-healing of vacancies in graphene due to the dissociation of hydrocarbon molecules [187]. The effects of the environment should also be considered in the 'shielding' of the irradiated 2D materials due to the unwanted but hardly-avoidable deposition and agglomeration of ubiquitous hydrocarbons on top of the 2D materials, which can also come during the transfer process.

In contrast to defect engineering of bulk materials, where the quality of the surface and the ambient conditions are often not the crucial factors, the surfaces of 2D materials have to be extremely clean and the structure of the material has to contain very few defects before the irradiation. Therefore, it is important to perform in-situ experiments under an ultra-high vacuum which requires dedicated experimental set-up including ion source joined with a chamber for the materials growth and characterization tools. A positive side of the sensitivity of the irradiated 2D materials to the environment is that post-irradiation treatment (e.g., exposure to gases) can be used to engineer the atomic structure of 2D systems. Impurities can be introduced by a post-bombardment deposition of the desired chemical species, which can interact with vacancies and other irradiation-induced defects and thus add new functionalities like magnetism or high catalytic activity to the irradiated 2D material. It is also evident that due to their atomically thin nature, 2D materials can be easily cut or patterned by focused ion beams, which opens many opportunities for the beam-mediated engineering of devices and nanomeshes for, e.g., DNA sequencing or molecule separation.

As for the theoretical aspects, based on the results obtained so far, one can conclude that the interaction of low-energy heavy ions with free-standing and supported 2D materials overall is well understood. The problem with the interpretation of the experimental results frequently lies in the difficulties arising from not having access to pure materials free of defects and adsorbed species. The adequate description of the evolution of the defects on a macroscopic time scale is another challenge. At the same time, the theoretical description of defect-creation mechanisms in 2D materials using swift and highly-charged ions are not fully understood. Note that contrary to bulk materials where the experimental results are the outcome of many scattering events, the interaction of an ion with a 2D material can be understood in terms of a single scattering event. This, on the one hand, makes the theoretical analysis more complicated. For example, the energy lost by the ion after passing a graphene sheet should depend on the impact parameter even in the electron stopping regime. However, on the other hand, the experiments on such a system can provide lots of insight into the fundamental aspects of ion-solid interaction. The theoretical tools used for the simulations of the interaction of highly charged and swift heavy ions with 2D systems should be further developed. One of the challenges here is to account for the ionization of the 2D system during the early stages of damage formation, as exited electrons will leave the 2D sheet, contrary to bulk materials.

It should be pointed out that the vast majority of the irradiation experiments have been carried out on graphene. Among inorganic 2D materials, effects of ion irradiation on MoS_2 have been mostly investigated, although some other TMDs, e.g., WS_2 have been studied. So the obvious direction for the future work is to explore the response of other 2D materials to ion bombardment. Specifically, a systematic study of the behavior of TMDs, which have essentially the same atomic structure, but are composed of atoms with different masses and also have different electronic properties, could provide insight into ion-solid interaction in these materials.

Acknowledgments

The Authors would like to thank T. Michely, M. Schleberger, S. Facsko, G. Hlawacek, R. Wilhelm, J. Kotakoski, C. Busse, K. Nordlund, N. Stenger and all the members of their groups for many years of successful collaboration on various aspects of ion irradiation of 2D materials.

References

[1] Nils Bohr, II. On the theory of the decrease of velocity of moving electrified particles on passing through matter, The London, Edinburgh, and Dublin Philosophical Magazine and Journal of Science 25 (145) (1913) 10–31, https://doi.org/10.1080/14786440108634305.

[2] Michael Nastasi, Mayer James, James K. Hirvonen, Ion-Solid Interactions: Fundamentals and Applications, Cambridge Solid State Science Series, Cambridge University Press, Cambridge, 1996.

[3] K.S. Novoselov, A.K. Geim, S.V. Morozov, D. Jiang, Y. Zhang, S.V. Dubonos, I.V. Grigorieva, A.A. Firsov, Electric field effect in atomically thin carbon films, Science 306 (5696) (2004) 666–669, https://doi.org/10.1126/science.1102896.

[4] B. Radisavljevic, A. Radenovic, J. Brivio, V.Giacometti, A.Kis, Single-layer MoS2 transistors, Nature Nanotechnology 6 (2011) 147–150, https://doi.org/10.1038/nnano.2010.279.

[5] J.N. Coleman, M. Lotya, A. O'Neill, S.D. Bergin, P.J. King, U. Khan, K. Young, A. Gaucher, S. De, R.J. Smith, I.V. Shvets, s.K. Arora, G. Stanton, H.-Y Kim, K. Lee, G.T. Kim, G.S. Duesberg, T. Hallam, J.J. Boland, J.J. Wang, J.F. Donegan, J.C. Grunlan, G. Moriarty, A. Shmeliov, R.J. Nicholls, J.M. Perkins, E.M. Grieveson, K. Theuwissen, D.W. McComb, P.D. Nellist, V. Nicolosi, Two-dimensional nanosheets produced by liquid exfoliation of layered materials, Science 331 (2011) 568–571.

[6] Se-Yang Kim, Jinsung Kwak, Cristian V. Ciobanu, Soon-Yong Kwon, Recent developments in controlled vapor-phase growth of 2D group 6 transition metal dichalcogenides, Advanced Materials 31 (20) (2019) 1804939, https://doi.org/10.1002/adma.201804939.

[7] Rui Zhang, Zaodi D. Zhang, Zesong S. Wang, Shixu Wang, Wei Wang, Dejun J. Fu, Jiarui Liu, Nonlinear damage effect in graphene synthesis by C-cluster ion implantation, Applied Physics Letters 101 (1) (2012) 011905, https://doi.org/10.1016/j.carbon.2018.01.098.

[8] Roger Smith (Ed.), Atomic and Ion Collisions in Solids and at Surfaces: Theory, Simulation and Applications, Cambridge University Press, Cambridge, 1997.

[9] R. Nair, M. Sepioni, I-Ling Tsai, O. Lehtinen, J. Keinonen, Arkady V. Krasheninnikov, T. Thomson, a.K. Geim, I.V. Grigorieva, Spin-half paramagnetism in graphene induced by point defects, Nature Physics 10 (2012) 199–202, https://doi.org/10.1038/nphys2183.

[10] Sunil Kumar, Ambuj Tripathi, Fouran Singh, Saif Ahmad Khan, Vikas Baranwal, Devesh Kumar Avasthi, Nanoscale Research Letters 9 (12) (2014) 126, https://doi.org/10.1186/1556-276X-9-126.

[11] A. Kumar, D.K. Avasthi, J.C. Pivin, P.M. Koinkar, Ordering of fullerene and carbon nanotube thin films under energetic ion impact, Applied Physics Letters 92 (2008) 221904.

[12] Alan Colli, Andrea Fasoli, Carsten Ronning, Simone Pisana, Stefano Piscanec, A.C. Ferrari, Ion beam doping of silicon nanowires, Nano Letters 8 (8) (2008) 2188–2193.

[13] A.V. Krasheninnikov, K. Nordlund, Ion and electron irradiation-induced effects in nanostructured materials, Journal of Applied Physics 107 (7) (2010) 071301, https://doi.org/10.1063/1.3318261.

[14] C-T. Pan, J. a Hinks, Q. Ramasse, G. Greaves, Ursel Bangert, S.E. Donnelly, S.J. Haigh, In-situ observation and atomic resolution imaging of the ion irradiation induced amorphisation of graphene, Scientific Reports 4 (i) (2014) 6334, https://doi.org/10.1038/srep06334.

[15] Ziqi Li, Feng Chen, Ion beam modification of two-dimensional materials: characterization, properties, and applications, Applied Physics Reviews 4 (1) (2017) 011103, https://doi.org/10.1063/1.4977087.

[16] Marika Schleberger, Jani Kotakoski, 2D material science: defect engineering by particle irradiation, Materials 11 (10) (2018) 1885, https://doi.org/10.3390/ma11101885.

[17] Guang-Yi Zhao, Hua Deng, Nathaniel Tyree, Michael Guy, Abdellah Lisfi, Qing Peng, Jia-An Yan, Chundong Wang, Yucheng Lan, Recent progress on irradiation-induced defect engineering of two-dimensional 2H-MoS2 few layers, Applied Sciences 9 (4) (2019) 678, https://doi.org/10.3390/app9040678.

[18] K. Nordlund, Historical review of computer simulation of radiation effects in materials, Journal of Nuclear Materials 520 (2019) 273–295, https://doi.org/10.1016/j.jnucmat.2019.04.028.

[19] Program TRIM, by J.F. Ziegler, J.P. Biersack, http://www.srim.org, 2008.

[20] Sadegh Ghaderzadeh, Mahdi Ghorbani-Asl, Silvan Kretschmer, Gregor Hlawacek, Arkady V. Krasheninnikov, Channeling effects in gold nanoclusters under He ion irradiation: insights from molecular dynamics simulations, Nanotechnology 31 (3) (2020), https://doi.org/10.1088/1361-6528/ab4847.

[21] O. Lehtinen, J. Kotakoski, A.V. Krasheninnikov, A. Tolvanen, K. Nordlund, J. Keinonen, Effects of ion bombardment on a two-dimensional target: atomistic simulations of graphene irradiation, Physical Review B 81 (15) (2010) 153401, https://doi.org/10.1103/PhysRevB.81.153401.

[22] Silvan Kretschmer, Mikhail Maslov, Sadegh Ghaderzadeh Mahdi Ghorbani-Asl, Gregor Hlawacek, Arkady V. Krasheninnikov, Supported two dimensional materials under ion irradiation: the substrate governs defect production, ACS Applied Materials & Interfaces 10 (36) (2018) 30827–30836, https://doi.org/10.1021/acsami.8b08471.

[23] Egor A. Kolesov, Choosing a substrate for the ion irradiation of twodimensional materials, Beilstein Journal of Nanotechnology 10 (2019) 531–539, https://doi.org/10.3762/BJNANO.10.54.

[24] Abhijit Chatterjee, Dionisios G. Vlachos, An overview of spatial microscopic and accelerated kinetic Monte Carlo methods, Journal of Computer-Aided Materials Design 14 (2) (2007) 253–308, https://doi.org/10.1007/s10820-006-9042-9.

[25] O. Lehtinen, J. Kotakoski, A.V. Krasheninnikov, J. Keinonen, Cutting and controlled modification of graphene with ion beams, Nanotechnology 22 (2011) 175306, https://doi.org/10.1088/0957-4484/22/17/175306.

[26] Roman Bttger, Karl-Heinz Heinig, Lothar Bischoff, Bartosz Liedke, Stefan Facsko, From holes to sponge at irradiated Ge surfaces with increasing ion energy – an effect of defect kinetics?, Applied Physics A 113 (2013) 53–59, https://doi.org/10.1007/s00339-013-7911-0.

[27] Zhangcan Yang, Michael A. Lively, Jean Paul Allain, Kinetic Monte Carlo simulation of self-organized pattern formation induced by ion beam sputtering using crater functions, Physical Review B 91 (12) (2015) 075427, https://doi.org/10.1103/PhysRevB.91.075427.

[28] Graeme Henkelman, Blas P. Uberuaga, Hannes Jonsson, A climbing image nudged elastic band method for finding saddle points and minimum energy paths, The Journal of Chemical Physics 113 (2000) 99019904, https://doi.org/10.1063/1.1329672.

[29] Frank H. Stillinger, Thomas A. Weber, Computer simulation of local order in condensed phases of silicon, Physical Review B 31 (8) (1985) 5262–5271, https://doi.org/10.1103/PhysRevB.31.5262.

[30] J. Tersoff, New empirical approach for the structure and energy of covalent systems, Physical Review B 37 (12) (1988) 6991–7000, https://doi.org/10.1103/PhysRevB.37.6991.

[31] James F. Ziegler, Jochen P. Biersack, The stopping and range of ions in matter, in: D. Allan Bromley (Ed.), Treatise on Heavy-Ion Science: Volume 6: Astrophysics, Chemistry, and Condensed Matter, Boston, MA, ISBN 978-1-4615-8103-1, 1985, pp. 93–129.

[32] J. Byggmstar, A. Hamedani, K. Nordlund, F. Djurabekova, Machine-learning interatomic potential for radiation damage and defects in tungsten, Physical Review B 100 (14) (2019) 144105, https://doi.org/10.1103/PhysRevB.100.144105.

[33] Hao Wang, Xun Guo, Linfeng Zhang, Han Wang, Jianming Xue, Deep learning inter-atomic potential model for accurate irradiation damage simulations, Applied Physics Letters 114 (2019) 244101, https://doi.org/10.1063/1.5098061.

[34] A. Meftah, F. Brisard, J.M. Costantini, E. Dooryhee, M. Hage-Ali, M. Hervieu, J.P. Stoquert, F. Studer, M. Toulemonde, Track formation in SiO_2 quartz and the thermal-spike mechanism, Physical Review B 49 (1994) 12457–12463, https://doi.org/10.1103/PhysRevB.49.12457.

[35] H. Hellmann, Einfhrung in die Quantenchemie, Angewandte Chemie 54 (11–12) (1941) 156, https://doi.org/10.1002/ange.19410541109.

[36] R.P. Feynman, Forces in molecules, Physical Review 56 (4) (1939) 340–343, https://doi.org/10.1103/PhysRev.56.340.

[37] J. Kotakoski, C.H. Jin, O. Lehtinen, K. Suenaga, A.V. Krasheninnikov, Electron knock-on damage in hexagonal boron nitride monolayers, Physical Review B 82 (11) (2010) 113404, https://doi.org/10.1103/PhysRevB.82.113404.

[38] Silvan Kretschmer, Tibor Lehnert, Ute Kaiser, Arkady V. Krashenin nikov, Formation of defects in two-dimensional MoS 2 in the transmission electron microscope at electron energies below the knock-on threshold: the role of electronic excitations, Nano Letters 20 (4) (2020) 2865–2870, https://doi.org/10.1021/acs.nanolett.0c00670.

[39] Arkady V. Krasheninnikov, Yoshiyuki Miyamoto, David Tomnek, Role of electronic excitations in ion collisions with carbon nanostructures, Physical Review Letters 99 (2007) 016104, https://doi.org/10.1103/PhysRevLett.99.016104.

[40] Weisen Li, Jianming Xue, Ion implantation of low energy Si into graphene: Insight from computational studies, RSC Advances 5 (121) (2015) 99920–99926, https://doi.org/10.1039/c5ra17250k.

[41] Mukesh Tripathi, Alexander Markevich, Roman Bttger, Stefan Facsko, Elena Besley, Jani Kotakoski, Toma Susi, Implanting germanium into graphene, ACS Nano 12 (5) (2018) 4641–4647, https://doi.org/10.1021/acsnano.8b01191.

[42] Ari Ojanpera, Arkady V. Krasheninnikov, Martti Puska, Electronic stopping power from first-principles calculations with account for core electron excitations and projectile ionization, Physical Review B 89 (3) (2014) 035120, https://doi.org/10.1103/PhysRevB.89.035120.

[43] Alfredo A. Correa, Jorge Kohanoff, Emilio Artacho, Daniel Snchez-Portal, Alfredo Caro, Nonadiabatic forces in ion-solid interactions: the initial stages of radiation damage, Physical Review Letters 108 (21) (2012) 213201, https://doi.org/10.1103/PhysRevLett.108.213201.

[44] Rafi Ullah, Emilio Artacho, Alfredo A. Correa, Core electrons in the electronic stopping of heavy ions, Physical Review Letters 121 (11) (2018) 116401, https://doi.org/10.1103/PhysRevLett.121.116401.

[45] Cheng-Wei Lee, Andr Schleife, Hot-electron-mediated ion diffusion in semiconductors for ion-beam nanostructuring, Nano Letters 19 (6) (2019) 3939–3947, https://doi.org/10.1021/acs.nanolett.9b01214.

[46] Tibor Lehnert, Mahdi Ghorbani-Asl, Janis Kster, Zhongbo Lee, Arkady V. Krasheninnikov, Ute Kaiser, Electron-beam-driven structure evolution of single-layer MoTe2 for quantum devices, ACS Applied Nano Materials 2 (5) (2019) 3262–3270, https://doi.org/10.1021/acsanm.9b00616.

[47] Xin Wu, Influence of Particle Beam Irradiation on the Structure and Properties of Graphene, Springer Nature, Singapore, 2018, pp. 51–72.

[48] J. Karthikeyan, H-P. Komsa, Matthias Batzill, Arkady V. Krasheninnikov, Which transition metal atoms can be embedded into two-dimensional molybdenum dichalcogenides and add magnetism?, Nano Letters 19 (2019) 4581–4587, https://doi.org/10.1021/acs.nanolett.9b01555.

[49] Xin Wu, Haiyan Zhao, Jiayun Pei, Dong Yan, Joining of graphene flakes by low energy N ion beam irradiation, Applied Physics Letters 110 (13) (2017) 133102, https://doi.org/10.1063/1.4979166.

[50] Edson P. Bellido, Jorge M. Seminario, Molecular dynamics simulations of ion-bombarded graphene, Journal of Physical Chemistry C 116 (3) (2012) 4044, https://doi.org/10.1021/jp208049t.

[51] Shijun Zhao, Jianming Xue, Tuning the band gap of bilayer graphene by ion implantation: insight from computational studies, Physical Review B 86 (16) (2012) 165428, https://doi.org/10.1103/PhysRevB.86.165428.

[52] ichul Yoon, Ali Rahnamoun, Jacob L. Swett, Vighter Iberi, David A. Cullen, Ivan V. Vlassiouk, Alex Belianinov, Stephen Jesse, Xiahan Sang, Olga S. Ovchinnikova, Adam J. Rondinone, Raymond R. Unocic, Adri C.T. Van Duin, Atomistic-scale simulations of defect formation in graphene under noble gas ion irradiation, ACS Nano 10 (9) (2016) 8376–8384, https://doi.org/10.1021/acsnano.6b03036.

[53] Martin Kalbac, Ossi Lehtinen, Arkady V. Krasheninnikov, Juhani Keinonen, Ion-irradiation-induced defects in isotopically-labeled two layered graphene: enhanced in-situ annealing of the damage, Advanced Materials 25 (2013) 1004–1009.

[54] ebastian Standop, Ossi Lehtinen, Charlotte Herbig, Georgia Lewes-Malandrakis, Fabian Craes, Jani Kotakoski, Thomas Michely, Arkady V. Krasheninnikov, Carsten Busse, Ion impacts on graphene/Ir(111): interface channeling, vacancy funnels, and a nanomesh, Nano Letters 13 (5) (2013) 1948–1955, https://doi.org/10.1021/nl304659n.

[55] Weisen Li, Li Liang, Shijun Zhao, Shuo Zhang, Jianming Xue, Fabrication of nanopores in a graphene sheet with heavy ions: a molecular dynamics study, Journal of Applied Physics 114 (23) (2013) 234304, https://doi.org/10.1063/1.4837657.

[56] Shihao Su, Jianming Xue, Facile fabrication of subnanopores in graphene under ion irradiation: molecular dynamics simulations, ACS Applied Materials & Interfaces 13 (10) (2021) 12366–12374, https://doi.org/10.1021/acsami.0c22288.

[57] O. Lehtinen, E. Dumur, J. Kotakoski, A.V. Krasheninnikov, K. Nordlund, J. Keinonen, Production of defects in hexagonal boron nitride monolayer under ion irradiation, Nuclear Instruments and Methods in Physics Research, Section B: Beam Interactions with Materials and Atoms 10 (11) (2011) 1327–1331, https://doi.org/10.1016/j.nimb.2010.11.027.

[58] Kelly A. Stephani, Iain D. Boyd, Molecular dynamics modeling of defect formation in many-layer hexagonal boron nitride, Nuclear Instruments and Methods in Physics Research, Section B: Beam Interactions with Materials and Atoms 365 (2015) 235–239, https://doi.org/10.1016/j.nimb.2015.10.020.

[59] Saransh Gupta, Prakash Periasamy, Badri Narayanan, Defect dynamics in two-dimensional black phosphorus under argon ion irradiation, Nanoscale 13 (18) (2021) 8575–8590, https://doi.org/10.1039/D1NR00567G.

[60] Leonardo C. Campos, Vitor R. Manfrinato, Javier D. Sanchez-Yamagishi, Jing Kong, Pablo Jarillo-Herrero, Anisotropic etching and nanoribbon formation in single-layer graphene, Nano Letters 9 (7) (2009) 2600–2604, https://doi.org/10.1021/nl900811r.

[61] T.J. Booth, Filippo Pizzocchero, Henrik Andersen, Thomas W. Hansen, Jakob B. Wagner, Joerg R. Jinschek, Rafal E. Dunin-Borkowski, Ole Hansen, P. Bggild, Discrete dynamics of nanoparticle channelling in suspended graphene, Nano Letters 11 (2011) 2689–2692, https://doi.org/10.1021/nl200928k.

[62] Xiao Yi Liu, Feng Chao Wang, Harold S. Park, Heng An Wu, Defecting controllability of bombarding graphene with different energetic atoms via reactive force field model, Journal of Applied Physics 114 (5) (2013) 054313, https://doi.org/10.1063/1.4817790.

[63] Zhitong Bai, Lin Zhang, Ling Liu, Bombarding graphene with oxygen ions: combining effects of incident angle and ion energy to control defect generation, Journal of Physical Chemistry C 119 (47) (2015) 26793–26802, https://doi.org/10.1021/acs.jpcc.5b09620.

[64] E.H. Ahlgren, J. Kotakoski, A.V. Krasheninnikov, Atomistic simulations of the implantation of low-energy boron and nitrogen ions into graphene, Physical Review B 83 (11) (2011) 115424, https://doi.org/10.1103/PhysRevB.83.115424.

[65] U. Bangert, Ion implantation of graphene – toward IC compatible technologies, Nano Letters 13 (10) (2013) 4902–4907, https://doi.org/10.1021/nl402812y.

[66] Cory D. Cress, Scott W. Schmucker, Adam L. Friedman, Pratibha Dev, James C. Culbertson, Joseph W. Lyding, Jeremy T. Robinson, Nitrogen-doped graphene and twisted bilayer graphene via hyperthermal ion implantation with depth control, ACS Nano 10 (3) (2016) 3714–3722, https://doi.org/10.1021/acsnano.6b00252.

[67] Yijun Xu, Kun Zhang, Christoph Brusewitz, Xuemei Wu, Hans Christian Hofsass, Investigation of the effect of low energy ion beam irradiation on mono-layer graphene, AIP Advances 3 (7) (2013) 072120, https://doi.org/10.1063/1.4816715.

[68] Mahdi Ghorbani-Asl, Silvan Kretschmer, Douglas E. Spearot, Arkady V. Krasheninnikov, Two-dimensional MoS2 under ion irradiation: from controlled defect production to electronic structure engineering, 2D Materials 4 (2) (2017) 25078, https://doi.org/10.1088/2053-1583/aa6b17.

[69] J.A. Stewart, D.E. Spearot, Atomistic simulations of nanoindentation on the basal plane of crystalline molybdenum disulfide (MoS2), Modelling and Simulation in Materials Science and Engineering 21 (4) (2013) 045003, https://doi.org/10.1088/0965-0393/21/4/045003.

[70] Jin-Wu Jiang, Harold S. Park, Timon Rabczuk, Molecular dynamics simulations of single-layer molybdenum disulphide (MoS2): Stillinger-Weber parametrization, mechanical properties, and thermal conductivity, Journal of Applied Physics 114 (6) (2013) 064307, https://doi.org/10.1063/1.48184140.

[71] Hannu-Pekka Komsa, Jani Kotakoski, Simon Kurasch, Ossi Lehtinen, Ute Kaiser, Arkady V. Krasheninnikov, Two-dimensional transition metal dichalcogenides under electron irradiation: defect production and doping, Physical Review Letters 109 (3) (2012) 035503, https://doi.org/10.1103/PhysRevLett.109.035503.

[72] Kedi Yin, Shengxi Huang, Xiaofei Chen, Xinwei Wang, Jing Kong, Yan Chen, Jianming Xue, Generating sub-nanometer pores in SingleLayer MoS2 by heavy-ion bombardment for gas separation: a theoretical perspective, ACS Applied Materials and Interfaces 10 (34) (2018) 28909–28917, https://doi.org/10.1021/acsami.8b10569.

[73] Weisen Li, Xinwei Wang, Xitong Zhang, Shijun Zhao, Huiling Duan, Jianming Xue, Mechanism of the defect formation in supported graphene by energetic heavy ion irradiation: the substrate effect, Scientific Reports 5 (2015) 9935, https://doi.org/10.1038/srep09935.

[74] Wolfhard Mller, W. Eckstein, Tridyn – a trim simulation code including dynamic composition changes, Nuclear Instruments and Methods in Physics Research B2 (2) (1984) 814–818.

[75] W. Mller, W. Eckstein, J.P. Biersack, Tridyn – binary collision simulation of atomic-collisions and dynamic composition changes in solids, Computer Physics Communications 51 (1988) 355–368.

[76] Shijun Zhao, Jianming Xue, Yugang Wang, Sha Yan, Effect of SiO2 substrate on the irradiation-assisted manipulation of supported graphene: a molecular dynamics study, Nanotechnology 23 (28) (2012) 285703, https://doi.org/10.1088/0957-4484/23/28/285703.

References 295

[77] D. Fox, Y.B. Zhou, A. O'Neill, S. Kumar, J.J. Wang, J.N. Coleman, G.S. Duesberg, J.F. Donegan, H.Z. Zhang, Helium ion microscopy of graphene: beam damage, image quality and edge contrast, Nanotechnology 24 (33) (2013) 335702, https://doi.org/10.1088/0957-4484/24/33/335702.

[78] Beidou Guo, Qian Liu, Erdan Chen, Hewei Zhu, Liang Fang, Jian Ru Gong, Controllable N-doping of graphene, Nano Letters 10 (12) (2010) 4975–4980, https://doi.org/10.1021/nl103079j.

[79] Grant Buchowicz, Peter R. Stone, Jeremy T. Robinson, Cory D. Cress, Jeffrey W. Beeman, Oscar D. Dubon, Correlation between structure and electrical transport in ion-irradiated graphene grown on Cu foils, Applied Physics Letters 98 (3) (2011) 032102, https://doi.org/10.1063/1.3536529.

[80] Daniel S. Fox, Yangbo Zhou, Pierce Maguire, Arlene ONeill, Cormac Coilein, Riley Gatensby, Alexey M. Glushenkov, Tao Tao, Georg S. Dues- berg, Igor V. Shvets, Mohamed Abid, Mourad Abid, Han-Chun Wu, Ying Chen, Jonathan N. Coleman, John F. Donegan, Hongzhou Zhang, Nanopatterning and electrical tuning of MoS2 layers with a subnanometer helium ion beam, Nano Letters 15 (8) (2015) 5307–5313, https://doi.org/10.1021/acs.nanolett.5b01673.

[81] Quan Ma, Patrick M. Odenthal, John Mann, Duy Le, Chen S. Wang, Yeming Zhu, Tianyang Chen, Dezheng Sun, Yamaguchi F, Tai Tran, Michelle Wurch, Jessica L. McKinley, Jonathan Wyrick, KatieMarie Magnone, Tony, Heinz Bartels, Talat S. Rahman, Roland Kawakami, Ludwig, Controlled argon beam-induced desulfurization of monolayer molybdenum disulfide, Journal of Physics: Condensed Matter 25 (25) (2013) 252201, https://doi.org/10.1088/0953-8984/25/25/252201.

[82] Cheng Lun Wu, Hsiang Ting Lin, Hsuan An Chen, Shih Yen Lin, Min Hsiung Shih, Chun Wei Pao, Defect formation and modulation during patterning supported graphene sheets using focused ion beams, Materials Today Communications 17 (2018) 60–68, https://doi.org/10.1016/j.mtcomm.2018.08.006.

[83] Huanyao Cun, Marcella Iannuzzi, Adrian Hemmi, Jrg Osterwalder, Thomas Greber, Ar implantation beneath graphene on Ru(0001): Nanotents and can-opener effect, Surface Science 634 (2015) 95–102, https://doi.org/10.1016/j.susc.2014.11.004.

[84] Yudi Rosandi, Herbert M. Urbassek, Subsurface and interface channeling of keV ions in graphene/SiC, Nuclear Instruments and Methods in Physics Research Section B: Beam Interactions with Materials and Atoms 340 (2014) 5–10, https://doi.org/10.1016/j.nimb.2014.07.031.

[85] Huanyao Cun, Marcella Iannuzzi, Adrian Hemmi, Jrg Osterwalder, T. Greber, Two-nanometer voids in single-layer hexagonal boron nitride: formation via the "Can-opener" effect and annihilation by self-healing, ACS Nano 8 (7) (2014) 7423–7431, https://doi.org/10.1021/nn502645w.

[86] Sergiy Bubin, Bin Wang, Sokrates Pantelides, Klmn Varga, Simulation of high-energy ion collisions with graphene fragments, Physical Review B 85 (23) (2012) 235435, https://doi.org/10.1103/PhysRevB.85.235435.

[87] Shijun Zhao, Wei Kang, Jianming Xue, Xitong Zhang, Ping Zhang, Comparison of electronic energy loss in graphene and BN sheet by means of time-dependent density functional theory, Journal of Physics: Condensed Matter 27 (2) (2015) 025401, https://doi.org/10.1088/0953-8984/27/2/025401.

[88] Zhi Wang, Shu-Shen Li, Lin-Wang Wang, Efficient real-time time-dependent density functional theory method and its application to a collision of an ion with a 2D material, Physical Review Letters 114 (6) (2015) 063004, https://doi.org/10.1103/PhysRevLett.114.063004.

[89] Yoshiyuki Miyamoto, Hong Zhang, Calculating interaction between a highly charged high-speed ion and a solid surface, Physical Review B 77 (4) (2008) 045433, https://doi.org/10.1103/PhysRevB.77.045433.

[90] Hong Zhang, Yoshiyuki Miyamoto, A. Rubio, Abinitioo simulation of helium-ion microscopy images: the case of suspended graphene, Physical Review Letters 109 (26) (2012) 265505, https://doi.org/10.1103/PhysRevLett.109.265505.

[91] H. Vazquez, E.H. Ahlgren, O. Ochedowski, A.A Leino, R. Mirzayev, R. Kozubek, H. Lebius, M. Karluic, M. Jakic, A.V. Krasheninnikov, J. Kotakoski, M. Schleberger, K. Nordlund, F. Djurabekova, Creating nanoporous graphene with Swift heavy ions, Carbon 114 (2017) 511–518, https://doi.org/10.1016/j.carbon.2016.12.015.

[92] Nadezhda Aleksandrovna Nebogatikova, Irina V. Antonova, Sergey V. Erohin, Dmitry G. Kvashnin, Andrzej Olejniczak, V.A. Volodin, A.V. Skuratov, Arkady V. Krasheninnikov, P.B. Sorokin, Leonid A. Chernozatonskii, Nanostructuring few-layer graphene films with Swift heavy ions for electronic application: tuning of electronic and transport properties, Nanoscale 10 (30) (2018) 14499–14509, https://doi.org/10.1039/C8NR03062F.

[93] Shijun Zhao, Jianming Xue, Modification of graphene supported on SiO2 substrate with Swift heavy ions from atomistic simulation point, Carbon 93 (2015) 169–179, https://doi.org/10.1016/j.carbon.2015.05.047.

[94] L. Tapaszt, G. Dobrik, P. Nemes-Incze, G. Vertesy, Ph. Lambin, L.P. Bir, Tuning the electronic structure of graphene by ion irradiation, Physical Review B 78 (2008) 233407.

[95] M. Ugeda, I. Brihuega, Fanny Hiebel, Pierre Mallet, Jean-Yves Veuillen, Jos Gmez-Rodrguez, Flix Yndurin, Electronic and structural characterization of divacancies in irradiated graphene, Physical Review B 85 (12) (2012) 121402, https://doi.org/10.1103/PhysRevB.85.121402.

[96] M.M. Lucchese, F. Stavale, E.H. Martins Ferreira, C. Vilani, M.V.O. Moutinho, Rodrigo B. Capaz, C.a. Achete, A. Jorio, Quantifying ion-induced defects and Raman relaxation length in graphene, Carbon 48 (5) (2010) 1592–1597, https://doi.org/10.1016/j.carbon.2009.12.057.

[97] Axel Eckmann, Alexandre Felten, Artem Mishchenko, Liam Britnell, Ralph Krupke, K.S. Novoselov, C. Casiraghi, Probing the nature of defects in graphene by Raman spectroscopy, Nano Letters 12 (8) (2012) 39253930, https://doi.org/10.1021/nl300901a.

[98] Kazunori Fujisawa, Bruno R. Carvalho, Tianyi Zhang, Nstor Perea-Lpez, Zhong Lin, Victor Carozo, Srgio L.L.M. Ramos, Ethan Kahn, Adam Bolotsky, He Liu, Ana Laura Elas, Mauricio Terrones, Quantification and healing of defects in atomically thin molybdenum disulfide: beyond the controlled creation of atomic defects, ACS Nano (2021), https://doi.org/10.1021/acsnano.0c10897.

[99] R.H. Telling, C.P. Ewels, A.A. El-Barbary, M.I. Heggie, Wigner defects bridge the graphite gap, Nature Materials 2 (2003) 333–337, https://doi.org/10.1038/nmat876.

[100] J.S. Williams, Ion implantation of semiconductors, Materials Science and Engineering: A 253 (1) (1998) 8–15, https://doi.org/10.1016/S0921-5093(98)00705-9.

[101] Toma Susi, Trevor P. Hardcastle, Hans Hofsss, Andreas Mittelberger, Timothy J. Pennycook, Clemens Mangler, Rik Drummond-Brydson, Andrew J. Scott, Jannik C. Meyer, Jani Kotakoski, Single-atom spectroscopy of phosphorus dopants implanted into graphene, 2D Materials 4 (2) (2017) 021013, https://doi.org/10.1088/2053-1583/aa5e78.

[102] W. Zhao, O. Hfert, K. Gotterbarm, J.F. Zhu, C. Papp, H.-P. Steinrck, Production of nitrogen-doped graphene by low-energy nitrogen implantation, The Journal of Physical Chemistry C 116 (8) (2012) 50625066, https://doi.org/10.1021/jp209927m.

[103] hilip Willke, Julian A. Amani, Anna Sinterhauf, Sangeeta Thakur, Thomas Kotzott, Thomas Druga, Steffen Weikert, Kalobaran Maiti, Hans Hofsss, Martin Wenderoth, Doping of graphene by low-energy ion beam implantation: structural, electronic, and transport properties, Nano Letters 15 (8) (2015) 5110–5115, https://doi.org/10.1021/acs.nanolett.5b01280.

[104] Hongtao Wang, Qingxiao Wang, Yingchun Cheng, Kun Li, Yingbang Yao, Qiang Zhang, Cezhou Dong, Peng Wang, U. Schwingenschlgl, Wei Yang, X.X. Zhang, Doping monolayer graphene with single atom substitutions, Nano Letters 12 (1) (2012) 141–144, https://doi.org/10.1021/nl2031629.

[105] Gaurav Nanda, Srijit Goswami, Kenji Watanabe, Takashi Taniguchi, Paul F.A. Alkemade, Defect control and n-doping of encapsulated graphene by helium-ion-beam irradiation, Nano Letters 15 (6) (2015) 40064012, https://doi.org/10.1021/acs.nanolett.5b00939.

[106] Slawomir Prucnal, Arsalan Hashemi, Mahdi Ghorbani-Asl, Ren Hbner, Juanmei Duan, Yidan Wei, Divanshu Sharma, Dietrich R.T. Zahn, Ren Ziegenrcker, Ulrich Kentsch, Arkady V. Krasheninnikov, Manfred Helm, Shengqiang Zhou, Chlorine doping of MoSe2 flakes by ion implantation, Nanoscale 13 (11) (2021) 5834–5846, https://doi.org/10.1039/D0NR08935D.

[107] Yu-Chuan Lin, Chenze Liu, Yiling Yu, Eva Zarkadoula, Mina Yoon, Alexander A. Puretzky, Liangbo Liang, Xiangru Kong, Yiyi Gu, Alex Strasser, Harry M. Meyer, Matthias Lorenz, Matthew F. Chisholm, Ilia N. Ivanov, Christopher M. Rouleau, Gerd Duscher, Kai Xiao, David B. Geohegan, Low energy implantation into transition-metal dichalcogenide monolayers to form Janus structures, ACS Nano 14 (4) (2020) 3896–3906, https://doi.org/10.1021/acsnano.9b10196.

[108] Liam H. Isherwood, Zachariah Hennighausen, Seok-Kyun Son, F. Ben Spencer, Paul T. Wady, Samir M. Shubeita, Swastik Kar, Cinzia Casiraghi, Aliaksandr Baidak, The influence of crystal thickness and interlayer interactions on the properties of heavy ion irradiated MoS 2, 2D Materials 7 (3) (2020) 035011, https://doi.org/10.1088/2053-1583/ab817b.

[109] S. Mathew, T.K. Chan, D. Zhan, K. Gopinadhan, a.-R. Barman, M.B.H. Breese, S. Dhar, Z.X. Shen, T. Venkatesan, John T.L. Thong, The effect of layer number and substrate on the stability of graphene under MeV proton beam irradiation, Carbon 49 (5) (2011) 1720–1726, https://doi.org/10.1016/j.carbon.2010.12.057.

[110] R. Nair, I.-L. Tsai, M. Sepioni, O. Lehtinen, J. Keinonen, Arkady V. Krasheninnikov, a.H. Castro Neto, M.I. Katsnelson, a.K. Geim, I.V. Grigorieva, Dual origin of defect magnetism in graphene and its reversible switching by molecular doping, Nature Communications 4 (1) (2013) 2010, https://doi.org/10.1038/ncomms3010.

[111] Gwangseok Yang, Byung Jae Kim, Kyeounghak Kim, Jeong Woo Han, Jihyun Kim, Energy and dose dependence of proton-irradiation damage in graphene, RSC Advances 5 (40) (2015) 31861–31865, https://doi.org/10.1039/c5ra03551a.

[112] Oleg Yazyev, Lothar Helm, Defect-induced magnetism in graphene, Physical Review B 75 (12) (2007) 125408, https://doi.org/10.1103/PhysRevB.75.125408.

[113] Oleg Yazyev, Emergence of magnetism in graphene materials and nanostructures, Reports on Progress in Physics 73 (5) (2010) 056501, https://doi.org/10.1088/0034-4885/73/5/056501.

[114] Tae-Young Kim, Kyungjune Cho, Woanseo Park, Juhun Park, Younggul Song, Seunghun Hong, Woong-Ki Hong, Takhee Lee, Irradiation effects of high-energy proton beams on MoS2 field effect transistors, ACS Nano 8 (3) (2014) 2774–2781, https://doi.org/10.1021/nn4064924.

[115] S. Mathew, K. Gopinadhan, T.K. Chan, X.J. Yu, D. Zhan, L. Cao, a. Rusydi, M.B.H. Breese, S. Dhar, Z.X. Shen, T. Venkatesan, John T.L. Thong, Magnetism in MoS2 induced by proton irradiation, Applied Physics Letters 101 (10) (2012) 102103, https://doi.org/10.1063/1.4750237.

[116] S. Akcltekin, H. Bukowska, T. Peters, O. Osmani, I. Monnet, I. Alzaher, B. Ban D'Etat, H. Lebius, M. Schleberger, Unzipping and folding of graphene by Swift heavy ions, Applied Physics Letters 98 (10) (2011) 103103, https://doi.org/10.1063/1.3559619.

[117] Oliver Ochedowski, Modification of 2D-Materials by Swift Heavy Ion Irradiation, PhD thesis, Universitt Duisburg-Essen, 2014.

[118] Oliver Ochedowski, Hanna Bukowska, Victor M. Freire Soler, Lara Brkers, Brigitte Ban-d'Etat, Henning Lebius, Marika Schleberger, Folding two dimensional crystals by Swift heavy ion irradiation, Nuclear Instruments and Methods in Physics Research Section B: Beam Interactions with Materials and Atoms 340 (2014) 39–43, https://doi.org/10.1016/j.nimb.2014.07.037.

[119] Jin Wu Jiang, The buckling of single-layer MoS2 under uniaxial compression, Nanotechnology 25 (35) (2014) 355402, https://doi.org/10.1088/0957-4484/25/35/355402.

[120] Lukas Madauss, Oliver Ochedowski, Henning Lebius, Brigitte Ban-D'Etat, Carl H. Naylor, A.T. Charlie Johnson, Jani Kotakoski, Marika Schleberger, Defect engineering of single- and few-layer MoS2 by Swift heavy ion irradiation, 2D Materials 4 (1) (2017) 015034, https://doi.org/10.1088/2053--1583/4/1/015034.

[121] Richard A. Wilhelm, Ayman S. El-Said, Franciszek Krok, René Heller, Elisabeth Gruber, Friedrich Aumayr, Stefan Facsko, Highly charged ion induced nanostructures at surfaces by strong electronic excitations, Progress in Surface Science 90 (3) (2015) 377–395, https://doi.org/10.1016/j.progsurf.2015.06.001.

[122] Richard A. Wilhelm, Elisabeth Gruber, Janine Schwestka, Roland Kozubek, Teresa I. Madeira, Jos P. Marques, Jacek Kobus, Arkady V. Krasheninnikov, Marika Schleberger, Friedrich Aumayr, Interatomic Coulombic decay: the mechanism for rapid deexcitation of hollow atoms, Physical Review Letters 119 (10) (2017) 103401, https://doi.org/10.1103/PhysRevLett.119.103401.

[123] Richard A. Wilhelm, Pedro L. Grande, Unraveling energy loss processes of low energy heavy ions in 2D materials, Communications Physics 2 (1) (2019) 89, https://doi.org/10.1038/s42005-019-0188-7.

[124] Janine Schwestka, Heena Inani, Mukesh Tripathi, Anna Niggas, Niall McEvoy, Florian Libisch, Friedrich Aumayr, Jani Kotakoski, Richard A. Wilhelm, Atomic-scale carving of nanopores into a van der Waals heterostructure with slow highly charged ions, ACS Nano 14 (8) (2020) 10536–10543, https://doi.org/10.1021/acsnano.0c04476.

[125] J. Hopster, R. Kozubek, J. Krmer, V. Sokolovsky, M. Schleberger, Ultra-thin MoS2 irradiated with highly charged ions, Nuclear Instruments and Methods in Physics Research Section B: Beam Interactions with Materials and Atoms 317 (2013) 165–169, https://doi.org/10.1016/j.nimb.2013.02.038.

[126] Elisabeth Gruber, Richard A. Wilhelm, Rmi Ptuya, Valerie Smejkal, Roland Kozubek, Anke Hierzenberger, Bernhard C. Bayer, Iigo Aldazabal, Andrey K. Kazansky, Florian Libisch, Arkady V. Krasheninnikov, Marika Schleberger, Stefan Facsko, Andrei G. Borisov, Andrs Arnau, Friedrich Aumayr, Ultrafast electronic response of graphene to a strong and localized electric field, Nature Communications 7 (1) (2016) 13948, https://doi.org/10.1038/ncomms13948.

[127] Roland Kozubek, Philipp Ernst, Charlotte Herbig, Thomas Michely, Marika Schleberger, Fabrication of defective single layers of hexagonal boron nitride on various supports for potential applications in catalysis and DNA sequencing, ACS Applied Nano Materials 1 (8) (2018) 3765–3773, https://doi.org/10.1021/acsanm.8b00903.

[128] Roland Kozubek, Mukesh Tripathi, Mahdi Ghorbani-Asl, Silvan Kretschmer, Lukas Madau, Erik Pollmann, Maria O'Brien, Niall McEvoy, Ursula Ludacka, Toma Susi, Georg S. Duesberg, Richard A. Wilhelm, Arkady V.

Krasheninnikov, Jani Kotakoski, Marika Schleberger, Perforating freestanding molybdenum disulfide monolayers with highly charged ions, Journal of Physical Chemistry Letters 10 (5) (2019) 904–910, https://doi.org/10.1021/acs.jpclett.8b03666.

[129] S. Creutzburg, J. Schwestka, A. Niggas, H. Inani, M. Tripathi, A. George, R. Heller, R. Kozubek, L. Madau, N. McEvoy, S. Facsko, J. Kotakoski, M. Schleberger, A. Turchanin, P.L. Grande, F. Aumayr, R.A. Wilhelm, Vanishing influence of the band gap on the charge exchange of slow highly charged ions in freestanding single-layer MoS_2, Physical Review B 102 (2020) 045408, https://doi.org/10.1103/PhysRevB.102.045408.

[130] Max C. Lemme, David C. Bell, James R. Williams, Lewis A. Stern, Britton W.H. Baugher, Pablo Jarillo-Herrero, Charles M. Marcus, Etching of graphene devices with a helium ion beam, ACS Nano 3 (9) (2009) 2674–2676, https://doi.org/10.1021/nn900744z.

[131] D.C. Bell, M.C. Lemme, L.A. Stern, J.R. Williams, C.M. Marcus, Precision cutting and patterning of graphene with helium ions, Nanotechnology 20 (45) (2009) 455301, https://doi.org/10.1088/0957-4484/20/45/455301.

[132] Y. Naitou, T. Iijima, S. Ogawa, Direct nano-patterning of graphene with helium ion beams, Applied Physics Letters 106 (3) (2015) 033103, https://doi.org/10.1063/1.4906415.

[133] Gaurav Nanda, Gregor Hlawacek, Srijit Goswami, Kenji Watanabe, Takashi Taniguchi, Paul F.A. Alkemade, Electronic transport in helium-ion-beam etched encapsulated graphene nanoribbons, Carbon 119 (2017) 419–425, https://doi.org/10.1016/j.carbon.2017.04.062.

[134] Marek Edward Schmidt, Takuya Iwasaki, Manoharan Muruganathan, Mayeesha Haque, Huynh Van Ngoc, Shinichi Ogawa, Hiroshi Mizuta, Structurally controlled large-area 10 nm pitch graphene nanomesh by focused helium ion beam milling, ACS Applied Materials & Interfaces 10 (2018) 10362–10368, https://doi.org/10.1021/acsami.8b00427.

[135] Gregor Hlawacek, Vasilisa Veligura, Raoul van Gastel, Bene Poelsema, Helium ion microscopy, Journal of Vacuum Science & Technology B 32 (2) (2014) 020801, https://doi.org/10.1116/1.4863676.

[136] J. Klein, M. Lorke, M. Florian, F. Sigger, L. Sigl, S. Rey, J. Wierzbowski, J. Cerne, K. Mller, E. Mitterreiter, P. Zimmermann, T. Taniguchi, K. Watanabe, U. Wurstbauer, M. Kaniber, M. Knap, R. Schmidt, J.J. Finley, A.W. Holleitner, Site-selectively generated photon emitters in monolayer MoS2 via local helium ion irradiation, Nature Communications 10 (1) (2019) 2755, https://doi.org/10.1038/s41467-019-10632-z.

[137] Vighter Iberi, Anton V. Ievlev, Ivan Vlassiouk, Stephen Jesse, Sergei V. Kalinin, David C. Joy, Adam J. Rondinone, Alex Belianinov, Olga S. Ovchinnikova, Graphene engineering by neon ion beams, Nanotechnology 27 (12) (2016) 125302, https://doi.org/10.1088/0957-4484/27/12/125302.

[138] Jani Kotakoski, Christian Brand, Yigal Lilach, Ori Cheshnovsky, Clemens Mangler, Markus Arndt, Jannik C. Meyer, Toward two-dimensional all-carbon heterostructures via ion beam patterning of single-layer graphene, Nano Letters 15 (9) (2015) 5944–5949, https://doi.org/10.1021/acs.nanolett.5b02063.

[139] Xin Wu, Haiyan Zhao, Dong Yan, Jiayun Pei, Investigation on gallium ions impacting monolayer graphene, AIP Advances 5 (6) (2015) 067171, https://doi.org/10.1063/1.4923395.

[140] Jothi Priyanka Thiruraman, Kazunori Fujisawa, Gopinath Danda, Paul Masih Das, Tianyi Zhang, Adam Bolotsky, Nstor Perea-Lpez, Adrien Nicola, Patrick Senet, Mauricio Terrones, Marija Drndi, Angstrom-size defect creation and ionic transport through pores in single-layer MoS2, Nano Letters 18 (3) (2018) 1651–1659, https://doi.org/10.1021/acs.nanolett.7b04526.

[141] Daehee Seol, Songkil Kim, Woo-Sung Jang, Yeongrok Jin, Seunghun Kang, Sera Kim, Dongyeun Won, Chanwoo Lee, Young-Min Kim, Jaekwang Lee, Heejun Yang, Mun Seok Jeong, Alex Belianinov, Alexander Tselev, Suhas Somnath, Christopher R. Smith, Olga S. Ovchinnikova, Nina Balke, Yunseok Kim, Selective patterning of out-of-plane piezoelectricity in MoTe2 via focused ion beam, Nano Energy 79 (2021) 105451, https://doi.org/10.1016/j.nanoen.2020.105451, issn: 2211-2855.

[142] Hu Li, Lakshya Daukiya, Soumyajyoti Haldar, Andreas Lindblad, Biplab Sanyal, Olle Eriksson, Dominique Aubel, Samar Hajjar-Garreau, Laurent Simon, Klaus Leifer, Site-selective local fluorination of graphene induced by focused ion beam irradiation, Scientific Reports 6 (1) (2016) 19719, https://doi.org/10.1038/srep19719.

[143] Chen Lin, Keivan Davami, Yijie Jiang, John Cortes, Michael Munther, Mehrdad Shaygan, Hessam Ghassemi, Jeremy T. Robinson, Kevin T. Turner, Igor Bargatin, Enhancing the stiffness of vertical graphene sheets through ion beam irradiation and fluorination, Nanotechnology 28 (29) (2017) 295701, https://doi.org/10.1088/1361-6528/aa75ac.

[144] Quan Ma, Miguel Isarraraz, Chen S. Wang, Edwin Preciado, Velveth Klee, Sarah Bobek, Koichi Yamaguchi, Emily Li, Patrick Michael odenthal, Ariana Nguyen, David Barroso, Dezheng Sun, Gretel von Son Palacio,

Michael Gomez, Andrew Nguyen, Duy Le, Greg Pawin, John Mann, Tony F. Heinz, Talat Shahnaz Rahman, Ludwig Bartels, Postgrowth tuning of the bandgap of single-layer molybdenum disulfide films by sulfur/selenium exchange, ACS Nano 8 (5) (2014) 4672–4677, https://doi.org/10.1021/nn5004327.

[145] Simone Bertolazzi, Sara Bonacchi, Guangjun Nan, Anton Pershin, David Beljonne, Paolo Samor, Engineering chemically active defects in monolayer MoS2 transistors via ion-beam irradiation and their healing via vapor deposition of alkanethiols, Advanced Materials 29 (18) (2017) 1606760, https://doi.org/10.1002/adma.201606760.

[146] Salah Elafandi, Robert Christiansen, Nurul Azam, Max Cichon, Minseo Park, Michael C. Hamilton, Masoud Mahjouri-Samani, Monolayer 2D quantum materials subjected to gamma irradiation in high-vacuum for nuclear and space applications, Applied Physics Letters 116 (21) (2020) 213105, https://doi.org/10.1063/5.0006919.

[147] O. Ochedowski, K. Marinov, G. Wilbs, G. Keller, N. Scheuschner, D. Severin, M. Bender, J. Maultzsch, F.J. Tegude, M. Schleberger, Radiation hardness of graphene and MoS2 field effect devices against Swift heavy ion irradiation, Journal of Applied Physics 113 (21) (2013) 214306, https://doi.org/10.1063/1.4808460.

[148] Tobias Vogl, Kabilan Sripathy, Ankur Sharma, Prithvi Reddy, James Sullivan, Joshua R. Machacek, Linglong Zhang, Fouad Karouta, C. Ben Buchler, Marcus W. Doherty, Yuerui Lu, Ping Koy Lam, Radiation tolerance of two-dimensional material-based devices for space applications, Nature Communications 10 (2019) 1, https://doi.org/10.1038/s41467-019-09219-5.

[149] Andrew J. Arnold, Tan Shi, Igor Jovanovic, Saptarshi Das, Extraordinary radiation hardness of atomically thin MoS2, ACS Applied Materials and Interfaces 11 (8) (2019) 8391–8399, https://doi.org/10.1021/acsami.8b18659.

[150] Yifan Zhang, Xiaofei Chen, Heshen Wang, Junfeng Dai, Jianming Xue, Xun Guo, Electronic properties of multilayer MoS2 field effect transistor with unique irradiation resistance, The Journal of Physical Chemistry C 125 (3) (2021) 2089–2096, https://doi.org/10.1021/acs.jpcc.0c09666.

[151] Mohnish Pandey, Filip A. Rasmussen, Korina Kuhar, Thomas Olsen, Karsten W. Jacobsen, Kristian S. Thygesen, Defect-tolerant monolayer transition metal dichalcogenides, Nano Letters 16 (4) (2016) 2234–2239, https://doi.org/10.1021/acs.nanolett.5b04513.

[152] Arkady V. Krasheninnikov, Are two-dimensional materials radiation tolerant?, Nanoscale Horizons 5 (11) (2020) 1447–1452, https://doi.org/10.1039/D0NH00465K.

[153] Wenqing Li, Xueying Zhan, Xianyin Song, Shuyao Si, Rui Chen, Jing Liu, Zhenxing Wang, Jun He, Xiangheng Xiao, A review of recent applications of ion beam techniques on nanomaterial surface modification: design of nanostructures and energy harvesting, Small 15 (31) (2019) 1901820, https://doi.org/10.1002/smll.201901820.

[154] Shu Nakaharai, Tomohiko Iijima, Shinichi Ogawa, Song Lin Li, Kazuhito Tsukagoshi, Shintaro Sato, Naoki Yokoyama, Current on-off operation of graphene transistor with dual gates and He ion irradiated channel, Physica Status Solidi (C) Current Topics in Solid State Physics 10 (11) (2013) 1608–1611, https://doi.org/10.1002/pssc.201300262.

[155] Nima Kalhor, Stuart A. Boden, Hiroshi Mizuta, Sub-10nm patterning by focused He-ion beam milling for fabrication of downscaled graphene nano devices, Microelectronic Engineering 114 (2014) 70–77, https://doi.org/10.1016/j.mee.2013.09.018.

[156] Philipp Valerius, Silvan Kretschmer, Boris V. Senkovskiy, Shilong Wu, Joshua Hall, Alexander Herman, Niels Ehlen, Mahdi Ghorbani-Asl, Alexander Grneis, Arkady V. Krasheninnikov, Thomas Michely, Reversible crystalline-to-amorphous phase transformation in monolayer MoS2 under grazing ion irradiation, 2D Materials 7 (2) (2020) 025005, https://doi.org/10.1088/2053-1583/ab5df4.

[157] Jakub Jadwiszczak, Darragh Keane, Pierce Maguire, Conor P. Cullen, Yangbo Zhou, Huading Song, Clive Downing, Daniel Fox, Niall McEvoy, Rui Zhu, Jun Xu, Georg S. Duesberg, Zhi Min Liao, John J. Boland, Hongzhou Zhang, MoS2 memtransistors fabricated by localized helium ion beam irradiation, ACS Nano 13 (12) (2019) 14262–14273, https://doi.org/10.1021/acsnano.9b07421.

[158] Zhihui Cheng, Hattan Abuzaid, Yifei Yu, Fan Zhang, Yanlong Li, Steven G. Noyce, Nicholas X. Williams, Yuh Chen Lin, James L. Doherty, Chenggang Tao, Linyou Cao, Aaron D. Franklin, Convergent ion beam alteration of 2D materials and metal-2D interfaces, 2D Materials 6 (3) (2019) 034005, https://doi.org/10.1088/2053-1583/ab1764.

[159] Kashif Shahzad, Kunpeng Jia, Chao Zhao, Dahai Wang, Muhammad Usman, Jun Luo, Effects of different ion irradiation on the contact resistance of Pd/Graphene contacts, Materials 12 (23) (2019) 3928, https://doi.org/10.3390/ma12233928.

[160] Yanran Liu, Zhibin Gao, Yang Tan, Feng Chen, Enhancement of out- of-plane charge transport in a vertically stacked two-dimensional heterostructure using point defects, ACS Nano 12 (2018) 1052910536, https://doi.org/10.1021/acsnano.8b06503.

[161] Michael G. Stanford, Pushpa Raj Pudasaini, Alex Belianinov, Nicholas Cross, Joo Hyon Noh, Michael R. Koehler, David G. Mandrus, Gerd Duscher, Adam J. Rondinone, Ilia N. Ivanov, T. Zac Ward, Philip D. Rack, Focused helium-ion beam irradiation effects on electrical transport properties of few-layer WSe2: enabling nanoscale direct write homo-junctions, Scientific Reports 6 (2016) 27276, https://doi.org/10.1038/srep27276.

[162] Yang Tan, Xiaobiao Liu, Zhiliang He, Yanran Liu, Mingwen Zhao, Han Zhang, Feng Chen, Tuning of interlayer coupling in large-area graphene/WSe2 van der Waals heterostructure via ion irradiation: optical evidences and photonic applications, ACS Photonics 4 (6) (2017) 1531–1538, https://doi.org/10.1021/acsphotonics.7b00296.

[163] L. Ma, Y. Tan, M. Ghorbani-Asl, R. Boettger, S. Kretschmer, S. Zhou, Z. Huang, A.V. Krasheninnikov, F. Chen, Tailoring the optical properties of atomically-thin WS2 via ion irradiation, Nanoscale 9 (31) (2017) 11027–11034, https://doi.org/10.1039/C7NR02025B.

[164] M. Fischer, J.M. Caridad, A. Sajid, S. Ghaderzadeh, M. Ghorbani-Asl, L. Gammelgaard, P. Bggild, K.S. Thygesen, A.V. Krasheninnikov, S. Xiao, M. Wubs, N. Stenger, Controlled generation of luminescent centers in hexagonal boron nitride by irradiation engineering, Science Advances 7 (8) (2021) eabe7138, https://doi.org/10.1126/sciadv.abe7138.

[165] M. Fischer, J.M. Caridad, A. Sajid, S. Ghaderzadeh, M. Ghorbani-Asl, L. Gammelgaard, P. Bggild, K.S. Thygesen, A.V. Krasheninnikov, S. Xiao, M. Wubs, N. Stenger, Controlled generation of luminescent centers in hexagonal boron nitride by irradiation engineering, Science Advances 7 (8) (2021), https://doi.org/10.1126/sciadv.abe7138.

[166] Sadegh Ghaderzadeh, Silvan Kretschmer, Mahdi Ghorbani-Asl, Gregor Hlawacek, Arkady V. Krasheninnikov, Atomistic simulations of defect production in monolayer and bulk hexagonal boron nitride under low- and high-fluence ion irradiation, Nanomaterials 11 (2021) 5, https://doi.org/10.3390/nano11051214.

[167] Hsu-Sheng Tsai, Ching-Hung Hsiao, Chia-Wei Chen, Hao Ouyang, Jenq-Horng Liang, Synthesis of nonepitaxial multilayer silicene assisted by ion implantation, Nanoscale 8 (18) (2016) 9488–9492, https://doi.org/10.1039/C6NR02274J.

[168] Hsu-Sheng Tsai, Jhe-Wei Liou, Icuk Setiyawati, Kuan-Rong Chiang, Chia-Wei Chen, Chi-Chong Chi, Yu-Lun Chueh, Hao Ouyang, Yu-Hui Tang, Wei-Yen Woon, Jenq-Horng Liang, Photoluminescence characteristics of multilayer HfSe2 synthesized on sapphire using ion implantation, Advanced Materials Interfaces 5 (8) (2018) 1701619, https://doi.org/10.1002/admi.201701619.

[169] Gang Wang, Miao Zhang, Su Liu, Xiaoming Xie, Guqiao Ding, Yongqiang Wang, Paul K. Chu, Heng Gao Wei Ren, Qinghong Yuan, Peihong Zhang, Xi Wang, Zengfeng Di, Synthesis of layer-tunable graphene: a combined kinetic implantation and thermal ejection approach, Advanced Functional Materials 25 (24) (2015) 3666–3675, https://doi.org/10.1002/adfm.201500981.

[170] Bishnupad Mohanty, Mahdi Ghorbani-Asl, Silvan Kretschmer, Arnab Ghosh, Puspendu Guha, Subhendu K. Panda, Bijayalaxmi Jena, Arkady V. Krashenin nikov, Bikash Kumar Jena, MoS2 quantum dots as efficient catalyst materials for the oxygen evolution reaction, ACS Catalysis 8 (3) (2018) 1683–1689, https://doi.org/10.1021/acscatal.7b03180.

[171] Bishnupad Mohanty, Yidan Wei, Mahdi Ghorbani-Asl, Arkady V. Krasheninnikov, Parasmani Rajput, Bikash Kumar Jena, Revealing the defect-dominated oxygen evolution activity of hematene, Journal of Materials Chemistry A 8 (2020) 6709–6716, https://doi.org/10.1039/D0TA00422G.

[172] Mahdi Ghorbani-Asl, Andrey N. Enyashin, Agnieszka Kuc, Gotthard Seifert, Thomas Heine, Defect-induced conductivity anisotropy in MoS_2 monolayers, Physical Review B 88 (24) (2013) 245440, https://doi.org/10.1103/PhysRevB.88.245440.

[173] Hong Li, Charlie Tsai, Ai Leen Koh, Lili Cai, Alex W. Contryman, Alex H. Fragapane, Jiheng Zhao, Hyun Soo Han, Hari C. Manoharan, Frank Abild- Pedersen, Jens K. Nrskov, Xiaolin Zheng, Activating and optimizing MoS2 basal planes for hydrogen evolution through the formation of strained sulphur vacancies, Nature Materials 15 (1) (2016) 48–53, https://doi.org/10.1038/nmat4465.

[174] Li Tao, Xidong Duan, Chen Wang, Xiangfeng Duan, Shuangyin Wang, Plasma-engineered MoS2 thin-film as an efficient electrocatalyst for hydrogen evolution reaction, Chemical Communications 51 (35) (2015) 7470–7473, https://doi.org/10.1039/C5CC01981H.

References

[175] Thomas G. Pedersen, Christian Flindt, Jesper Pedersen, Niels Asger Mortensen, Antti-Pekka Jauho, Kjeld Pedersen, Graphene antidot lattices: designed defects and spin qubits, Physical Review Letters 100 (13) (2008) 136804, https://doi.org/10.1103/PhysRevLett.100.136804.

[176] Bjarke S. Jessen, Lene Gammelgaard, Morten R. Thomsen, David M.A. Mackenzie, Joachim D. Thomsen, Jos M. Caridad, Emil Duegaard, Kenji Watanabe, Takashi Taniguchi, T.J. Booth, Thomas G. Pedersen, Antti-pekka Jauho, Peter Bggild, Lithographic band structure engineering of graphene, Nature Nanotechnology 14 (2019) 340–346, https://doi.org/10.1038/s41565-019-0376-3.

[177] Luda Wang, Michael S.H. Boutilier, Piran R. Kidambi, Doojoon Jang, Nicolas G. Hadjiconstantinou, Rohit Karnik, Fundamental transport mechanisms, fabrication and potential applications of nanoporous atomically thin membranes, Nature Nanotechnology 12 (6) (2017) 509–522, https://doi.org/10.1038/nnano.2017.72.

[178] Gopinath Danda, Marija Drndi, Two-dimensional nanopores and nanoporous membranes for ion and molecule transport, Current Opinion in Biotechnology 55 (2019) 124–133.

[179] Meni Wanunu, Nanopores: a journey towards DNA sequencing, Physics of Life Reviews 9 (2) (2012) 125–158, https://doi.org/10.1016/j.plrev.2012.05.010.

[180] Slaven Garaj, W. Hubbard, A. Reina, J. Kong, D. Branton, J.A. Golovchenko, Graphene as a subnanometre trans-electrode membrane, Nature 467 (7312) (2010) 190–193.

[181] Bala Murali Venkatesan, Rashid Bashir, Nanopore sensors for nucleic acid analysis, Nature Nanotechnology 6 (10) (2011) 615–624, https://doi.org/10.1038/nnano.2011.129.

[182] James Clarke, Hai-Chen Wu, Lakmal Jayasinghe, Alpesh Patel, Stuart Reid, Hagan Bayley, Continuous base identification for single-molecule nanopore DNA sequencing, Nature Nanotechnology 4 (4) (2009) 265.

[183] Sumedh P. Surwade, Sergei N. Smirnov, Ivan V. Vlassiouk, Raymond R. Unocic, Gabriel M. Veith, Sheng Dai, Shannon M. Mahurin, Water desalination using nanoporous single-layer graphene, Nature Nanotechnology 10 (5) (2015) 459–464, https://doi.org/10.1038/nnano.2015.37.

[184] Sadegh Ghaderzadeh, Vladimir Ladygin, Mahdi Ghorbani-Asl, Gregor Hlawacek, Marika Schleberger, Arkady V. Krasheninnikov, Freestanding and supported MoS2 monolayers under cluster irradiation: insights from molecular dynamics simulations, ACS Applied Materials & Interfaces 12 (33) (2020) 37454–37463, https://doi.org/10.1021/acsami.0c09255.

[185] M.C. Clochard, G. Melilli, G. Rizza, B. Madon, M. Alves, J.E. Wegrowe, M.E. Toimil-Molares, M. Christian, L. ortolani, R. Rizzoli, V. Morandi, V. Palermo, S. Bianco, F. Pirri, M. Sangermano, Large area fabrication of self-standing nanoporous graphene-on-PMMA substrate, Materials Letters 184 (2016) 47–51, https://doi.org/10.1016/j.matlet.2016.07.133.

[186] C. Herbig, T. Michely, Graphene: the ultimately thin sputtering shield, 2D Materials 3 (2) (2016) 025032, https://doi.org/10.1088/2053-1583/3/2/025032.

[187] Recep Zan, Q. Ramasse, Ursel Bangert, K.S. Novoselov, Graphene reknits its holes, Nano Letters 12 (8) (2012) 3936–3940, https://doi.org/10.1021/nl300985q.

10

Tailoring defects in 2D materials for electrocatalysis

Leping Yang, Yuchi Wan, and Ruitao Lv
State Key Laboratory of New Ceramics and Fine Processing, School of Materials Science and Engineering, Tsinghua University, Beijing, China

10.1 Introduction

Electrocatalysis is crucial for developing sustainable clean energy techniques in the future to alleviate environmental and energy crisis via conversion of small energy-related molecules (H_2O, O_2, CO_2, N_2, etc.) to value-added chemical feedstocks or fuels (H_2, CH_4, NH_3, etc.) [1–3]. In general, electrocatalysis undergoes a series of complicated multi-electron reaction pathways and occurs at the interface of the electrocatalysts, where electrons, ions and molecules can participate in reactions on specific active sites [4,5]. Therefore, the factors of mass transfer, chemical components, sizes, porosity and interface properties are intimately associated with the catalytic performance of the electrocatalysts [6]. Recently, 2D materials have ignited extensive interests in electrocatalysis due to their large surface to bulk ratios, robust structure and controllable band-gap states, which not only endow the electrocatalysts with ideal platforms to design accessible active sites, but also guarantee the catalytic durability and efficient ion/electron transfer [7]. However, the anisotropy induces 2D materials with active edges, while the dominating basal planes are usually electrochemically inert [6]. In addition, weak out-of-plane van der Waals interactions suppress electron transport in vertical direction, especially for multilayer or bulk materials, causing poor vertical conductivity and sluggish kinetic processes [8]. Hence, the catalytic properties are significantly below expectations in practical applications. So far, numerous effective strategies have been proposed to enhance their intrinsic activity such as defect engineering, strain engineering and composite engineering. Among them, defect engineering with various defect species, defect numbers, defect distributions and defect-reactant chemisorption modes emerges as hot topics, which can regulate the local coordination environment at the atomic scale. Defect engineering will enable achieving the following targets: (1) tuning band structures to optimize the reaction energy barriers; (2) enhancing the electrical conductivity to accelerate electron

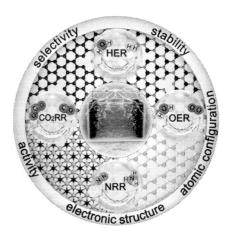

SCHEME 10.1 The illustration of defect-tailored two-dimensional materials for various electrocatalytic processes.

transport; and (3) incorporation of accessible active sites to improve the surface electrochemical reactivity [9]. At present, great progress has been made in tailoring proper defects to satisfy various catalytic reactions, such as oxygen evolution reaction (OER), hydrogen evolution reaction (HER), carbon dioxide reduction reaction (CO_2RR) and nitrogen reduction reaction (NRR), as shown in Scheme 10.1. In this chapter, we aim to summarize the latest progress on defect engineering in 2D materials for electrocatalysis, in particular focusing on vacancies, dopants, grain boundaries and heterojunctions. We emphasize their advantages for structural modification and discuss the defect-performance correlations in depth. Finally, the challenges and opportunities of defect engineering in 2D electrocatalysts are proposed.

10.2 Defect-tailored 2D electrocatalysts for hydrogen evolution reaction (HER)

10.2.1 Fundamental principles of electrocatalytic HER

Hydrogen evolution reaction is a two-electron transfer process with H_2 as the product. Given the source of proton donors, the reaction mechanisms are different in acidic (H_3O^+) and alkaline (H_2O) conditions, as shown in Table 10.1 [10,11]. Typically, HER goes through the Volmer-Heyrovsky pathway or the Volmer-Tafel pathway. For the Volmer step, acidic media is more favorable because of the adequate H^+ in electrolyte to participate in reactions. But for alkaline media, it requires a higher activation energy to break the H-OH bond before adsorbing hydrogen (H_{ad}). Subsequently, the H_{ad} will transform to H_2 by a Heyrovsky process or a Tafel process. Finally, the generated H_2 will be released from the surface of the electrocatalysts.

TABLE 10.1 Hydrogen evolution reaction (HER) Mechanisms in acidic and alkaline conditions. Adapted from Ref. [10] with permission from American Chemical Society. Data were collected from Ref. [11].

	Acidic	Alkaline	Tafel slope (b)
Volmer	$H_3O^+ + e^- \rightarrow H_{ad} + H_2O$	$H_2O + e^- \rightarrow H_{ad} + OH^-$	$b = \frac{2.3RT}{\alpha F} \approx 118$ mV dec^{-1}
Heyrovsky	$H^+ + e^- + H_{ad} \rightarrow H_2$	$H_2O + e^- + H_{ad} \rightarrow H_2 + OH^-$	$b = \frac{2.3RT}{(1+\alpha)F} \approx 39$ mV dec^{-1}
Tafel	$2H_{ad} \rightarrow H_2$	$2H_{ad} \rightarrow H_2$	$b = \frac{2.3RT}{2F} \approx 30$ mV dec^{-1}

H_{ad} is the adsorbed hydrogen; b is the Tafel slope; α is the transfer coefficient; R is the universal gas constant; T is the temperature; F is the Faraday constant.

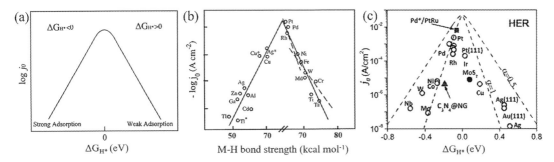

FIGURE 10.1 (a) Relationship between exchange current density (j_0) and hydrogen adsorption Gibbs free energy (ΔG_{H^*}) under assumption of a Langmuir adsorption model. (b) Electrochemically measured j_0 values on metals versus the strengths of M-H bond (M represents the different metals). (c) Dependence of j_0 on ΔG_{H^*} for hydrogen evolution reaction (HER) on the surface of various metals, alloy compounds and non-metallic materials. Reproduced from Ref. [13] with permission from Wiley-VCH.

10.2.2 Catalytic activity descriptors of electrocatalytic HER

In 1958, Parsons proposed a volcano-type plot to describe the correlation of exchange current density (j_0) and the thermodynamically derived hydrogen adsorption Gibbs free energy (ΔG_{H^*}) [12]. As seen in Fig. 10.1a, an ideal HER electrocatalyst should have a ΔG_{H^*} near 0 eV, meanwhile delivering the largest j_0 [13]. Both overly weak (right) or strong (left) hydrogen adsorption will result in a poor HER activity. Inspired by this, Trasatti acquired another volcano-type plot, which adopted M-H bond strength (M represents the different metals, such as Zn, Pt, and Pd) to reflect the catalytic activity of polycrystalline metals [14]. As shown in Fig. 10.1b, only the metals with modest M-H bond strengths, such as Pt, Pd, and Rh, can produce large values of j_0 [13]. Using density functional theory (DFT) simulations, Nørskov and co-workers calculated the ΔG_{H^*} for different metals [15]. Subsequently, the DFT simulation has been extensively used for many other electrocatalysts including alloys, metals carbides, sulfides and non-metallic materials [13]. As described in Fig. 10.1c, when the experimentally obtained j_0 is plotted as a function of the theoretical ΔG_{H^*}, a volcano-type plot can also be obtained, which is a guide for the prediction and selection of potential HER electrocatalysts.

Experimental parameters are used to evaluate the HER performance of an electrocatalyst as follows [16–18]:

(1) *Overpotential*

Theoretically, the Nernst potential of HER versus normal hydrogen electrode (vs. NHE) is zero volt under the standard conditions. However, the practical HER process requires a larger potential to overcome the sluggish kinetics. Overpotential (η) represents the deviation degree from theoretical decomposition voltage, which can be calculated by Eq. (10.1) [16]:

$$\eta = E_{\text{electrocatalysis}} - E_{\text{reversible}} - iR \tag{10.1}$$

Where $E_{\text{reversible}}$ is the theoretical decomposition voltage; iR is the ohmic drop caused by electrolyte, wires and contact points; $E_{\text{electrocatalysis}}$ is the applied voltage. Commonly, to quantitatively compare the HER performance of different electrocatalysts, the η value at a current density of 10 mA cm^{-2} is regarded as an important descriptor to evaluate the catalytic performance. An ideal electrocatalyst should deliver a smaller η with larger j_0.

(2) *Charge transfer resistance*

Charge transfer resistance (R_{ct}) reflects the interfacial charge transfer process. It can be obtained from the electrochemical impedance spectroscopy (EIS) measurement. A small R_{ct} indicates a rapid charge transfer.

(3) *Tafel slope and exchange current density*

Tafel slope (b) is an important parameter to unveil reaction kinetics, which can be acquired by fitting linear sweep voltammetry (LSV) curves from the Tafel Equation (10.2) [19]:

$$\eta = b \log(j/j_0) \tag{10.2}$$

Where j is the current density; j_0 is the exchange current density; b is the Tafel slope. It represents the requirement of a higher η to increase the catalytic current by an order of magnitude. The unit is mV decade^{-1} (or mV dec^{-1}, for short). In addition, according to the Butler-Volmer kinetics, when the Volmer, Heyrovsky and Tafel processes are employed as rate-determining steps, respectively, the theoretical Tafel slopes are 118 mV dec^{-1}, 39 mV dec^{-1} and 30 mV dec^{-1}, as seen in Table 10.1. Therefore, the Tafel slope can serve as an indicator of the rate-determining steps for different electrocatalysts [11].

Exchange current density (j_0) is also a key descriptor to assess the catalytic efficiency, which is proportional to the electroactive surface area. The corresponding value can be calculated by extrapolating the Tafel plot to x-axis. The large j_0 indicates a large surface area, fast electron transfer and favorable HER kinetics.

(4) *Turnover frequency*

Turnover frequency (*TOF*) shows the number of reactant molecules evolved per second per surface site, which is employed as a metric to describe the intrinsic activity of electrocatalysts at the molecular level. The value can be calculated through Eq. (10.3) [16]:

$$TOF = \frac{J}{2nF} \tag{10.3}$$

10.2.3 Defect-tailored 2D electrocatalysts for hydrogen evolution reaction (HER)   307

Where J is the measured current (A); F is the Faraday constant (96485 C mol^{-1}); n is the number of the active sites (mol). The factor of $1/2$ indicates that two electrons are required to generate one H_2 molecule. The large TOF indicates an excellent catalytic ability.

(5) *Faradaic efficiency*

Faradaic efficiency (FE) denotes the electron transfer efficiency in an electrocatalytic process, which is the ratio of the utilized charge by the reactants and the total charge from the external circuit. In general, the experimentally quantified H_2 volume ($V_{experimental}$) and theoretically calculated H_2 volume ($V_{theoretical}$) can be used to calculate FE, as shown in Eq. (10.4) [20]:

$$FE = \frac{V_{experimental}}{V_{theoretical}}. \tag{10.4}$$

10.2.3 Defect-tailored 2D electrocatalysts for HER

2D transition metal dichalcogenides (TMDs), such as MoS_2, WS_2, $MoSe_2$, and NbS_2, exhibit superior activity, selectivity and stability for hydrogen generation [21]. Among them, MoS_2 is one of the most popular candidates due to its thermo-neutral hydrogen adsorption energy in the volcano plot [22]. However, both experiments and computations identify that the catalytic activity of MoS_2 relies mainly on the defects of under-coordinated edge sites, whereas the basal planes are electrochemically inert [23–25]. Earlier reports have been devoted to creating plenty of nanopores, cracks and grain boundaries in MoS_2 to expose more active edge sites. For instance, Ajayan and co-workers introduced cracks and holes in MoS_2 via oxygen or hydrogen etching, forming additional edge active sites for more efficient HER process [26]. Jaramillo and co-workers synthesized a mesoporous MoS_2 network structure with double-gyroid morphology [27]. The nanoscale surface curvature can hinder the growth of MoS_2 basal planes and expose a large fraction of edge defects, exhibiting an improved η of 150–200 mV vs. reversible hydrogen electrode (RHE) at 10 mA cm^{-2} and a low b of 50 mV dec^{-1}. Zhu and co-workers created two kinds of different boundaries (2H-2H domain boundaries and 2H-1T phase boundaries) in monolayer MoS_2 [28]. These linear boundary defects effectively increased active edges sites, which achieved a large j_0 of 0.57×10^{-4} A cm^{-2}, a low b of 73 mV dec^{-1}, and a long-term operating stability of over 200 h. However, the above strategies can only mechanically increase the total amount of edge sites per unit geometric area, but do not alter the poor electronic properties and insufficient utilization of the basal planes. Recently, Jin and co-workers chemically exfoliated bulk MoS_2 to MoS_2 nanosheets by Li$^+$ intercalation, which not only endowed a high density of defects, but also tailored the electrical conductivity by transforming semiconducting 2H-MoS_2 to metallic 1T-MoS_2 [29]. As a result, an enhanced HER activity with a low η of 187 mV vs. RHE at 10 mA cm^{-2} and a small b of 43 mV dec^{-1} was achieved.

Defect engineering in basal planes is another constructive route to activate inert components through the perturbation of the local density of states (DOS) and creation of additional energy levels, so as to optimize ΔG_{H^*} [30,31]. Typically, S-vacancy engineering is beneficial for HER process, where the new gap states around the Fermi level allow hydrogen to bond directly with the exposed Mo atoms [32,33]. For instance, Zheng and co-workers introduced S vacancies into basal plane of monolayer 2H-MoS_2 (see Fig. 10.2a) [33]. Note that the ΔG_{H^*}

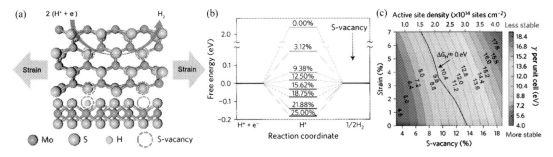

FIGURE 10.2 (a) Schematic illustration of the top (upper panel) and side (lower panel) views of MoS$_2$ with strained S vacancies on the basal plane. (b) Free energy versus the reaction coordinate of HER for the S vacancies range of 0–25%. (c) Colored contour plot of the surface energy per unit cell γ (with respect to the bulk MoS$_2$) as a function of S-vacancy and strain. Color bar represents the value of γ. The black lines indicate the combination of S vacancy and strain that yield $\Delta G_{H^*} = 0$ eV. Reproduced from Ref. [33] with permission from Springer Nature.

decreases significantly with the increasing concentration of S vacancies and the optimal value is achieved for a S vacancy concentration between 12.50 and 15.62% (see Fig. 10.2b). By theoretical analysis, a major reason to do this was that the S vacancies introduce new gap states around the Fermi level, which are favorable for hydrogen binding. After combining with uniaxial strain, the ΔG_{H^*} can be further tuned closer to 0 eV, as shown in Fig. 10.2c, allowing the catalyst to achieve the highest intrinsic HER activity among the molybdenum sulphide-based electrocatalysts. Beyond that, a similar conclusion was also reported by Pedersen's group, who reduced the S atoms on 2H-MoS$_2$ basal planes by electrochemical desulfurization [34]. Simply, the desulfurization degree can be regulated under various desulfurization potentials, which realized stable active sites and showed consistent HER activity with the above results.

Nevertheless, S vacancies are easy to be poisoned or etched because of the presence of a high density of dangling bonds [28]. Hence, in-plane heteroatom doping with a stable co-ordination environment as another alternative strategy has been widely studied in recent years, which can enhance charge transfer, introduce new active sites and adjust hydrogen adsorption/desorption equilibrium. Up to now, both non-metal dopants (e.g., C [35], O [36], P [37], N [38]) and metal dopants (e.g., Co [39], Ni [40], Pt [41], Sn [42]) have been successfully incorporated into the basal planes of MoS$_2$ to ameliorate the electronic structure and reactivity. For instance, DFT calculations verified that C doping can generate empty p orbitals perpendicular to the basal planes of MoS$_2$ for water adsorption and dissociation, which were essential for HER in alkaline conditions [35]. P-doped MoS$_2$ with a range of expanded interlayer spacing (see Fig. 10.3a) can not only enhance the electrical conductivity (see Fig. 10.3b), but also serve as new active sites by virtue of a moderate ΔG_{H^*} (0.04 eV) (see Fig. 10.3c) [37]. The different interlayer spacings introduced by P doping also play an important role in tuning ΔG_{H^*}, as shown in Fig. 10.3d. Moreover, P and N dual-doped MoS$_2$ has also been designed by Liu and co-workers [43]. The resultant new active site of Mo-N-P with appropriate ΔG_{H_2O} (see Fig. 10.3e) and ΔG_{H^*} (see Fig. 10.3f) results in water dissociation and proton adsorption, thus dramatically enhancing the catalytic kinetics. Except for non-metal doping, Bao and co-workers implanted Pt atoms into in-plane domain of few-layer MoS$_2$, which exhibited highly Pt-like activity for hydrogen evolution [41]. DFT calculations revealed that the doped Pt atoms can increase the electronic states around the Fermi level, thus tuning the Volmer and

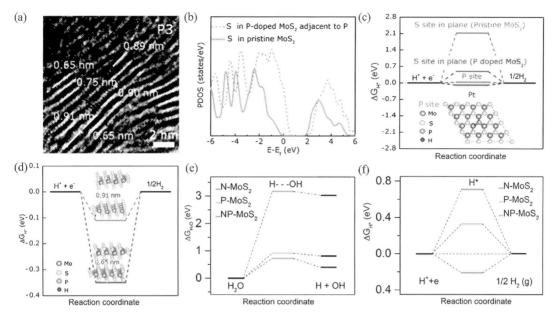

FIGURE 10.3 (a) High-resolution transmission electron microscopy (HRTEM) image taken at the edges of P-doped MoS$_2$ nanosheets. (b) The partial density of states (PDOS) for in-plane S atoms for pristine MoS$_2$ and P-doped monolayer MoS$_2$. (c) HER free energy diagram for P sites and S sites in basal planes of P-doped MoS$_2$ and pristine MoS$_2$. Inset is schematic illustration of P-doped MoS$_2$ basal plane with H adatom on P site. (d) Calculated free energy diagram for HER over P-doped MoS$_2$ with different interlayer spacings. Reproduced from Ref. [37] with permission from American Chemical Society. (e) The calculated water dissociation Gibbs free energy (ΔG_{H_2O}) value. (f) The calculated hydrogen adsorption Gibbs free energy (ΔG_{H^*}) value. Reproduced from Ref. [43] with permission from Elsevier.

Heyrovsky pathways with lower reaction barriers. Moreover, other metal dopants were also systematically investigated, as shown in Fig. 10.4. The dopants such as Pt, Zn, Ag, Pd, and Co can serve as effective candidates to tailor MoS$_2$ because of the proper ΔG_{H^*} near 0 eV, offering a theoretical guidance in designing HER electrocatalysts. Experimentally, Xu and co-workers selected zinc atoms as dopants to manipulate S vacancies in MoS$_2$, which showed an improved η of 194 mV vs. RHE at 10 mA cm^{-2} and a low b of 78 mV dec^{-1} [44]. DFT calculations revealed that zinc atoms can lower the formation energy of S vacancies and decrease the ΔG_{H^*} of S vacancies near the Zn atoms. Further, Qi and co-workers incorporated an atomic cobalt array in the MoS$_2$ basal plane through electrochemical cyclic voltammetry leaching, which induced the phase transformation from 2H to distorted 1T phase, causing an optimal hydrogen binding energy and Pt-like electrocatalytic activity for HER [45].

The above strategies are also applicable to other TMDs. Shaijumon and co-workers reported spiral WS$_2$ domains via a screw dislocation-driven growth [46]. These spiral structures can create many active edge sites, which are connected by dislocation lines in the vertical direction and result in exceptional catalytic properties for HER. Lv and co-workers synthesized WS$_2$/WO$_{2.9}$/C hybrid structures grown on graphite foils (see Fig. 10.5a) via a spin-coating process and a subsequent thermal treatment [47]. Benefitting from the abundant sulfur vacancies in 2D WS$_2$ (see Fig. 10.5b) and oxygen vacancies in 1D WO$_{2.9}$ (see Fig. 10.5c) and

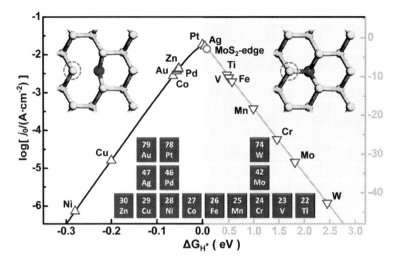

FIGURE 10.4 The relationship between log(j_0) and ΔG_{H^*}. The inserted graphs represent the different configurations of metal-doped MoS$_2$ as coordinated with four (left) and six (right) S atoms. The adsorption sites for H atom are marked by the red dashed circles. The relevant metal atoms are shown in the Periodic Table in the inset. Green ball: Mo; yellow ball: S; blue and purple balls: doped metal atoms. Reproduced from Ref. [41] with permission from the Royal Society of Chemistry.

the fast electron transport in graphite foils, the hybrid WS$_2$/WO$_{2.9}$/C catalyst exhibited outstanding HER activity with a small onset potential of −20 mV vs. RHE (see Fig. 10.5d) and a low b of 36 mV dec^{-1} (see Fig. 10.5e). Chen and co-workers simulated various defects in MoSe$_2$ for hydrogen generation by DFT calculations, as shown in Fig. 10.6 [48]. The results indicated that thermodynamically stable defects such as monoselenium vacancy (V$_{Se}$), diselenium vacancy (V$_{Se2}$), Se antisite (Se$_{Mo}$), hole model (V$_{Mo3Se2}$), and edge configurations, i.e., the reconstructed Mo and Se edges (Mo-R and Se-R), were electrochemically active for HER, which were attributed to the modified electronic structures and hydrogen adsorption behavior (see Fig. 10.6a,b). Furthermore, they pointed out that the Volmer-Tafel mechanism was preferred for V$_{Se}$-enriched and V$_{Se2}$-enriched MoSe$_2$, whereas the Volmer-Heyrovsky mechanism was more favorable for other types of defects in MoSe$_2$, especially for edge sites (see Fig. 10.6c,d).

Recently, MXenes, a fast-growing family of 2D materials with multiple surface chemical environments, robust frameworks and high hydrophilicity are anticipated to act as good candidates for HER [49]. In particular, theoretical calculations show that the basal planes of MXenes are electrochemically active for HER, which are different from that of the widely reported 2D TMDs [50]. However, the use of MXenes for practical applications is very limited due to the sluggish ion/mass transport and poor surface reactivity. To address the above issues, Wang and co-workers reported double transition metal MXene (Mo$_2$TiC$_2$T$_x$) nanosheets with abundant surface Mo vacancies (V$_{Mo}$) by electrochemical exfoliation (Mo$_2$TiC$_2$T$_x$-V$_{Mo}$), in which the Mo vacancies can act as the subsequent anchoring sites for Pt-atom doping (Mo$_2$TiC$_2$T$_x$-Pt$_{SA}$) [51]. After immobilizing Pt atoms, the Mo$_2$TiC$_2$T$_x$-Pt$_{SA}$ catalyst exhibited Pt-like properties towards HER with a low η of 30 mV vs. RHE at 10 mA cm^{-2} and a small b of 30 mV dec^{-1}. An and co-workers doped P atoms into V$_2$CT$_x$ by heat treatment of V$_2$CT$_x$

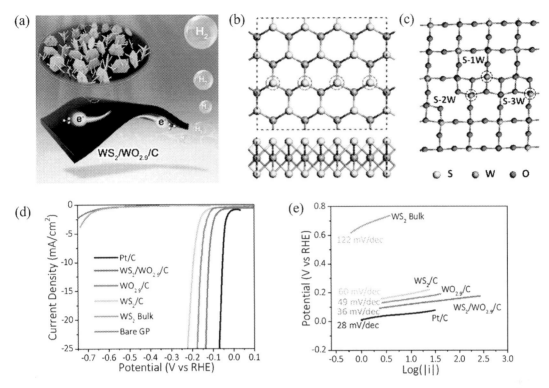

FIGURE 10.5 (a) Schematic illustration of a flexible WS$_2$/WO$_{2.9}$/C hybrid membrane. (b) Top and side views of monolayer WS$_2$, the possible S vacancies are shown in red dashed circles. (c) Top view of WO$_{2.9}$ (100) surface, three adsorption sites are denoted as S-1W, S-2W and S-3W. (d) Polarization curves for HER of bare graphite paper (GP), WS$_2$/WO$_{2.9}$/C hybrid, WS$_2$/C, WO$_{2.9}$/C on GP, 10% Pt/C and WS$_2$ bulk on glassy carbon (GC) electrode. All the measurements are carried out in 0.5 M H$_2$SO$_4$ at a sweep rate of 5 mV s^{-1}. (e) Corresponding Tafel plots for different electrocatalysts. Reproduced from Ref. [47] with permission from Wiley-VCH.

with triphenylphosphine (TPP) as P source [52]. Theoretical calculations identified that the P-C bond weakened the H binding strength and balanced the energy barriers of H$^+$ reduction and H$_{ad}$ desorption. Finally, the P3-V$_2$CT$_x$ with the highest density of P-C bands exhibited excellent catalytic reactivity. Therefore, defect engineering by heteroatoms doping opens the possibility of improving electrocatalytic properties of many MXenes.

10.3 Defect-tailored 2D electrocatalysts for oxygen evolution reaction (OER)

10.3.1 Fundamental principles of electrocatalytic OER

Oxygen evolution reaction is an energy-intensive process involving four-electron transfers. Similar to HER, it can also occur in both acidic or alkaline conditions. The fundamental mechanisms of electrocatalytic OER are shown in Fig. 10.7 [53]. There are two main routes to generate O$_2$ from M-O intermediates (M represents the electrocatalyst surface). One is

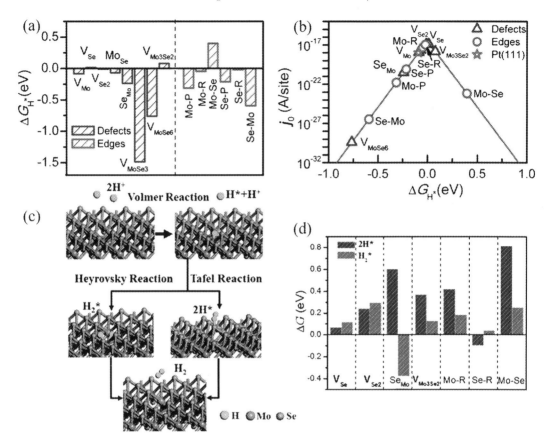

FIGURE 10.6 (a) Calculated ΔG_{H^*} values of various point defects and edges on the monolayer MoSe$_2$. (b) A volcano plot between j_0 and ΔG_{H^*}. The star, triangular, and circle symbols represent the data of Pt (111), MoSe$_2$ defects and edge structures, respectively. (c) Schematic illustration for the HER processes following the Volmer-Heyrovsky pathway and the Volmer-Tafel pathway on MoSe$_2$ surface. Green ball: H; yellow ball: Se; blue ball: Mo. (d) Gibbs free energies (ΔG) of 2H* and H$_2$* on various MoSe$_2$ catalytic sites. Reproduced from Ref. [48] with permission from American Chemical Society.

through direct combination of two M-O intermediates to produce O$_2$, the other involves the formation of M-OOH intermediate and subsequent production of O$_2$. The more visual reaction steps are shown in Table 10.2 [53]. As such, oxygen evolution experiences a series of consecutive reaction steps with only one electron transfer per step, which has to overcome large energy barriers and sluggish kinetics.

10.3.2 Catalytic activity descriptors of electrocatalytic OER

For OER, many critical parameters, such as overpotential (η), Tafel slope (b), turnover frequency (*TOF*), and exchange current density (j_0) are adopted to evaluate the catalytic activity of the electrocatalysts.

10.3. Defect-tailored 2D electrocatalysts for oxygen evolution reaction (OER)

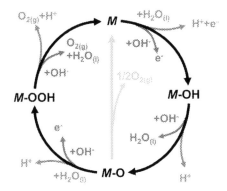

FIGURE 10.7 Schematic illustration of oxygen evolution reaction (OER) in alkaline (red route inside the circle) and acidic (blue route outside the circle) conditions, where M represents the electrocatalyst surface. The black line indicates the formation of M-OOH intermediate. The green line represents the direct combination of M-O intermediates to produce O_2. Reproduced from Ref. [53] with permission from the Royal Society of Chemistry.

TABLE 10.2 The mechanisms of OER in acidic and alkaline conditions [53].

Acidic	Alkaline
$M + H_2O_{(l)} \rightarrow M\text{-}OH + H^+ + e^-$	$M + OH^- \rightarrow M\text{-}OH$
$M\text{-}OH + OH^- \rightarrow M\text{-}O + H_2O_{(l)} + e^-$	$M\text{-}OH + OH^- \rightarrow M\text{-}O + H_2O_{(l)}$
$2M\text{-}O \rightarrow 2M + O_{2(g)}$	$2M\text{-}O \rightarrow 2M + O_{2(g)}$
$M\text{-}O + H_2O_{(l)} \rightarrow M\text{-}OOH + H^+ + e^-$	$M\text{-}O + OH^- \rightarrow M\text{-}OOH + e^-$
$M\text{-}OOH + H_2O_{(l)} \rightarrow M + O_{2(g)} + H^+ + e^-$	$M\text{-}OOH + OH^- \rightarrow M + O_{2(g)} + H_2O_{(l)}$

M is the electrocatalyst surface.

(1) *Overpotential*

Overpotential (η) is the potential difference between the theoretical potential and the applied potential achieving a specific current density. Usually, the η value at a current density of 10 mA cm^{-2} is regarded as an important descriptor to evaluate the catalytic performance of an electrocatalyst. A good OER electrocatalyst should have a small η at a specific current density.

(2) *Tafel slope*

Tafel slope (b) can help to reveal the reaction kinetics of the electrocatalysts, which can be calculated from Tafel Equation (10.2). A small b means that the current density can increase rapidly with smaller η change, indicating excellent reaction kinetics.

(3) *Exchange current density*

Exchange current density (j_0) is another important indicator to evaluate the catalytic efficiency of the electrocatalysts, which can be calculated by extrapolating the Tafel plot to x-axis. The large j_0 indicates a large surface area, fast electron transfer and favorable OER kinetics.

(4) *Turnover frequency*

Turnover frequency (*TOF*) is a crucial descriptor to evaluate the catalytic ability of the catalysts. It represents that the conversion of the number of reactant molecules on per active site in unit time. The value can be calculated through Eq. (10.5) [54]:

$$TOF = \frac{J}{4nF} \tag{10.5}$$

Where J is the measured current (A); F is the Faraday constant (96485 C mol^{-1}); n is the number of the active sites (mol). The factor of $1/4$ indicates that four electrons are transferred to form one O_2 molecule. The large *TOF* indicates an excellent catalytic activity.

In addition to the above experimental parameters, some computational data, such as the adsorption energies of various reaction intermediates OH*, O*, and OOH*, can also be employed to evaluate catalytic performance of the electrocatalysts.

10.3.3 Defect-tailored 2D electrocatalysts for OER

Layered double hydroxides (LDHs, $M^{II}_{1-x} M^{III}_x$ (OH)$_2$ (A$^{n-}_{x/n}$)·mH$_2$O, where M^{II} is a divalent metal, M^{III} is a trivalent metal, A^{n-} is a intercalated anion), based on Fe, Co, Ni derivatives, as a category of inexpensive 2D materials have drawn a lot of attention for OER in alkaline conditions owing to their steerable chemical components, large surface-to-bulk ratios and electroactive surface areas [54,55]. However, in terms of the limited active sites, inferior stability and poor electrical conductivity, the exploration of highly reactive LDHs is still in its infancy. In recent years, research has focused on the defect-tailored LDHs for OER, which regulates the chemical components, porosity, electronic structure and interface properties of the electrocatalysts, causing abundant active sites and enhancing multiple-electron transfer and mass transport [56–60]. Given that the M^{III} is usually isolated by M^{II}, the formation of various metal vacancies or dopants is thermodynamically favorable to modulate OER performance. For example, Wang and co-workers exfoliated bulk CoFe LDHs into lamellar CoFe LDHs nanosheets by Ar plasma etching, which generated multiple vacancies (including O, Co, Fe vacancies) on the fresh exposed surfaces [61]. These new vacancies tuned the surface electron structure, degree of disorder and coordination numbers of ultrathin CoFe LDHs nanosheets, which significantly promoted the intrinsic activity towards OER. In addition, they also utilized N_2 plasma to etch the bulk CoFe LDHs, as shown in Fig. 10.8a [62]. After a 60 min treatment the bulk CoFe LDHs was exfoliated into lamellar CoFe LDHs nanosheets with numerous atomic-sized holes and edges, causing an abundance of dangling bonds and unsaturated surface atoms. Meanwhile, N atoms were implanted into ultrathin CoFe LDHs nanosheets, which not only formed a supportive coordination environment to absorb intermediates, but also decreased the R_{ct} to accelerate kinetic processes. When tested as work electrode, the N-doped CoFe LDHs nanosheets delivered a low η of 233 mV vs. RHE at 10 mA cm^{-2} and a small b of 40.03 mV dec^{-1} (see Fig. 10.8b,c). Apart from gas plasma etching, Yang and co-workers employed ionic reductive complexation extraction method to disperse Cu (II) vacancies in NiCu LDHs nanosheets (SAV-NiCu$_x$ LDHs), which efficiently increased the oxidation states of the neighboring Ni ions and tuned the rate-determining step

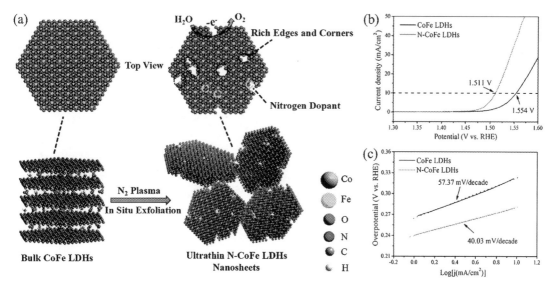

FIGURE 10.8 (a) Schematic illustration of the exfoliation of bulk CoFe layered double hydroxides (LDHs) to ultrathin CoFe LDHs nanosheets by N_2 plasma treatment. (b) Linear sweep voltammetry (LSV) curves of bulk CoFe LDHs and ultrathin N-doped CoFe LDHs nanosheets (N-CoFe LDHs). (c) The corresponding Tafel plots. Reproduced from Ref. [62] with permission from Wiley-VCH.

with reduced free energy barriers (from O* to OOH*) for OER [63]. Notably, the current density of the SAV-NiCu$_x$ LDHs catalyst was 8.4 times higher than that of the pristine NiCu LDHs at 1.6 V vs. RHE, revealing the significant role of Cu vacancies in promoting the OER activity. Sun and co-workers activated basal planes of NiFe LDHs by Mn^{2+} doping [64]. Because of the strong reducing ability and weak electronegativity of Mn^{2+}, the electrons transferred from Mn^{2+} to neighboring Fe and Ni sites in hydroxide matrix, causing an electron-rich environment of Fe and Ni sites. As a result, the d band center shifted closer to the Fermi level and enhanced the binding strength between Mn^{2+}-doped NiFe LDHs and −OH. In addition, Mn^{2+} doping also promoted deprotonation step by reducing uphill energy barriers, exhibiting an enhanced OER activity with a small b of 68.6 mV dec^{-1} and outstanding stability over 40 h.

Besides the various vacancies and dopants, designing heterostructures of LDHs with the appropriate 2D materials is another reliable route to improve the OER performance due to strong synergistic effects of the different components [59,65]. Recently, Yao and co-workers reported a heterostructure of exfoliated NiFe LDH nanosheet and defective graphene (NiFe LDH-NS@DG) by electrostatic stacking [59]. The negatively charged defective graphene (DG) provided sufficient anchoring sites to immobilize the Ni atoms and/or Fe atoms in NiFe LDH nanosheet, causing a prominent electron transfer at local domains. Moreover, the restacking of exfoliated LDH nanosheets was avoided, leading to improved electrical conductivity and a higher electroactive surface area. The charge distribution on this heterostructure was also investigated by DFT calculations. Note that the charge transferred partly from NiFe LDH-NS to DG, especially around the defect sites, resulting in the holes and electrons accumulation on the NiFe LDH-NS and DG respectively. This heterostructure can serve as a

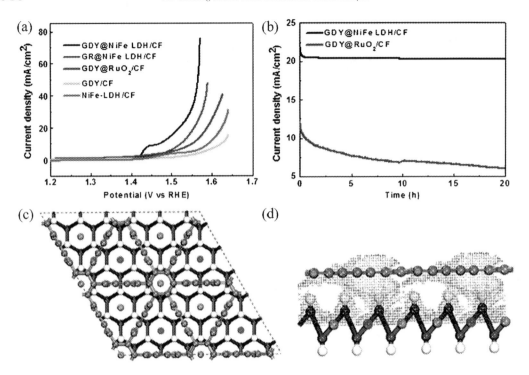

FIGURE 10.9 (a) LSV curves of different electrocatalysts in 1 M KOH. (b) Time-dependent current density curves of the graphdiyne@NiFe layered double hydroxide/copper foam (GDY@NiFe LDH/CF) and GDY@RuO$_2$/CF catalysts. (c) The two-dimensional side view of GDY@NiFe LDH interface. (d) The three-dimensional side view of interfacial electron density difference (the isosurface value is 0.001eÅ^{-3}) between a GDY sheet and a NiFe LDH layer. Blue isosurface conveys charge accumulation between GDY and NiFe LDH. Grey, purple, blue, red and white balls represent C, Fe, Ni, O and H atoms, respectively. Reproduced from Ref. [69] with permission from Elsevier.

bifunctional electrocatalyst to promote OER and HER simultaneously. As expected, a prototype electrolytic cell was assembled to split water at room temperature, which exhibited superior performance as compared to NiFe LDH-NS@graphene (NiFe LDH-NS@G), NiFe LDH-NS@N-doped graphene (NiFe LDH-NS@NG) and other non-noble metal bifunctional catalysts. Furthermore, it could also be used in solar-water electrolysis, achieving oxygen and hydrogen production by powering with a 1.5 V solar panel. Even so, DG just physically adsorbs transition metal atoms, which greatly decreases the stability and activity of electrocatalysts for long-term use. More recently, graphdiyne (GDY) as an ideal substrate has received a lot of attention owing to the strong chemical bonds between the transition metal atoms and surrounding acetylenic bonds [66–68]. Zhang and co-workers prepared a GDY@NiFe LDH heterojunction grown on copper foam (GDY@NiFe LDH/CF) for water splitting [69]. For OER, the GDY@NiFe LDH/CF exhibited a small η of 0.22 V at 10 mA cm^{-2} and a stable time-dependent current density over a time period of 20 h (see Fig. 10.9a,b). Theoretical calculations indicated that the charge density accumulated around the interface between the Ni atoms and/or Fe atoms and the acetylenic bonds. Such synergistic effects facilitated electron transfer and shortened the diffusion length by maximizing interfacial contact, thus tuning electronic properties and reaction kinetics towards OER and HER (see Fig. 10.9c,d). Except for the graphene and GDY mentioned above, other 2D materials, such as g-C_3N_4 [70], Ti_3C_2 MXene [71], and MoS_2 [72,73], can also be employed as promising candidates to construct heterostructures with LDHs, showing excellent electrocatalytic performance for OER.

Apart from LDHs, other 2D materials can also be modified to improve OER activity. For example, $CoSe_2$ with its unique electronic configuration ($t_{2g}^6 e_g^1$) and metallic behaviors would be a promising catalyst for water oxidation [74]. However, because it has a low concentration of active sites, the expected catalytic efficiency is very low. Recently, Xie and co-workers has overcome the shortcomings of limited active sites by exfoliating the bulk $CoSe_2$ to atomically thin $CoSe_2$ nanosheets, which generated abundant Co vacancies (V_{Co}) on $CoSe_2$ surface [75]. DFT calculations showed that the Co sites and V_{Co} of ultrathin $CoSe_2$ nanosheets exhibited higher H_2O adsorption energy than that of the bulk $CoSe_2$, resulting in a small b of 44 mV dec^{-1} and a large TOF of 0.33 s^{-1} at an η of 0.5 V in alkaline conditions. In addition, exfoliated 2D TMDs with abundant edge defects also perform excellent OER activity in acidic conditions, as confirmed by Ajayan and co-workers [76]. They proposed that 1T-MoS_2 is the best candidate for OER followed by 1T-TaS_2, 2H-MoS_2 and 2H-TaS_2. Theoretical studies indicate that edge sites are the dominant active centers instead of in-plane atoms, revealing a polymorphic dependence of electrocatalytic activity. Phosphorene with high carrier mobility and fully exposed active sites can also act as a potential electrocatalyst for OER. In order to further enhance its catalytic activity, Feng and co-workers studied the OER performance of Te-doped phosphorene by DFT calculations [77]. They found that Te atoms in phosphorene tended to form clusters (Te_P^x) stabilized by a series of intrinsic defects, such as Stone-Wales, single vacancy defects, and zigzag nanoribbons. Due to the synergy of both Te atoms and various defects, the binding strength of O* at Te sites dramatically decreased, showing better OER catalytic activity than pristine phosphorene. Moreover, defect-tailored carbon-based electrocatalysts, such as graphene and graphitic carbon nitride nanosheets, have been reported in many studies for OER. For example, Dai and co-workers explored N and S co-doped graphitic sheets with a high concentration of stereoscopic holes (SHG), which not only tuned electronic and chemical characteristics, but also provided abundant accessible active sites and rich mesopores for high-efficiency electron and electrolyte transport [78]. As a consequence, the SHG catalyst delivered a comparable OER performance to the state-of-the-art RuO_2 electrode. Qiao and co-workers developed molecular-level g-C_3N_4 coordinated Co atoms (Co-C_3N_4) as OER electrocatalysts [79]. Theoretical studies showed that Co-N_2 coordination in the g-C_3N_4 matrix is the active center to trigger OER activity with low reaction barriers in agreement with experimental validation. In addition, other 3d transition metal-centered C_3N_4 (M-C_3N_4, M = Fe, Ni, Cu, Zn, Mn, Cr) were also studied, providing a valuable guidance for catalyst design.

10.4 Defect-tailored 2D electrocatalysts for nitrogen reduction reaction (NRR)

10.4.1 Fundamental principles of electrocatalytic NRR

In contrast with the conventionally accepted high-polluting and energy-intensive Haber-Bosch method, electrocatalytic nitrogen reduction has been viewed as a promising substitu-

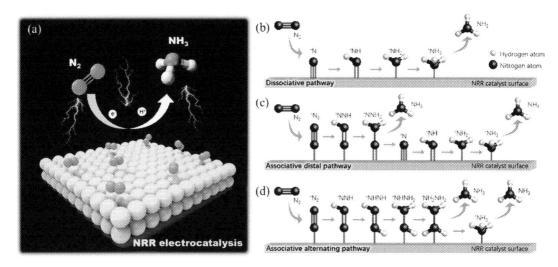

FIGURE 10.10 (a) Schematic illustration of the electrocatalytic nitrogen reduction reaction (NRR). (b–c) Possible reaction mechanisms including the dissociative pathway, the associative distal pathway and the associative alternating pathway. Reproduced from Ref. [82,83] with permission from Elsevier.

tion for ammonia production due to its high efficiency and sustainability [80]. However, due to the chemical inertness of nitrogen molecules (N_2), the practical conversion from nitrogen to ammonia is suppressed by a series of sluggish steps, especially for the N_2 adsorption and the breaking of the N≡N triple bond, which induce large overpotentials, sluggish kinetics and low NH_3 yield [81]. As reported in previous studies, the fundamental principles of electrocatalytic NRR involve two different pathways: the dissociative pathway and the associative pathway, as shown in Fig. 10.10a–d [82,83]. As for the dissociative pathway, the breaking of the N≡N triple bond occurs before the hydrogenation process, which requires a high energy to overcome the strong N≡N triple bond energy (see Fig. 10.10b). While in the associative pathway, the separation of two N atoms in dinitrogen is accompanied by the release of the first NH_3 molecule. Given the hydrogenation sequences, the associative pathway could be further classified as the associative distal pathway and the associative alternating pathway [83]. For the associative distal pathway, three H atoms adsorb on the remotely distal N atom to generate one NH_3 molecule, and then repeating the above similar steps on the proximal N atom to release the second NH_3 molecule (see Fig. 10.10c). While in the associative alternating pathway, H atoms adsorb on both N atoms, releasing two NH_3 molecules one by one in the final step (see Fig. 10.10d). The elementary steps for NRR are listed in Table 10.3 [82]. Hence, complex reaction pathways and slow kinetics make the requirements of the overpotentials for nitrogen electroreduction much larger than that of HER, leading to the poor selectivity for ammonia production owing to the competing HER process.

10.4.2 Catalytic activity descriptors of electrocatalytic NRR

Similar to HER, volcano diagrams for NRR on various transition-metal surfaces have been established through DFT calculations, as shown in Fig. 10.11 [84]. Note that the rate-

10.4. Defect-tailored 2D electrocatalysts for nitrogen reduction reaction (NRR)

TABLE 10.3 The mechanisms of the electrocatalytic NRR. Reproduced from Ref. [82] with permission from Elsevier.

Mechanism	Elementary reaction steps
Dissociative pathway	$N_2 + 2^* \rightarrow 2^*N$
	$2^*N + 2e^- + 2H^+ \rightarrow 2^*NH$
	$2^*NH + 2e^- + 2H^+ \rightarrow 2^*NH_2$
	$2^*NH_2 + 2e^- + 2H^+ \rightarrow 2NH_3 + 2^*$
Associative distal pathway	$N_2 + {}^* \rightarrow {}^*N_2$
	${}^*N_2 + e^- + H^+ \rightarrow {}^*NNH$
	${}^*NNH + e^- + H^+ \rightarrow {}^*NNH_2$
	${}^*NNH_2 + e^- + H^+ \rightarrow {}^*N + NH_3$
	${}^*N + e^- + H^+ \rightarrow {}^*NH$
	${}^*NH + e^- + H^+ \rightarrow {}^*NH_2$
	${}^*NH_2 + e^- + H^+ \rightarrow NH_3 + {}^*$
Associative alternating pathway	$N_2 + {}^* \rightarrow {}^*N_2$
	${}^*N_2 + e^- + H^+ \rightarrow {}^*NNH$
	${}^*NNH + e^- + H^+ \rightarrow {}^*NHNH$
	${}^*NHNH + e^- + H^+ \rightarrow {}^*NHNH_2$
	${}^*NHNH_2 + e^- + H^+ \rightarrow {}^*NH_2NH_2$
	${}^*NH_2NH_2 + e^- + H^+ \rightarrow {}^*NH_2 + NH_3$
	${}^*NH_2 + e^- + H^+ \rightarrow NH_3 + {}^*$

* is an adsorption site on catalyst surface.

determining steps of the metals on the left branches are the protonation process from $^*NH$ to $^*NH_2$ and the removal of $^*NH_2$ as NH_3. While for the metals on the right branches, the breaking of $N{\equiv}N$ triple bond is the rate-limiting step for the dissociative mechanism. In the case of the associative mechanism, N_2 adsorption and the first protonation process from $^*N_2$ to $^*N_2H$ greatly restrict their activity. Hence, the reaction barriers of the rate-determining steps could be employed as theoretical descriptors to evaluate the electrocatalytic activity of the NRR electrocatalysts. Based on the above results, the desired NRR electrocatalysts should meet the following two requirements: (1) the electrocatalysts can accept lone-pair electrons to adsorb N_2 molecules, while donating electrons to the antibonding orbitals of N_2 to decrease the activation energy of the inert $N{\equiv}N$ triple bond; (2) the electrocatalysts should stabilize the $^*N_2H$ intermediate and facilitate the removal of $^*NH_2$ to produce NH_3.

In addition, some electrochemical parameters are also employed as important descriptors. For example, Faradaic efficiency (FE) intuitively unveils the selectivity of NH_3 synthesis due to the disturbance of hydrogen or nitrogen by-products, which can be defined as the ratio of the charge used for N_2 reduction ($N_2 \rightarrow NH_3$) to the total charge provided by the external circuit. The value can be calculated using Eq. (10.6) [82]:

$$FE = \frac{n \cdot F \cdot C \cdot V}{M \cdot Q} \tag{10.6}$$

Where n is the number of electron transfer for producing one NH_3 molecule ($n = 3$); F is the Faraday constant (96485 C mol^{-1}); C is the concentration of NH_3; V is the volume of the

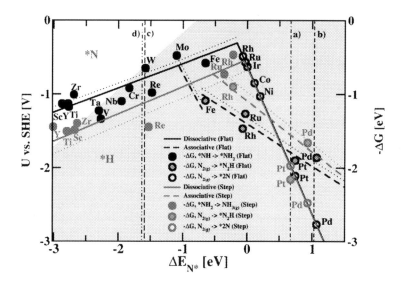

FIGURE 10.11 Volcano diagrams of electrocatalytic NRR via a dissociative mechanism on a flat surface (black solid line), associative mechanism on a flat surface (black dotted line), dissociative mechanism on a stepped surface (red solid line) and associative mechanism on a stepped surface (red dotted line) for a selected series of metals. The redox potential limiting steps in the catalytic cycle for every metal were calculated and highlighted by circles. In the grey area, the adsorption of *H is favored over *N. Reproduced from Ref. [84] with permission from the Royal Society of Chemistry.

electrolyte; M is the relative molecular mass of NH_3 ($M = 17$); Q is the total charge provided by the external circuit.

The NH_3 yield rate (R) is another descriptor to evaluate the NRR rate for NH_3 synthesis, which represents the mole number of the NH_3 generation per unit time and unit catalyst loading mass (or unit geometric area/electrochemically active surface area of the electrode), as shown in Eq. (10.7) [82]:

$$R = \frac{C \cdot V}{t \cdot s} \quad (10.7)$$

Where t is the reaction time; s is the catalyst loading mass or the geometric area of the electrode or the electrochemically active surface area of the electrode.

10.4.3 Defect-tailored 2D electrocatalysts for NRR

Transition metal nitrides (TMNs), such as W_2N_3[85] and MoN_2[86] with abundant N vacancies show conspicuous superiority for electrocatalytic NRR under suitable operating potentials. By virtue of the electron-deficient environment, N vacancy is capable of accommodating lone-pair electrons of N_2 molecule and activate N≡N triple bond, resulting in low uphill barriers for electrocatalytic NRR [85]. Recently, Qiao and co-workers deliberately created plenty of N vacancies on 2D layered W_2N_3 surface (NV-W_2N_3), which was regarded as efficient electrocatalyst to achieve a stable ammonia production rate of 11.66 ±

0.98 $\mu g\,h^{-1}\,mg^{-1}$ catalyst and a higher FE of $11.67 \pm 0.93\%$ at -0.2 V vs. RHE than pristine W_2N_3 electrocatalyst [85]. Fourier transform extended X-ray adsorption fine structure (FTEXAFS) spectra characterization revealed that N coordination environment of NV-W_2N_3 was highly stable after NRR process due to the high valence state of tungsten atoms and 2D confinement effect. The reaction pathways through a distal mechanism were calculated by DFT calculations, which indicated that N vacancies served as active sites can stabilize and activate N_2 molecule, causing a reduced thermodynamic limiting potential and low reaction barriers for improving NRR performance. For 2D Mo-N nanosheets, N vacancies also play an important role in adsorbing and activating N_2 molecule [86]. However, because of the strong interactions between Mo and N atoms, the high energy is required to break Mo-N bond. Fe doping in 2D Mo-N nanosheets can weaken Mo-N interactions and promote the formation of the second NH_3 molecule, effectively improving NRR selectivity and offering higher FE. Extending to other nitrides, the strategy of nitrogen-defective engineering is still universally applicable. For example, Liu and co-workers reported N-defective carbon nitride layers with a thickness of 42 nm grown on carbon paper as a highly active NRR electrocatalyst, as shown in Fig. 10.12a,b [87]. X-ray photoelectron spectroscopy (XPS) analysis indicated that the ratio of C=N-C/N-C_3 for CN/C_{600} was smaller than that of CN/C_{500}, revealing a higher concentration of C=N-C N_{2C} vacancies in CN/C_{600} (see Fig. 10.12c). The formation of the C=N-C N_{2C} vacancies efficiently facilitated N_2 adsorption and increased NH_3 yield up to 2.87 $\mu g\,h^{-1}\,mg^{-1}$ catalyst at -0.3 V vs. RHE ($\mu g\,h^{-1}\,mg^{-1}_{cat.}$ for short, see Fig. 10.12d). In fact, previous study has proposed that N_2 molecule prefers to adsorb on the C=N−C N_{2C} vacancy via an end-on mode, which can stretch N≡N triple bond and decrease N_2 adsorption energy [88]. Hence, the unpaired electrons in carbon nitride layers are facile to donate back to the adsorbed N_2 molecule and finally impel electrocatalytic NRR process. Beside N vacancies, S vacancies can also be adopted to facilitate electrocatalytic NRR by tuning local electron structure of the electrocatalysts. Recently, Lou and co-workers incorporated the sulfur vacancies to MoS_{2-x} basal plane by Co doping, revealing polycrystalline domains with a superior N_2-to-NH_3 conversion ability [89]. Theoretical calculations showed that the sulfur vacancies were favorable for the N_2 chemisorption process and the dissociation process of N≡N triple bond. In addition, the presence of Co atoms could further accelerate the above processes by modulating the configuration of Mo-N bond.

Apart from vacancy engineering, heteroatom doping including metal-atom and non-metal-atom dopants is of great concern for NRR electrocatalysts to tailor their electronic structures and physicochemical adsorption/desorption properties for better nitrogen reduction [90]. In terms of metal-atom doping, atomic-scale dispersion in NRR electrocatalysts is one of the most popular defect engineering strategies, which can maximize the atomic utilization and reduce the material costs. Additionally, the homogeneous atomic coordination environments enable high selectivity for NH_3 production, while suppressing the competition of HER process [91]. For example, Chen and co-worker established the model of a single transition metal atom (e.g. Zn, Mo, Ru, Pd) supported on a defective boron nitride monolayer (see Fig. 10.13a) as electrocatalysts for NH_3 production [92]. As can be seen in Fig. 10.13b,c, Mo-doped boron nitride monolayer was the only electrocatalyst satisfying screening requirements. Mo doping can not only promote the chemisorption and activation of N_2 molecule on Mo sites, but also stabilize *N_2H and destabilize *NH_2 to optimize the rate-determining steps for electrocatalytic NRR. The potential applications of single transition metal atom anchored

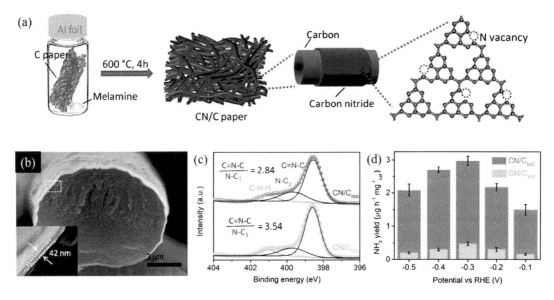

FIGURE 10.12 (a) The synthesis of N-defective carbon nitride on carbon paper (CN/C$_{600}$, synthesized at 600 °C). The enlarged image is N vacancy-containing carbon nitride structure. Carbon and nitrogen atoms are represented by the gray and blue balls, respectively. (b) Cross-sectional scanning electron microscopy (SEM) image of an individual carbon nitride coated carbon fiber. Inset is the zoomed-in SEM image of the squared area. (c) N 1s X-ray photoelectron spectroscopy (XPS) images of CN/C$_{500}$ (synthesized at 500 °C) and CN/C$_{600}$ catalysts. (d) NH$_3$ yield rate of CN/C$_{600}$ and CN/C$_{500}$ catalysts at different potentials. Reproduced from Ref. [87] with permission from American Chemical Society.

on N-doped graphene[93,94] and g-C$_3$N$_4$[95] for electrocatalytic NRR were also explored by DFT calculations. Yet to date, there is no experimental evidence, suggesting further investigation. Interestingly, metal-atom-doped MoS$_2$ nanosheets have been successfully synthesized for electrocatalytic NRR in recent years [96,97]. For instance, Zhang and co-workers immobilized Fe atoms onto MoS$_2$ nanosheets (see Fig. 10.13e), which induced the local electric fields on Fe-doped MoS$_2$ interface [97]. The resultant interfacial polarization promoted the electrons injection into N$_2$ antibonding orbitals, appreciably increasing the length of N≡N triple bond, as confirmed in Fig. 10.13d. Therefore, the cleavage of N≡N triple bond can occur at a low energy barrier, accelerating the NH$_3$-evolving rate (36.1 ± 3.6 mmol g^{-1} h^{-1}) and improving the FE (31.6% ± 2%) at a low potential of −0.2 V vs. RHE (see Fig. 10.13f). Qian and co-workers explored the Fe-doped MoS$_2$/carbon cloth (FMS/CC) catalyst by substituting Mo atoms for NRR [96]. Compared with un-doped MoS$_2$/carbon cloth (MS/CC) catalyst, the FMS/CC catalyst increased the FE 5 times. By theoretical analysis, the improvement of NRR performance contributed to the reduction of the N$_2$ adsorption energy by affecting the chemical states of Mo and S atoms and the stacking interlayer instance of MoS$_2$.

As for non-metal-atom doping, elements such as B, N, S, and O are usually selected as dopants to promote electrocatalytic NRR. In the case of graphene, B doping can suppress HER process. Because of boron's relatively low electronegativity it can induce electron transfer from B to adjacent C atoms. Hence, the electron-deficient B atoms can prevent the binding

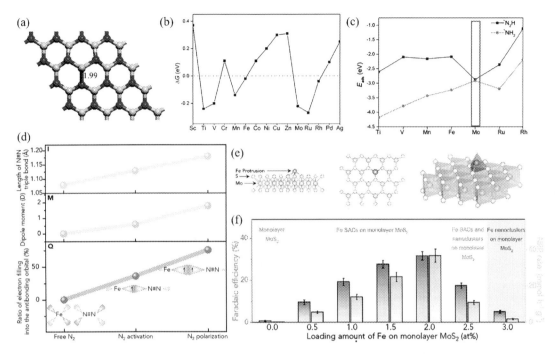

FIGURE 10.13 (a) Optimized structure of Mo-doped boron nitride monolayer. The unit of bond length is Å. Pink, blue and green balls represent B, N and Mo atoms, respectively. (b) Gibbs free energies of N_2 adsorption on various single transition metal atoms supported by defective boron nitride nanosheet. (c) Adsorption energies of *N_2H and *NH_2 species on various single transition metal atoms supported by defective boron nitride nanosheet. Reproduced from Ref. [92] with permission from American Chemical Society. (d) The change of the N≡N triple bond length (Å), dipole moment (D) and the ratio of electron filling into the antibonding orbital (%) during the N_2 adsorption, activation and polarization processes. (e) Side view, top view and perspective view of Fe-doped MoS_2 crystalline structure. (f) Performance comparison of single-atom catalysts (SACs)-MoS_2-Fe-Y (Y = 0.5, 1.0, 1.5, 2.0, 2.5, and 3.0 at. %) at −0.2 V vs. RHE in flow cell. Reproduced from Ref. [97] with permission from Elsevier.

of Lewis acid H^+ and strengthen the binding stability to N_2 molecules. Recently, Zheng and co-workers developed a series of B-doped graphene electrocatalysts (BG-1, BOG and BG-2) for electrochemical NRR, where BG-1 with its high-content BC_3-type structure showed an excellent NH_3 production rate of 9.8 μg h^{-1} cm^{-2} and an enhanced selectivity of 10.8% FE at −0.5 V vs. RHE (see Fig. 10.14a-d) [98]. DFT calculations supported the experimental results, where BC_3-type structure was found to exhibit the lowest energy barriers for NH_3 production among several B-doped graphene structures, providing a useful guideline for B-doped carbon-based materials (see Fig. 10.14e). In contrast, N-doped or S-doped graphene, due to their high electronegativity, S (2.58) and N (3.04), they can induce positive charge densities on the adjacent carbon atoms. For instance, Liu and co-workers reported S and N co-doped graphene (NSG) for efficient and durable electrocatalytic NRR, which dramatically outperformed N- or S-doped graphene (NG and SG) (see Fig. 10.14f) [99]. DFT calculations indicated that N and S co-doping with enhanced conductivity can facilitate N_2 adsorption

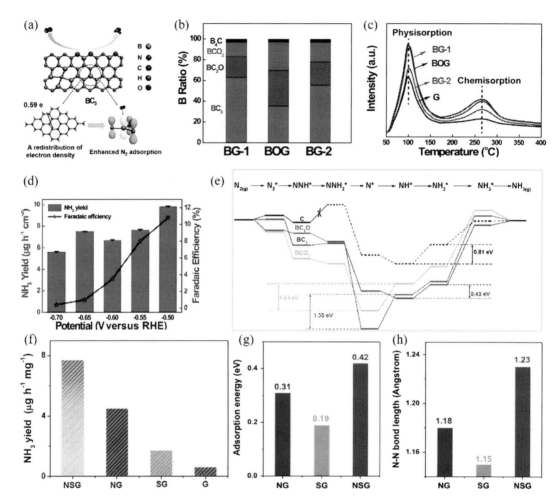

FIGURE 10.14 (a) Schematic illustration of B-doped graphene (BG). The images below are the enlarged image and the atomic orbital image of BC$_3$ for binding N$_2$. (b) Percentages of different B types in different BG samples. Three different BG samples are denoted as BG-1, BG-2 and BOG, where BG-1 and BG-2 were prepared by annealing the H$_3$BO$_3$/graphene oxide mixture in H$_2$/Ar gasses with different ratios of 5:1 and 1:10, respectively. BOG was synthesized by annealing a H$_3$BO$_3$/graphene oxide mixture (5:1) in Ar gas. (c) N$_2$ temperature-programmed desorption (N$_2$-TPD) curves of BG-1, BOG, BG-2 and undoped graphene (G). (d) The NH$_3$ production rates (left y axis) and FE_{NH_3} (right y axis) of BG-1. The error bars represent the average of three independent measurements. (e) Free energy diagram of NRR on different B types of BC$_3$, BC$_2$O and BCO$_2$, respectively. Undoped C was also calculated as the control group. Reproduced from Ref. [98] with permission from Elsevier. (f) NH$_3$ yields of N and S co-doped graphene (NSG), N-doped graphene (NG), S-doped graphene (SG) and undoped graphene (G) in N$_2$-saturated solution at −0.6 V vs. RHE. (g) Calculated adsorption energies of N$_2$ on NG, SG and NSG. (h) Calculated N≡N triple bond elongations of N$_2$ absorbed on NG, SG and NSG. Reproduced from Ref. [99] with permission from Springer Nature.

and N≡N triple bond elongation thus making it easy to break the N≡N triple bond and stimulate the following NRR processes (see Fig. 10.14g,h).

Apart from above point defects, assembling two or more kinds of 2D materials to form heterostructures has also stimulated interests for electrocatalytic NRR in past few years [100,101]. Due to the mismatch of the interfacial atoms, the heterostructures deliver strong interface interactions, which modulate the interfacial physicochemical properties by electron coupling and charge transfer. Recently, Tian and co-workers explored a 2D/2D MoS_2/C_3N_4 heterostructure by pyrolysis and subsequent hydrothermal process, which exhibited a high NH_3 yield of 18.5 µg h^{-1} mg^{-1} and an improved FE of 17.8% at -0.3 V vs. RHE, far better than those of individual MoS_2 or C_3N_4 (see Fig. 10.15a-c) [102]. Theoretical calculations indicated that the interface engineering of C_3N_4 and MoS_2 can significantly activate N_2 molecule and stabilize the $^*N_2H$ intermediate on Mo edge sites (see Fig. 10.15d-f). However, it can also cause the desorption obstacle of formed NH_3 (see Fig. 10.15g). Nevertheless, MoS_2/C_3N_4 heterostructure protected the NRR-active Mo edge sites from the competition of HER process and reduced reaction energy barriers, which efficiently improved FE for NH_3 production. In addition, a MnO_2-$Ti_3C_2T_x$ MXene heterostructure was also prepared with excellent electrocatalytic NRR activity and durability [100]. DFT computations showed the unsaturated Mn atoms act as active sites can adsorb and activate the inert N_2 molecules. Meanwhile, the presence of $Ti_3C_2T_x$ further improves the bonding between N and Mn atoms.

10.5 Defect-tailored 2D electrocatalysts for carbon dioxide reduction reaction (CO_2RR)

10.5.1 Fundamental principles of electrocatalytic CO_2RR

Electrocatalytic CO_2RR is a promising approach to alleviate the ever-increasing global CO_2 emission, meanwhile, it can generate a series of value-added chemical feedstocks or fuels, such as HCOOH, HCHO, CH_4, CH_3OH, C_2H_4, CH_3CH_2OH, and $H_2C_2O_4$ [103,104]. However, CO_2 molecule is thermodynamically stable with the highest oxidation state, inducing high energy barriers, poor selectivity and sluggish kinetics of CO_2RR [3,105]. Depending on the total number of proton and electron transfer, CO_2 can be reduced to various products, as shown in Table 10.4 [106]. According to the Gibbs free energy Eq. (10.8):

$$\Delta G = nFE^0 \tag{10.8}$$

Where ΔG is the Gibbs free energy; n is the number of electron transfer for CO_2RR (mol); F is the Faraday constant (96485 C mol^{-1}); E^0 is the standard redox potential. A more negative ΔG can be favor CO_2 reduction. Hence, the formation of some multi-carbon (C_n, n>1) hydrocarbons and alcohol products are more thermodynamically favorable than some C_1 products (CO, HCOOH, HCHO, etc.). However, the formation of C-C bond is kinetically sluggish and must compete with the formation of C-H bond and C-O bond, thereby causing a larger energy barrier. Taken together, the simple C_1 products possess a more satisfactory FE than multi-carbon products, especially for CO and HCOOH. A summary of the possible CO_2RR mechanisms towards some typical products is shown in Fig. 10.16 [107].

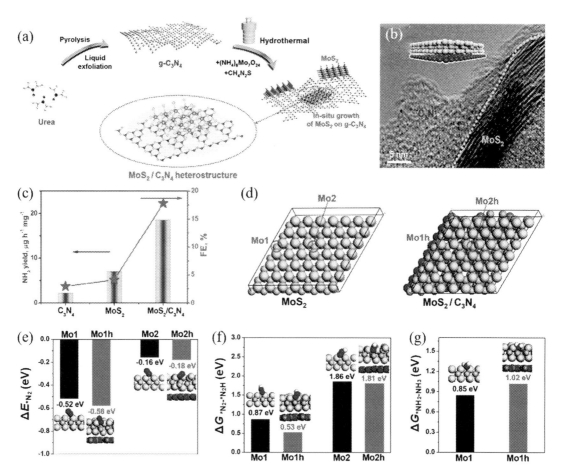

FIGURE 10.15 (a) Synthesis process of the MoS$_2$/C$_3$N$_4$ heterostructure. (b) HRTEM image of the MoS$_2$/C$_3$N$_4$ heterostructure. (Inset is the schematic illustration of MoS$_2$/C$_3$N$_4$ heterostructure. Green, yellow, blue and gray balls represent Mo, S, N and C atoms, respectively.) (c) NRR performance of MoS$_2$, C$_3$N$_4$ and MoS$_2$/C$_3$N$_4$ catalysts at -0.3 V vs. RHE. (d) Optimized structures of MoS$_2$ and MoS$_2$/C$_3$N$_4$ and their potential NRR active sites. (edge Mo1 and Mo1h sites; basal-plane Mo2 and Mo2h sites) (e) Adsorption energies of N$_2$ (ΔE_{*N_2}) for different sites. (f) Free energy changes of the first hydrogenation step ($\Delta G_{*N_2-*N_2H}$) of Mo1, Mo1h, Mo2 and Mo2h sites. (g) Free energy changes of the last hydrogenation step ($\Delta G_{*NH_2-NH_3}$) of Mo1 and Mo1h sites. Reproduced from Ref. [102] with permission from American Chemical Society.

10.5.2 Catalytic activity descriptors of electrocatalytic CO$_2$RR

Some critical parameters, such as onset potential, overpotential (η), Tafel slope, Faradaic efficiency (FE), and turnover frequency (TOF), can be used to evaluate the activity of the electrocatalysts for electrocatalytic CO$_2$RR, which are similar to the electrocatalytic HER, OER and NRR. In addition, energy efficiency (EE) is also essential, which can be calculated under

10.5. Defect-tailored 2D electrocatalysts for carbon dioxide reduction reaction (CO$_2$RR)

TABLE 10.4 Standard redox potentials and the number of electron transfer of CO$_2$RR for different products [106]. (CO$_2$ reduction potential versus standard hydrogen electrode (SHE) at pH $= 7$).

The number of electron transfer	Reaction pathway	E^0 (V vs. standard hydrogen electrode)
2e$^-$	$CO_2 + 2H^+ + 2e^- \rightarrow HCOOH$	-0.61
	$CO_2 + 2H^+ + 2e^- \rightarrow CO + H_2O$	-0.53
	$2CO_2 + 2H^+ + 2e^- \rightarrow H_2C_2O_4$	-0.913
4e$^-$	$CO_2 + 4H^+ + 4e^- \rightarrow HCHO + H_2O$	-0.48
6e$^-$	$CO_2 + 6H^+ + 6e^- \rightarrow CH_3OH + H_2O$	-0.38
8e$^-$	$CO_2 + 8H^+ + 8e^- \rightarrow CH_4 + 2H_2O$	-0.24
12e$^-$	$2CO_2 + 12H^+ + 12e^- \rightarrow C_2H_5OH + 3H_2O$	-0.329
14e$^-$	$2CO_2 + 14H^+ + 14e^- \rightarrow C_2H_6 + 4H_2O$	-0.27
18e$^-$	$3CO_2 + 18H^+ + 18e^- \rightarrow C_3H_7OH + H_2O$	-0.31

E^0 is the standard redox potential.

a certain η through the following Eq. (10.9) [106]:

$$EE = \frac{E^0}{E^0 + \eta} \times FE \tag{10.9}$$

Where E^0 is the standard redox potential.

10.5.3 Defect-tailored 2D electrocatalysts for CO$_2$RR

It is well known that the complex reaction pathways and various reduction products, at least 16 different C$_1$-C$_3$ species, greatly degrade the selectivity towards CO$_2$RR [103]. This is the case for the synthesis of multi-carbon products, where undesired by-products and slow kinetics greatly depress the FE. Hence, it is urgent to design suitable electrocatalysts towards specific reduction products. Defect tailoring in 2D electrocatalysts is a common avenue to regulate the local charge distribution of the electrocatalysts and the absorption/desorption behaviors of reaction intermediates, leading to higher activity and selectivity for CO$_2$RR.

Recently, heteroatom-doped graphene-based materials have emerged as potential catalysts towards electrocatalytic CO$_2$RR. For example, N defects in the graphene lattice (NG) can promote the conversion of CO$_2$ to CO in an aqueous electrolyte, which yield a maximum FE of 85% at a low η of -0.47 V vs. RHE and an improved stability of over 5 h [108]. Theoretical calculations supported by experimental data show that pyridinic-N defect is the most selective site toward CO generation due to the relatively low free energy barrier to form adsorbed COOH*. Therefore, by deliberately increasing pyridinic-N content in graphene-based materials can lead to high selectivity, i.e. from CO$_2$ to CO. Phani and co-workers incorporated B atoms into graphene to exclusively produce formate, leading to an improved current density with a FE of 66% at -1.4 V vs. saturated calomel electrode (SCE) [109]. Further DFT calculations showed that boron doping impeded symmetrically distributed electron delocalization and introduced asymmetric charge and spin density distribution throughout the

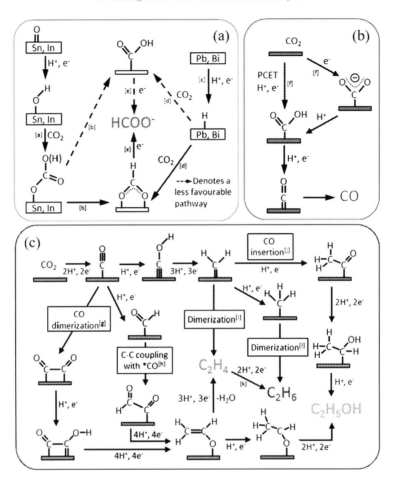

FIGURE 10.16 The mechanisms of carbon dioxide reduction reaction (CO$_2$RR) towards various products. (a) CO$_2$ reduction to formate. (b) CO$_2$ reduction to CO. (c) CO$_2$ reduction to C$_2$H$_4$, C$_2$H$_6$ and C$_2$H$_5$OH. Reproduced from Ref. [107] with permission from Elsevier.

ground state geometry. The positive spin density on B and C atoms makes chemisorption of *COOH feasible. Fluorine doping with its larger electronegativity behaves differently from nitrogen and boron doping. Recently, Yao's group studied a fluorine-doped carbon catalyst (FC) for electrocatalytic CO$_2$RR, which achieved a FE of 89% for CO generation with 68% EE [110]. The enhanced activity is mainly associated with a positive charge density and asymmetrical spin of the neighbor C atoms, causing a stronger adsorption of COOH* intermediate and better suppression of HER process. Except for non-metal doping, metal-doped graphene enables better catalytic performance for CO$_2$RR. For instance, Tour and co-workers synthesized atomic Fe anchored nitrogen-doped graphene (Fe/NG) for high selectivity and activity of CO$_2$ reduction to CO [111]. The Fe atom coordinated with adjacent four nitrogen atoms (Fe-N$_4$-C) were confirmed as active centers to form CO, which dramatically lowered the for-

10.5. Defect-tailored 2D electrocatalysts for carbon dioxide reduction reaction (CO_2RR)

mation energy of *COOH from 0.63 eV to 0.29 eV and promoted CO desorption, resulting in a low overpotential and high selectivity of 80% FE for CO production.

Apart from graphene, other 2D materials with abundant defects have been extensively developed for CO_2RR. Recently, Xie and co-workers deliberately created oxygen vacancies (V_o) on the surface of Co_3O_4 atomic layers (with the average thickness of 0.84 nm) and attempted to uncover the inherent link between the oxygen vacancies and electrocatalytic CO_2RR from atomic-level insights [112]. As shown in Fig. 10.17a, V_o-rich Co_3O_4 atomic layers exhibited higher selectivity (87.6% FE) for formate formation at moderate applied potentials with a negligible H_2 content. While for V_o-poor Co_3O_4 atomic layers, although the FEs of formate formation were modulated at different applied potentials, the maximum value (67.3% FE) was still not comparable with that of V_o-rich Co_3O_4 atomic layers (see Fig. 10.17b). Therefore, the presence of V_o can increase the activity and selectivity to form formate. Free energy diagrams (see Fig. 10.17c,d) indicated that V_o can stabilize the $HCOO^{-*}$ intermediate and favor the hydrogenation process, lowering the overall activation energy for CO_2RR. Hence, V_o-rich Co_3O_4 atomic layers showed a small Tafel slope of 37 mV dec^{-1} and robust stability over 40 h at −0.87 V vs. SCE. Salehi-Khojin and co-workers reported the defect structure of Nb-doped MoS_2 for electrocatalytic CO_2RR, in which 5 at.% Nb was successfully introduced into the MoS_2 lattice without structural modification or other types of defects [113]. Theoretical analysis of the reaction pathways from CO_2 to CO and the trends of $COOH^*$ formation energies and CO desorption energies in Nb-doped MoS_2 showed that inserting Nb atoms at the Mo edges can efficiently reduce the binding strength of CO on Mo sites, while maintaining the formation of $COOH^*$ and CO^* exergonic, which give rise to one order of magnitude higher TOF for CO formation than pristine MoS_2 (see Fig. 10.17e,f). Nørskov and co-workers showed the performance of other metal-doped MoS_2 for electrocatalytic CO_2RR by DFT calculations, further verifying that dopants can stabilize the key intermediates, such as $COOH^*$, CHO^*, and COH^*, independent of CO^* [114]. The results demonstrated that Ni atom is the most selective and reactive dopant among all of the non-noble metal dopants, and this can help us design better electrocatalysts in the future (see Fig. 10.17g).

Currently, several pure metals (Cu, Ag, Au, etc.) have shown potential catalytic performance for CO_2RR [115–117]. This is especially true for the lamellar poly-crystalline 2D metals, the exposed facets and unique grain boundaries with low-coordinated surface atoms have been viewed as probable sites adjusting the electrocatalytic activity towards electrocatalytic CO_2RR. For instance, Li and co-workers synthesized the ultrathin mono-crystalline Bi nanosheets (BiNS, approximately 8.7 nm) for formate production [118]. The BiNS with a large specific area with an abundant under-coordinated Bi surface delivered excellent FE (over 90% in a wide voltage window), activity (large current density of 24 mA cm^{-2} at −1.74 V vs. SCE) and durability (at least 10 h). Xu's group discovered the thickness-dependent CO_2RR performance of Bi nanosheets [119]. The thinner the Bi nanosheets, the better the catalytic activity. When the thickness was ultimately decreased to form 2D Bi monolayer (Bismuthene, with the thickness of about 0.6 nm, as seen in Fig. 10.18a,b), the selectivity and stability can be sharply enhanced to 99% FE at a lower overpotential of −0.58 V vs. RHE and 75 h, respectively (see Fig. 10.18d,e). According to Fourier transform image analysis, the exposed facet occurred fundamental changes from (011) of thicker Bi nanosheets to (111) of monolayer Bismuthene (see Fig. 10.18c), which weakened the adsorption of $COOH^*$, $OCHO^*$ and OH^*

330   10. Tailoring defects in 2D materials for electrocatalysis

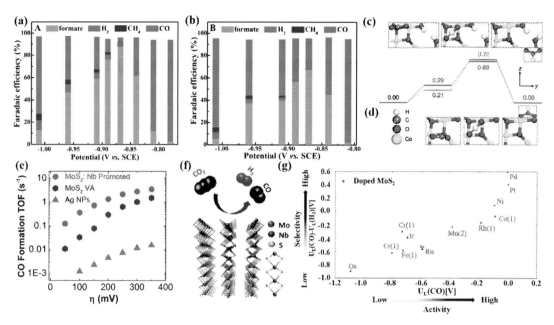

FIGURE 10.17 (a–b) FEs of formate, H_2, CH_4 and CO at different applied potentials for V_O-rich and V_O-poor Co_3O_4 atomic layers, respectively. (c–d) Calculated free energy diagrams for the electrochemical reduction of CO_2 to formate on the Co_3O_4 atomic layer with oxygen vacancies and intact Co_3O_4, respectively. White, red, grey and light blue balls represent H, O, C and Co atoms, respectively. Reproduced from Ref. [112] with permission from Springer Nature. (e) Calculated $TOFs$ of CO formation at different applied overpotentials for different catalysts. (f) Schematic illustration of Nb-doped MoS_2. Reproduced from Ref. [113] with permission from American Chemical Society. (g) Difference between limiting potentials of CO and H_2 versus the limiting potential of CO for CO_2 reduction on the variously doped sulfur edge of MoS_2. Reproduced from Ref. [114] with permission from American Chemical Society.

intermediates and avoided poisoning due to the unique compressive strain (6% compression relative to bulk Bi structure) (see Fig. 10.18f).

10.6 Challenges and perspectives of defect engineering for 2D electrocatalysts

In summary, different defects, such as vacancies, dopants, grain boundaries, and heterojunctions are generally applicable to trigger the intrinsic electrocatalytic activity of 2D materials for various electrochemical reactions, e.g., OER, HER, NRR, and CO_2RR. Defects can not only tailor electronic configurations of the adjacent atoms by inducing charge localizations to harmonize the reaction barriers of key reaction steps, but also serve as new active centers to increase the density of catalytic sites. Up to now, a lot of progress regarding defect-tailored 2D electrocatalysts has been proposed. The defective structure and corresponding defect-property correlations at atomic scale have been unveiled by characterization techniques, such as X-ray absorption fine spectroscopy, spherical aberration corrected transmission elec-

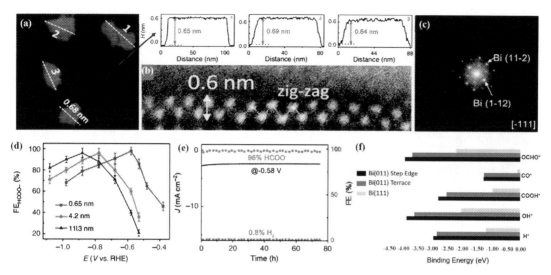

FIGURE 10.18 (a) Typical atomic force microscopy (AFM) image and corresponding height profiles for three Bismuthene nanosheets. (b) Typical lateral high angle annular dark field-scanning transmission electron microscopy (HAADF-STEM) image of a Bismuthene nanosheet, directly showing the single-atom thickness of the layer with zig-zag structure. (c) Fast Fourier transform (FFT) image of Bismuthene. (d) Comparison of FEs for formate at each applied potential for Bi nanosheets with different thicknesses. (e) Long-term stability of Bismuthene nanosheets at a potential of −0.58 V vs. RHE and the corresponding FEs for HCOO$^-$ and H$_2$. (f) Binding energy of OCHO*, CO*, COOH*, OH* and H* on the Bi(111) monolayer (green bar), on the step edge sites of the Bi(011) surface (black bar), and on the terrace sites of the Bi(011) surface (red bar). The more negative binding energy reflects stronger interaction. Zero corresponds to the solid model at infinite separation from the gas phase adsorbate species. Reproduced from Ref. [119] with permission from Springer Nature.

tron microscope, high-resolution scanning transmission electron microscope, and electron paramagnetic resonance. Nevertheless, a series of inevitable challenges still impede their development for practical applications. For example, defects are reactive and easy to be poisoned or removed under certain conditions. In fact, most of the reaction conditions are either strong acids or strong alkalis. If the catalysts do not have a wide pH tolerance, the catalytic active sites will be rapidly corroded during the electrocatalytic process. In addition, high oxidation potentials or reduction potentials significantly induce surface reconstruction of the electrocatalysts, especially for OER and NRR, causing large deviation in composition from the pristine stoichiometry. Thus, defect evolution at extreme conditions is a non-negligible issue for defect-tailored 2D electrocatalysts. Another significant issue is the controlled synthesis of specific defects. Variations in coordination environments cause them to exhibit different electronic properties, thus affecting the performance of electrocatalysis. Due to the complexity and diversity of defect structures involving defect types, defect concentrations and defect locations, it is very challenging to obtain specific defects without other by-products. In most situations, several different defects are concomitant in the lattice of active materials. Some of them could produce "a whole greater than the sum of the individual components" by synergistic effects, others will be harmful for electrocatalysis due to their huge incompatibility. In addition, a multitude of defects can disturb the study of defect-property relationships. At last, current researchers focus more on the fine characterization of the original or final defect structure rather than dynamic evolution throughout the catalytic process. Hence, it is challenging to uncover the real active sites and catalytic mechanisms. In addition, many studies have adopted DFT calculations to explore the variations of the electronic structure and reaction energy barriers. Even so, the simulated models are highly controversial due to lack of statistically significant experimental data. Hence, the defects and corresponding catalytic mechanisms require more comprehensive understanding.

Finally, a rational design of high-efficiency electrocatalysts on basis of these critical issues can be used to fabricate defect-chemistry-based electrocatalysts. With the advancement in synthetic processes and well-developed characterization tools during growth and post growth, high-efficiency and cost-effective electrocatalysis could be achieved in the future.

Acknowledgments

The authors acknowledge the financial support from the National Natural Science Foundation of China (Grant Nos. 51972191, 51722207). We also appreciate the fruitful discussions with Dr. Hongwei Zhang.

References

[1] C.C. McCrory, S. Jung, I.M. Ferrer, S.M. Chatman, J.C. Peters, T.F. Jaramillo, Benchmarking hydrogen evolving reaction and oxygen evolving reaction electrocatalysts for solar water splitting devices, Journal of the American Chemical Society 137 (2015) 4347–4357.

[2] G.J.K. Acres, J.C. Frost, G.A. Hards, R.J. Potter, T.R. Ralph, D. Thompsett, et al., Electrocatalysts for fuel cells, Catalysis Today 38 (1997) 339–400.

[3] Q. Lu, F. Jiao, Electrochemical CO_2 reduction: electrocatalyst, reaction mechanism, and process engineering, Nano Energy 29 (2016) 439–456.

[4] Y. Jia, K. Jiang, H. Wang, X. Yao, The role of defect sites in nanomaterials for electrocatalytic energy conversion, Chem 5 (2019) 1371–1397.

[5] C. Tang, H.F. Wang, Q. Zhang, Multiscale principles to boost reactivity in gas-involving energy electrocatalysis, Accounts of Chemical Research 51 (2018) 881–889.

[6] X. Chia, M. Pumera, Characteristics and performance of two-dimensional materials for electrocatalysis, Nature Catalysis 1 (2018) 909–921.

[7] D. Deng, K.S. Novoselov, Q. Fu, N. Zheng, Z. Tian, X. Bao, Catalysis with two-dimensional materials and their heterostructures, Nature Nanotechnology 11 (2016) 218–230.

[8] Y. Yu, S.Y. Huang, Y. Li, S.N. Steinmann, W. Yang, L. Cao, Layer-dependent electrocatalysis of MoS_2 for hydrogen evolution, Nano Letters 14 (2014) 553–558.

[9] C. Xie, D. Yan, W. Chen, Y. Zou, R. Chen, S. Zang, et al., Insight into the design of defect electrocatalysts: from electronic structure to adsorption energy, Materials Today 31 (2019) 47–68.

[10] L. Zhang, J. Xiao, H. Wang, M. Shao, Carbon-based electrocatalysts for hydrogen and oxygen evolution reactions, ACS Catalysis 7 (2017) 7855–7865.

[11] M. Zeng, Y. Li, Recent advances in heterogeneous electrocatalysts for the hydrogen evolution reaction, Journal of Materials Chemistry A 3 (2015) 14942–14962.

[12] R. Parsons, The rate of electrolytic hydrogen evolution and the heat of adsorption of hydrogen, Transactions of the Faraday Society 54 (1958) 1053–1063.

[13] Y. Zheng, Y. Jiao, M. Jaroniec, S.Z. Qiao, Advancing the electrochemistry of the hydrogen-evolution reaction through combining experiment and theory, Angewandte Chemie. International Edition in English 54 (2015) 52–65.

[14] S. Trasatti, Work function, electronegativity, and electrochemical behaviour of metals: III. Electrolytic hydrogen evolution in acid solutions, Journal of Electroanalytical Chemistry 39 (1972) 163–184.

[15] J.K. Nørskov, T. Bligaard, A. Logadottir, J. Kitchin, J.G. Chen, S. Pandelov, et al., Trends in the exchange current for hydrogen evolution, Journal of the Electrochemical Society 152 (2005) J23–J26.

References

[16] J. Wang, F. Xu, H. Jin, Y. Chen, Y. Wang, Non-noble metal-based carbon composites in hydrogen evolution reaction: fundamentals to applications, Advanced Materials 29 (2017) 1605838.

[17] A.P. Murthy, J. Madhavan, K. Murugan, Recent advances in hydrogen evolution reaction catalysts on carbon/carbon-based supports in acid media, Journal of Power Sources 398 (2018) 9–26.

[18] G. Zhao, K. Rui, S.X. Dou, W. Sun, Heterostructures for electrochemical hydrogen evolution reaction: a review, Advanced Functional Materials 28 (2018) 1803291.

[19] J. Zhu, L. Hu, P. Zhao, L.Y.S. Lee, K.Y. Wong, Recent advances in electrocatalytic hydrogen evolution using nanoparticles, Chemical Reviews 120 (2020) 851–918.

[20] X. Zou, Y. Zhang, Noble metal-free hydrogen evolution catalysts for water splitting, Chemical Society Reviews 44 (2015) 5148–5180.

[21] D. Voiry, J. Yang, M. Chhowalla, Recent strategies for improving the catalytic activity of 2D TMD nanosheets toward the hydrogen evolution reaction, Advanced Materials 28 (2016) 6197–6206.

[22] T.F. Jaramillo, K.P. Jørgensen, J. Bonde, J.H. Nielsen, S. Horch, I. Chorkendorff, Identification of active edge sites for electrochemical H_2 evolution from MoS_2 nanocatalysts, Science 317 (2007) 100–102.

[23] G. Li, D. Zhang, Q. Qiao, Y. Yu, D. Peterson, A. Zafar, et al., All the catalytic active sites of MoS_2 for hydrogen evolution, Journal of the American Chemical Society 138 (2016) 16632–16638.

[24] Y. Huang, Y. Sun, X. Zheng, T. Aoki, B. Pattengale, J. Huang, et al., Atomically engineering activation sites onto metallic 1T-MoS_2 catalysts for enhanced electrochemical hydrogen evolution, Nature Communications 10 (2019) 982.

[25] X.-L. Fan, Y. Yang, P. Xiao, W.-M. Lau, Site-specific catalytic activity in exfoliated MoS_2 single-layer polytypes for hydrogen evolution: basal plane and edges, Journal of Materials Chemistry A 2 (2014) 20545–20551.

[26] G. Ye, Y. Gong, J. Lin, B. Li, Y. He, S.T. Pantelides, et al., Defects engineered monolayer MoS_2 for improved hydrogen evolution reaction, Nano Letters 16 (2016) 1097–1103.

[27] J. Kibsgaard, Z. Chen, B.N. Reinecke, T.F. Jaramillo, Engineering the surface structure of MoS_2 to preferentially expose active edge sites for electrocatalysis, Nature Materials 11 (2012) 963–969.

[28] J. Zhu, Z.C. Wang, H. Dai, Q. Wang, R. Yang, H. Yu, et al., Boundary activated hydrogen evolution reaction on monolayer MoS_2, Nature Communications 10 (2019) 1348.

[29] M.A. Lukowski, A.S. Daniel, F. Meng, A. Forticaux, L. Li, S. Jin, Enhanced hydrogen evolution catalysis from chemically exfoliated metallic MoS_2 nanosheets, Journal of the American Chemical Society 135 (2013) 10274–10277.

[30] J. Hong, Z. Hu, M. Probert, K. Li, D. Lv, X. Yang, et al., Exploring atomic defects in molybdenum disulphide monolayers, Nature Communications 6 (2015) 6293.

[31] W. Zhou, X. Zou, S. Najmaei, Z. Liu, Y. Shi, J. Kong, et al., Intrinsic structural defects in monolayer molybdenum disulfide, Nano Letters 13 (2013) 2615–2622.

[32] A.Y. Lu, X. Yang, C.C. Tseng, S. Min, S.H. Lin, C.L. Hsu, et al., High-sulfur-vacancy amorphous molybdenum sulfide as a high current electrocatalyst in hydrogen evolution, Small 12 (2016) 5530–5537.

[33] H. Li, C. Tsai, A.L. Koh, L. Cai, A.W. Contryman, A.H. Fragapane, et al., Corrigendum: Activating and optimizing MoS_2 basal planes for hydrogen evolution through the formation of strained sulphur vacancies, Nature Materials 15 (2016) 48–53.

[34] C. Tsai, H. Li, S. Park, J. Park, H.S. Han, J.K. Norskov, et al., Electrochemical generation of sulfur vacancies in the basal plane of MoS_2 for hydrogen evolution, Nature Communications 8 (2017) 15113.

[35] Y. Zang, S. Niu, Y. Wu, X. Zheng, J. Cai, J. Ye, et al., Tuning orbital orientation endows molybdenum disulfide with exceptional alkaline hydrogen evolution capability, Nature Communications 10 (2019) 1217.

[36] J. Xie, J. Zhang, S. Li, F. Grote, X. Zhang, H. Zhang, et al., Controllable disorder engineering in oxygen-incorporated MoS_2 ultrathin nanosheets for efficient hydrogen evolution, Journal of the American Chemical Society 135 (2013) 17881–17888.

[37] P. Liu, J. Zhu, J. Zhang, P. Xi, K. Tao, D. Gao, et al., P dopants triggered new basal plane active sites and enlarged interlayer spacing in MoS_2 nanosheets toward electrocatalytic hydrogen evolution, ACS Energy Letters 2 (2017) 745–752.

[38] W. Xiao, P. Liu, J. Zhang, W. Song, Y.P. Feng, D. Gao, et al., Dual-functional N dopants in edges and basal plane of MoS_2 nanosheets toward efficient and durable hydrogen evolution, Advanced Energy Materials 7 (2017) 1602086.

[39] N. Huang, R. Peng, Y. Ding, S. Yan, G. Li, P. Sun, et al., Facile chemical-vapour-deposition synthesis of vertically aligned Co-doped MoS_2 nanosheets as an efficient catalyst for triiodide reduction and hydrogen evolution reaction, Journal of Catalysis 373 (2019) 250–259.

[40] J. Zhang, T. Wang, P. Liu, S. Liu, R. Dong, X. Zhuang, et al., Engineering water dissociation sites in MoS_2 nanosheets for accelerated electrocatalytic hydrogen production, Energy & Environmental Science 9 (2016) 2789–2793.

[41] J. Deng, H. Li, J. Xiao, Y. Tu, D. Deng, H. Yang, et al., Triggering the electrocatalytic hydrogen evolution activity of the inert two-dimensional MoS_2 surface via single-atom metal doping, Energy & Environmental Science 8 (2015) 1594–1601.

[42] C. Du, H. Huang, J. Jian, Y. Wu, M. Shang, W. Song, Enhanced electrocatalytic hydrogen evolution performance of MoS_2 ultrathin nanosheets via Sn doping, Applied Catalysis. A, General 538 (2017) 1–8.

[43] K. Sun, L. Zeng, S. Liu, L. Zhao, H. Zhu, J. Zhao, et al., Design of basal plane active MoS_2 through one-step nitrogen and phosphorus co-doping as an efficient pH-universal electrocatalyst for hydrogen evolution, Nano Energy 58 (2019) 862–869.

[44] W. Wu, C. Niu, C. Wei, Y. Jia, C. Li, Q. Xu, Activation of MoS_2 basal planes for hydrogen evolution by Zinc, Angewandte Chemie. International Edition in English 58 (2019) 2029–2033.

[45] K. Qi, X. Cui, L. Gu, S. Yu, X. Fan, M. Luo, et al., Single-atom cobalt array bound to distorted 1T MoS_2 with ensemble effect for hydrogen evolution catalysis, Nature Communications 10 (2019) 5231.

[46] P.V. Sarma, A. Kayal, C.H. Sharma, M. Thalakulam, J. Mitra, M.M. Shaijumon, Electrocatalysis on edge-rich spiral WS_2 for hydrogen evolution, ACS Nano 13 (2019) 10448–10455.

[47] X. Wang, X. Gan, T. Hu, K. Fujisawa, Y. Lei, Z. Lin, et al., Noble-metal-free hybrid membranes for highly efficient hydrogen hvolution, Advanced Materials 29 (2017) 1603617.

[48] H. Shu, D. Zhou, F. Li, D. Cao, X. Chen, Defect engineering in $MoSe_2$ for the hydrogen evolution reaction: from point defects to edges, ACS Applied Materials & Interfaces 9 (2017) 42688–42698.

[49] M. Yu, S. Zhou, Z. Wang, J. Zhao, J. Qiu, Boosting electrocatalytic oxygen evolution by synergistically coupling layered double hydroxide with MXene, Nano Energy 44 (2018) 181–190.

[50] T.A. Le, Q.V. Bui, N.Q. Tran, Y. Cho, Y. Hong, Y. Kawazoe, et al., Synergistic effects of nitrogen doping on MXene for enhancement of hydrogen evolution reaction, ACS Sustainable Chemistry & Engineering 7 (2019) 16879–16888.

[51] J. Zhang, Y. Zhao, X. Guo, C. Chen, C.-L. Dong, R.-S. Liu, et al., Single platinum atoms immobilized on an MXene as an efficient catalyst for the hydrogen evolution reaction, Nature Catalysis 1 (2018) 985–992.

[52] Y. Yoon, A.P. Tiwari, M. Choi, T.G. Novak, W. Song, H. Chang, et al., Precious-metal-free electrocatalysts for activation of hydrogen evolution with nonmetallic electron donor: chemical composition controllable phosphorous doped Vanadium Carbide MXene, Advanced Functional Materials 29 (2019) 1903443.

[53] N.T. Suen, S.F. Hung, Q. Quan, N. Zhang, Y.J. Xu, H.M. Chen, Electrocatalysis for the oxygen evolution reaction: recent development and future perspectives, Chemical Society Reviews 46 (2017) 337–365.

[54] Z. Cai, X. Bu, P. Wang, J.C. Ho, J. Yang, X. Wang, Recent advances in layered double hydroxide electrocatalysts for the oxygen evolution reaction, Journal of Materials Chemistry A 7 (2019) 5069–5089.

[55] F. Song, X. Hu, Exfoliation of layered double hydroxides for enhanced oxygen evolution catalysis, Nature Communications 5 (2014) 4477.

[56] J. Li, R. Lian, J. Wang, S. He, S.P. Jiang, Z. Rui, Oxygen vacancy defects modulated electrocatalytic activity of iron-nickel layered double hydroxide on Ni foam as highly active electrodes for oxygen evolution reaction, Electrochimica Acta 331 (2020) 135395.

[57] H. Xu, B. Wang, C. Shan, P. Xi, W. Liu, Y. Tang, Ce-doped NiFe-layered double hydroxide ultrathin nanosheets/nanocarbon hierarchical nanocomposite as an efficient oxygen evolution catalyst, ACS Applied Materials & Interfaces 10 (2018) 6336–6345.

[58] R. Liu, Y. Wang, D. Liu, Y. Zou, S. Wang, Water-plasma-enabled exfoliation of ultrathin layered double hydroxide nanosheets with multivacancies for water oxidation, Advanced Materials 29 (2017) 1701546.

[59] Y. Jia, L. Zhang, G. Gao, H. Chen, B. Wang, J. Zhou, et al., A heterostructure coupling of exfoliated Ni-Fe hydroxide nanosheet and defective graphene as a bifunctional electrocatalyst for overall water splitting, Advanced Materials 29 (2017) 1700017.

[60] X. Zhang, Y. Zhao, Y. Zhao, R. Shi, G.I.N. Waterhouse, T. Zhang, A simple synthetic strategy toward defect-rich porous monolayer NiFe-layered double hydroxide nanosheets for efficient electrocatalytic water oxidation, Advanced Energy Materials 9 (2019) 1900881.

[61] Y. Wang, Y. Zhang, Z. Liu, C. Xie, S. Feng, D. Liu, et al., Layered double hydroxide nanosheets with multiple vacancies obtained by dry exfoliation as highly efficient oxygen evolution electrocatalysts, Angewandte Chemie. International Edition in English 56 (2017) 5867–5871.

References

[62] Y. Wang, C. Xie, Z. Zhang, D. Liu, R. Chen, S. Wang, In situ exfoliated, N-doped, and edge-rich ultrathin layered double hydroxides nanosheets for oxygen evolution reaction, Advanced Functional Materials 28 (2018) 1703363.

[63] Y.S. Xie, Z. Wang, M. Ju, X. Long, S. Yang, Dispersing transition metal vacancies in layered double hydroxides by ionic reductive complexation extraction for efficient water oxidation, Chemical Science 10 (2019) 8354–8359.

[64] D. Zhou, Z. Cai, Y. Jia, X. Xiong, Q. Xie, S. Wang, et al., Activating basal plane in NiFe layered double hydroxide by Mn(2+) doping for efficient and durable oxygen evolution reaction, Nanoscale Horizons 3 (2018) 532–537.

[65] J. Su, G.D. Li, X.H. Li, J.S. Chen, 2D/2D heterojunctions for catalysis, Advanced Science 6 (2019) 1801702.

[66] Y. Xue, J. Li, Z. Xue, Y. Li, H. Liu, D. Li, et al., Extraordinarily durable graphdiyne-supported electrocatalyst with high activity for hydrogen production at all values of pH, ACS Applied Materials & Interfaces 8 (2016) 31083–31091.

[67] Y. Fang, Y. Xue, L. Hui, H. Yu, Y. Liu, C. Xing, et al., In situ growth of graphdiyne based heterostructure: toward efficient overall water splitting, Nano Energy 59 (2019) 591–597.

[68] Y. Xue, Z. Zuo, Y. Li, H. Liu, Y. Li, Graphdiyne-supported $NiCo_2S_4$ nanowires: a highly active and stable 3D bifunctional electrode material, Small 13 (2017) 1700936.

[69] H.-Y. Si, Q.-X. Deng, L.-C. Chen, L. Wang, X.-Y. Liu, W.-S. Wu, et al., Hierarchical graphdiyne@NiFe layered double hydroxide heterostructures as a bifunctional electrocatalyst for overall water splitting, Journal of Alloys and Compounds 794 (2019) 261–267.

[70] M. Shakeel, M. Arif, G. Yasin, B. Li, H.D. Khan, Layered by layered Ni-Mn-LDH/g-C_3N_4 nanohybrid for multi-purpose photo/electrocatalysis: Morphology controlled strategy for effective charge carriers separation, Applied Catalysis. B, Environmental 242 (2019) 485–498.

[71] C. Hao, Y. Wu, Y. An, B. Cui, J. Lin, X. Li, et al., Interface-coupling of CoFe-LDH on MXene as high-performance oxygen evolution catalyst, Materials Today Energy 12 (2019) 453–462.

[72] P. Xiong, X. Zhang, H. Wan, S. Wang, Y. Zhao, J. Zhang, et al., Interface modulation of two-dimensional superlattices for efficient overall water splitting, Nano Letters 19 (2019) 4518–4526.

[73] M.S. Islam, M. Kim, X. Jin, S.M. Oh, N.-S. Lee, H. Kim, et al., Bifunctional 2D superlattice electrocatalysts of layered double hydroxide–transition metal dichalcogenide active for overall water splitting, ACS Energy Letters 3 (2018) 952–960.

[74] J. Suntivich, K.J. May, H.A. Gasteiger, J.B. Goodenough, Y. Shao-Horn, A perovskite oxide optimized for oxygen evolution catalysis from molecular orbital principles, Science 334 (2011) 1383–1385.

[75] Y. Liu, H. Cheng, M. Lyu, S. Fan, Q. Liu, W. Zhang, et al., Low overpotential in vacancy-rich ultrathin $CoSe_2$ nanosheets for water oxidation, Journal of the American Chemical Society 136 (2014) 15670–15675.

[76] J. Wu, M. Liu, K. Chatterjee, K.P. Hackenberg, J. Shen, X. Zou, et al., Exfoliated 2D transition metal disulfides for enhanced electrocatalysis of oxygen evolution reaction in acidic medium, Advanced Materials Interfaces 3 (2016) 1500669.

[77] J. Zhu, X. Jiang, Y. Yang, Q. Chen, X.X. Xue, K. Chen, et al., Synergy of tellurium and defects in control of activity of phosphorene for oxygen evolution and reduction reactions, Physical Chemistry Chemical Physics 21 (2019) 22939–22946.

[78] C. Hu, L. Dai, Multifunctional carbon-based metal-free electrocatalysts for simultaneous oxygen reduction, oxygen evolution, and hydrogen evolution, Advanced Materials 29 (2017) 1604942.

[79] Y. Zheng, Y. Jiao, Y. Zhu, Q. Cai, A. Vasileff, L.H. Li, et al., Molecule-level g-C_3N_4 coordinated transition metals as a new class of electrocatalysts for oxygen electrode reactions, Journal of the American Chemical Society 139 (2017) 3336–3339.

[80] X. Wu, Z. Wang, Y. Han, D. Zhang, M. Wang, H. Li, et al., Chemically coupled NiCoS/C nanocages as efficient electrocatalysts for nitrogen reduction reactions, Journal of Materials Chemistry A 8 (2020) 543–547.

[81] N. Cao, G. Zheng, Aqueous electrocatalytic N_2 reduction under ambient conditions, Nano Research 11 (2018) 2992–3008.

[82] Y. Wan, J. Xu, R. Lv, Heterogeneous electrocatalysts design for nitrogen reduction reaction under ambient conditions, Materials Today 27 (2019) 69–90.

[83] C.J. van der Ham, M.T. Koper, D.G. Hetterscheid, Challenges in reduction of dinitrogen by proton and electron transfer, Chemical Society Reviews 43 (2014) 5183–5191.

[84] E. Skulason, T. Bligaard, S. Gudmundsdottir, F. Studt, J. Rossmeisl, F. Abild-Pedersen, et al., A theoretical evaluation of possible transition metal electro-catalysts for N_2 reduction, Physical Chemistry Chemical Physics 14 (2012) 1235–1245.

[85] H. Jin, L. Li, X. Liu, C. Tang, W. Xu, S. Chen, et al., Nitrogen vacancies on 2D layered W_2N_3: a stable and efficient active site for nitrogen reduction reaction, Advanced Materials 31 (2019) 1902709.

[86] Q. Li, L. He, C. Sun, X. Zhang, Computational study of MoN_2 monolayer as electrochemical catalysts for nitrogen reduction, Journal of Physical Chemistry C 121 (2017) 27563–27568.

[87] G. Peng, J. Wu, M. Wang, J. Niklas, H. Zhou, C. Liu, Nitrogen-defective polymeric carbon nitride nanolayer enabled efficient electrocatalytic nitrogen reduction with high Faradaic efficiency, Nano Letters 20 (2020) 2879–2885.

[88] C. Ren, Y. Zhang, Y. Li, Y. Zhang, S. Huang, W. Lin, et al., Whether corrugated or planar vacancy graphene-like carbon nitride (g-C_3N_4) is more effective for nitrogen reduction reaction?, The Journal of Physical Chemistry C 123 (2019) 17296–17305.

[89] J. Zhang, X. Tian, M. Liu, H. Guo, J. Zhou, Q. Fang, et al., Cobalt-modulated molybdenum-dinitrogen interaction in MoS_2 for catalyzing ammonia synthesis, Journal of the American Chemical Society 141 (2019) 19269–19275.

[90] D. Yan, H. Li, C. Chen, Y. Zou, S. Wang, Defect engineering strategies for nitrogen reduction reactions under ambient conditions, Small Methods 3 (2018) 1800331.

[91] Y. Qiu, X. Peng, F. Lü, Y. Mi, L. Zhuo, J. Ren, et al., Single-atom catalysts for the electrocatalytic reduction of nitrogen to ammonia under ambient conditions, Chemistry-An Asian Journal 14 (2019) 2770–2779.

[92] J. Zhao, Z. Chen, Single Mo atom supported on defective boron nitride monolayer as an efficient electrocatalyst for nitrogen fixation: a computational study, Journal of the American Chemical Society 139 (2017) 12480–12487.

[93] P. Liu, C. Fu, Y. Li, H. Wei, Theoretical screening of single atoms anchored on defective graphene for electrocatalytic N_2 reduction reactions: a DFT study, Physical Chemistry Chemical Physics 22 (2020) 9322–9329.

[94] C. Choi, S. Back, N.-Y. Kim, J. Lim, Y.-H. Kim, Y. Jung, Suppression of hydrogen evolution reaction in electrochemical N_2 reduction using single-atom catalysts: a computational guideline, ACS Catalysis 8 (2018) 7517–7525.

[95] Z. Chen, J. Zhao, C.R. Cabrera, Z. Chen, Computational screening of efficient single-atom catalysts based on graphitic carbon nitride (g-C_3N_4) for nitrogen electroreduction, Small Methods 3 (2018) 1800368.

[96] J. Guo, T. Tadesse Tsega, I. Ul Islam, A. Iqbal, J. Zai, X. Qian, Fe doping promoted electrocatalytic N_2 reduction reaction of 2H MoS_2, Chinese Chemical Letters 31 (2020) 2487–2490.

[97] J. Li, S. Chen, F. Quan, G. Zhan, F. Jia, Z. Ai, et al., Accelerated dinitrogen electroreduction to ammonia via interfacial polarization triggered by single-atom protrusions, Chem 6 (2020) 885–901.

[98] X. Yu, P. Han, Z. Wei, L. Huang, Z. Gu, S. Peng, et al., Boron-doped graphene for electrocatalytic N_2 reduction, Joule 2 (2018) 1610–1622.

[99] Y. Tian, D. Xu, K. Chu, Z. Wei, W. Liu, Metal-free N, S co-doped graphene for efficient and durable nitrogen reduction reaction, Journal of Materials Science 54 (2019) 9088–9097.

[100] W. Kong, F. Gong, Q. Zhou, G. Yu, L. Ji, X. Sun, et al., An MnO_2-$Ti_3C_2T_x$ MXene nanohybrid: an efficient and durable electrocatalyst toward artificial N_2 fixation to NH_3 under ambient conditions, Journal of Materials Chemistry A 7 (2019) 18823–18827.

[101] X. Li, X. Ren, X. Liu, J. Zhao, X. Sun, Y. Zhang, et al., A MoS_2 nanosheet–reduced graphene oxide hybrid: an efficient electrocatalyst for electrocatalytic N_2 reduction to NH_3 under ambient conditions, Journal of Materials Chemistry A 7 (2019) 2524–2528.

[102] K. Chu, Y.P. Liu, Y.B. Li, Y.L. Guo, Y. Tian, Two-dimensional (2D)/2D interface engineering of a MoS_2/C_3N_4 heterostructure for promoted electrocatalytic nitrogen fixation, ACS Applied Materials & Interfaces 12 (2020) 7081–7090.

[103] Y. Zheng, A. Vasileff, X. Zhou, Y. Jiao, M. Jaroniec, S.Z. Qiao, Understanding the roadmap for electrochemical reduction of CO_2 to multi-carbon oxygenates and hydrocarbons on copper-based catalysts, Journal of the American Chemical Society 141 (2019) 7646–7659.

[104] T.-T. Zhuang, Z.-Q. Liang, A. Seifitokaldani, Y. Li, P. De Luna, T. Burdyny, et al., Steering post-C–C coupling selectivity enables high efficiency electroreduction of carbon dioxide to multi-carbon alcohols, Nature Catalysis 1 (2018) 421–428.

[105] L. Sun, V. Reddu, A.C. Fisher, X. Wang, Electrocatalytic reduction of carbon dioxide: opportunities with heterogeneous molecular catalysts, Energy & Environmental Science 13 (2020) 374–403.

[106] Z. Sun, T. Ma, H. Tao, Q. Fan, B. Han, Fundamentals and challenges of electrochemical CO_2 reduction using two-dimensional materials, Chem 3 (2017) 560–587.

Reference 337

[107] L.R.L. Ting, B.S. Yeo, Recent advances in understanding mechanisms for the electrochemical reduction of carbon dioxide, Current Opinion in Electrochemistry 8 (2018) 126–134.

[108] J. Wu, M. Liu, P.P. Sharma, R.M. Yadav, L. Ma, Y. Yang, et al., Incorporation of nitrogen defects for efficient reduction of CO_2 via two-electron pathway on three-dimensional graphene Foam, Nano Letters 16 (2016) 466–470.

[109] N. Sreekanth, M.A. Nazrulla, T.V. Vineesh, K. Sailaja, K.L. Phani, Metal-free boron-doped graphene for selective electroreduction of carbon dioxide to formic acid/formate, Chemical Communications 51 (2015) 16061–16064.

[110] J. Xie, X. Zhao, M. Wu, Q. Li, Y. Wang, J. Yao, Metal-free fluorine-doped carbon electrocatalyst for CO_2 reduction outcompeting hydrogen evolution, Angewandte Chemie. International Edition in English 130 (2018) 9788–9792.

[111] C. Zhang, S. Yang, J. Wu, M. Liu, S. Yazdi, M. Ren, et al., Electrochemical CO_2 reduction with atomic iron-dispersed on nitrogen-doped graphene, Advanced Energy Materials 8 (2018) 1703487.

[112] S. Gao, Z. Sun, W. Liu, X. Jiao, X. Zu, Q. Hu, et al., Atomic layer confined vacancies for atomic-level insights into carbon dioxide electroreduction, Nature Communications 8 (2017) 14503.

[113] P. Abbasi, M. Asadi, C. Liu, S. Sharifi-Asl, B. Sayahpour, A. Behranginia, et al., Tailoring the edge structure of molybdenum disulfide toward electrocatalytic reduction of carbon dioxide, ACS Nano 11 (2017) 453–460.

[114] X. Hong, K. Chan, C. Tsai, J.K. Nørskov, How doped MoS_2 breaks transition-metal scaling relations for CO_2 electrochemical reduction, ACS Catalysis 6 (2016) 4428–4437.

[115] X. Feng, K. Jiang, S. Fan, M.W. Kanan, Grain-boundary-dependent CO_2 electroreduction activity, Journal of the American Chemical Society 137 (2015) 4606–4609.

[116] K. Jiang, R.B. Sandberg, A.J. Akey, X. Liu, D.C. Bell, J.K. Nørskov, et al., Metal ion cycling of Cu foil for selective C-C coupling in electrochemical CO_2 reduction, Nature Catalysis 1 (2018) 111–119.

[117] Q. Lu, J. Rosen, Y. Zhou, G.S. Hutchings, Y.C. Kimmel, J.G. Chen, et al., A selective and efficient electrocatalyst for carbon dioxide reduction, Nature Communications 5 (2014) 3242.

[118] N. Han, Y. Wang, H. Yang, J. Deng, J. Wu, Y. Li, et al., Ultrathin bismuth nanosheets from in situ topotactic transformation for selective electrocatalytic CO_2 reduction to formate, Nature Communications 9 (2018) 1320.

[119] F. Yang, A.O. Elnabawy, R. Schimmenti, P. Song, J. Wang, Z. Peng, et al., Bismuthene for highly efficient carbon dioxide electroreduction reaction, Nature Communications 11 (2020) 1088.

CHAPTER
11

Devices and defects in two-dimensional materials: outlook and perspectives

Amritesh Rai[a], Anupam Roy[a], Amithraj Valsaraj[a], Sayema Chowdhury[a], Deepyanti Taneja[a], Yaguo Wang[b], Leonard Frank Register[a], and Sanjay K. Banerjee[a]

[a]Microelectronics Research Center and Department of Electrical and Computer Engineering, The University of Texas at Austin, Austin, TX, United States [b]Department of Mechanical Engineering and Texas Materials Institute, The University of Texas at Austin, Austin, TX, United States

11.1 Introduction

Two-dimensional (2D) layered materials range from insulators to semiconductors and metals and among these 2D materials, the family of transition metal dichalcogenides (TMDs) has garnered the most attention. Transition metal dichalcogenides are characterized by the general formula MX_2 where M represents a transition metal (M = Mo, W, Re, etc.) and X is a chalcogen (X = S, Se, Te). These layered 2D TMDs can be isolated down to a single atomic layer from their bulk form. A TMD monolayer can be visualized as a layer of transition metal atoms sandwiched in-between two layers of chalcogen atoms (of the form X-M-X) with strong intra-layer covalent bonding, whereas the inter-layer bonding between two adjacent TMD layers is of the van der Waals (vdW) type. Moreover, depending on the specific crystal structure and atomic layer stacking sequence 1T, 2H or 3R, these TMDs can have metallic, semiconducting, or superconducting phases. Of particular interest is the subset of semiconducting 2D TMDs as they offer several promising advantages over conventional 3D semiconductors (Si, Ge and III-Vs) such as: (i) inherent ultra-thin bodies enabling enhanced electrostatic gate control and carrier confinement versus 3D bulk semiconductors (this can help mitigate SCE in ultra-scaled FETs based on 2D TMDs as their ultra-thin bodies can allow significant reduc-

Defects in Two-Dimensional Materials
https://doi.org/10.1016/B978-0-12-820292-0.00017-3

339

Copyright © 2022 Elsevier Inc. All rights reserved.

tion of the so-called characteristic "channel length (L_{CH}) scaling" factor "λ"); (ii) availability of a wide range of sizable bandgaps and diverse band-alignments; and (iii) lack of surface "dangling bonds" unlike conventional 3D semiconductors allowing for the formation of pristine defect-free interfaces (especially 2D/2D vdW interfaces). These attributes make the semiconducting 2D TMDs extremely promising for future "ultra-scaled" and "ultra-low-power" devices. Among the semiconducting 2D TMDs, MoS_2 has been the most popular and widely pursued material by the research community owing to its natural availability and environmental/ambient stability. Like most semiconducting TMDs, MoS_2 is characterized by a thickness-dependent bandgap and this bandgap variability, together with high carrier mobilities, mechanical flexibility, and optical transparency, makes 2D MoS_2 extremely attractive for practical nano- and optoelectronic device applications on both rigid and flexible platforms.

MoS_2 can also be combined with other 2D materials (e.g., TMDs or graphene), and 1D and 0D materials to form various 2D/3D, 2D/2D, 2D/1D and 2D/0D vdW heterostructure devices, enabling a wide gamut of functionalities. Several device applications such as ultra-scaled FETs, digital logic, memory, analog/RF, conventional diodes, photodetectors, light emitting diodes (LEDs), lasers, photovoltaics, sensors, ultra-low-power tunneling-devices such as tunnel-FETs (TFETs), and piezotronics, among several others, have been demonstrated using 2D MoS_2 (either on exfoliated MoS_2 flakes or synthesized MoS_2 films), highlighting its promise and versatility. Concurrently, massive research efforts have been devoted to solving various key technical challenges, such as large-area wafer-scale synthesis using techniques like chemical vapor deposition (CVD) and its variants (e.g. metal–organic CVD or MOCVD), van der Waals (vdW) epitaxy, reduction of parasitic contact resistance (R_C) and defect density, as well as enhancement of charge carrier mobility (μ), that can improve the operational efficiency of these devices and allow 2D MoS_2-based circuits and systems to become technologically and commercially relevant. While much of the ensuing discussion in this chapter is centered around 2D MoS_2, the results and concepts presented are mostly applicable and extendable to other members of the 2D semiconducting TMD family as well.

The following section will first focus on providing an overview of a novel defect characterization technique using ultrafast laser spectroscopy that can determine defect density and reveal the defect-carrier interaction in 2D semiconductors. Extraction of the defect-modulated carrier lifetime for exfoliated, CVD-grown and MBE-grown 2D materials as well as real-time detection of defect trapping-induced carrier behavior using ultrafast laser spectroscopy will be discussed. Next, the latest progress, current challenges, and future opportunities in the device applications of 2D TMD materials grown *via* proven large-area growth techniques, such as CVD and molecular beam epitaxy (MBE), will be elucidated. The focus will then shift to 2D van der Waals (vdW) heterostructures wherein their latest fabrication methods and prominent experimental reports of heterostructure devices based on 2D TMDs will be highlighted. The emerging field of 'Twistronics', wherein the angle or 'twist' between the two constituent layers of a 2D heterostructure can be manipulated to realize novel electronic properties, will also be discussed. The penultimate section of the chapter will shed light on some major fundamental bottlenecks to realizing low-power, high-performance and ultra-scaled devices based on 2D semiconducting TMDs and will talk about some promising defect engineering techniques and how they can be utilized to tailor the properties of 2D TMDs for realizing devices with enhanced functionalities. Finally, the chapter will conclude with a theoretical analysis of the effects of metal and chalcogen vacancies in 2D TMDs, such as MoS_2, and the influ-

11.2 Defect characterization in 2D TMDs using ultrafast pump-probe spectroscopy

ence of interfacial oxygen vacancies at TMD/High-κ dielectric interfaces on the 2D material properties.

11.2 Defect characterization in 2D TMDs using ultrafast pump-probe spectroscopy

11.2.1 Motivation

As we learned from the semiconductor industry that has evolved for more than half a century, defects play a key role in tailoring material properties to achieve novel and versatile functions [1–3]. As new members of layered 2D material after graphene, TMDs have attracted extensive research interests ever since their re-discovery over a decade ago [4]. Few-layer TMDs have a common chemical formula MX_2, where, as shown in Fig. 11.1a, one layer of M atoms (transition metal atom, Mo, W, etc.) is sandwiched between two layers of X atoms (chalcogen atom, S, Se, or Te) [5]. Many types of defects have been identified in 2D TMD materials [6,7], such as X/M vacancies (Fig. 11.1b and 11.1e), impurities (Fig. 11.1c), grain boundaries (Fig. 11.1d and 11.1e), line defects (Fig. 11.1d), antisites (Mo atom replacing S atom, or vice versa) and so on. The presence of defects impacts the electronic band structure, optical and transport properties, all of which are governed by the defect-carrier interaction in TMDs. Even though most efforts by the crystal growth community are devoted to the production of large-scale, defect-free samples, the presence of defects is not always negative. Defects may provide us novel routes to tailor the properties of TMDs. For example, theoretical calculations predict that antisite Mo_S in monolayer MoS_2 can generate a magnetic moment [8]. Also, vacancies generated by irradiation can enhance the ferromagnetism in bulk MoS_2 [9].

When utilizing defect engineering to tune the physical properties of TMD materials, we need to know not only the defect species and density, but also how these defects interact with carriers, both of which are important to realize the desired functions. Therefore, understanding defect-carrier interaction and its effects on carrier dynamics and carrier transport is necessary to design and fabricate high-performance TMD-based devices with versatile functions. Our current understanding about defects in TMDs can be summarized as follows: (a) X/M vacancies typically can introduce mid-gap states in the band gap [10–12]; form bound excitons (neutral excitons bound to defects) in the materials with detectable PL peak at low temperature [10,13,14]; decrease the intrinsic exciton PL efficiency [10]; n-dope (X vacancy) [15] or p-dope the material [16] *via* the introduction of unpaired electrons; and induce a hopping behavior in electron transport [12]. (b) Grain boundaries/edges can hinder carrier transport by introducing localized charge-carrier states, and endow the material with an insulator-like property [17]. (c) When oxygen impurities occupy X vacancy sites, mid-gap states induced by the vacancies can be removed, which enhances carrier mobility [18]. (d) Charged impurity scattering, either from ionized impurities inside TMDs [19] or charge traps at the TMDs/substrate interface [20], has been proposed to be the dominant factor for the low mobility observed in TMDs devices at room-temperature. Charge traps at the TMD-gate insulator interface have also been suggested as the cause of hysteresis observed in TMD-based field effect transistors (FET) [21].

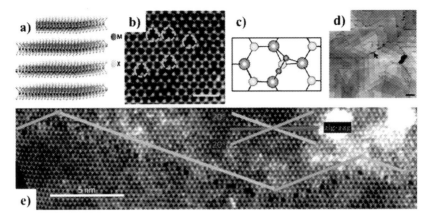

FIGURE 11.1 (a) Lattice structure of TMDs. (b) ADF-STEM image of mono- and bi-sulfur vacancies in exfoliated MoS$_2$ [8]. (c) Schematic picture of oxygen molecules bonded onto a perfect MoS$_2$ lattice [7]. (d) STM image of CVD grown MoS$_2$ film, showing grain boundary constructed by 1D defect line (red arrow), twin-mirror boundary (black arrow) and tilt boundary (green arrow) [60]. (e) ADF-STEM image showing ring defects at the grain boundary [16].

Due to the complex nature of defects in TMDs, as well as the large variety of defect species and density in samples synthesized with different approaches, combining various techniques to characterize the defects is necessary. Examples of conventional characterization techniques include PL spectroscopy and mapping to reveal the defect bound excitons, XPS to characterize the film stoichiometry and chemical composition, Raman spectroscopy to probe the defect-induced change on vibrational properties, and S/TEM to reveal the atomic scale defect morphology. Even though a lot of progress has been achieved, these steady state characterization approaches fail to provide information about the carrier dynamics in TMDs, which are important for opto-electronic devices and can be strongly influenced by the presence of defects. For example, PL can only probe the radiative recombination process, cannot reveal the non-radiative recombination that emit phonons, or the defects' role as electron/hole scatters. Transport properties, such as mobility and electrical resistance, have also been characterized extensively. Nevertheless, phenomena observed in transport measurements are still not being fully understood. There is an urgent need to conduct a thorough and systematic research to unveil defect-carrier interactions and defect-associated carrier dynamics in TMDs, which will advance fundamental physics and facilitate defect engineering in TMD materials and devices.

11.2.2 Pump-probe spectroscopy

For applications of TMDs in opto-electronic devices, photons are often used as external stimuli to excite electrons from the valence band to the conduction band [22–24]. Excited electrons can then form excitons or trions, emit photons or phonons, or recombine with holes, and eventually relax back to the valence band. Relaxation mechanisms of excited carriers (electrons, excitons, trions, etc.) determine the performance of TMD-based devices. **Excitons** are electron-hole pairs bound *via* the Coulomb force. In monolayer TMDs, the exciton binding

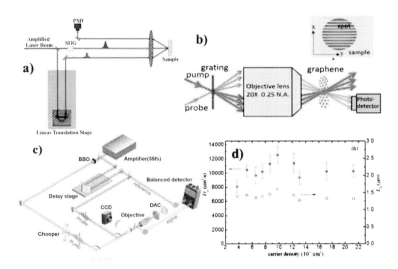

FIGURE 11.2 (a) Principle of ULS. (b) Schematics of ULS integrating a transmission grating for measuring diffusion coefficient. (c) Schematics of ULS setup. (d) Extracted carrier diffusion coefficient and diffusion length in monolayer graphene [36].

energies are found to be several hundred meV [25], corresponding to a temperature of several thousand Kelvin. Hence, the exciton recombination always dominates the photon emission process in monolayer TMDs [26,27]. If an extra electron or hole is bound with an exciton, a "trion" is formed. **Carrier lifetime** (τ) is the time taken by excited carriers to relax back to valence band, which determines response time of electronic devices. For some optoelectronics, e.g. ultrafast detectors [28], short carrier lifetimes are preferred. In most semiconductors, carrier lifetimes fall into picosecond to nanosecond regime, and cannot be determined by transport measurements. For non-gated 2D materials in the absence of an electrical field, carrier transport is dominated by diffusion. The **carrier diffusion length**, $L^2 = D\tau$, determines how far carriers can travel before recombination, where D is carrier diffusion coefficient. The exciton lifetime and exciton diffusion coefficient in monolayer TMDs are reported to be around 100 ps and about 10–20 cm^2/s [29–31].

Ultrafast laser spectroscopy (ULS) is a powerful time-resolved technique to measure carrier dynamics [32–35]. Principles of ULS are illustrated in Fig. 11.2a. An 800-nm laser beam is divided into two parts: a strong pump and a much weaker probe. The pump beam is converted to 400 nm *via* a second harmonic crystal (SHG). The pump beam prepares excited electron states. The probe beam detects the transient states in the material. The optical path of probe beam is controlled by a step-motorized linear translation stage, through which a time delay between pump and probe pulses is created. With a transmission grating integrated in ULS (Fig. 11.2b); density of excited carriers can be modulated periodically in 2D materials. The carrier decay is then governed by both carrier recombination and carrier diffusion. With the carrier/exciton recombination time pre-measured with the standard ULS, the carrier/exciton diffusion coefficient can be extracted with this grating imaging technique. We have recently demonstrated the feasibility of using the grating imaging technique to

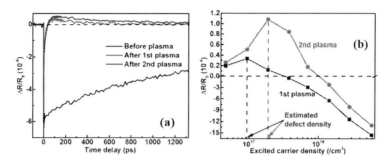

FIGURE 11.3 (a) Transient differential reflection signals of exfoliated MoSe$_2$ before and after plasma irradiation. (b) $\Delta R/R_0$ values at certain delay times as a function of excited carrier density [45].

measure the hot carrier diffusion in monolayer CVD-grown graphene (Fig. 11.2d) [36] and GaAs/AlAs quantum wells [37]. Fig. 11.2c shows our custom-built pump-probe spectrometer for ULS and imaging grating measurements. Our Ti:Sapphire amplifier system (Spitfire ACE, Spectra Physics) generates laser pulses with a 35-fs pulse width (full width at half maximum (FWHM)), 5kHz repetition rate, 1.2 mJ pulse energy, and 800 nm central wavelength. The long working distance (30.5 mm) of our objective lens (Mitutoyo) is specifically chosen to accommodate an optical cryostat (Janis ST500) for temperature-dependent measurements. The reflectivity change of the probe at different delay times is detected with a computer-controlled data acquisition system consisting of high-speed photodetectors and a lock-in amplifier. A mechanical chopper is placed in the path of the pump beam, to improve signal-to-noise ratio. A CCD camera is used to monitor sample surface.

11.2.3 Point defects

Among all of the point defects, chalcogen vacancies are the most common ones in both exfoliated and CVD grown TMDs [11]. Chemical vapor deposition is a relatively fast synthesis technique of TMDs where transition metal precursors are chemically reduced by gaseous chalcogen precursors. During CVD growth of TMDs it is important to control the metal to chalcogen ratio to ensure a stoichiometric film with a vacancy concentration. This is important because the vapor pressure of the chalcogen atoms is much higher than that of the metals. DFT calculations have shown that chalcogen vacancies can introduce mid-gap defect states [38], which can trap carrier and thus reduce the carrier mobility. Electrically active defects in 2D TMDs can typically form bound excitons (neutral excitons bound to defects). If the recombination of the bound-excitons is radiative, the bound-excitons will also contribute to the PL spectrum and form the bound-exciton (X_B) PL peak at the relevant energy positions. In both CVD grown and plasma irradiated exfoliated TMDs, X_B PL peaks are observed with an energy below the band gap value [10,39–43].

Even though the point defect morphology and its effects on optical properties have been extensively investigated, the effect of point defects on carrier dynamics has not been carefully studied yet. ULS has revealed the carrier trapping process by oxygen impurities in CVD grown MoSe$_2$ [44] and plasma irradiated exfoliated MoSe$_2$ [45]. Fig. 11.3a shows our results of the transient differential reflection $\Delta R/R_0$ signals in $MoSe_2$ before and after Ar plasma irradiation. Before plasma treatment, the signal just shows a simple exponential decay, which reflects the recombination of the pump-excited carriers. However, after plasma treatment, the signal changes significantly, and is composed of a sharp negative peak followed by a slow positive decay. The rapid sign change of $\Delta R/R_0$ signals after excitation indicates that the absorption of the sample first decreases and then increases, which has been well accepted as a signature of defect-capturing-carrier phenomenon. The decrease of absorption is due to Pauli repulsion just after the carrier injection, and the increase of absorption is due to the additional available transition channel from the mid-gap defect states to high energy states in the conduction/valence band. Therefore, the measured transient $\Delta R/R_0$ signals suggest that oxygen impurities can induce mid-gap state in TMD materials and serve as effective carrier trapping sites.

We can also estimate the defect density from ULS measurements. Since the total $\Delta R/R_0$ signal is a result of the competition between the trapped carriers and the excess free carriers, with large enough laser power, all the mid-gap states can be filled up. Fig. 11.3b shows $\Delta R/R_0$ values at certain delay times as a function of excited carrier density. The trend-turning points indicated by the dashed lines in Fig. 11.3b correspond to the density of the available trapping-defects in the sample. The estimated defect density after the 1st and 2nd plasma treatments are $9.4 \times 10^{16}/cm^3$ and $1.8 \times 10^{17}/cm^3$, respectively. These results are consistent with the observation from the XPS data [45] where with longer irradiation times, the oxygen impurities density increases. Therefore, ULS can also be used to estimate the density of oxygen impurities in TMDs.

Argon plasma treatment creates both Mo and Se vacancies. Theoretical calculations with density functional theory (DFT) have shown that both molybdenum and chalcogen vacancies can induce a mid-gap state [18,46–49], (Fig. 11.4a and b). However, the occupation of the chalcogen vacancy by oxygen atom (Mo-O bonding) can remove the mid-gap state and restore the band structure to a clean defect-free band gap [18,46,47] (Fig. 11.4c). In oxygen treated MoS_2 samples, substitution of chalcogen vacancy defects with oxygen atoms (Mo-O bonding) has been shown experimentally to improve the PL efficiency [18] and enhance carrier mobility [50]. Therefore, the question that remains is whether introducing oxygen atoms into the Mo vacancy sites to form Se-O bonding will also passivate the mid-gap states. We performed DFT calculations to obtain the band structure of bulk $MoSe_2$ with O atoms occupying Mo vacancy sites. As can be seen in Fig. 11.4d and e, with O atoms sitting in the Mo vacancy sites, there are additional defect states located deep inside the original band gap of bulk $MoSe_2$. These mid-gap states can serve as effective carrier trapping sites [51]. Therefore, carrier trapping in plasma irradiated $MoSe_2$ arises from the mid-gap defect states generated by oxygen impurities occupying Mo vacancy sites which form Se-O bonds.

11.2.4 Edges/grain boundaries

Due to the limited size achievable with exfoliated samples, considerable research effort has been devoted towards developing large area growth techniques, such as CVD and MBE. Even though CVD and MBE TMDs have been grown with a comparable concentration of chalcogen vacancies with that in exfoliated samples [8], transport measurements in CVD and MBE grown films consistently show poorer performance [52–59]. Previous studies have found the

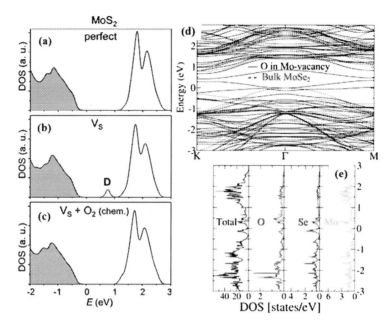

FIGURE 11.4 (a)–(c) Density of states (DOS) for perfect MoS_2, and those containing sulfur vacancies (Vs), and oxygen impurities on the Vs sites. The defect states generated by S vacancies (labeled as "D") are cleared out after introducing oxygen impurities [7]. (d) and (e) Our DFT calculation of band structure of bulk $MoSe_2$ without defects and with oxygen atoms sitting in Mo sites, and atom-projected density of states for $MoSe_2$ with oxygen atoms in Mo vacancy. Mid-gap states can be seen clearly for the O in Mo-vacancy case [45].

following defects in CVD and MBE grown TMDs, such as chalcogen vacancy lines, edges, and grain boundaries [17,60–62]. Our results on CVD grown MoS_2 reveal that sulfur vacancy lines can interconnect into triangular loops (Fig. 11.1d) and encompass inversion domains [60], which can introduce mid-gap states [61]. Stoichiometry variations across different domains also form mirror-twin and tilt boundaries (Fig. 11.1d and e). Ring defects consisting of multiple chalcogen vacancies are also observed near grain boundaries (Fig. 11.1e) [16]. Dangling bonds and vacancies on edges are more reactive forming bonds with foreign atoms (e.g., oxygen). These defects have shown significant impacts on the carrier properties in TMDs [16,62–65], but their effects on carrier dynamics and carrier diffusion are still unknown. **For simplicity, we call all these types of defects as edges/grain boundaries.** High-performance devices with versatile functionalities will require simultaneous control over edges/grain boundaries and point defects.

We have successfully grown large area continuous films of $MoTe_2$ and $MoSe_2$ by MBE on insulating sapphire substrates [66]. As shown in Fig. 11.5a and b, even though there are no edges, considerable disorder and grain boundaries in TEM image are observed, these arise from a corresponding propensity for chalcogen defects and a small grain size. These results suggest that the choice of substrate, growth conditions, and post-growth treatment play a major role on the film quality. Moreover, the large difference in vapor pressure between the Mo and chalcogen species (Te and Se) tightens the available window of growth. Even small

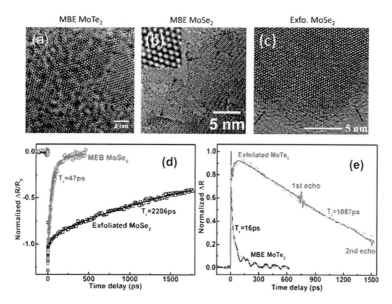

FIGURE 11.5 (a) and (b) TEM images of MBE grown MoTe$_2$ and MoSe$_2$. Abundant nano-size grains and plenty of grain boundary/edge defects are present. The arrows indicate some of the grain boundaries and edges. (c) TEM images of exfoliated MoSe$_2$. Very few grain boundary/edge defects can be seen. The arrows indicate the border of the exfoliated sheet. (d) and (e) Comparison of normalized $\Delta R/R_0$ signals in MBE and exfoliated samples. The fitted value of carrier lifetime τ_r is 2206 ps for exfoliated MoSe$_2$ and 47 ps for MBE MoTe$_2$; 1087 ps for exfoliated MoTe$_2$ and 16 ps for MBE MoTe$_2$ [66].

variations in these growth parameters can cause large variations in the grain size and vacancy density, thereby providing an opportunity to have some control of grain size and grain boundaries.

With respect to carrier relaxation/recombination, defects, especially mid-gap defects, can sometimes serve as recombination centers, which can accelerate or assist the carrier recombination process. Note that recombination centers are a special case of carrier trapping with similar electron- and hole-capture rates. In our results, we have observed this acceleration effect from grain boundaries in MBE grown MoTe$_2$ and MoSe$_2$ films. As shown in Fig. 11.5a and b, large number of nanometer-size grains and thus plenty of grain boundary/edge sites can be seen in the MBE samples. Similar grain boundaries have also been reported in MBE MoSe$_2$ by other groups [67,68]. For comparison, Fig. 11.5c shows the TEM image of exfoliated MoSe$_2$ depicting very few gran boundaries/edge defects. The large number of boundary/edge sites in MBE TMDs provides us with a good physical model to study the effect of grain boundaries on the photoexcited carrier dynamics by direct comparison of the transient differential reflection signals between MBE and the exfoliated samples, as shown in Fig. 11.5d and e. The transient differential reflection signals in MBE samples decay much faster than those in the exfoliated ones. Fittings with exponential decay functions suggest that carrier lifetimes in MBE samples are about 50 times shorter than those in exfoliated samples. This giant difference in carrier lifetime suggests that grain boundary defects in TMDs are effective recombination centers for the photoexcited carriers. To our knowledge, this study represents the first at-

348   11. Devices and defects in two-dimensional materials: outlook and perspectives

tempt to understand the effects of grain boundaries on carrier dynamics [66]. We expect that numerous exciting new physics in this field can be explored.

11.3 Devices fabricated on 2D CVD-grown TMDs

Chemical vapor deposition (CVD) of TMD materials is the principal deposition/growth process pursued by the scientific community for industrial applications because of its ability to scale to large areas and acceptance by the electronics industry. CVD grown TMDs have shown excellent promise in several fields of electronic and optoelectronic devices. Here, we focus on device applications of MoS_2 based field effect transistors (FETs). The performance of FETs depends largely on the type of gate dielectrics which protect and passivate the channel region [69], reduce coulombic scattering [70], and have the potential to introduce dopants and improve mobility [71]. Several dielectrics have been used in this approach, some of which are explained in the following subsections.

11.3.1 Effect of top gate dielectrics

11.3.1.1 Al_2O_3

One of the most widely used top gate dielectric material for MoS_2 channel FETs is Al_2O_3 grown by atomic layer deposition (ALD). Qian *et al.* [72] reports the use of an ~18 nm thick ALD-grown Al_2O_3 as top gate dielectric for CVD-MoS_2. Due to the absence of dangling bonds on the inert flat surface of the MoS_2, ALD becomes challenging and hence a thin layer of AlN was used as an interfacial buffer. Moreover, this interfacial buffer layer greatly helps in minimizing the oxidation of the chalcogen rich surface of the MoS_2 (or TMD) layer which enhances the nucleation of the dielectric to enable conformal deposition and full coverage. Furthermore, the buffer layer helps enhance the electrical stability of the MoS_2 transistors. Devices with AlN/Al_2O_3 dielectric stack show a hysteresis as low as 0.1 V as seen from the transfer characteristics in Fig. 11.6a. The output curves, as shown in Fig. 11.6b, saturate at high V_D indicating good control of gate. The use of a buffer layer to deposit ALD alumina is also reported by Liu *et al.* [73], where a 0.8–1 nm Al seed layer was deposited, and the samples were left in air overnight for the Al layer to be oxidized to form Al_2O_3. Following this, a 15 nm of Al_2O_3 was deposited *via* ALD at 200 °C. The dual gated FETs were fabricated on Si/SiO_2 substrate which also served as the back gate and the 16 nm Al_2O_3 as the top gate. The output characteristics of the MoS_2 FET operated under a back gate bias (V_{BG}) and top gate bias (V_{TG}) are shown in Fig. 11.6c and Fig. 11.6d, respectively. At 2 V drain voltage, a drain current of 14.9 mA/mm is achieved under a back gate bias (V_{BG}) of 100 V, with a top gate bias (V_{TG}) of 0 V, and 2.71 mA/mm for $V_{TG} = 2$ V and $V_{BG} = 0$ V for the same device. This difference in ON current can be attributed to different contact resistances. For a back gated configuration, higher gate voltages will increase the carrier density in MoS_2 under the source drain region. Unlike conventional semiconductors, the source drain regions are not heavily doped in MoS_2 based FETs. Hence, the contact resistance (R_c) mainly depends upon the Schottky barrier height at the metal-semiconductor region. The large contact resistance in case of 2D TMDs results from Fermi level pinning at metal-semiconductor interface, due to high interface traps, giving rise to a notable Schottky barrier height. A higher gate voltage facilitates current injection from metal to MoS_2, thereby reducing contact resistance. On the other hand, the top gate does not influence the carrier density of the TMD under the source drain electrodes. Thus, the contact resistance remains high for the top gated configuration. Due to their atomically thin body, traditional approaches such as ion implantation cannot be applied here to reduce contact resistance. Among various approaches, one method with notable success is doping introduced by sub stoichiometric high-k dielectric deposition, which is discussed in detail in the next subsection.

11.3.1.2 HfO$_2$

Several approaches have been used to deal with the contact resistance issue of 2D TMDs which includes work function engineering [74,75], chemical and electrical doping of TMDs [76] and interface engineering [77]. Among these, the dielectric mediated doping is the most technologically realistic approach and has successfully been implemented for exfoliated MoS_2 samples [71]. Alharbi *et al.* [78] demonstrated the effect of different doping levels of HfO_x dielectric layer on CVD grown MoS_2. HfO_x thin films with varying oxygen content were deposited on CVD-MoS_2 *via* ALD. Fig. 11.6e shows the Raman spectra of four devices S-1 to S-4 with varying O/Hf ratios of 2.0, 1.9, 1.75 and 1.5. As the oxygen content in HfO_x decreases, the doping in MoS_2 increases as evident from the A_{1g} peak shift and broadening. Additionally, the E_{2g} peak position remains unchanged which means that there is no dielectric induced strain and that the shifts in the A_{1g} peak is solely due to increased doping level due to high-k deposition. Fig. 11.6f shows the variation of R_c with gate overdrive voltage for different samples. The data shows that R_c decreases by two orders of magnitude due to injection of carriers into the channel region as the doping levels increases. The mobility for all the devices lies within the same range, 50 to 65 cm^2/V-s at room temperature. In addition, μ_{4pt} shows a power law dependence on temperature, $\mu_{4pt} \propto T^{-\gamma}$ with $\gamma = 1$ for $T > 200$ K. This leads to the conclusion that the dominant scattering mechanism is acoustic phonon scattering and that the dielectric mediated doping process had no effect on the mobility limited scattering mechanism [79]. Therefore, the reduction in R_c was solely because of doping induced Schottky barrier height reduction at the metal-semiconductor interface.

11.3.1.3 ZrO$_2$

Apart from the most widely used top gate dielectrics Al_2O_3 and HfO_2, ZrO_2 has also been explored as a potential gate dielectric. Hu *et al.* report the use of ALD ZrO_2 on CVD-MoS_2 transistors [80]. Fig. 11.6g shows the transfer characteristics of MoS_2 device without and with ZrO_2, respectively, under back gated configuration. The device shows an increase in field effect mobility from 6.9 cm^2/V-s to 11.5 cm^2/V-s after ZrO_2 deposition due to suppression of the charge impurity scattering *via* dielectric encapsulation. Fig. 11.6h shows the transfer characteristics of a top gated FET with ZrO_2 gate dielectric. With the back-gate floating, the device shows ON/OFF ratio of 10^6. At higher V_{TG} ($V_{BG}=0$) the drain current reduces which can be due to injection and trapping of electrons in gate dielectric. Measuring the devices under a positive V_{BG} induces electron accumulation within the source drain access regions and reduces both the access resistance and the contact resistance. The dotted line in Fig. 11.6h, left, shows that the application of 100 V back gate voltage leads to a 20-fold increase in drive current resulting in higher ON/OFF ratios $\sim 10^7$. Fig. 11.6h, right, further shows that the leakage

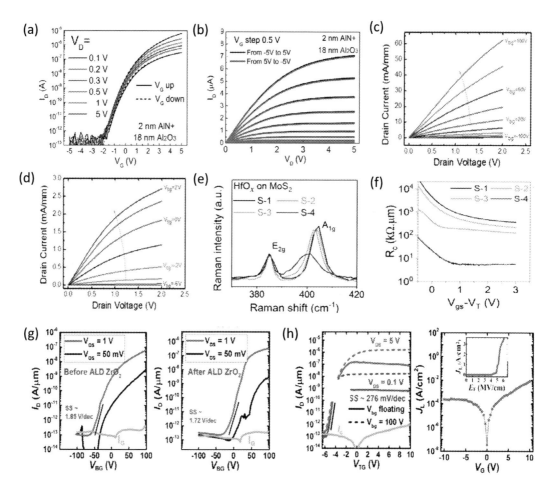

FIGURE 11.6 (a–b) For transistor with 2-nm AlN/18-nm Al$_2$O$_3$ dielectric stack: (a) Transfer curves (I_D vs. V_G) at different drain biases with the gate voltage swept from −5 V to 5 V (solid lines, up-sweep) then back to −5 V (dashed line, down-sweep). (b) Output curves by stepping V_G up (black) and down (red) with a V_G step 0.5 V (adapted from Qian et al. [72]). (c–d) For a transistor with 1 nm Al/15 nm Al$_2$O$_3$ dielectric stack: (c) Output curves for a device under different back-gate bias showing maximum drain current to be over 60 mA/mm. (d) Output curves of a device under top-gate bias (adapted from Liu et al. [73]). (e–f) Effect of HfO$_2$ top gate dielectric: (e) Raman spectra of the monolayer MoS$_2$ at regions near the source/drain electrodes for the devices S-1 to S-4. The data indicate the noticeable broadening of the out-of-plane Raman peak accompanied with the change in its peak position with increasing the doping level. (f) Contact resistance R_C of devices S-1 to S-4 using four-point measurements, indicating marked reduction in R_C with increasing the doping level (adapted from Alharbi et al. [78]). (g–h) Effect of ZrO$_2$ top gate dielectric: (g) Transfer characteristics of the same MoS$_2$/SiO$_2$ back-gate transistor device before (left) and after (right) the deposition of a 27 nm ALD ZrO$_2$ overcoat (h) Transfer characteristics of ZrO$_2$/MoS$_2$ top-gate transistor (left) and leakage current characteristics of a MOSFET structure (Ti/ALD-ZrO$_2$/MoS$_2$) (right). The inset of the figure shows the breakdown field of 27 nm ALD ZrO$_2$ on MoS$_2$ (adapted from Hu et al. [80]). (i) Schematic showing gate-first process used for MoS$_2$ local gate devices and circuits, and (j) Transfer characteristics in linear (right y-axis) and log scale (left y-axis) for a gate-first FET (adapted from Yu et al. [83]). (k) Transfer curves, Ids–Vgs, for a 150 nm gate length CVD MoS$_2$ device. The device shows enhancement mode operation with an I_{ON}/I_{OFF} ratio of 10^8 at 100 mV of drain bias (adapted from Sanne et al. [84]).

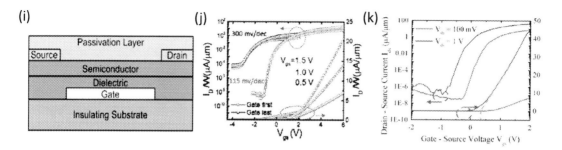

FIGURE 11.6 (*continued*)

current lies within the range of 10^{-4} to 10^{-2} A/cm^2 for V$_g$ below 10 V, which is significantly lower than that of ALD deposited HfO$_2$ [81]. The breakdown field for 27 nm of ALD ZrO$_2$ is calculated to be 4.9 MV/cm and is comparable to similar reports in literature for ALD Al$_2$O$_3$ [82].

11.3.2 Embedded gate FETs

In case of conventional FETs with gate last processes, dielectric integration is often preceded with lithography steps and seeding layers which introduce fixed charge and trapped states inside the dielectric. This often causes an undesirable negative shift in threshold voltage (V$_t$). Yu et al. [83] demonstrate an enhancement mode "gate first" device (Fig. 11.6i) which exhibits larger V$_t$, 10X larger ON current, 100X smaller OFF current and a steeper subthreshold swing (Fig. 11.6j). Another recent report uses this embedded gate structure to fabricate short channel CVD-MoS$_2$ radio frequency transistors with a largest high-field saturation velocity, v$_{sat}$ = 1.88 × 10^6 cm/s, in MoS$_2$ [84]. This device also shows enhancement mode operation, I$_{ON}$/I$_{OFF}$ ratio of 10^8, an a transconductance (g$_m$) of 70 µS/ µm (Fig. 11.6k).

11.3.3 Effect of growth substrates

11.3.3.1 Al$_2$O$_3$

In an alternate approach to achieve resist free channel TMDs are directly grown TMDs on the high-k dielectric. Since CVD growth usually involves high temperature processes, using ALD deposited materials as a substrate runs the risk of dielectric degradation. Bergeron et al. [85] report the effects of the MoS$_2$ growth conditions on the electronic quality of Al$_2$O$_3$ dielectric *via* metal-insulator-semiconductor (MIS) capacitance-voltage (C-V) measurements on an Al$_2$O$_3$/Si substrate annealed in a temperature cycle identical to that of the MoS$_2$ growth process. Fig. 11.7a shows a histogram of the capacitance values at V = 4 V for 121 devices with a narrow distribution and average value of 328 nF/cm^2. The effective dielectric constant was found to be 8.45, corresponding to an effective silicon oxide thickness ~10.5 nm. These are comparable to as-deposited ALD Al$_2$O$_3$ which suggest that the dielectric quality does not degrade after CVD growth at high temperatures [86]. The room temperature transfer characteristics of the same device exhibit an I$_{ON}$/I$_{OFF}$ ratio >10^4, sub-threshold swing of ~220 mV/dec, and $V_T = 2.1V$ as shown in Fig. 11.7b (note that these devices were not encapsulated, and electrical measurements were done at 25 °C below a pressure of 5×10^{-5} Torr). The field effect mobility was calculated in the range of 0.4 to 4.1 cm^2/V-s at $V_D = 1V$. The devices also show low hysteresis in comparison to other reports on exfoliated 2D MoS$_2$ back-gated FETs [87].

11.3.3.2 ZrO$_2$

Liu *et al.* [88] reported a low power consumption back gated transistor based on CVD-MoS$_2$ grown on ALD ZrO$_2$ dielectric substrate. Fig. 11.7c and d shows the transfer and output characteristics of the MoS$_2$ devices. The graphs show low operation voltage of 4 V, threshold voltage of 1.2 V and an ON/OFF ratio of 10^8. Mobility was calculated to be ~65 cm^2/V-s along with a subthreshold swing of 100 mV/dec. These characteristics ensure the high performance of the device with low energy consumption.

11.3.4 Effect of encapsulation and protective layer

An encapsulation layer helps improve device characteristics since it protects the TMD surface from air. In Fig. 11.7e, Illaniorov *et al.* [69] compare the I_D–V_G characteristics for encapsulated CVD-grown MoS$_2$/SiO$_2$ (25 nm) devices, bare devices fabricated on CVD-grown MoS$_2$ using the same method, and bare exfoliated single-layer MoS$_2$/SiO$_2$(90 nm) FETs. Significant hysteresis is present for devices made on bare exfoliated flakes which decreases in the case of bare CVD-grown devices. The hysteresis, however, is much suppressed for their encapsulated counterparts. This indicates that the Al$_2$O$_3$ encapsulation layer efficiently protects the devices from fast adsorbent-type trapping sites which leads to hysteresis. Amani *et al.* [89] reports similar encapsulation effect on CVD MoS$_2$ FETs [Fig. 11.7f] showing significant improvement in the channels on-state conductance and an increase in the back-gated mobility by a factor of ~3 at room temperature, despite the large contact resistance.

11.3.5 Defects in CVD MoS$_2$

One of the major problems associated with CVD growth of TMDs are the grain boundaries (GBs) that form when two or more domains merge. (*e.g.*, mirror twins when two GBs merge at 180°, tilt twins when merging happens at random angles, etc.). These GBs can potentially impact the electrical performance of TMD devices [90,91]. Zande *et al.* [92] demonstrated electrical performance of FETs fabricated in three different configurations: within a grain (pristine), across a GB (perpendicular) and along a GB (parallel). Although the perpendicular device (cyan curve) shows nearly identical performance to the pristine devices (magenta and black curves) indicating little effect of mirror twin boundary on channel conductivity, the transfer curves for the parallel device show 60% larger OFF state conductivity (Fig. 11.7g). In contrast to mirror twins, FETs on tilt boundaries show a decrease in conductance for both parallel and perpendicular configuration with inconsistent degree of variability. This can be attributed to the fact that the variability is strictly dependent on the structure of the boundary as well as tilt angle [93] as seen also in the case of graphene [94]. Kim *et al.* [95] reported the current level of MoS$_2$ FET without a GB to be about five times higher and mobility values to be 4 times higher than that of with GB. Ly *et al.* [96] reported a similar trend where they show

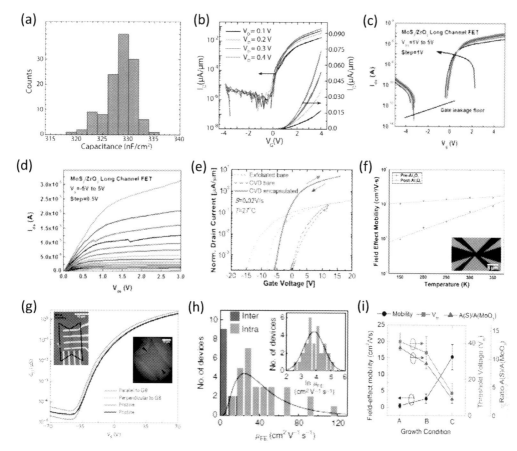

FIGURE 11.7 (a–b) Al$_2$O$_3$ growth substrate: (a) Histogram of the capacitance at V = 4 V (accumulation regime) of 121 Au/Al$_2$O$_3$/Si MIS capacitors where the ALD Al$_2$O$_3$/Si substrate was annealed at CVD growth temperatures (b) Linear and semi-log transfer characteristics of the same device at different drain biases, including forward and backward sweeps (normalized by the device channel width) (adapted from Bergeron et al. [85]). (c–d) ZrO$_2$ growth substrate: (c) Transfer I–V characteristics of MoS$_2$/ZrO$_2$ long channel FETs under different V$_{ds}$ (from 1 V–5 V). (d) Output I–V characteristics of the same device under different V$_g$ (from −5 V to 5 V and 0.5 V per step) (adapted from Liu et al. [88]). (e) The I$_D$ −V$_G$ characteristics of exfoliated MoS$_2$/SiO$_2$ (90 nm) FET (L=1.5 μm, W =7 μm) and CVD-grown MoS$_2$/SiO$_2$ (25 nm) devices (L = 6 μm, W = 10 μm for bare and L = 8 μm, W = 20 μm for encapsulated device) measured using both sweep directions and a constant sweep rate S = 0.02 V/s. The current is normalized by width (adapted from Illarionov et al. [69]). (f) Field effect mobility measured for the same L/W= 800 nm/1000 nm at V$_{DS}$=0.5V before and after the deposition of a 15 nm ALD Al$_2$O$_3$ overcoat; the inset shows an optical micrograph of the finished device (adapted from Amani et al. [89]). (g) Logarithmic electrical transport transfer curves of 4 FETs fabricated from the mirror twin MoS$_2$ island shown in the inset. The curves correspond with pristine regions (magenta and black), and regions containing a grain boundary running perpendicular (cyan) and parallel (orange) to the flow of electrons. All data were measured at room temperature, using the Si growth substrate as a back-gate and a source–drain bias of 500 mV (adapted from Zande et al. [92]). (h) Statistical distribution of the intra- and inter-domain mobilities with $\mu_{FE(inter)} < 16$ cm^2/V-s $< \mu_{FE(intra)}$. The intra-domain μ_{FE} displays a log-normal distribution (black line) with expectation value $\mu_{FE} = 44$ cm^2/V-s, as exemplified by ln(μ_{FE}) in the inset (adapted from Ly et al. [96]). (i) Comparison of field-effect mobility, threshold voltage (V$_{th}$), and the normalized ratio of the S-2p and MoO$_3$ 3d XPS core level areas with growth condition (adapted from Kim et al. [98]).

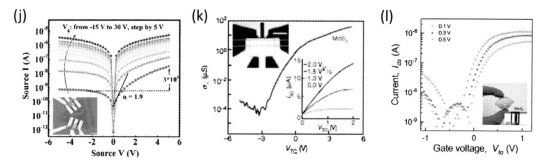

FIGURE 11.7 (*continued*) (j) (Inset): OM image of the MoS$_2$ FETs, the areas of original, processed, and homojunction channels are marked with blue, red, and green arrows, respectively, scale bar, 10 μm. Rectification behavior of the homojunction in different gate voltages (tested in green arrows indicated electrodes in inset) (adapted from Gao *et al.* [99]). (k) Top gate (V_{TG})-dependent $\sigma_\square$ for dual-gate monolayer MoS$_2$ FET (device shown in the upper inset). Lower inset: V_{TG}-dependent I_{SD}–V_{SD} curves showing current saturation and ohmic electrode contact. Scale bar, 10 mm (adapted from Kang *et al.* [100]). (l) Electrical performance of flexible MoS$_2$ FET (Inset). Digital photograph of flexible MoS$_2$ FET. The image shows its original color and flexibility (adapted from Mun *et al.* [103]).

a statistical analysis on 43 devices (29 intra domain and 14 inter domain devices), as shown in Fig. 11.7h, and have obtained the following trend: For low angle GBs, $\mu_{FE(inter)} < 16$ cm^2/V-s $< \mu_{FE(intra)}$. For high angle GB, more complicated defects exist, along with a greater number of dangling bonds in a zigzag shape or containing holes which are not fully understood yet.

The electrical stability and reliability of MoS$_2$ FETs are also dependent on the thickness of the MoS$_2$ material that is used as the channel material. In a recent study, Park *et al.* [97] investigated the variations in transistor properties by applying a positive and negative bias stress. Although FETs based on 3 nm thick films demonstrated stable behavior under both bias stress polarities, the threshold voltage and mobility values of FETs based on thicker 9 nm films greatly differed under positive and negative bias stresses. These shifts can be attributed to sulfur interstitials, oxygen partially dissociated from the MoO$_3$ precursor, and Mo completely dissociated from the MoO$_3$ precursor, thereby, providing metallic Mo defects. In addition to defects associated with GBs, the stoichiometry of the grown film also effects electrical performance. Kim *et al.* [98] discusses the effect of stoichiometry on the electrical performance of MoS$_2$. Among three different samples grown with different stoichiometries, a reduction in S content is seen to bring about an increase in the average field-effect mobility, n-type conductivity and ON/OFF ratio and a decrease in the threshold voltage (Fig. 11.7i). The sulfur vacancies act as electron donors, thereby increasing the field effect mobilities and contributing positively to the conductivity. Apart from defects inherently associated with CVD growth, there have been reports of intentionally engineering defects to modulate the electronic structure of MoS$_2$ for applications in logic circuits. Gao *et al.* [99] report H$_2$O$_2$ aqueous solution to form sulfur vacancies within a homogenous MoS$_2$ flake, which reduces the electronic concentration and increases work function by 100 meV. This knowledge is then used to create an in-plane homogenous monolayer MoS$_2$ logic inverter which shows a rectification ratio of about 3×10^2 with an ideal rectification factor of 1.9 under the bottom gate at −15 V with a voltage gain reaching 4 (Fig. 11.7j).

11.3.6 MOCVD MoS$_2$

To gain a more precise control over the layer thickness, gas phase precursors are now used to grow large area MoS$_2$ by metal organic chemical vapor deposition (MOCVD). Several reports of successful MoS$_2$ wafer scale growth *via* MOCVD have already been demonstrated with high electrical performances [100–104]. Kang *et al.* [100] demonstrate wafer scale growth of MOCVD MoS$_2$ with excellent spatial homogeneity of the electrical properties. Field effect transistors showed high ON/OFF ratios (10^6), current saturation at relatively lower bias, large transconductance and μ_{FE} of around 30 cm^2/V-s (Fig. 11.7k). A different approach used halide assisted MoS$_2$ grown on *c*-plane sapphire and transferred it onto degenerately doped Si/SiO$_2$ and fabricated FETs using polymer electrolyte gating to build electron double layer transistors (EDLTs) [101]. The devices show a SS of 125 mV/dec and 4-probe electron mobility of 95 cm^2/V-s at T = 12 K. Smets *et al.* [102] further show ultra-scaled devices with channel length as low as 29 nm, low SS of 80 mV/dec and maximum ON current of 250 µA/ µm. MOCVD growth has also been used to grow films directly on polymer substrates as demonstrated by Mun *et al.* [103]. The two-point flexible FETs fabricated directly on polyimide substrate show an ON-state current of 10^{-6} A and OFF-state current of 10^{-9} A, and a calculated field effect mobility is 1 cm^2/V-s (Fig. 11.7l). A recent photolithographic fabrication technique enabled wafer scale patterning of MoS$_2$ on Si/SiO$_2$ as well as polymer substrates followed by lift-off of the deposited metals using an Irgacure 651-doped polymethyl methacrylate resist and a water free developer. This is especially important because water often intercalates between the MoS$_2$ and substrate interface and causes detachment of MoS$_2$ during the photoresist development process [104]. These devices show ON/OFF ratios of 10^5 and field effect mobilities of 6 cm^2/V-s at V$_{DS}$ of 3 V in the back gated configuration.

11.4 Devices fabricated on MBE-grown TMDs

Besides CVD and MOCVD growth techniques, there have also been efforts to grow large-area TMD films using molecular beam epitaxy (MBE), a proven method for the growth of ultra-pure thin films of commercial electronic materials such as III-V semiconductors and their hetero-structures. Growth of traditional semiconductors, non-2D materials, require careful lattice matching conditions between the films and the substrates. Koma *et al.* first proposed that epitaxial growth of Van der Waals materials did not require the same lattice matching conditions as for example cubic materials, this process is referred to as Van der Waals epitaxy (vdWE) [105,106]. Recently, MBE growth methods have drawn significant attention to grow different TMDs on various substrates under ultra-high vacuum environment [107–119]. However, MBE-grown films are primarily limited only to structural, spectroscopic, and optical characterization, and the detail nature of charge transport in MBE-grown films merits more studies. Here, we will focus on the electrical properties of group-VIB based TMDs (with 'Mo' and 'W' as transition metals and S, Se, Te as chalcogens) that are 2D semiconductors in 2H phase.

TMD films grown by MBE have structural and spectroscopic properties similar to their exfoliated counterparts. However, the electrical properties of as-grown films reveal insulating transport behavior through Mott variable-range hopping (VRH) mechanism due to a high degree of carrier localization, possibly arising from disorder due to chalcogen vacancies and/or their nanocrystalline nature [Fig. 11.8a] [113]. Presence of defects in MBE-grown few-layer $MoSe_2$ thin films on c-Al_2O_3(0001) substrates is evident in the high-resolution transmission electron microscopy (HR-TEM) and is further confirmed using X-ray photoelectron spectroscopy studies [113]. The high density of grain boundaries and other potential defects mandates further optimization of the MBE growth process to improve the electrical quality of TMD films. Similar insulating VRH transport characteristic of MBE grown $MoSe_2$ [Fig. 11.8b] was also observed by Dau *et al.* [115]. In addition, an unexpected negative magnetoresistance (MR) was evidenced at high temperatures and high magnetic fields. The negative MR is likely arising due to the hopping nature of the vacancy induced VRH transport at high temperatures.

Several works are reported on TMD-based devices using an electric double layer transistors (EDLT) structure where the ionic liquid gate helps to achieve a higher carrier density. Chen *et al.* [120] demonstrated an ambipolar behavior close to 280 K in MBE-grown $MoSe_2$ on GaAs(111)B substrates using an EDLT that enables a wide range of electrostatically induced doping levels and reduces the Schottky barrier heights at contacts. In these devices an I_{on}/I_{off} ratio of $\sim10^4$ (for n sides) and $\sim10^2$ (for p sides), and a subthreshold swing ~ 390 mV/dec for both sides, are extracted [Fig. 11.8c]. However, the field-effect mobility values ($\mu_e \sim 0.05$ cm^2/V-s and $\mu_h \sim 0.28$ cm^2/V-s) are significantly lower than those of CVD-grown $MoSe_2$. Devices without a polymer electrolyte show an extremely poor transistor behavior. These results indicate that charge carrier transport is strongly influenced by the disorder in the film and follows a VRH transport mechanism, as also observed by others [113,115].

Recently, Poh *et al.* [121] demonstrated MBE growth of mono and bilayers of $MoSe_2$ on hBN and fabricated devices directly on the grown films. Field effect transistors were fabricated using standard electron-beam lithography to pattern the samples with thermally evaporated Cr(2 nm)/Au(50 nm) metal electrodes and with the hBN substrate as the dielectric layer. All the electrical transport measurements were performed in a nitrogen-filled glovebox and the contacts are ohmic as observed from the linear I_D–V_D curves at different gate voltages. I_D increases monotonically with increasingly positive V_G, indicating a typical n-type FET behavior, and the on/off ratio is of the order of 10^4 [Fig. 11.8d]. Use of hBN as the substrate can reduce the Coulomb scattering and thereby improve the electrical performance. This is observed as the average μ_e of ~15 cm^2/V-s is achieved at room temperature (RT).

In the recent few years, TMD-based FETs have attracted significant attention as promising active materials in optics because they have an optimum band gap in the range of the visible to infrared (IR), high mobility and ability to absorb light (as high as 15%) in ultra-thin film limit [122–125]. MBE-grown bilayer $MoSe_2$ on SiO_2 has been used to investigate the characteristics of the photo-FET with gate bias and illumination, resulting in superior photoresponsivity (~242 A/W), suggesting $MoSe_2$ to be a promising material for the next-generation phototransistors [Fig. 11.8e] [126]. The mechanism of photoresponsivity was understood in terms of the change in quasi-Fermi levels, defect states, mobility, and metal contact. Especially, capturing the electrons (holes) generated in the defect states increases the probability that the untrapped holes (electrons) reach the source and drain electrode, resulting in the high photoresponsivity in these TMD-based photo-FETs [126].

11.4. Devices fabricated on MBE-grown TMDs

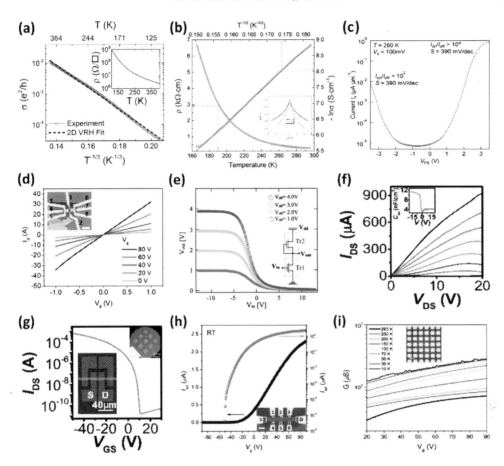

FIGURE 11.8 (a) Electrical transport measurements of MoSe$_2$ grown on Al$_2$O$_3$(0001) substrate using MBE. Conductivity on a semilog scale vs T$^{-1/3}$ shows a linear trend, which can be fit to a 2D Mott VRH transport model (dotted lines). The temperature dependence of the measured resistivity (inset) follows an insulating trend (adapted from ref. [113]). (b) Resistivity vs temperature and the logarithm of the conductivity vs T$^{-1/3}$ following 2D-VRH model (solid red line fit). Variations of magnetoresistance with magnetic field applied parallel and perpendicular to the sample surface at 250 K are shown in inset (adapted from ref. [115]). (c) Channel current as a function of polymer electrolyte voltage (V$_{PE}$) showing ambipolar behavior in MoSe$_2$ EDLT close to RT (T = 280 K) (adapted from ref. [120]). (d) Output characteristics of MBE-MoSe$_2$ FET at various back-gate voltages (Scale bar of inset: 5 µm) (adapted from ref. [121]). (e) Voltage-transport characteristics with drain voltage of a photo-inverter based on MBE-grown MoSe$_2$. The circuit schematic is illustrated in the inset (adapted from ref. [126]). Electrical characteristics of the MBE-grown wafer-scale monolayer 2H-MoTe$_2$ film: (f) Output and (g) transfer characteristics of 2H-MoTe$_2$ FET measured at RT in ambient. Inset in (f): capacitance vs voltage measured from the FET device. Inset in (g): (top) batch-fabricated MoTe$_2$ FET on a 2-inch SiO$_2$/Si wafer and (bottom) FET device with a channel length of 4 µm and width of 8 µm, respectively (adapted from ref. [127]). Transport properties of monolayer MoS$_2$ FETs made from MBE-grown film on exfoliated hBN: (h) Two-probe transfer curve of a back-gated MoS$_2$ FET measured at RT. Source-drain bias was fixed at 0.5 V. Inset shows the optical micrograph of the multi-terminal FET device. The channel is capped by an hBN thin flake, and the electrodes are contacted by graphene. (i) Four-probe transfer curves measured at different temperatures. Voltage was recorded between probe leads 1 and 2, as marked in inset of (h). Inset shows the optical image of a FET device array fabricated from a wafer-scale monolayer MoS$_2$ film that was transferred onto a 285 nm SiO$_2$/Si substrate before device fabrication (adapted from ref. [128]).

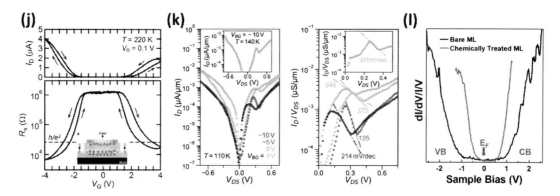

FIGURE 11.8 (*continued*) (j) Ambipolar transistor operation in top-gated electric-double-layer transistor based on WSe$_2$ epitaxial thin film grown on Al$_2$O$_3$ substrate: (Top) Response of the drain current (I$_D$) as a function of gate voltage (V$_G$) taken at T = 220 K and (Bottom) corresponding variation of the four-terminal sheet resistance (R$_S$) as a function of V$_G$. Inset shows the schematic of the device structure (adapted from ref. [129]). (k) Esaki tunneling and conductance slope in devices based on MBE-grown WSe$_2$: (Left) Current-voltage characteristics at T = 110 K and at T = 140 K (inset). (Right) Absolute conductance I$_D$/V$_{DS}$ vs V$_{DS}$ for different back-gate biases. Dashed-dotted lines indicate the linear fit of the conductance slope in the NDR region (adapted from ref. [130]). (l) Averaged STS curves of ML WSe$_2$ before (black curve) and after (red curve) (NH$_4$)$_2$S(aq) chemical treatment (adapted from ref. [131]).

Transport studies of MoTe$_2$ have received much less attention than other TMDs. Electrical measurement directly on few-layer MoTe$_2$ grown on Al$_2$O$_3$(0001) using MBE shows an insulating behavior dominated by a VRH transport mechanism [113]. Only recently, device applications of MBE-grown wafer-scale continuous monolayer 2H-MoTe$_2$ films on SiO$_2$/Si substrate are reported by He *et al.* [127]. Monolayer 2H-MoTe$_2$ film is observed to be a p-type semiconductor with a carrier density $\sim 10^{11}$ cm^{-2}, and the Fermi level located in the lower half of the bandgap ($\sim$0.15 eV above the valence-band maximum), as confirmed from photoluminescence and ultraviolet/x-ray photoelectron spectroscopy studies. FETs fabricated on this sample show an on/off ratio of the order of 10^7 with a hole mobility of $\sim$ 23.4 cm^2/V-s [Fig. 11.8f and g].

Transport properties of the FETs fabricated on MBE-grown MoS$_2$ films on exfoliated hBN and with hBN capping was evaluated by Fu *et al.* [128]. The film shows n-type conduction and an extracted device mobility of $\sim$45 cm^2/V-s at RT, which increases to $\sim$375 cm^2/V-s when the temperature is decreased to 20 K. The temperature-dependent mobility above $\sim$100 K follows a power law $\mu \approx T^{-\gamma}$, where $\gamma \approx 1.8$, suggesting a phonon scattering mechanism [Fig. 11.8h and i].

Monolayer WSe$_2$ possesses the largest spin splitting amongst all group VI 2H-TMDs making it a promising material for potential spintronic and valleytronic based devices. However, transport properties of MBE-grown WSe$_2$ thin films still need to be addressed in detail. Nakano *et al.* [129] demonstrated ambipolar transistor operation of EDLTs of MBE-grown WSe$_2$ thin films grown on Al$_2$O$_3$ substrates [Fig. 11.8j]. The maximum field-effect mobilities (μ) obtained are $\sim$ 3 cm^2/V-s for holes and 1 cm^2/V-s for electrons, respectively. Paletti *et al.* [130] on the other hand presented the experimental demonstration of an EDL Esaki junction in MBE-WSe$_2$ thin films grown on high-*k* dielectric [Al$_2$O$_3$(30 nm)/SiO$_2$(10 nm)/Si, grown *via* atomic layer deposition (ALD)]. Degenerate and abrupt doping profiles are obtained by modulating the electron/cation and hole/anion EDLs formed at the interface between a 40 nm long WSe$_2$ channel and a solid polymer electrolyte. The EDL-doped tunnel diode exhibits repeatable, gate-tunable band-to-band tunneling with negative differential resistance in the forward bias regime at temperatures up to 140 K, and strong conduction in reverse bias with a maximum peak-to-valley current ratio of 3.5 at 110 K [Fig. 11.8k].

To enhance its p-type electrical performance, band structure engineering of MBE-grown WSe$_2$ *via* chemical treatment using ammonium sulfide solution was demonstrated. Scanning tunneling spectroscopy (STS) reveals that chemical treatment reduces the electronic bandgap of monolayer WSe$_2$ from 2.1 to 1.1 eV, and the Fermi level shifts toward the VBE [Fig. 11.8l]. As proof of concept, this p-doping effect of the chemical treatment was demonstrated on back-gated few-layer ambipolar WSe$_2$ FETs, with the p-branch ON-currents enhanced by about 2 orders of magnitude (from 10^7 to 10^9) following the treatment. The p-doping induces narrowing of the Schottky barrier width leading to an enhanced hole injection at the WSe$_2$/contact metal interface thereby improving the hole field-effect mobility. The two-point field-effect mobility for holes showed a 6-fold increase, from $\sim$3.5 cm^2/V-s to 22.7 cm^2/V-s, following the chemical treatment [131].

11.5 2D van der Waals (vdW) heterostructures

Two-dimensional materials, being layered in nature, offer an excellent platform for fabricating on-demand, designer heterostructures. Termed as van der Waals (vdW) heterostructures, these represent an epitome of *'materials by design'* where any species and number of TMDs, regardless of the commensurability relationship between the layers, could be picked up and placed atop each other. This is realized due to the presence of weak van der Waals forces between two layers of 2D materials, which enable an easy 'peel-off' of the layers followed by 'drop-off' at specified locations. With the vast variety of 2D materials possessing different electronic band structures, including semiconducting, insulating, metallic and semi-metallic, available today, there exists a phenomenal opportunity to create innumerable types of vdW heterostructures.

As one would imagine, the synergetic effects that emerge when different quantum confined 2D crystals are stacked vertically in close proximity with each other are plentiful [132]. Charge redistribution could occur between near and distant neighbors; structural changes, as well as stress and strain, could be induced between layers; surface reconstruction could occur; charge transport in one layer could start to affect that in another, leading to tunneling and Coulomb drag effects. When 2D materials with not-so-different lattice constants are overlaid on one another with a small rotation/twist between them, several compelling physical properties emerge which are different from the individual layers [132]. Such studies now constitute a part of an emerging research field called *'twistronics'*, a very hot topic in the scientific community. In addition to the several remarkable changes in properties, the assembly of 2D devices also enables the fabrication of unique devices with a multitude of functionalities. These will be reviewed later in this section, together with the topic of *'twistronics'*.

11.5.1 Device fabrication

The most common method employed for the fabrication of vdWs heterostructures involves a direct stamping/stacking of 2D materials (mostly exfoliated crystals) *via* a deterministic pick and place mechanism. Several different state-of-the-art deterministic transfer setups exist in laboratories worldwide, however, they all comprise of the following two central components: a long working distance microscope and a couple of micro-positioning systems. The basic mechanism first involves aligning a polymeric (typically PDMS, polydimethylsiloxane and PPC, polypropylene carbonate) stamp connected to a micromanipulator in an overhanging cantilever configuration onto a substrate containing the 2D flake that needs to be picked up. The alignment is facilitated using long working distance objectives. The stamp is then brought in contact with the 2D flake on the substrate and a combination of heat and pressure enables the polymeric stamp to pick up the desired flake. Once the flake is re-positioned on the stamp, it can then be dropped off at a user-specified location (e.g., atop another flake or contacts) using the same procedure of alignment, followed by pressure contact albeit at a relatively higher temperature than the pick-up. Fig. 11.9a shows a schematic and an optical image of a stamp and Fig. 11.9b depicts the steps employed in one variation of the transfer process which utilizes hBN [133]. For an excellent review of the several existing variations of the technique used for deterministic transfer, please refer to Ref. [134]. Thermal annealing steps and/or chemical treatments are typically incorporated between consecutive layer transfers to remove polymer and other organic residues. The pick and place assembly method for 2D heterostructures is known to give rise to atomically sharp interfaces and, for certain 2D materials, could produce contamination-free large-scale areas between the two 2D layers, due to the 'self-cleaning' mechanism [132,135].

With the surge in *twistronics* related research, the deterministic transfer setups now also incorporate the capability for sample rotation. Furthermore, it is also possible to work with several highly reactive 2D materials if the transfer setup is housed in an inert glove box environment. Fig. 11.9c and d show images of a fully automated deterministic transfer setup comprising of a highly precise sample rotation stage, installed in an Argon glovebox such as the one at the Microelectronics and Engineering Research Center at UT Austin. Research efforts in other labs are also underway to build a Quantum Press (QPress), an automated system comprising of an exfoliator, a cataloger, a library, a stacker and a characterizer for 2D materials [136].

An intrinsic limitation of the heterostructures composed of exfoliated crystals is their small lateral size. For large-area heterostructures, CVD grown films are transferred utilizing the method of growth substrate etching while the 2D material is supported by a polymer, as demonstrated since the early days of graphene research [137–139], followed by stamping/transferring on a target substrate. As such, these large-area heterostructures include those comprising of only TMDs [140–142], as well as combinations of TMDs with graphene [143–145]. Finally, a direct growth of vertical heterostructures (bottom-up synthesis) has also been demonstrated through the technique of van der Waals epitaxy [146–148]. The technique relies on the existence of van der Waals force between 2D materials, which in essence has a two-edged character. Its weakly interacting nature, in general, renders the growth of large-area heterostructures of 2D materials challenging, however, at the same time, the absence of dangling bonds in these materials leads to minimal strain in between two layers and provides

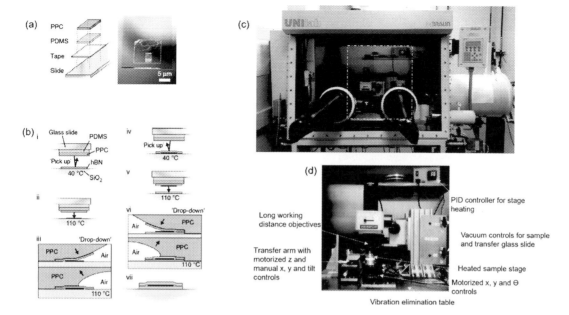

FIGURE 11.9 (a) Schematic and optical image of a polymeric stamp on a glass slide used for the pick up and drop off of 2D flakes. (b) Schematic process flow for one variation of the deterministic pick and place technique involving hBN. Reproduced from Ref. [133]. (c) Fully automated deterministic transfer set-up with a precise sample rotation stage installed in an Argon glove box at the Microelectronics and Engineering Research Center at The University of Texas at Austin. Close-up view of the transfer set-up (marked with white dot-dashed rectangle) is shown in (d).

the opportunity to explore heteroepitaxy of 2D materials with a significant lattice mismatch between them [146,149].

11.5.2 Applications

Amongst one of the early applications of vdW heterostructures was the utilization of ultra-flat and chemically clean hexagonal boron nitride (hBN) as a substrate for graphene, in an attempt to reduce undesirable substrate effects such as surface roughness and charge scattering leading to improved device performance [150]. Forging ahead, hBN also found use as an encapsulation layer for graphene [151] and TMDs [152,153] lending protection against environmental adsorbates and leading to significant improvements in carrier mobility. Recently, it has also been used for encapsulating highly reactive 2D materials, such as black phosphorus [154], WTe_2 [155], CrI_3 [156] etc.

VdW heterostructures have been used for the demonstration of numerous electronic devices, including vertical tunneling transistors, vertical diodes, memory devices, light-emission and light-detection devices etc. While a discussion of all the devices is beyond the scope of this work, we briefly review vertical tunneling transistors and photo-diodes, and refer the readers to Ref. [157], which forms an excellent review of vdW heterostructure devices. Vertical tunneling transistors are field-effect transistors relying on quantum tunneling from

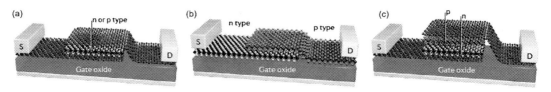

FIGURE 11.10 Schematic diagrams of vdWs vertical photodiode structures (a) Vertical graphene-TMD-graphene configuration, (b) 2D semiconductor p-n diode with lateral contacts configuration, (c) 2D semiconductor p-n diode with vertical graphene contacts configuration. Reproduced from Ref. [157].

one 2D material to another through a thin insulating barrier. They were first demonstrated in graphene-hBN-graphene and graphene-MoS$_2$-graphene configurations, with switching ratios of ~ 50 and $\sim 10{,}000$ respectively [158]. Furthermore, when the work functions of the two electrodes align, resonant tunneling of carriers can take place through the barrier, leading to the demonstration of resonant tunneling transistors and the accompanying negative differential resistance [159,160].

With intrinsic bandgaps and high photon absorption [161,162], TMDs constitute as good candidates for photo-detectors. Three basic types of vdW heterostructure platforms have been explored so far [157]. The first one has a vertical graphene-TMD-graphene configuration [163,164] (Fig. 11.10a), where the TMD functions as the primary absorption layer for the photons, and graphene, capacitively coupled to a gate, as the active contact. These devices exhibit a finite quantum efficiency, due to an efficient photon absorption in the atomically thin TMDs. The second configuration involves vertically stacking n- and p-type 2D semiconducting materials (Fig. 11.10b), e.g., MoS$_2$-WSe$_2$ [165–167], MoS$_2$-black phosphorus [168,169], MoS$_2$-GaTe [170]. In this configuration too, an external electric field from a gate can be used to modulate the diode characteristics. To overcome the large series resistance leading to longer diffusion times in a lateral channel, graphene electrodes could be used on the top and the bottom of the vertical p-n junction [171] (see Fig. 11.10c). This constitutes the third configuration of photo-diode devices, where vertical charge-transfer leads to increased charge-collection efficiency.

One of the most intriguing areas of study for vdW heterostructures is the investigation of properties of heterostructures where the constituent 2D materials are twisted with respect to each other. Garnering significant attention from the physics community in the early years, twisted heterostructures now constitute a popular research theme amongst electrical engineers, materials scientists, and experimental and theoretical physicists alike. The field is commonly referred to as *'twistronics'*, the study of how the twist between layers of 2D materials can alter their electrical properties and holds the potential of ushering in a new era of technology. We review some of the major research highlights of twisted vdW heterostructures below.

As is common in most research fields, theoretical investigations of twisted heterostructures preceded experimental discoveries of intriguing phenomena in these structures. Utilizing ab-initio calculations to develop a tight-binding model, it was shown that when two layers of graphene are rotated with respect to each other at an angle of $\Theta \sim 1.5°$, flat bands close to the Fermi level (K point) emerge and the Fermi velocity approaches zero [172]. Shortly afterwards, R. Bistritzer and A.H. Macdonald showed that a moire pattern [173–176]

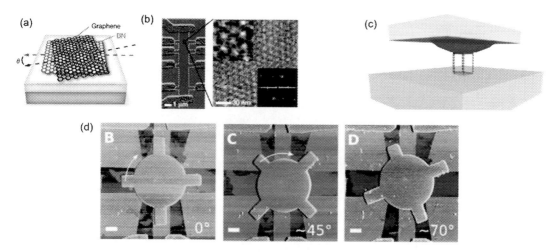

FIGURE 11.11 (a) Sketch of graphene on hBN showing the emergence of a Moire pattern. (b) Left: an AFM image of a multiterminal Hall bar. Right: A high resolution image of a magnified region. The moire pattern is evident as a triangular lattice (upper inset shows a further magnified region). A fast Fourier transform of the scan area (lower inset) confirms a triangular lattice symmetry wit periodicity 15.5 ± 0.9 nm. Reproduced from Ref. [182]. (c) Illustration of the hemispherical handle substrate used for selective flake pick-up and transfer. Reproduced from Ref. [185]. (d) AFM images of a graphene/BN/graphene device where the top BN layer could be freely rotated by pushing on one of the arms of the uppermost BN using the AFM. Dynamically rotatable heterostructures could be obtained using this technique. Images are shown for three different orientations of the top BN. The angle identified in each panel is the absolute angle referenced to the AFM coordinate system (Φ_A). Scale bars, 1 μm. Reproduced from Ref. [187].

is formed in twisted double-layer graphene and predicted the existence of a discrete set of magic angles ($0.18° < \Theta < 1.2°$) for which the Dirac velocity vanishes [177]. Subsequent studies of twisted bilayer graphene in the presence of magnetic field [178] showed the existence of a fractal energy spectrum, called 'Hofstader butterfly' [179]. This spectrum arises in a two-dimensional electron system subjected to a magnetic field due to an interplay of the characteristic length scales associated with the quantizing magnetic field and the periodic electrostatic potential. The first significant experimental discovery in the field came with the fabrication of graphene on hBN vdW heterostructures (for which the existence of moire patterns was demonstrated previously [174,175]) with a very small rotational misalignment ($\Theta < 2°$) [180–182] (Fig. 11.11a and b). In these works, with the demonstration of 'Hofstader butterfly', the band structure of twisted graphene-hBN heterostructures was shown to vary significantly from that of bare graphene.

While resonant tunneling transistors were demonstrated previously [159,160], with the surge in twisted heterostructures research, twist controlled resonant tunneling [183,184] was achieved where, both energy and in-plane momentum were conserved. In terms of twisted heterostructures' fabrication, traditionally, the rotational alignment between constituent layers of vdW heterostructures was achieved by using the straight edges of the flakes as crystallographic references. However, the limited precision of this method was overcome by K. Kim *et al.* by developing a technique for high rotational accuracy based on sectioning a single exfoliated flake into two using a hemispherical handle substrate (see Fig. 11.11c) made from a polymer [185]. Resonant tunneling in graphene-hBN-graphene [185] and WSe_2-hBN-WSe_2 [186] heterostructures was also demonstrated using this method of fabrication.

The above examples of investigations in twisted vdW heterostructures irrefutably demonstrate the potential of such heterostructures in studying fundamental physical effects, as well as for applications-based devices. Recently, dynamically rotatable heterostructures for twistable electronics [187] were demonstrated by Ribeiro-Palau *et al.* In their method, they achieved rotational alignment with high accuracy in hBN-encapsulated graphene structures by using an AFM tip to 'push' on the top hBN layer to achieve twist angles as little as $0.2°$ (Fig. 11.11d).

Six years after the prediction of magic angles in twisted bilayer graphene, the existence of flat bands in such a system at the first magic angle of $\Theta \sim 1 - 1.1°$ was experimentally demonstrated [188,189]. These investigations revealed the presence of electron-electron interactions in small angle twisted bilayer graphene. Moreover, remarkably, unconventional superconductivity with a critical temperature of 1.7 K upon electrostatic doping of the material away from the insulating states, arising due to flat bands, has been demonstrated [190]. Furthermore, optical studies of twisted heterostructures providing evidence of moiré excitons [191] have also been demonstrated. While still quite fresh, the field of *twistronics*, with potential applications in quantum computation, electrical transmission [192] and several other arenas, coupled with the opportunities that it presents for investigating fundamental physical phenomena, could lead to unprecedented advancements.

11.6 Enhancing 2D device performance using defect engineering

One of the biggest issues confronting MoS_2-based devices is the presence of a Schottky barrier (SB) at the interface between MoS_2 and the contact metal electrode. This results in a "non-Ohmic" or a Schottky electrical contact characterized by an energy barrier, called the Schottky barrier height (SBH or Φ_{SB}), that hinders the injection of charge carriers into the device channel [193]. Consequently, this notable SBH leads to a large R_C and performance degradation (e.g., low field-effect mobilities) in two-terminal MoS_2 devices since a large portion of the applied drain bias gets dropped across this R_C [194,195]. These Schottky barriers are thought to be formed due to strong Fermi level pinning (FLP) effects at the contact metal/MoS_2 interface [193,196,197] and have been attributed to sulfur vacancies (SVs) and subsurface metal-like impurities, which are thought to be responsible for the strong FLP [198–201]. These SV defects/impurities lead to a large background n-doping in the MoS_2 and introduce unwanted energy levels or "mid-gap states" closer to the conduction band edge (CBE) within its bandgap that ultimately governs the location of the charge neutrality level where the metal Fermi level gets pinned resulting in fixed barrier heights at the contact/MoS_2 interface [202–204]. Further insight on the possible origin of this FLP effect was shed by Kang *et al.* who, using theoretical calculations based on density functional theory (DFT), reported that interactions between certain metals and MoS_2 can lead to the formation of a "metal/MoS_2 alloy" at the contact interface with a much lower work function than unalloyed MoS_2. This leads to an abnormal FLP as if the MoS_2 is contacted to a low work function metal [205]. Gong *et al.*, on the other hand, claimed that the FLP at the metal/MoS$_2$ interface is a result of two simultaneous effects: first, a modification of the metal work function by interface dipole formation due to the charge redistribution at the interface and, second, by the formation of mid-gap states originating from Mo d-orbitals, that result from the weakening of the intralayer S-Mo bonds due to the interfacial interaction, and the degree thereof, between the metal and the S atom orbitals [206]. A qualitatively similar result was obtained by Farmanbar *et al.* [207]. Additionally, Guo *et al.* suggested that the strongly pinned SBHs at the metal/2D MoS$_2$ interface arises due to strong bonding between the contact metal atoms and the TMD chalcogen atoms [208], in accordance with the age-old theory of metal-induced gap states (MIGS) [209–211].

Regardless of the exact underlying physical mechanism involved, FLP is an undesired effect as it leads to fixed SBHs at metal/MoS$_2$ interfaces. It is for this very pinning effect that most metal-contacted MoS$_2$ FETs typically show unipolar n-type behavior as the metal Fermi level gets pinned near the conduction band edge (CBE) of MoS$_2$ irrespective of the metal work function [212–215]. In addition to degrading the device performance due to large R_C, the reduced tunability of the SBH due to FLP is detrimental towards realizing both n-type and p-type Ohmic contacts to MoS$_2$ desirable for CMOS applications [196]. Besides SBH, another relevant parameter associated with these Schottky barriers is the width of its depletion region in the semiconductor channel or, simply, the Schottky barrier width (SBW). The SBW is largely dependent on the extent of semiconductor "band-bending" in the 2D TMD/MoS$_2$ channel under the electrode contacted region [216]. Both the SBH and the SBW together determine the charge injection in the 2D MoS$_2$ channel. While SBH governs the extent of thermionic emission of carriers "over" the barrier, SBW determines the extent of thermionic field emission (i.e., thermally assisted tunneling) and/or field emission (i.e., direct tunneling) "through" the width of this barrier due to the quantum mechanical tunneling of charge carriers (Fig. 11.12 shows the band-alignment at the metal/2D TMD interface under different gating conditions and illustrates the different charge carrier injection mechanisms) [193,196,217,218]. Hence, both the SBH and SBW must be minimized to achieve efficient injection of charge carriers (electrons or holes) from the contact into the semiconducting MoS$_2$ channel.

For ultra-short-channel FETs (targeting the sub-10 nm node) based on 2D MoS$_2$, it is extremely important to minimize the specific contact resistivity, ρ_C, to achieve low R_C. This is important because the R_C of any FET can only be a small fraction ($\sim$20%) of the total FET resistance (i.e., $R_{CH} + 2R_C$) for the transistor to operate properly while ensuring that its current-voltage (I–V) behavior is primarily determined by the intrinsic channel resistance R_{CH} [196,219]. Hence, it is imperative that R_C must scale (i.e., reduce) together with both channel length and contact length before MoS$_2$-based FETs can come anywhere close to rivaling the performance of state-of-the-art Si and III-V device analogs [193,196,220]. The ρ_C is strongly dependent on the SBH among other factors, hence minimizing or eliminating the SBH is a guaranteed way to alleviate the R_C issue in MoS$_2$ FETs. Other major challenges that serve as roadblocks for TMD-based devices are the ultra-thin nature of the 2D TMDs and the existence of extrinsic charge carrier scattering mechanisms. The ultra-thin nature of the 2D MoS$_2$ makes it incredibly challenging to employ conventional CMOS-compatible doping techniques to perform controlled and area-selective doping to control the carrier type (n or p) and carrier concentration (ranging from degenerate in the source/drain contact regions to non-degenerate in the channel region) in MoS$_2$ FETs, especially at the monolayer limit [221].

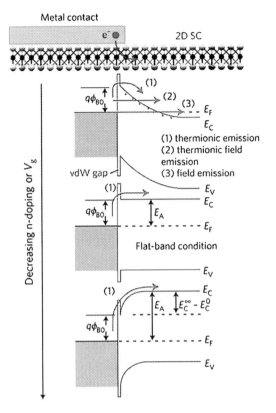

FIGURE 11.12 Energy band diagram of the n-type contact/MoS$_2$ interface under different gating (electrostatic n-doping) conditions depicting the different charge injection mechanisms from the metal into the MoS$_2$ channel across the SB. qΦ_{B0} represents the SBH. Thermionic emission is represented by Path (1), thermionic field emission by Path (2) and field emission by Path (3) as shown in the top band diagram for the case of maximum n-doping or maximum gate voltage V_g (that causes maximum downward band-bending). The additional tunnel barrier due to the vdW gap is also shown (marked by the red text). The lateral distance through which the carriers "tunnel" through in Paths (2) and (3) represents the SBW. As V_g decreases (i.e., n-doping decreases), the band-bending decreases and charge injection is governed by thermionic emission only, as shown by Path (1) in the middle and bottom energy band diagrams. [196].

Extrinsic charge carrier scattering mechanisms, such as substrate remote phonons, surface roughness, charged impurities, intrinsic structural defects (e.g., SVs), interface charge traps (D_{it}) and grain boundary (GB) defects, that can severely degrade the mobility in MoS$_2$-based devices [222–231].

To achieve low-power, high-performance and ultra-scaled devices based on 2D MoS$_2$, it is necessary to come up with effective solutions to alleviate the various problems, as mentioned above, that have an adverse effect on key device performance metrics. There has been an extensive research effort in the past few years to explore effective solutions for mitigating the challenges associated with the contact, doping and mobility engineering of 2D MoS$_2$ devices. This section focuses on some of the most promising 'defect engineering' techniques that have

11.6.1 Defect passivation techniques

In addition to all the intrinsic phonon scattering mechanisms that inevitably set an upper bound on the MoS_2 charge carrier mobility [232–236], the performance/mobility of MoS_2 FETs is often dominated by several extrinsic carrier scattering factors, such as structural defects, interface traps and surface states, leading to carrier localization and lower experimental mobilities than the predicted phonon-limited values [223–231,237,238]. Moreover, the scattering problem becomes worse for monolayer or ultra-thin MoS_2 devices (due to lack of efficient screening) since charge carriers in it are more susceptible to getting scattered by impurities present at the MoS_2/dielectric interface(s) as well as those residing within the MoS_2 lattice. One of the main structural defects in MoS_2 are the sulfur vacancies (SVs) that can act as charged impurities (CIs), charge trapping as well as short-range scattering centers by introducing localized gap states [239–241]. Moreover, SVs are also responsible for the strong contact FLP effect that leads to uncontrolled and large SBH for carrier injection [199,200,204,242]. Therefore, passivating these SVs is important to enhance the MoS_2 device performance. Yu *et al.* reported a facile, low-temperature thiol (-SH) chemistry to repair these SVs and improve the monolayer MoS_2/dielectric interface, resulting in significant reduction of charged impurities and interfacial traps [243]. They treated both sides of the monolayer MoS_2 using the chemical (3-mercaptopropyl) trimethoxysilane, abbreviated as "MPS", under mild annealing. In this approach, not only the sulfur atoms from the MPS molecules passivate the chemically reactive SVs, but the trimethoxysilane groups in MPS react with the SiO_2 substrate to form a self-assembled-monolayer (SAM) that can effectively passivate the MoS_2/SiO_2 interface (Fig. 11.13a and b illustrate the chemical structure of the MPS molecule, and the SV passivation process in MoS_2 by the MPS sulfur atom, respectively). Moreover, using theoretical modeling, the authors were able to extract the densities of CIs and interface trap states (D_{it}) that showed lower values in MPS-treated samples than the as-exfoliated ones (CIs reduced from 0.7×10^{12} to 0.24×10^{12} cm^{-2}, and D_{it} reduced from 8.1×10^{12} to 5.22×10^{12} cm^{-2}, after MPS treatment). These results highlight the importance of passivating the SV defects in MoS_2 to minimize carrier scattering and to improve the MoS_2 device performance (Fig. 11.13c compares the monolayer MoS_2 FET conductivity before and after MPS chemical treatment at RT) [243]. Another promising approach to passivate the SVs in monolayer MoS_2 was reported by Amani *et al.* using an air-stable, solution-based chemical treatment by an organic superacid, namely, bis(trifluoromethane) sulfonamide or TFSI, resulting in a near-unity photoluminescence (PL) yield and minority carrier lifetime enhancement, showing great promise for MoS_2-based optoelectronic devices [244]. Giannazzo *et al.* showed that D_{it} at the MoS_2/oxide interface can be reduced by carrying out a "temperature-bias" annealing process on MoS_2 FETs [245]. Forming gas annealing could be another effective way to passivate D_{it} and enhance the MoS_2 device performance (due to lowering of the SS and increase in the μ_{FE}) as shown by both Bolshakov *et al.* [246,247] and Young *et al.* [248].

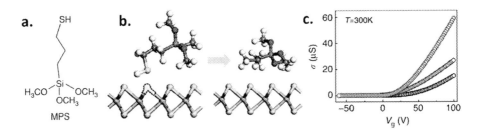

FIGURE 11.13 (a) Chemical structure of an MPS molecule showing the thiol (-SH) end-termination. (b) Schematic illustration of the MoS$_2$ SV passivation mechanism by the sulfur atom (yellow balls) derived from the -SH group of the MPS molecule. The SV in the MoS$_2$ layer is marked by the black-dashed contour. (c) Comparison of conductivity versus back gate voltage V$_G$ for FETs made on as-exfoliated (black curve), top side-treated (blue curve) and double side-treated (red curve) monolayer MoS$_2$ by MPS at RT. It is evident that the double-side treated FET shows the highest conductivity due to SV passivation on both the top and bottom surfaces of the monolayer MoS$_2$ [243].

Besides sulfur treatments, oxygen/ozone treatment has also been explored to repair the structural defects/SVs in MoS$_2$. Nan *et al.* used a mild oxygen (O$_2$) plasma treatment to improve the mobility of MoS$_2$ devices by an order of magnitude. This was attributed to the passivation of localized states originating from the SVs (that serve as carrier scattering centers) by the incorporation of oxygen ions that chemically bond with the MoS$_2$ at these SV sites. However, the plasma power and exposure time must be carefully controlled as excessive treatment may damage the MoS$_2$ lattice (either by physical damage or by excessive MoO$_3$ formation due to oxidation) and deteriorate the material quality [249]. Another novel report of mobility enhancement in multilayer MoS$_2$ NFETs was by Guo *et al.* where they used the synergistic effects of ultraviolet (UV) exposure and ozone (O$_3$) plasma treatment [250]. The authors showed an abnormal enhancement, up to an order of magnitude, in the FET mobility (from 2.76 cm^2/V-s to 27.63 cm^2/V-s) and attributed this to the passivation of interface traps/scattering centers as well as to an n-doping effect arising due to the photo-generated excess carriers during the UV/ozone plasma treatment. In this approach, negatively charged oxygen ions get incorporated in the MoS$_2$ lattice at the SV sites during the O$_3$ plasma treatment which are simultaneously neutralized by the excess photo-generated holes due to the UV exposure. This results in an aggregate of electrons in the MoS$_2$ lattice effectively causing n-doping and downward band-bending in the MoS$_2$ near the contact regions, thereby, narrowing the SBW and reducing the n-type R$_C$ [250]. The reader should note that, while controlled O$_2$/O$_3$ treatment can passivate the structural defects and interface traps in MoS$_2$ devices, excessive O$_2$/O$_3$ exposure can be harmful for the device performance. While chemical passivation of structural defects and traps are beneficial for improving the MoS$_2$ device performance, introduction of controlled physical damage in the contact regions *via* argon (Ar) plasma treatment has also been shown to be beneficial for improving the R$_C$ at the metal/MoS$_2$ interface (possibly due to the creation of unsaturated/reactive MoS$_2$ edge sites that can bond better with the contact metal atoms and/or due to the etching/removal of any surface oxide layers) [251,252].

It is also instructive to note that in addition to the commonly observed structural defects in 2D MoS$_2$ films derived from either naturally occurring or synthetically grown (*via* chemical vapor transport, CVT, etc.) bulk MoS$_2$ crystals, there also exists a set of intrinsic structural defects uniquely associated with synthetically grown large-area 2D MoS_2 nanosheets. From a commercial viewpoint, it is imperative to realize wafer-scale growth of uniform 2D MoS_2 films (either on rigid or flexible substrates) with tunable thicknesses for any MoS_2-based technology to become scalable and practically viable [253]. Hence, various synthetic routes, such as CVD [254,255], ALD [256], and vdW epitaxy [257] among others, utilizing a diverse set of precursor materials and growth conditions, have been explored to grow wafer-scale MoS_2 films [258]. However, these synthesized MoS_2 films are typically polycrystalline (i.e., they are formed by the coalescence of many MoS_2 domains) and typically contain a rich variety of unique point defects (e.g., antisite defects, vacancy complexes of Mo with three nearby sulfurs or disulfur pairs, etc.) and a diverse set of inter-domain dislocations and grain boundaries (GBs) that can introduce localized mid-gap states which, in turn, can scatter and/or trap the charge carriers. These defects can have dire consequences on the electrical performance of devices derived from synthesized MoS_2 films [259,260]. While several of the defect-passivation techniques described earlier can also be applied to these synthetic MoS_2 films, it is necessary to optimize the synthetic growth process itself to achieve defect-free, single-crystalline and pure MoS_2 films on commercial wafer-scale substrates that will enable the integration of large-scale 2D MoS_2-based devices and circuits.

11.6.2 Doping & defect engineering using dielectrics

The role of dielectrics is of paramount importance in the development and integration of high-performance MoS_2 devices. Dielectrics can play a critical role in doping the MoS_2, thereby, enhancing its carrier mobility, and in passivating/protecting the device channel against ambient exposure. In an interesting experimental and theoretical study, Rai *et al.* demonstrated the use of sub-stoichiometric high-κ oxides, such as TiO_x ($x < 2$), HfO_x ($x < 2$) and Al_2O_x ($x < 3$), as air-stable n-type charge transfer dopants on monolayer MoS_2 [261]. This high-κ oxide doping effect, arising due to interfacial-oxygen-vacancies in the high-κ oxide, could be used as an effective way to fabricate high-κ-encapsulated top-gated MoS_2 FETs with selective doping of the S/D access regions to alleviate the R_C issue, merely by adjusting the interfacial high-κ oxide stoichiometry. The underlying doping mechanism is similar for all high-κ oxides and involves the creation of donor states/bands near the conduction band edge (CBE) of MoS_2 by the uncompensated interfacial metal atoms of the sub-stoichiometric high-κ oxides. Moreover, this doping effect is absent in the case of purely stoichiometric high-κ oxides as has been verified both experimentally and theoretically using DFT calculations [261–263]. Fig. 11.14a and b show the R_C of a back-gated monolayer MoS_2 FET, extracted using the transfer length measurement or "TLM" method, as a function of back gate bias before and after sub-stoichiometric TiO_x doping and the DFT-calculated band structures. Using this doping technique, the authors reported low contact resistance, and an enhancement in both the μ_{FE} and intrinsic mobility in TiO_x-encapsulated monolayer MoS_2 FETs, strongly indicating that this high-κ doping effect plays an important role in boosting the electron mobility in high-κ-encapsulated MoS_2 FETs. In similar reports, both Alharbi *et al.* [264] and McClellan *et al.* [265] demonstrated the efficacy of n-doping by sub-stoichiometric high-κ oxides in improving the performance of MoS_2 FETs. Alharbi *et al.* used sub-stoichiometric HfO_x as the top gate dielectric in FETs fabricated on CVD-grown monolayer MoS_2 and achieved an R_C as low as $\sim 480 \, \Omega \cdot \mu m$ under heavy HfO_x doping ($>100\times$ improvement than the light HfO_x

doping case) and a mobility of $\sim$64 cm^2/V-s [264]. McClellan *et al.*, on the other hand, utilized AlO$_x$ encapsulation to n-dope back-gated monolayer MoS$_2$ FETs and achieved an R$_C$ of $\sim$480 $\Omega \cdot$ μm, a μ_{FE} of $\sim$34 cm^2/V-s and a record ON-current of 700 μA/μm. A key step in their approach was annealing of the MoS$_2$ devices in an N$_2$ ambient after AlO$_x$ encapsulation, which helped restore the SS and μ_{FE} by converting the "deep-level traps" at or near the AlO$_x$/MoS$_2$ interface into "shallow-level donors" (Fig. 11.14c and d show the back-gated MoS$_2$ FET schematic, and the effect of AlO$_x$ doping, as well as N$_2$ post-annealing on the FET transfer curves, respectively) [265].

High-κ dielectrics can also help mitigate the deleterious Coulombic interaction between charge carriers in low-dimensional (i.e., 2D and 1D) semiconductors and their surrounding charged impurities (CIs) [266]. This is so even without considering the high-k dielectric propensity for doping the MoS$_2$ *via* sub-stoichiometric surface charge transfer or their capability to provide enhanced electrostatically-induced carrier densities in the device channel (due to the much higher gate capacitances they offer) [266]. For 2D MoS$_2$ devices, the CIs typically reside at the MoS$_2$/dielectric interface and can originate from various kinds of incorporated residues and adsorbates (gaseous or chemical residue) during device processing (note that the widely used SiO$_2$ substrate for MoS$_2$ devices is highly prone to the adsorption of these CIs due to its highly reactive/hydrophilic surface). Moreover, the CIs can also originate from the intrinsic structural defects such as SVs and/or trapped ionic species in the MoS$_2$ host lattice. Charged impurities serve as major scattering centers by giving rise to localized electric fields that can interact strongly with, and perturb the motion of, the MoS$_2$ charge carriers. The high-κ dielectrics, thanks to their large ionic polarizability, can effectively cancel out or "screen" the local electric fields generated by these CIs, thereby, minimizing the scattering effect of CIs on the charge carriers. Note that higher "κ" values increased polarizability of the high-κ dielectric, leading to improved dielectric screening of CIs [230,237,267,268].

Employing high-κ dielectrics to offset the effect of CIs, Li *et al.* [269] demonstrated the use of an HfO$_2$/Al$_2$O$_3$ high-κ dielectric stack to fabricate dual-gated MoS$_2$ NFETs on SiO$_2$ substrates. They showed that the high-κ stack enhanced the RT electron mobility (from 55 to 81 cm^2/V-s) as well as enabled high drain currents at low-T ($\sim$660 μA/ μm at 4.3 K), while effectively eliminating the self-heating-induced negative differential resistance (NDR) effect owing to its higher thermal conductivity as compared to SiO$_2$ [269]. Similarly, Yu *et al.* [270] reported monolayer MoS$_2$ back-gated FETs on HfO$_2$ and Al$_2$O$_3$ substrates showing RT mobilities of 148 and 113 cm^2/V-s, representing an 85% and a 41% improvement over FETs on SiO$_2$ substrates, respectively. Through experimental and rigorous theoretical modeling, they demonstrated the efficacy of high-κ dielectrics over SiO$_2$, and of higher-κ HfO$_2$ ($\kappa \sim 17$) over Al$_2$O$_3$ ($\kappa \sim 10$), in providing improved screening against CI scattering. The top plot of Fig. 11.15 compares the temperature-dependent mobility of MoS$_2$ FETs on SiO$_2$, Al$_2$O$_3$ and HfO$_2$ substrates, at a fixed sheet carrier density, showing a good match between experimental data and the theoretical model employed by the authors. Moreover, the authors theoretically calculated the dependence of the CI-limited mobility (i.e., after subtracting the contribution of phonon scattering) on the MoS$_2$ sheet carrier density (n_{2D}) for MoS$_2$ FETs on SiO$_2$, Al$_2$O$_3$ and HfO$_2$ substrates and found that while the mobility increased with increasing n_{2D} for all dielectric substrates, the mobility values were highest in the case of HfO$_2$, followed by Al$_2$O$_3$, and lowest for SiO$_2$ (Bottom plot of Fig. 11.15 shows the calculated CI-limited mobility as a function of n_{2D}). A similar dependence was observed by the authors in their experimental

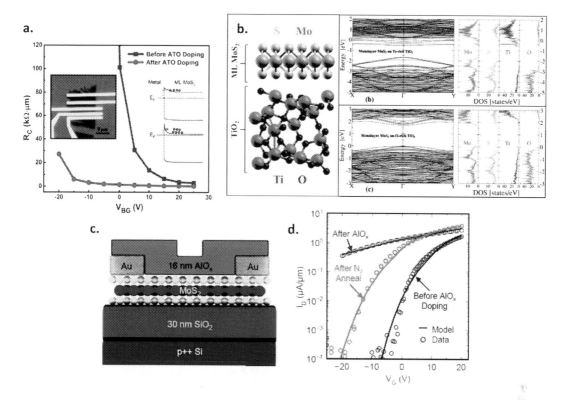

FIGURE 11.14 (a) Extracted R_C versus V_{BG} before (blue) and after (red) amorphous TiO$_x$ (ATO) doping measured using a TLM structure. The R_C shows a strong gate dependence before doping (Schottky behavior) and a weak gate dependence after doping (Ohmic behavior). Left inset: Optical micrograph of the as-fabricated transfer line method (TLM) structure. Right inset: Qualitative band diagrams of the metal/MoS$_2$ interface before (**top**) and after (**bottom**) ATO doping showing increased band-bending in the MoS$_2$ after doping leading to SBW reduction and enhanced electron injection into the channel *via* tunneling [262]. (b) **Left schematic**: Supercell showing the composite crystal structure of the monolayer MoS$_2$/TiO$_2$ interface used in the DFT simulations. **Top right**: Band structure and atom-projected-density-of-states (AP-DOS) plots for MoS$_2$-on-sub-stoichiometric TiO$_x$ case. In the presence of interfacial oxygen "O" vacancies, electronic states/bands from the uncompensated Ti atoms are introduced near the CBE of monolayer MoS$_2$ causing the Fermi level (represented by the 0 eV energy level on the y-axis) to get pinned above the conduction band indicating strong n-doping. **Bottom right**: Band structure and AP-DOS plots for the MoS$_2$-on-stoichiometric TiO$_2$ case. No doping effect is seen in this case and the Fermi level remains pinned at the VBE of MoS$_2$. [262] (c) Schematic of a back-gated CVD-grown monolayer MoS$_2$ FET with a top AlO$_x$ doping layer and Au contacts. [265] (d) Semilog transfer curves of the MoS$_2$ FET before and after AlO$_x$ deposition, and after N$_2$ anneal. A significant n-doping effect is seen after AlO$_x$ deposition accompanied with an SS degradation. The N$_2$ anneal helps restore the SS while maintaining the n-doping effect of AlO$_x$ due to conversion of "deep-level traps" into "shallow-level donors" [265].

data, further validating their theoretical model [270]. These results make sense as increased sheet carrier densities in the MoS$_2$ channel would provide additional screening against the CI scattering centers, in addition to the high-κ screening effect of the various dielectrics (with HfO$_2$ being more effective than Al$_2$O$_3$ due to its higher κ value). In addition to the com-

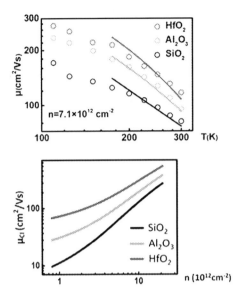

FIGURE 11.15 Top: Mobility versus temperature (at a fixed carrier density) for MoS$_2$ FETs on HfO$_2$, Al$_2$O$_3$ and SiO$_2$ substrates. While FETs with both high-κ dielectrics show mobility enhancement over SiO$_2$, highest mobilities are achieved in the case of HfO$_2$ due to the enhanced CI screening effect of higher-κ HfO$_2$ ($\kappa \sim 17$) than Al$_2$O$_3$ ($\kappa \sim 10$). **Bottom**: Log-log plot showing the calculated RT CI-limited mobility versus sheet carrier density (at a fixed CI density) for monolayer MoS$_2$ FETs on the three different dielectrics. Highest mobilities are achieved in the case of HfO$_2$ which outperforms Al$_2$O$_3$ which, in turn, outperforms SiO$_2$ [270].

monly used ALD-deposited high-κ dielectrics on MoS$_2$ such as HfO$_2$ and Al$_2$O$_3$, researchers have also resorted to integrating several other high-κ dielectrics, such as tantalum pentoxide (Ta$_2$O$_5$) [271] and zirconium oxide (ZrO$_2$) [272], in TMD devices using novel integration approaches showing interesting device results.

An extremely important point to note is that the mobility enhancement due to high-κ screening is possible only when the MoS$_2$ carrier mobility is strongly limited by charged impurities (i.e., the CI density is high, typically $>10^{12}$ cm^{-2}, which limits the MoS$_2$ carrier mobility to values well below 100 cm^2/V-s due to Coulombic scattering). In other words, in the absence or dearth of CIs, high-κ dielectrics would no longer be useful for improving the device performance of clean MoS$_2$ samples any further. This is because high-κ dielectrics are a major source of surface optical (SO) phonons or remote optical phonons, which can serve as a major extrinsic carrier scattering source. These SO-phonon modes originate from the oscillations of the polarized metal-oxide bonds (note that these are the same polarized bonds that provide the screening against CIs) in the high-κ dielectric and typically have low activation energies such that these SO-phonon modes can easily be activated at RT. Moreover, the ultra-thin 2D channel of TMDs such as MoS$_2$ (especially in the monolayer limit) is highly susceptible to its surrounding dielectric environment. These SO-phonon modes can easily couple to the MoS$_2$ channel and scatter the charge carriers [223,237,270,273]. At the MoS$_2$/high-κ dielectric interface, therefore, there is always a competition between the 'detrimental' SO-phonon scattering effect and the 'advantageous' CI screening effect on the carrier mobility, and SO-phonon scattering ultimately becomes the dominant mobility-limiting factor for 2D MoS_2 devices in the limit of decreasing charged impurities. The magnitude of SO-phonon scattering in MoS_2 is directly (inversely) proportional to the κ-value (SO-phonon energy) of the surrounding dielectric media. Now, for SiO_2 and other commonly used high-κ dielectrics, their SO-phonon energies are as follows (listed in ascending order in units of meV): HfO_2 (12.4) < ZrO_2 (16.67) < Al_2O_3 (48.18) < SiO_2 (55.6) [274,275]. In general, the magnitude of the SO-phonon energy of a dielectric is inversely related to its κ-value. Thus, HfO_2 would cause the worst SO-phonon scattering of the MoS_2 charge carriers, and SiO_2 the least, at RT (i.e., when the thermal energy $kT/q \sim 26$ meV). Calculation of RT field-effect mobility for mono-layer MoS_2 devices on different dielectric substrates as a function of CI density by Yu *et al.* revealed a "critical" CI density of $\sim 0.3 \times 10^{12}$ cm^{-2} above which the mobility was strongly limited by the CIs (this corresponds to the "impure" regime in which high-κ dielectrics can be beneficial for enhancing the mobility by effectively screening the scattering effect of these CIs) and below which the mobility was limited by phonons (this corresponds to the "ultra-clean" regime where high-κ dielectrics are no longer useful due to the detrimental effect of their SO-phonons) [270].

It is clear that to achieve the maximum MoS_2 device performance, ultra-clean samples (i.e., those having CI densities well below 10^{12} cm^{-2}) and low-κ dielectrics (as opposed to high-κ HfO_2, ZrO_2, Al_2O_3, etc.) having higher SO-phonon activation energies (to minimize SO-phonon scattering) must be integrated together. To achieve ultra-clean samples, the structural and electronic quality of synthetically grown large-area MoS_2 as well as the device processing/fabrication steps must be carefully optimized to minimize the CI density. Regarding choice of dielectrics, nitride-based wide-bandgap dielectrics such as 2D hexagonal boron nitride (hBN) and aluminum nitride (AlN), both having $E_g \sim 6$ eV [276], hold the most promise since they are both medium-κ dielectrics ($\kappa \sim 5$ for hBN and ~ 9 for AlN) and have much higher SO-phonon energies (hBN: ~ 93 meV, AlN: ~ 81 meV) compared to the other high-κ dielectrics discussed above. Therefore, both hBN and AlN can offer an optimized combination of high gate capacitances (required for better electrostatic control over the device channel) and high carrier mobilities (required for achieving high ON-state currents) that are essential for realizing high-performance FETs based on MoS_2 [237].

Indeed, there have been several reports demonstrating MoS_2 FETs with hBN dielectrics. Cui *et al.* [277] reported a vdW heterostructure device platform wherein the MoS_2 layers were fully encapsulated within hBN to minimize external scattering due to charged impurities and remote SO-phonons, while gate-tunable graphene was used as the contacts to MoS_2. Using this approach, the authors extracted a low-temperature mobility of 1020 cm^2/V-s for monolayer MoS_2 and 34,000 cm^2/V-s for 6-layer MoS_2 at 4 K, with the values being up to two orders of magnitude higher than what was reported previously. Theoretical fit to the experimental data revealed the interfacial long-range CI density of $\sim 6 \times 10^9$ cm^{-2} that was about two orders of magnitude lower than the CI density typically obtained for graphene devices on SiO_2. However, even with such clean samples and hBN dielectrics, the maximum RT MoS_2 electron mobility was only 120 cm^2/V-s [277]. Xu *et al.* reported sandwiched hBN/MoS_2/hBN devices with low-T Ohmic contacts which resulted in a high low-T μ_{FE} of 14,000 cm^2/V-s and Hall mobility (μ_{Hall}) of 9,900 cm^2/V-s at 2 K. The extracted RT device mobilities in this study, however, were only ~ 50 cm^2/V-s [278].

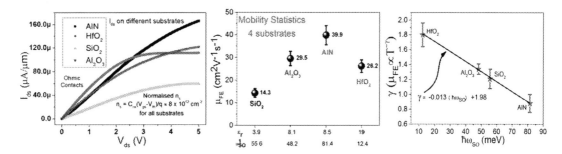

FIGURE 11.16 Left: Normalized output characteristics of MoS$_2$ FETs on SiO$_2$ (green curve), Al$_2$O$_3$ (blue curve), HfO$_2$ (red curve) and AlN (black curve) substrates showing the highest saturation drain current in the case of AlN. **Middle**: Extracted average RT μ_{FE} for MoS$_2$ FETs on four different dielectric substrates as a function of their SO-phonon energies. The highest μ_{FE} is achieved for FETs on AlN substrates which has the lowest scattering effect due to its relatively high SO-phonon energy (~81 meV). **Right**: Extracted mobility degradation factor "γ" as a function of the dielectric SO-phonon energy for MoS$_2$ FETs on different dielectrics. As expected, γ is highest for HfO$_2$ (which has the lowest SO-phonon energy ~12 meV) and lowest for AlN (which has the highest SO-phonon energy) [279].

It is interesting to observe that in most of the experimental studies on hBN-encapsulated MoS$_2$ devices, while record-high low-T electron mobilities well in excess of 1000 cm^2/V-s have been achieved, the technologically relevant RT mobility still lags behind the best reported values for MoS$_2$ FETs fabricated using other dielectrics. This could be due to a multitude of factors such as differences in the material or electronic quality of the starting MoS$_2$ used in these isolated experiments, the innate material quality of the hBN itself (with perhaps lower than expected SO-phonon energies etc.), and other differences such as quality of the S/D electrical contacts and processing-induced impurities/defects. Nonetheless, these theoretical and experimental results clearly highlight the advantages of using pristine 2D hBN as an ideal dielectric for MoS$_2$-based electronics over commonly used SiO$_2$ and other high-κ dielectrics, as hBN can afford much lower densities of interface traps and charged impurities, lower surface roughness scattering and much better immunity against SO-phonon scattering.

In addition to hBN, experimental evidence of the benefits of using aluminum nitride (AlN) as an alternative nitride-based dielectric for MoS$_2$ devices was also reported by Bhattacharjee et al. [279]. They compared the performance of identical MoS$_2$ FETs fabricated on SiO$_2$, Al$_2$O$_3$, HfO$_2$ and AlN substrates, with the MoS$_2$-on-AlN FETs outperforming its counterparts. Temperature-dependent μ_{FE} calculations revealed the mobility degradation factor (γ) to be lowest for the FET on AlN ($\gamma = 0.88$) as compared to all other dielectrics ($\gamma = 1.21, 1.32,$ and 1.80 for SiO$_2$, Al$_2$O$_3$, and HfO$_2$, respectively). Since phonon scattering is the dominant scattering mechanism at high temperatures and $\mu \propto T^{-\gamma}$, it is no surprise that AlN affords the lowest phonon scattering owing to its relatively high SO-phonon activation energy of ~ 81 meV. (Fig. 11.16 compares the output characteristics as well as the extracted μ_{FE} and γ values as a function of the SO-phonon energy for MoS$_2$ FETs fabricated on the four different dielectrics used in this study, with FETs on AlN showing the best performance) [279].

11.6.3 Substitutional doping & alloying

There have also been several reports of "substitutional doping" of MoS_2 wherein both the Mo cation and the S anion atoms have been substituted by appropriate "donor" or "acceptor" dopant atoms to yield either n-type or p-type MoS_2, respectively. In conventional CMOS technology, substitutional doping using ion implantation is the method of choice for controllably doping selected areas of the semiconductor wafer (either Si, Ge or III-Vs) to fabricate complementary FETs and realize complex circuits with desired performances. The ion implantation technique is also used to selectively and degenerately dope the S/D regions of the FET to realize Ohmic n- and p-type contacts for NMOS (i.e., n^+-p-n^+) and PMOS (i.e., p^+-n-p^+) device configurations, respectively, as well as to realize various bipolar devices, such as LEDs and photodetectors, for optoelectronic applications [280–293]. Owing to the atomically thin nature of 2D MoS_2 (recall that an MoS_2 monolayer is only ~0.65 nm thick), it is extremely challenging to employ the conventional ion implantation technique to dope MoS_2 (or 2D materials in general) as the process can induce irreparable surface damage and etching of the MoS_2 layers. However, practical and stable doping requires "substitution" of a given fraction of the host lattice atoms by the dopant atoms wherein the latter covalently bonds with other atoms in the host lattice. While little progress has been made in the controlled and area-selective "top-down" substitutional doping of MoS_2, most substitutional doping efforts on MoS_2 have relied on the incorporation of dopant atoms during the "bottom-up" or in-situ synthetic growth process (e.g., CVD) which may provide controlled, but inevitably unselective, doping of the entire MoS_2 film [221,294,295].

The first report demonstrating the in-situ CVD substitutional doping of MoS_2 was by Laskar *et al.* [296] where they p-doped MoS_2 using niobium (Nb) atoms. In this approach, the Nb atoms replace the Mo cations in the MoS_2 host lattice and act as efficient electron acceptors because they have one less valence electron than the Mo atoms (note that Nb lies to the left of Mo in the periodic table). The authors showed that the crystalline nature of MoS_2 is preserved after Nb doping and reported reasonable RT mobilities (8.5 cm^2/V-s) at high hole doping densities (3.1×10^{20} cm^{-3}), and a low R_C (0.6 $\Omega \cdot$ mm) for p-type conduction (since the high hole doping concentration would help reduce the SBW for hole injection, favoring hole tunneling at the contacts) [296]. A similar in-situ Nb doping approach was reported by Suh *et al.* where they obtained a degenerate hole density of 3×10^{19} cm^{-3} in their MoS_2 thin films and confirmed the p-doping *via* XPS, TEM and electrical measurements among others [297] (Fig. 11.17a shows the 3D schematic of a Nb-doped MoS_2 lattice along with the XPS spectra depicting the shift of the Mo 3d core level peaks associated with p-type doping). Mirabelli *et al.* studied the back-gated FET behavior of Nb-substituted highly p-doped 10 nm thick MoS_2 flakes and highlighted the importance of high Nb doping levels in improving the metal/MoS_2 contact resistance [298]. The hole concentration after doping was extracted to be 4.3×10^{19} cm^{-3} from Hall-effect measurements. Although the FET ON/OFF ratio was compromised due to the uniform high hole doping throughout the MoS_2 contact and channel regions, the authors extracted the specific contact resistivity (ρ_C) for holes to be 1.05×10^{-7} $\Omega \cdot cm^2$ from TLM measurements which was even lower than the ρ_C for electrons as reported by English *et al.* (5×10^{-7} $\Omega \cdot cm^2$ using UHV metal deposition [299]) and Kang *et al.* (2.2×10^{-7} $\Omega \cdot cm^2$ using hybridized Mo contacts [300]), confirming that heavy doping of MoS_2 can be an effective way to drive down the ρ_C (due to SBW thinning in the contact regions) [298].

In contrast to the in-situ p-doping of MoS_2 by Mo cation substitution, Gao *et al.* [301] reported the in-situ n-doping of monolayer MoS_2 by substitution of Mo with rhenium (Re) atoms. Note that Re has seven valence electrons as compared to six valence electrons in Mo and, hence, donates an extra electron to the MoS_2 lattice when substituted at the Mo atom site. The authors confirmed the n-doping *via* XPS and PL measurements as well as DFT calculations (Fig. 11.17b shows the DFT-calculated band structure of Re-doped MoS_2 depicting the presence of Re donor bands near the MoS_2 CBE and the XPS spectra depicting the shift of the Mo 3d core level peaks associated with n-type doping). Their Re-doped MoS_2 NFETs displayed linear output characteristics even at temperatures as low as 10 K implying Ohmic nature of the S/D contacts (due to the doping-induced reduction of the SBW that facilitated efficient carrier injection into the MoS_2 channel *via* tunneling). Hence, these results clearly demonstrate the efficacy of Re doping on the n-type R_C reduction in MoS_2 FETs [301]. However, the electron mobility of their Re-doped MoS_2 films was low, ranging between 0.1 to 0.7 cm^2/V-s, implying the presence of various carrier scattering sources (including scattering from the incorporated Re dopant atoms).

In addition to substitution of Mo atoms, doping *via* substitution of the sulfur (S) anion has also been investigated for MoS_2. Yang *et al.* [302] reported a novel and simple chloride-based molecular doping technique wherein the chlorine (Cl) atoms covalently attach to the Mo atoms at the sulfur vacancy (SV) sites in the MoS_2 lattice upon treatment with 1,2-dichloroethane (DCE) at RT. Since the Cl atom has an extra valence electron than the S atom (note that Cl lies to the right of S in the periodic table), it donates its extra electron to the MoS_2 lattice when substituted at the S atom sites resulting in n-type doping. Using this doping approach, an R_C as low as 500 $\Omega \cdot$ μm was extracted for Ni-contacted few-layer MoS_2 NFETs *via* TLM analysis, and the low R_C was attributed to the high electron doping density in the MoS_2 ($\sim 2.3 \times 10^{19}$ cm^{-3}) that causes increased band-bending at the contact/MoS_2 interface leading to a significant reduction of the SBW, thereby, facilitating electron tunneling. Significant improvements in the extracted transfer length (L_T) and specific contact resistivity (ρ_C) were also observed after Cl doping, with the L_T reducing from 590 nm to 60 nm and the ρ_C reducing from 3×10^{-5} to 3×10^{-7} $\Omega \cdot cm^2$. The authors demonstrated high-performance 100 nm channel Cl-doped MoS_2 NFETs with a high ON-current of 460 μA/ μm, μ_{FE} of 50–60 cm^2/V-s, high ON/OFF ratio of 6.3×10^5, and long-term environmental stability [302]. Moreover, the Cl doping technique can also be applied to other 2D semiconducting TMDs such as WS_2 [303]. Incorporation of oxygen (O) in the MoS_2 lattice was also shown to cause a p-type doping effect by Neal *et al.* and was attributed to the formation of acceptor states, about ~ 214 meV above the MoS_2 VBE, by the formation of high work function MoS_xO_{3-x} clusters in the MoS_2 lattice. Going against the notion that oxygen exposure only helps in passivating the SVs in MoS_2 leading to a decrease in the background electron concentration causing an 'apparent' p-type doping effect, this work provided evidence that oxygen atoms can independently cause p-doping of the MoS_2 when substituted at the S atom sites [304]. A very similar work by Giannazzo *et al.* further confirmed the local substitutional p-doping effect of O atoms *via* conductive atomic force microscopy (CAFM) measurements. They used "soft" O_2 plasma treatments to modify the top surface of multilayer MoS_2 resulting in the formation of high work function MoO_xS_{2-x} localized alloy clusters. [305].

A breakthrough in the controlled and area-selective p-doping of few-layer MoS_2 was reported by Nipane *et al.* who used a novel and CMOS-compatible plasma immersion ion

11.6. Enhancing 2D device performance using defect engineering

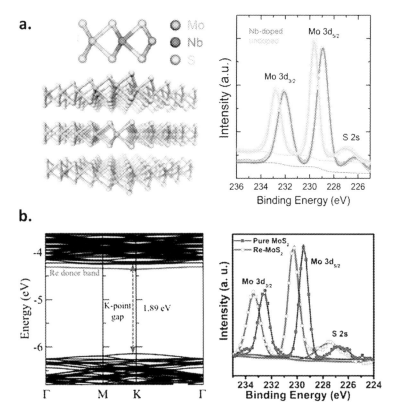

FIGURE 11.17 (a) **Left schematic**: 3D cross-sectional illustration of Nb-doped few-layer MoS$_2$ wherein the Nb dopant atoms replace the Mo host atoms in the MoS$_2$ lattice. **Right**: XPS spectra of the Mo 3d core level peaks as a function of electron binding energy as measured from the Nb-doped (red) and undoped MoS$_2$ (light blue). A clear shift in the Mo 3d peaks towards lower binding energies is observed after Nb doping confirming the lowering of the MoS$_2$ Fermi level due to p-type doping. [297] (b) **Left**: DFT-calculated electronic band structure of Re-doped MoS$_2$ showing the presence of Re donor bands/levels close to the CBE of MoS$_2$ confirming the n-type substitutional doping. **Right**: XPS spectra of the Mo 3d core level peaks as a function of electron binding energy measured from the Re-doped (red) and undoped MoS$_2$ (blue). In this case, the Mo 3d peaks shift towards higher binding energies after Re doping (opposite to the case of Nb-doped p-type MoS$_2$) confirming the upshift of the MoS$_2$ Fermi level due to n-type doping. [301]

implantation (PIII) process using phosphorous (P) atoms [306]. P lies to the left of S in the periodic table and, hence, acts as an acceptor in the MoS$_2$ lattice due to its electron deficient nature. In this method, the MoS$_2$ flakes were exposed to an inductively coupled phosphine (PH$_3$)/He plasma inside a PIII chamber either before or after S/D contact patterning to achieve area-selective P implants. Various characterization techniques were employed to identify suitable PIII processing conditions to achieve low surface damage and minimal etching of the MoS$_2$. Back-gated FETs fabricated on P-implanted MoS$_2$ with varying implant energies and doses showed clear evidence of p-type conduction, ranging from non-degenerate to degenerate behavior, and the p-doping was also verified experimentally *via* XPS. The peak

μ_{FE} for holes was extracted to be 8.4 cm^2/V-s and 137.7 cm^2/V-s for the degenerate and non-degenerate doping cases, respectively. Furthermore, using rigorous DFT analysis, the authors confirmed the substitutional p-doping of MoS$_2$ by incorporation of P atoms at the S atom sites, and found that pre-existing SVs could enhance this doping effect by providing "empty" sites for the P atoms to latch onto [306,307]. Another promising result for substitutional doping of MoS$_2$ using high energy ion implantation was reported by Xu *et al.* [308] wherein they directly utilized traditional ion-implanters to p-dope few-layer MoS$_2$ films using P atoms. In their approach, a thin layer of poly(methyl methacrylate) or PMMA resist (200 nm or 1000 nm thick) was spin-coated onto the ultrathin MoS$_2$ flakes (<10 nm thick) as a "protective masking layer" which helped decelerate the P dopant ions and led to the successful retention of a portion of these ions inside the 2D MoS$_2$ lattice [308]. Raman, TEM and HRTEM characterization revealed negligible damage to the MoS$_2$ crystal structure upon removal of the PMMA layer, highlighting an advantage of this PMMA-coated ion implantation technique over the PIII process described above (where the plasma can cause unintended etching of the 2D MoS$_2$ layers). This ion implantation technique is highly promising for achieving large-scale, controlled and area-selective doping of MoS$_2$ (both n- and p-type) and is compatible with existing infrastructures in the semiconductor industry.

Another extremely promising experimental approach was demonstrated by Chuang *et al.* [309] where they utilized a "2D/2D" vertical heterostructure contact strategy to realize low-resistance p-type Ohmic contacts to MoS$_2$ FETs. In their approach, the undoped semiconducting MoS$_2$ channel is contacted in the S/D regions by degenerately p-doped Mo$_{0.995}$Nb$_{0.005}$S$_2$ (the degenerately p-doped MoS$_2$ was obtained by substitutional doping of MoS$_2$ using Nb during the crystal growth process). The work function difference between the undoped and the degenerately p-doped MoS$_2$ creates a band offset across the 2D/2D vdW interface which can be electrostatically tuned by a back gate voltage owing to the weak interlayer vdW interaction at the 2D/2D junction, essentially resulting in a negligible SBH in the ON-state of the FET. Note that the vdW interface also promotes weaker FLP by suppressing the formation of interface gap states. The authors reported field-effect hole mobilities as high as 180 cm^2/V-s at RT and linear output characteristics down to 5 K in their p-type MoS$_2$ FETs with these low-resistance 2D/2D contacts [309]. An alloyed 2D metal/TMD contact scheme, similar to recent reports on 2D tungsten diselenide (WSe$_2$) where a NbSe$_2$/W$_x$Nb$_{1-x}$Se$_2$/WSe$_2$ contact interface was realized (here NbSe$_2$ is a metallic 2D TMD), could also be used to realize p-type TMD FETs having a seamless contact interface (as opposed to creating stacked 2D/2D heterostructure contacts). Such alloyed 2D junctions have been shown to have atomically sharp vdW interfaces with reduced interface traps and SBH, and they can help maximize the electrical reliability of 2D devices [310].

11.7 Theoretical investigation of defects in 2D TMDs

The unique electrical and optical properties of monolayer (ML) TMDs [311,312] have spurred intense research interest towards development of nanoelectronic devices utilizing these novel materials [313–320]. The atomically thin form of ML TMDs translates to excellent electrostatic gate control even at nanoscale channel length dimensions [321–323]. However, the two-dimensional (2D) nature of ML TMDs makes their properties susceptible to the surrounding environment, as evidenced by the mobility enhancement of ML MoS_2 when surrounded by a high-k dielectric such as hafnia (HfO_2) [313]. This mobility improvement in 2D materials was attributed to the damping of Columbic impurity scattering by high-k dielectrics [324]. Theoretical calculations of HfO_2 interfaces have indicated that band offsets can be altered chemically by utilizing different interface terminations [325]. The conductive characteristics of MoS_2 deposited on SiO_2 have been shown to be dependent on the interface structure [326]. Controllable n-type doping of graphene transistors with extended air stability have been demonstrated by using self-encapsulated doping layers of titanium sub-oxide (TiO_x) thin films [327]. These results put into stark focus the need to consider the effect of surrounding materials and the interfaces with them on the characteristics of ML TMDs.

In an earlier preliminary study [328], we considered MoS_2 on O- and Hf (Al)- terminated HfO_2 (Al_2O_3) *via* density functional theory (DFT). Those results suggest that O-terminated and H-passivated HfO_2 and Al_2O_3 exhibit potential as good substrates or gate insulators for ML MoS_2, with a straddling gap (Type 1) band structure for the composite system devoid of any defect states within the band gap of the MoS_2. However, ML MoS_2 on Hf-terminated HfO_2 shows a staggered gap alignment (Type 2), such that holes would be localized in the oxide rather than the MoS_2 layer. But in the case of ML MoS_2 on Al-terminated Al_2O_3 slab, we found that there is a straddling gap (Type 1) alignment to again provide localization of holes to the MoS_2 layer, although with some spill over into the nearby surface O states.

In subsequent work, we focused on the effects of O vacancies (O deficiency) in MoS_2 on HfO_2 and on Al_2O_3, and the effects of Mo and S vacancies in MoS_2 on HfO_2. We have used both density functional theory (DFT) and experimental analysis. For the O deficient systems, two possible terminations for the HfO_2 (Al_2O_3) slab are considered using DFT: An O-terminated HfO_2 (Al_2O_3) slab with H passivation and an Hf (Al)-terminated HfO_2 (Al_2O_3). The naming of two possible terminations is indicative of the initial structures used as starting point in our atomistic relaxations. The effects of O-vacancies in the first few layers of oxide on the band structure of the MoS_2-oxide system were simulated, with results for vacancies in the topmost/MoS_2-adjacent O layer shown here. Among our findings, O vacancies can lead to modulation-like doping of the MoS_2 from donor states in the oxide depending on the oxide terminations.

11.7.1 Computational details

The DFT calculations were performed using the projector-augmented wave method with a plane-wave basis set as implemented in the Vienna *ab initio* simulation package (VASP) [329,330]. A kinetic energy cutoff of 400 eV was chosen. The k-mesh grid of $7 \times 7 \times 1$ for the sampling of the first Brillouin zone of the supercell was selected according to Monkhorst-Pack type meshes with the origin being at the Γ point for all calculations except the band structure calculation. The local density approximation (LDA) [331] was employed primarily for the exchange-correlation potential as LDA has been shown to reproduce the apparent experimental band gap ($E_g = 1.8$ eV) [9] of ML MoS_2 well [332,333]. The calculated lattice constant for the MoS_2 layer after volume relaxation, $a = 3.122$ Å, is also a good match to the experimental value [334]. We have also re-checked some of the DFT results using the generalized gradient approximation (GGA) [335]. We note, however, that both the LDA and the GGA underesti-

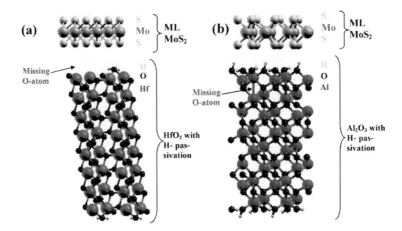

FIGURE 11.18 (a) Supercell of ML MoS$_2$ on an H-passivated, O-terminated HfO$_2$ slab of approximately 2 nm thickness with O-vacancy (side view). (b) Supercell of ML MoS$_2$ on an H-passivated, O-terminated Al$_2$O$_3$ slab of approximately 2 nm thickness with O-vacancy (side view). The monolayer of MoS$_2$ belongs to the space group P-6m2 (point group D3h). [341].

mate the band gap of at least the bulk HfO$_2$ and Al$_2$O$_3$, which makes the prediction of band offsets from theoretical calculations unreliable. With approximately 150 atoms per supercell, use of presumably more accurate hybrid functionals or GW methods for atomistic relaxations was not practical. However, we have utilized hybrid functionals, namely HSE06 [336], to perform band structure calculations using the relaxed structures from our GGA simulations to further check our key conclusions. However, the primary objective of this theoretical work is to explore possible pathways to insulating and doping MoS$_2$ MLs qualitatively toward device applications, ultimately for experimental follow-up for promising cases. Similarly, we did not include spin orbit coupling here, which causes substantial spin splitting in the valence band, for similar reason. However, only conduction band doping is observed in our results, mitigating the impact of this latter approximation. Van der Waal's forces also were simulated due to the absence of covalent bonding between the TMD and the oxides [337]. In our computations, we have adopted the DFT-D2 scheme to model the non-local dispersive forces wherein a semi-empirical correction is added to the conventional Kohn-Sham DFT theory [338].

The two representative dielectrics, HfO$_2$ and Al$_2$O$_3$, were chosen for high-k value and minimal lattice mismatch, respectively. The MoS$_2$ ML of principle interest, with its hexagonal lattice, was taken to be unstrained with its above-noted volume-relaxed lattice constant of $a = 3.122$ Å. For the dielectric oxide, the energetically stable crystalline phases of bulk HfO$_2$ and Al$_2$O$_3$ at ambient conditions, namely, monoclinic HfO$_2$ [339] and hexagonal Al$_2$O$_3$ [340], respectively, were utilized. Our simulations were performed by constructing a supercell of ML MoS$_2$ on an approximately 2 nm thick oxide slab. For HfO$_2$, atomic relaxation was performed within a rectangular supercell ($a = 9.366$ Å, $b = 5.407$ Å) chosen to reduce the lattice mismatch between ML MoS$_2$ and monoclinic HfO$_2$. However, a roughly 6% strain remains along the in-plane directions in the HfO$_2$ (Fig. 11.18a). For Al$_2$O$_3$, atomic relaxation was performed in a (rotated) hexagonal supercell ($a = 8.260$ Å) with a strain of only about

0.2% (Fig. 11.18b). The systems were relaxed until the Hellmann-Feynman forces on the atoms were less than 0.02 eV/Å. During relaxation, all the MoS_2 ML atoms and the top half of the layers of the dielectric oxide were allowed to move in all three spatial dimensions. Oxygen vacancies were modeled by removing a single O atom from an O-layer of the supercell. Since we have periodic supercells, the O vacancy is repeated in each instance of the supercell. The system is then allowed to relax again with the introduced O-vacancy. A similar procedure was followed in the modeling of Mo and S vacancies in the MoS_2-HfO_2 system. In the latter case, the S atom vacancy was introduced in the layer adjacent to the oxide surface. All simulations were performed at a temperature of 0 K.

11.7.2 Results and discussion

The band structure and atom-projected density of states (AP-DOS) have been calculated for the ML MoS_2-oxide system considering different possible terminations of the oxide at the interface in the presence of O vacancies in the oxide or Mo and S vacancies in the MoS_2. It is to be noted that the discussion of the ML MoS_2-oxide system is limited to the MoS_2-HfO_2 system in this section. Results for the MoS_2-Al_2O_3 system are qualitatively similar and details can be found in [341]. We compared (overlaid) the band structures for the ideal MoS_2-oxide systems to those for freestanding ML MoS_2, and the MoS_2-oxide systems with vacancies to the ideal MoS_2-oxide results. In all cases, the highest occupied state of the system with vacancies serves as the zero-energy reference in these 0 K simulations. However, the reference band structures absent vacancies are shifted up or down to provide a rough fit to the former in terms of band structure and the atom projected densities of states (AP-DOS) of the Mo and S atoms (otherwise, the zero-energy reference for the latter would be the valence band edge).

11.7.2.1 Monolayer MoS_2 on HfO_2 slab

The simulated band structure of monolayer MoS_2 on O-terminated HfO_2 with H-passivation is shown in Fig. 11.19a (solid black lines). The band structure of free-standing MoS_2 is provided for reference (red dashed lines). Since a rectangular supercell was used in these simulations of MoS_2 on HfO_2, the corresponding Brillouin zone (BZ) is smaller and the K point of the primitive unit cell—where the monolayer MoS_2 band edges are located—folds into the Γ point in the supercell's BZ. The O-terminated and H-passivated HfO_2 has little effect on the band structure of monolayer MoS_2. The band gap is devoid of any defect states, suggesting the potential for HfO_2 to act as an excellent gate insulator or substrate. In addition, the direct band gap is preserved, which is crucial for optoelectronic applications. The absence of band gap states also is reflected in the AP-DOS for the system (Fig. 11.19b). The AP-DOS also indicates a straddling gap (Type-1) band alignment between the MoS_2 and the HfO_2.

In contrast, our results indicate a significant reduction in the effective band gap of the combined Hf-terminated HfO_2 slab-MoS_2 system (energy gap E_g of approximately 0.9 eV) as compared to that of freestanding monolayer MoS_2, although a direct band gap is preserved (Fig. 11.20a). In MoS_2, the DOS at the conduction and valence band edges are dominated by d_{xz} and d_{z^2} orbitals from the Mo atoms. The lower portion of what otherwise would be the bandgap is now filled with states arising from the contribution of Hf (d orbitals) and O atoms (p orbitals), as shown in the AP-DOS of Fig. 11.20b. Indeed, there now appears to be a

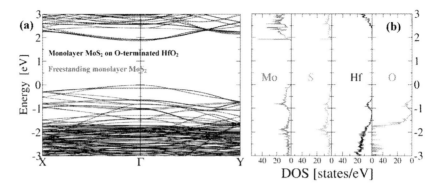

FIGURE 11.19 (a) Band structure of monolayer MoS$_2$ on an O-terminated HfO$_2$ slab with H passivation, plotted along the high symmetry directions of the BZ (black solid lines). The 0 eV reference corresponds to the highest occupied state in these 0 K simulations. The band structure of freestanding monolayer MoS$_2$ is superimposed for comparison (red dashed lines). This latter band structure, however, is shifted up or down to provide a reasonable fit to the former. (b) Atom-projected density of states (AP-DOS) per eV per supercell for the monolayer MoS$_2$ and O-terminated HfO$_2$ system. A straddling gap alignment is observed, devoid of defect states [328].

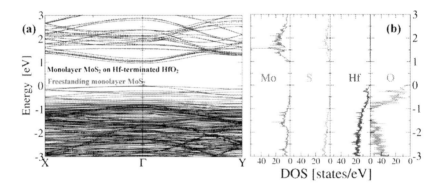

FIGURE 11.20 (a) Band structure of monolayer MoS$_2$ on Hf-terminated HfO$_2$ slab, plotted along the high symmetry directions of the BZ (solid black lines). The band structure of freestanding monolayer MoS$_2$ is superimposed for comparison (red dashed lines). The reference band structure is shifted up or down to provide a reasonable fit to primary band structure. (b) Atom-projected density of states (AP-DOS) for the monolayer MoS$_2$ and Hf-terminated HfO$_2$ system. A staggered gap alignment is observed, such that holes would be localized in the oxide layer rather than the MoS$_2$ [328].

staggered gap (Type II) alignment, where holes would be localized in the HfO$_2$ instead of the MoS$_2$ layer.

11.7.2.2 Monolayer MoS$_2$ on HfO$_2$ slab with O vacancy

When an O vacancy is introduced into the top layer of the O-terminated and H-passivated HfO$_2$ slab, in these 0 K simulations, an occupied defect state (band) is introduced within the band gap of ML MoS$_2$ (Fig. 11.21a), which is associated primarily with Hf atoms in the oxide. Analogous Hf-associated defect states also arise in an isolated O-terminated and H-

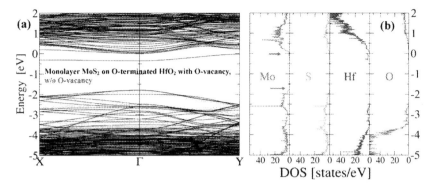

FIGURE 11.21 (a) Band structure of ML MoS$_2$ on an H-passivated, O-terminated HfO$_2$ slab with an O-vacancy in the top layer, plotted along the high symmetry directions of the BZ (black solid lines). The 0 eV reference corresponds to the highest occupied state in these 0 K simulations. The band structure of vacancy-free ML MoS$_2$-HfO$_2$ system (O-terminated) is superimposed for comparison (red dashed lines). However, this latter band structure, which otherwise would have its zero-reference energy at the upper edge of the valence band, is shifted up or down to provide a reasonable fit to the former (b) Atom-projected density of states for the ML MoS$_2$ and O-terminated HfO$_2$ system with an O-vacancy. Red arrows indicate the conduction and valence band edges. An occupied defect state (band) is introduced within the band gap of ML MoS$_2$ [341].

passivated HfO$_2$ slab (Fig. 11.22a and b). In this latter case (and for analogous cases below) we simply removed the MoS$_2$ layer from the combined system, while otherwise holding the crystal structure fixed as a control. However, the close proximity of the occupied defect band to the conduction band (of the reference band structure) suggests that these states might be able to act as donors. As can be seen from the AP-DOS (Fig. 11.21b), the conduction band edge for MoS$_2$ is pinned at the Fermi level indicating n-type doping. However, the defect band formation due to the limited supercell size and associated very large (1.97×10^{14}/cm^2) O-vacancy density in these simulations leaves the binding energy for lower defect densities uncertain. Alternatively, these interface states could function as relatively shallow charge traps, leading to degradation of device performance. Since a rectangular supercell was used in these simulations of MoS$_2$ on HfO$_2$, the corresponding Brillouin zone (BZ) is smaller and the K point of the primitive unit cell—where the ML MoS$_2$ band edges are located—folds into the Γ point in the supercell's BZ.

In the case of Hf-terminated HfO$_2$-MoS$_2$ system with an O vacancy in the top layer of oxide, there is a straddling gap alignment (Type-1) as seen in the AP-DOS (Fig. 11.23b) for this large O-vacancy density, much as for O-terminated HfO$_2$. Moreover, there are now two partially occupied bands at the bottom of the conduction band (Fig. 11.23a), both of which are largely localized to the MoS$_2$ layer, resulting in a system that now appears metallic. Calculation of the band structure for a freestanding Hf-terminated HfO$_2$ slab with an O vacancy exhibits occupied conduction band states associated with the Hf atoms (Fig. 11.24a and b). In the combined HfO$_2$-MoS$_2$ system, these electrons are then transferred into the lower conduction-band-edge MoS$_2$ layer, in a modulation-doping-like process. In MoS$_2$, the DOS at the conduction and valence band edges are dominated by d_{xz} and d_{z^2} orbitals from the Mo atoms while in the HfO$_2$ the band edge states arise mainly from the contribution of Hf- d orbitals and O- p orbitals.

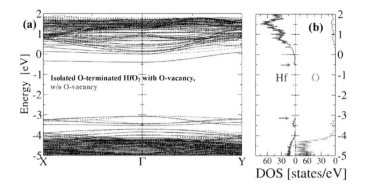

FIGURE 11.22 (a) Band structure of a freestanding H-passivated, O-terminated HfO_2 slab with an O-vacancy in the top layer, plotted along the high symmetry directions of the BZ (black solid lines). The energy-shifted band structure of vacancy-free HfO_2 slab (O-terminated) is superimposed for comparison (red dashed lines). (b) Atom-projected density of states for the O-terminated HfO_2 system with an O-vacancy. Defect states associated with Hf-atoms are observed [341].

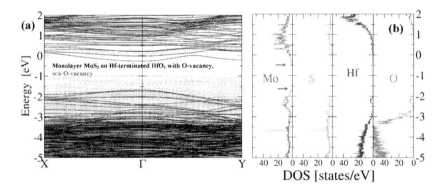

FIGURE 11.23 (a) Band structure of ML MoS_2 on Hf-terminated HfO_2 slab with an O-vacancy in the top layer, plotted along the high symmetry directions of the BZ (black solid lines). The energy-shifted band structure of vacancy free ML MoS_2-HfO_2 system with Hf-termination is superimposed for comparison (red dashed lines). (b) Atom-projected density of states for the ML MoS_2 and Hf-terminated HfO_2 system with an O-vacancy. A straddling gap band alignment is now observed along with two partially occupied bands at the conduction band edge both of which are largely localized to the MoS_2 layer, resulting in a system that now appears metallic [341].

For the HfO_2-MoS_2 with O vacancy systems, we also repeated the simulations with the GGA approximation for comparison with the above LDA results. Fig. 11.25a shows the band structure of ML MoS_2 on Hf-terminated HfO_2 with an O-vacancy, as obtained using both the GGA and the LDA approximations. The same nominal crystal structure was used, but a separate relaxation was performed for the LDA and GGA calculations (the latter, however, starting with the former for computational efficiency). As can be seen, the results match closely, including the degree of degenerate doping. A similar comparison (not shown) was performed for MoS_2 on O-terminated HfO_2, again with good agreement between the results

11.7. Theoretical investigation of defects in 2D TMDs

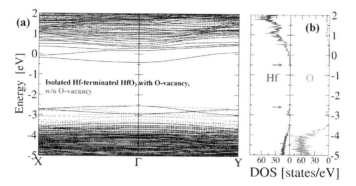

FIGURE 11.24 (a) Band structure of a freestanding Hf-terminated HfO$_2$ slab with an O-vacancy in the top layer, plotted along the high symmetry directions of the BZ. The energy-shifted band structure of vacancy-free HfO$_2$ slab (Hf-terminated) is superimposed for comparison (red dashed lines). (b) Atom-projected density of states for the Hf-terminated HfO$_2$ system with an O-vacancy. An occupied conduction band edge state associated with the Hf atoms is observed [341].

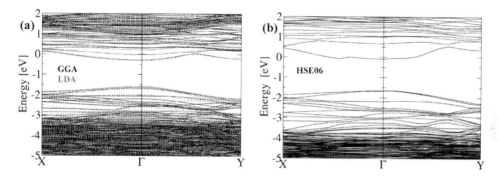

FIGURE 11.25 (a) Band structure of ML MoS$_2$ on an Hf-terminated HfO$_2$ with an O vacancy obtained using the GGA (solid lines). The band structure obtained using the LDA is overlaid on top for comparison (dashed lines). Both results exhibit n-type doping, and essentially the same degree of degeneracy. (The zero-energy reference remains the Fermi level in each case). (b) Band structure of ML MoS$_2$ on an Hf-terminated HfO$_2$ with an O vacancy obtained using the HSE06. The conduction band edge is pulled below the Fermi level indicating n-type doping of MoS$_2$ [341].

obtained with the GGA and with the LDA including the location of the occupied defect band just below the conduction band.

Finally, in Fig. 11.25b, we have used hybrid functionals, specifically HSE06, which provide a more accurate value for the band gap of bulk HfO$_2$ to simulate the band structure of ML MoS$_2$ on Hf-terminated HfO$_2$ to further check key results. The much larger computational demands required for hybrid method combined with the large supercell size constrained us to use a coarse k-point grid for evaluation of the band structure and precluded us from running any relaxations of the structure using the hybrid method. Instead, we reused the structure obtained from the GGA relaxations. As shown in Fig. 11.25b, with the hybrid method, the conduction band edge is again pulled below the Fermi level as in our previous GGA and LDA results, indicating the n-type doping of ML MoS$_2$ modulated by dielectric oxide.

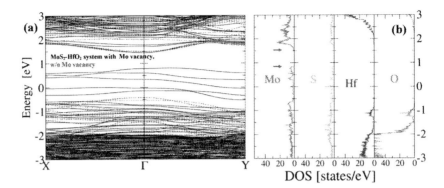

FIGURE 11.26 (a) Band structure of ML MoS$_2$ on an H-passivated, O-terminated HfO$_2$ slab with Mo-vacancy in the ML, plotted along the high symmetry directions of the BZ (black solid lines). The energy-shifted band structure of vacancy free ML MoS$_2$ – HfO$_2$ system (O-terminated) is superimposed for comparison (red dashed lines). (b) Atom-projected density of states for the ML MoS$_2$ and O-terminated HfO$_2$ system with Mo vacancy. Several states are introduced within the nominal band gap and significant distortion of the valence band edge structure is observed [341].

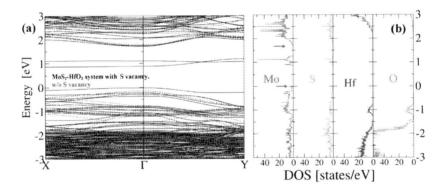

FIGURE 11.27 (a) Band structure of ML MoS$_2$ on an H-passivated, O-terminated HfO$_2$ slab with S-vacancy in the ML, plotted along the high symmetry directions of the BZ (black solid lines). The energy-shifted band structure of vacancy free ML MoS$_2$-HfO$_2$ system (O-terminated) is superimposed for comparison (red dashed lines). (b) Atom-projected density of states for the ML MoS$_2$ and O-terminated HfO$_2$ system with S vacancy. Two unoccupied defect states (bands) are introduced near mid gap, and significant distortion of the valence band edge structure is observed [341].

11.7.2.3 Mo and S vacancies in MoS$_2$

We introduced a single Mo or S atom vacancy in the supercell of the MoS$_2$ on O-terminated and H-passivated HfO2 system, with the S atom vacancy introduced in the layer adjacent to the oxide surface. The corresponding vacancy density is $1.97 \times 10^{14}/cm^2$ in either case. The resultant band structure and AP-DOS are plotted in Fig. 11.26 and Fig. 11.27 for MoS$_2$-HfO$_2$ systems with Mo and S vacancies, respectively. In both cases, a straddling gap alignment is retained, but defect states are introduced into the band gap of the MoS$_2$ that are localized to the MoS$_2$ layer (Fig. 11.26b and Fig. 11.27b). With the Mo vacancy, several states are introduced within the nominal band gap, both occupied and empty, although at this vacancy concentration, the valence band edge is difficult to define (Fig. 11.26a). In the case of S vacancies in the MoS_2 ML, there are two unoccupied defect states (bands) introduced near mid gap, as well as significant distortion of the valence band edge structure at these vacancy concentrations (Fig. 11.27a).

References

[1] D. Caughey, R. Thomas, Carrier mobilities in silicon empirically related to doping and field, Proceedings of the IEEE 55 (12) (1967) 2192–2193.

[2] W. Spear, P. Le Comber, Substitutional doping of amorphous silicon, Solid State Communications 17 (9) (1975) 1193–1196.

[3] Y. Cui, X. Duan, J. Hu, C.M. Lieber, Doping and electrical transport in silicon nanowires, The Journal of Physical Chemistry B 104 (22) (2000) 5213–5216.

[4] K.S. Novoselov, D. Jiang, F. Schedin, T.J. Booth, V.V. Khotkevich, S.V. Morozov, A.K. Geim, Two-dimensional atomic crystals, Proceedings of the National Academy of Sciences 102 (30) (2005 Jul. 26) 10451–10453.

[5] Q.H. Wang, K. Kalantar-Zadeh, A. Kis, J.N. Coleman, M.S. Strano, Electronics and optoelectronics of two-dimensional transition metal dichalcogenides, Nature Nanotechnology 7 (11) (2012) 699–712.

[6] Z. Lin, B.R. Carvalho, E. Kahn, R. Lv, R. Rao, H. Terrones, M.A. Pimenta, M. Terrones, Defect engineering of two-dimensional transition metal dichalcogenides, 2D Materials 3 (2) (2016) 022002.

[7] Y. Liu, P. Stradins, S.H. Wei, Air passivation of chalcogen vacancies in two-dimensional semiconductors, Angewandte Chemie International Edition 55 (2016) 965–968.

[8] J. Hong, Z. Hu, M. Probert, K. Li, D. Lv, X. Yang, L. Gu, N. Mao, Q. Feng, L. Xie, Exploring atomic defects in molybdenum disulphide monolayers, Nature Communications 6 (2015).

[9] S. Mathew, K. Gopinadhan, T. Chan, X. Yu, D. Zhan, L. Cao, A. Rusydi, M. Breese, S. Dhar, Z. Shen, Magnetism in MoS_2 induced by proton irradiation, Applied Physics Letters 101 (10) (2012) 102103.

[10] S. Tongay, J. Suh, C. Ataca, W. Fan, A. Luce, J.S. Kang, J. Liu, C. Ko, R. Raghunathanan, J. Zhou, Defects activated photoluminescence in two-dimensional semiconductors: interplay between bound, charged, and free excitons, Scientific Reports 3 (2013) 2657.

[11] W. Zhou, X. Zou, S. Najmaei, Z. Liu, Y. Shi, J. Kong, J. Lou, P.M. Ajayan, B.I. Yakobson, J.-C. Idrobo, Intrinsic structural defects in monolayer molybdenum disulfide, Nano Letters 13 (6) (2013) 2615–2622.

[12] H. Qiu, T. Xu, Z. Wang, W. Ren, H. Nan, Z. Ni, Q. Chen, S. Yuan, F. Miao, F. Song, G. Long, Y. Shi, L. Sun, J. Wang, X. Wang, Hopping transport through defect-induced localized states in molybdenum disulphide, Nature Communications 4 (2013) 2642.

[13] P.K. Chow, R.B. Jacobs-Gedrim, J. Gao, T.-M. Lu, B. Yu, H. Terrones, N. Koratkar, Defect-induced photoluminescence in monolayer semiconducting transition metal dichalcogenides, ACS Nano 9 (2) (2015) 1520–1527.

[14] V. Carozo, Y. Wang, K. Fujisawa, B.R. Carvalho, A. McCreary, S. Feng, Z. Lin, C. Zhou, N. Perea-López, A.L. Elías, Optical identification of sulfur vacancies: bound excitons at the edges of monolayer tungsten disulfide, Science Advances 3 (4) (2017) 1602813.

[15] S. McDonnell, R. Addou, C. Buie, R.M. Wallace, C.L. Hinkle, Defect-dominated doping and contact resistance in MoS_2, ACS Nano 8 (3) (2014) 2880–2888.

[16] A.M. Van Der Zande, P.Y. Huang, D.A. Chenet, T.C. Berkelbach, Y. You, G.-H. Lee, T.F. Heinz, D.R. Reichman, D.A. Muller, J.C. Hone, Grains and grain boundaries in highly crystalline monolayer molybdenum disulphide, Nature Materials 12 (6) (2013) 554–561.

[17] A. Roy, H.C. Movva, B. Satpati, K. Kim, R. Dey, A. Rai, T. Pramanik, S. Guchhait, E. Tutuc, S.K. Banerjee, Structural and electrical properties of $MoTe_2$ and $MoSe_2$ grown by molecular beam epitaxy, ACS Applied Materials & Interfaces 8 (11) (2016) 7396–7402.

[18] W. Su, L. Jin, X. Qu, D. Huo, L. Yang, Defect passivation induced strong photoluminescence enhancement of rhombic monolayer MoS_2, Physical Chemistry Chemical Physics 18 (20) (2016) 14001–14006.

[19] S. Ghatak, A.N. Pal, A. Ghosh, Nature of electronic states in atomically thin MoS_2 field-effect transistors, ACS Nano 5 (10) (2011) 7707–7712.

[20] N. Ma, D. Jena, Charge scattering and mobility in atomically thin semiconductors, Physical Review X 4 (1) (2014) 011043.

[21] Y.Y. Illarionov, G. Rzepa, M. Waltl, T. Knobloch, A. Grill, M.M. Furchi, T. Mueller, T. Grasser, The role of charge trapping in MoS_2/SiO_2 and MoS_2/hBN field-effect transistors, 2D Materials 3 (3) (2016) 035004.

[22] L. Britnell, R. Ribeiro, A. Eckmann, R. Jalil, B. Belle, A. Mishchenko, Y.-J. Kim, R. Gorbachev, T. Georgiou, S. Morozov, Strong light-matter interactions in heterostructures of atomically thin films, Science 340 (6138) (2013) 1311–1314.

[23] W.J. Yu, Y. Liu, H. Zhou, A. Yin, Z. Li, Y. Huang, X. Duan, Highly efficient gate-tunable photocurrent generation in vertical heterostructures of layered materials, Nature Nanotechnology 8 (12) (2013) 952–958.

[24] X. Hong, J. Kim, S.-F. Shi, Y. Zhang, C. Jin, Y. Sun, S. Tongay, J. Wu, Y. Zhang, F. Wang, Ultrafast charge transfer in atomically thin MoS_2/WS_2 heterostructures, Nature Nanotechnology (2014).

[25] A. Chernikov, T.C. Berkelbach, H.M. Hill, A. Rigosi, Y. Li, O.B. Aslan, D.R. Reichman, M.S. Hybertsen, T.F. Heinz, Exciton binding energy and nonhydrogenic Rydberg series in monolayer WS_2, Physical Review Letters 113 (7) (2014) 076802.

[26] A.M. Jones, H. Yu, N.J. Ghimire, S. Wu, G. Aivazian, J.S. Ross, B. Zhao, J. Yan, D.G. Mandrus, D. Xiao, Optical generation of excitonic valley coherence in monolayer WSe_2, Nature Nanotechnology 8 (9) (2013) 634–638.

[27] H. Zeng, J. Dai, W. Yao, D. Xiao, X. Cui, Valley polarization in MoS_2 monolayers by optical pumping, Nature Nanotechnology 7 (8) (2012) 490–493.

[28] F. Xia, T. Mueller, Y.-m. Lin, A. Valdes-Garcia, P. Avouris, Ultrafast graphene photodetector, Nature Nanotechnology 4 (12) (2009) 839–843.

[29] Q. Cui, F. Ceballos, N. Kumar, H. Zhao, Transient absorption microscopy of monolayer and bulk WSe_2, ACS Nano 8 (3) (2014) 2970–2976.

[30] R. Wang, B.A. Ruzicka, N. Kumar, M.Z. Bellus, H.-Y. Chiu, H. Zhao, Ultrafast and spatially resolved studies of charge carriers in atomically thin molybdenum disulfide, Physical Review B 86 (4) (2012) 045406.

[31] N. Kumar, Q. Cui, F. Ceballos, D. He, Y. Wang, H. Zhao, Exciton diffusion in monolayer and bulk $MoSe_2$, Nanoscale 6 (9) (2014) 4915–4919.

[32] a. Woutersen, U. Emmerichs, H. Bakker, Femtosecond mid-IR pump-probe spectroscopy of liquid water: evidence for a two-component structure, Science 278 (5338) (1997) 658–660.

[33] C. Stamm, T. Kachel, N. Pontius, R. Mitzner, T. Quast, K. Holldack, S. Khan, C. Lupulescu, E. Aziz, M. Wietstruk, Femtosecond modification of electron localization and transfer of angular momentum in nickel, Nature Materials 6 (10) (2007) 740–743.

[34] S. Sundaram, E. Mazur, Inducing and probing non-thermal transitions in semiconductors using femtosecond laser pulses, Nature Materials 1 (4) (2002) 217–224.

[35] J. Fujimoto, J. Liu, E. Ippen, N. Bloembergen, Femtosecond laser interaction with metallic tungsten and nonequilibrium electron and lattice temperatures, Physical Review Letters 53 (19) (1984) 1837.

[36] K. Chen, M.N. Yogeesh, Y. Huang, S. Zhang, F. He, X. Meng, S. Fang, N. Sheehan, T.H. Tao, S.R. Bank, Non-destructive measurement of photoexcited carrier transport in graphene with ultrafast grating imaging technique, Carbon 107 (2016) 233–239.

[37] K. Chen, N. Sheehan, F. He, X. Meng, S.C. Mason, S.R. Bank, Y. Wang, Measurement of ambipolar diffusion coefficient of photoexcited carriers with ultrafast reflective grating-imaging technique, ACS Photonics 4 (6) (2017) 1440–1446.

[38] H.-P. Komsa, A.V. Krasheninnikov, Native defects in bulk and monolayer MoS_2 from first principles, Physical Review B 91 (12) (2015) 125304.

[39] A. McCreary, A. Berkdemir, J. Wang, M.A. Nguyen, A.L. Elías, N. Perea-López, K. Fujisawa, B. Kabius, V. Carozo, D.A. Cullen, Distinct photoluminescence and Raman spectroscopy signatures for identifying highly crystalline WS_2 monolayers produced by different growth methods, Journal of Materials Research 31 (7) (2016) 931–944.

[40] Z. Wu, Z. Luo, Y. Shen, W. Zhao, W. Wang, H. Nan, X. Guo, L. Sun, X. Wang, Y. You, Defects as a factor limiting carrier mobility in WSe_2: a spectroscopic investigation, Nano Research 9 (12) (2016) 3622–3631.

[41] A. Zafar, H. Nan, Z. Zafar, Z. Wu, J. Jiang, Y. You, Z. Ni, Probing the intrinsic optical quality of CVD grown MoS_2, in: Nano Research, 2016, pp. 1–10.

[42] H.R. Gutiérrez, N. Perea-López, A.L. Elías, A. Berkdemir, B. Wang, R. Lv, F. López-Urías, V.H. Crespi, H. Terrones, M. Terrones, Extraordinary room-temperature photoluminescence in triangular WS_2 monolayers, Nano Letters 13 (8) (2012) 3447–3454.

[43] N. Peimyoo, J. Shang, C. Cong, X. Shen, X. Wu, E.K. Yeow, T. Yu, Nonblinking, intense two-dimensional light emitter: monolayer WS_2 triangles, ACS Nano 7 (12) (2013) 10985–10994.

References

[44] K. Chen, R. Ghosh, X. Meng, A. Roy, J.-S. Kim, F. He, S.C. Mason, X. Xu, J.-F. Lin, D. Akinwande, Experimental evidence of exciton capture by mid-gap defects in CVD grown monolayer $MoSe_2$, npj 2D Materials and Applications 1 (1) (2017) 15.

[45] K. Chen, A. Roy, A. Rai, A. Valsaraj, X. Meng, F. He, X. Xu, L.F. Register, S. Banerjee, Y. Wang, Carrier trapping by oxygen impurities in molybdenum diselenide, ACS Applied Materials & Interfaces (2017).

[46] B. Akdim, R. Pachter, S. Mou, Theoretical analysis of the combined effects of sulfur vacancies and analyte adsorption on the electronic properties of single-layer MoS_2, Nanotechnology 27 (18) (2016) 185701.

[47] A.V. Krivosheeva, V.L. Shaposhnikov, V.E. Borisenko, J.-L. Lazzari, C. Waileong, J. Gusakova, B.K. Tay, Theoretical study of defect impact on two-dimensional MoS_2, Journal of Semiconductors 36 (12) (2015) 122002.

[48] A. Valsaraj, J. Chang, A. Rai, L.F. Register, S.K. Banerjee, Theoretical and experimental investigation of vacancy-based doping of monolayer MoS_2 on oxide, 2D Materials 2 (4) (2015) 045009.

[49] S. Zhang, C.-G. Wang, M.-Y. Li, D. Huang, L.-J. Li, W. Ji, S. Wu, Defect structure of localized excitons in a WSe_2 monolayer, Physical Review Letters 119 (4) (2017) 046101.

[50] H. Nan, Z. Wu, J. Jiang, A. Zafar, Y. You, Z. Ni, Improving the electrical performance of MoS_2 by mild oxygen plasma treatment, Journal of Physics D: Applied Physics 50 (15) (2017) 154001.

[51] S.M. Sze, K.K. Ng, Physics of Semiconductor Devices, 3rd ed., John Wiley & Sons, 2006.

[52] M.M. Perera, M.-W. Lin, H.-J. Chuang, B.P. Chamlagain, C. Wang, X. Tan, M.M.-C. Cheng, D. Tománek, Z. Zhou, Improved carrier mobility in few-layer MoS2 field-effect transistors with ionic-liquid gating, ACS Nano 7 (5) (2013) 4449–4458.

[53] Y. Zhang, J. Ye, Y. Matsuhashi, Y. Iwasa, Ambipolar MoS_2 thin flake transistors, Nano Letters 12 (3) (2012) 1136–1140.

[54] M.S. Fuhrer, J. Hone, Measurement of mobility in dual-gated MoS_2 transistors, Nature Nanotechnology 8 (3) (2013) 146–147.

[55] Z. Yu, Z. Pan, Y. Shen, Z. Wang, Z.-Y. Ong, T. Xu, R. Xin, L. Pan, B. Wang, L. Sun, J. Wang, G. Zhang, Y.W. Zhang, Y. Shi, X. Wang, Towards intrinsic charge transport in monolayer molybdenum disulfide by defect and interface engineering, Nature Communications 5 (2014) 5290.

[56] B.W. Baugher, H.O. Churchill, Y. Yang, P. Jarillo-Herrero, Intrinsic electronic transport properties of high-quality monolayer and bilayer MoS_2, Nano Letters 13 (9) (2013) 4212–4216.

[57] W. Bao, X. Cai, D. Kim, K. Sridhara, M.S. Fuhrer, High mobility ambipolar MoS_2 field-effect transistors: substrate and dielectric effects, Applied Physics Letters 102 (4) (2013) 042104.

[58] H. Schmidt, S. Wang, L. Chu, M. Toh, R. Kumar, W. Zhao, A. Castro Neto, J. Martin, S. Adam, B. Özyilmaz, Transport properties of monolayer MoS_2 grown by chemical vapor deposition, Nano Letters 14 (4) (2014) 1909–1913.

[59] H. Liu, M. Si, S. Najmaei, A.T. Neal, Y. Du, P.M. Ajayan, J. Lou, P.D. Ye, Statistical study of deep submicron dual-gated field-effect transistors on monolayer chemical vapor deposition molybdenum disulfide films, Nano Letters 13 (6) (2013) 2640–2646.

[60] A. Roy, R. Ghosh, A. Rai, A. Sanne, K. Kim, H.C. Movva, R. Dey, T. Pramanik, S. Chowdhury, E. Tutuc, Intra-domain periodic defects in monolayer MoS_2, Applied Physics Letters 110 (20) (2017) 201905.

[61] A.N. Enyashin, M. Bar-Sadan, L. Houben, G. Seifert, Line defects in molybdenum disulfide layers, The Journal of Physical Chemistry C 117 (20) (2013) 10842–10848.

[62] O. Lehtinen, H.-P. Komsa, A. Pulkin, M.B. Whitwick, M.-W. Chen, T. Lehnert, M.J. Mohn, O.V. Yazyev, A. Kis, U. Kaiser, Atomic scale microstructure and properties of Se-deficient two-dimensional $MoSe_2$, ACS Nano 9 (3) (2015) 3274–3283.

[63] Y.L. Huang, Y. Chen, W. Zhang, S.Y. Quek, C.-H. Chen, L.-J. Li, W.-T. Hsu, W.-H. Chang, Y.J. Zheng, W. Chen, Bandgap tunability at single-layer molybdenum disulphide grain boundaries, Nature Communications (2015).

[64] M. Ghorbani-Asl, A.N. Enyashin, A. Kuc, G. Seifert, T. Heine, Defect-induced conductivity anisotropy in MoS_2 monolayers, Physical Review B 88 (24) (2013) 245440.

[65] V.K. Sangwan, D. Jariwala, I.S. Kim, K.-S. Chen, T.J. Marks, L.J. Lauhon, M.C. Hersam, Gate-tunable memristive phenomena mediated by grain boundaries in single-layer MoS_2, Nature Nanotechnology 10 (5) (2015) 403–406.

[66] K. Chen, A. Roy, A. Rai, H.C. Movva, X. Meng, F. He, S.K. Banerjee, Y. Wang, Accelerated carrier recombination by grain boundary/edge defects in MBE grown transition metal dichalcogenides, APL Materials 6 (5) (2018 May 1) 056103.

[67] S. Vishwanath, X. Liu, S. Rouvimov, P.C. Mende, A. Azcatl, S. McDonnell, R.M. Wallace, R.M. Feenstra, J.K. Furdyna, D. Jena, Comprehensive structural and optical characterization of MBE grown $MoSe_2$ on graphite, CaF_2 and graphene, 2D Materials 2 (2) (2015) 024007.

[68] Z. Lei, Y. Zhou, P. Wu, Simultaneous exfoliation and functionalization of $MoSe_2$ nanosheets to prepare "smart" nanocomposite hydrogels with tunable dual stimuli-responsive behavior, Small 12 (23) (2016) 3112–3118.

[69] Y.Yu.Illarionov, K.K.H. Smithe, M. Waltl, T. Knobloch, E. Pop, T. Grasser, Improved hysteresis and reliability of MoS_2 transistors with high-quality CVD growth and Al_2O_3 encapsulation, IEEE Electron Device Letters 38 (12) (2017) 1763–1766, https://doi.org/10.1109/LED.2017.2768602.

[70] A. Rai, H.C.P. Movva, A. Roy, D. Taneja, S. Chowdhury, S.K. Banerjee, Progress in contact, doping and mobility engineering of MoS_2: an atomically thin 2D semiconductor, Crystals 8 (8) (2018) 316, https://doi.org/10.3390/cryst8080316.

[71] A. Rai, A. Valsaraj, H.C.P. Movva, A. Roy, R. Ghosh, S. Sonde, S. Kang, J. Chang, T. Trivedi, R. Dey, S. Guchhait, S. Larentis, L.F. Register, E. Tutuc, S.K. Banerjee, Air stable doping and intrinsic mobility enhancement in monolayer molybdenum disulfide by amorphous titanium suboxide encapsulation, Nano Letters 15 (7) (2015) 4329–4336, https://doi.org/10.1021/acs.nanolett.5b00314.

[72] Q. Qian, B. Li, M. Hua, Z. Zhang, F. Lan, Y. Xu, R. Yan, K.J. Chen, Improved gate dielectric deposition and enhanced electrical stability for single-layer MoS_2 MOSFET with an AlN interfacial layer, Scientific Reports 6 (1) (2016) 27676, https://doi.org/10.1038/srep27676.

[73] H. Liu, M. Si, S. Najmaei, A.T. Neal, Y. Du, P.M. Ajayan, J. Lou, P.D. Ye, Statistical study of deep submicron dual-gated field-effect transistors on monolayer chemical vapor deposition molybdenum disulfide films, Nano Letters 13 (6) (2013) 2640–2646, https://doi.org/10.1021/nl400778q.

[74] D. Shahrjerdi, A.D. Franklin, S. Oida, J.A. Ott, G.S. Tulevski, W.Haensch, High-performance air-stable n-type carbon nanotube transistors with erbium contacts, ACS Nano 7 (9) (2013) 8303–8308, https://doi.org/10.1021/nn403935v.

[75] S. Das, H.-Y. Chen, A.V. Penumatcha, J.Appenzeller, High performance multilayer MoS_2 transistors with scandium contacts, Nano Letters 13 (1) (2013) 100–105, https://doi.org/10.1021/nl303583v.

[76] A. Valsaraj, J. Chang, A. Rai, L.F. Register, S.K. Banerjee, Theoretical and experimental investigation of vacancy-based doping of monolayer MoS_2 on oxide, 2D Materials 2 (4) (2015) 045009, https://doi.org/10.1088/2053-1583/2/4/045009.

[77] J. Wang, Q. Yao, C.-W. Huang, X. Zou, L. Liao, S. Chen, Z. Fan, K. Zhang, W. Wu, X. Xiao, C. Jiang, W.-W.Wu, High mobility MoS_2 transistor with low Schottky barrier contact by using atomic thick h-BN as a tunneling layer, Advanced Materials 28 (37) (2016) 8302–8308, https://doi.org/10.1002/adma.201602757.

[78] A. Alharbi, D. Shahrjerdi, Analyzing the effect of high-k dielectric-mediated doping on contact resistance in top-gated monolayer MoS_2 transistors, I.E.E.E. Transactions on Electron Devices 65 (10) (2018) 4084–4092, https://doi.org/10.1109/TED.2018.2866772.

[79] H.C.P. Movva, A. Rai, S. Kang, K. Kim, B. Fallahazad, T. Taniguchi, K. Watanabe, E. Tutuc, S.K. Banerjee, High-mobility holes in dual-gated WSe_2 field-effect transistors, ACS Nano 9 (10) (2015) 10402–10410, https://doi.org/10.1021/acsnano.5b04611.

[80] Y. Hu, H. Jiang, K.M. Lau, Q. Li, Chemical vapor deposited monolayer MoS_2 top-gate MOSFET with atomic-layer-deposited ZrO_{ga} as gate dielectric, Semiconductor Science and Technology 33 (4) (2018) 045004, https://doi.org/10.1088/1361-6641/aaaa5f.

[81] X. Zou, J. Wang, C.-H. Chiu, Y. Wu, X. Xiao, C. Jiang, W.-W. Wu, L. Mai, T. Chen, J. Li, J.C. Ho, L. Liao, Interface engineering for high-performance top-gated MoS_2 field-effect transistors, Advanced Materials 26 (36) (2014) 6255–6261, https://doi.org/10.1002/adma.201402008.

[82] J.W. Liu, M.Y. Liao, M. Imura, Y. Koide, High k ZrO_2/Al_2O_3 bilayer on hydrogenated diamond: band configuration, breakdown field, and electrical properties of field-effect transistors, Journal of Applied Physics 12 (2016) 120, https://doi.org/10.1063/1.4962851, p. 124504.

[83] L. Yu, D. El-Damak, S. Ha, X. Ling, Y. Lin, A. Zubair, Y. Zhang, Y.-H. Lee, J. Kong, A. Chandrakasan, T. Palacios, Enhancement-mode single-layer CVD MoS_2 FET technology for digital electronics, in: 2015 IEEE International Electron Devices Meeting (IEDM), 2015, pp. 32.3.1–32.3.4.

[84] A. Sanne, S. Park, R. Ghosh, M.N. Yogeesh, C. Liu, L. Mathew, R. Rao, D. Akinwande, S.K. Banerjee, Embedded gate CVD MoS_2 microwave FETs, npj 2d Materials and Applications 1 (1) (2017) 26, https://doi.org/10.1038/s41699-017-0029-z.

[85] H. Bergeron, V.K. Sangwan, J.J. McMorrow, G.P. Campbell, I. Balla, X. Liu, M.J. Bedzyk, T.J. Marks, M.C. Hersam, Chemical vapor deposition of monolayer MoS_2 directly on ultrathin Al_2O_3 for low-power electronics, Applied Physics Letters 110 (5) (2017) 053101, https://doi.org/10.1063/1.4975064.

[86] M.D. Groner, F.H. Fabreguette, J.W. Elam, S.M. George, Low-temperature Al_2O_3 atomic layer deposition, Chemistry of Materials 16 (4) (2004) 639–645, https://doi.org/10.1021/cm0304546.

[87] D.J. Late, B. Liu, H.S.S.R. Matte, V.P. Dravid, C.N.R. Rao, Hysteresis in single-layer MoS_2 field effect transistors, ACS Nano 6 (6) (2012) 5635–5641, https://doi.org/10.1021/nn301572c.

[88] X. Liu, Y. Chai, Z. Liu, Investigation of chemical vapour deposition MoS_2 field effect transistors on SiO_2 and ZrO_2 substrates, Nanotechnology 28 (16) (2017) 164004, https://doi.org/10.1088/1361-6528/aa610a.

[89] M. Amani, M.L. Chin, A.G. Birdwell, T.P. O'Regan, S. Najmaei, Z. Liu, P.M. Ajayan, J. Lou, M. Dubey, Electrical performance of monolayer MoS_2 field-effect transistors prepared by chemical vapor deposition, Applied Physics Letters 102 (19) (2013) 193107, https://doi.org/10.1063/1.4804546.

[90] A. Roy, R. Ghosh, A. Rai, A. Sanne, K. Kim, H.C.P. Movva, R. Dey, T. Pramanik, S. Chowdhury, E. Tutuc, S.K. Banerjee, Intra-domain periodic defects in monolayer MoS_2, Applied Physics Letters 110 (20) (2017) 201905, https://doi.org/10.1063/1.4983789.

[91] S. Chowdhury, A. Roy, I. Bodemann, S.K. Banerjee, Two-dimensional to three-dimensional growth of transition metal diselenides by chemical vapor deposition: interplay between fractal, dendritic, and compact morphologies, ACS Applied Materials & Interfaces 12 (13) (2020) 15885–15892, https://doi.org/10.1021/acsami.9b23286.

[92] A.M. van der Zande, P.Y. Huang, D.A. Chenet, T.C. Berkelbach, Y. You, G.-H. Lee, T.F. Heinz, D.R. Reichman, D.A. Muller, J.Hone, J.C.Hone, Grains and grain boundaries in highly crystalline monolayer molybdenum disulphide, Nature Materials 12 (6) (2013) 554–561, https://doi.org/10.1038/nmat3633.

[93] X. Zou, Y. Liu, B.I. Yakobson, Predicting dislocations and grain boundaries in two-dimensional metal-disulfides from the first principles, Nano Letters 13 (1) (2013) 253–258, https://doi.org/10.1021/nl3040042.

[94] A.W. Tsen, L. Brown, M.P. Levendorf, F. Ghahari, P.Y. Huang, R.W. Havener, C.S. Ruiz-Vargas, D.A. Muller, P. Kim, J. Park, Tailoring electrical transport across grain boundaries in polycrystalline graphene, Science 336 (6085) (2012) 1143–1146, https://doi.org/10.1126/science.1218948.

[95] J.-K. Kim, Y. Song, T.-Y. Kim, K. Cho, J. Pak, B.Y. Choi, J. Shin, S. Chung, T. Lee, Analysis of noise generation and electric conduction at grain boundaries in CVD-grown MoS_2 field effect transistors, Nanotechnology 28 (47) (2017) 47LT01, https://doi.org/10.1088/1361-6528/aa9236.

[96] T.H. Ly, D.J. Perello, J. Zhao, Q. Deng, H. Kim, G.H. Han, S.H. Chae, H.Y. Jeong, Y.H. Lee, Misorientation-angle-dependent electrical transport across molybdenum disulfide grain boundaries, Nature Communications 7 (1) (2016) 10426, https://doi.org/10.1038/ncomms10426.

[97] J.-I. Park, Y. Jang, J.-S. Bae, J.-H. Yoon, H.U. Lee, Y. Wakayama, J.-P. Kim, Y. Jeong, Effect of thickness-dependent structural defects on electrical stability of MoS_2 thin film transistors, Journal of Alloys and Compounds 814 (2020) 152134, https://doi.org/10.1016/j.jallcom.2019.152134.

[98] I.S. Kim, V.K. Sangwan, D. Jariwala, J.D. Wood, S. Park, K.-S. Chen, F. Shi, F. Ruiz-Zepeda, A. Ponce, M. Jose-Yacaman, V.P. Dravid, T.J. Marks, M.C. Hersam, L.J. Lauhon, Influence of stoichiometry on the optical and electrical properties of chemical vapor deposition derived MoS_2, ACS Nano 8 (10) (2014) 10551–10558, https://doi.org/10.1021/nn503988x.

[99] L. Gao, Q. Liao, X. Zhang, X. Liu, L. Gu, B. Liu, J. Du, Y. Ou, J. Xiao, Z. Kang, Z. Zhang, Y. Zhang, Defect-engineered atomically thin MoS_2 homogeneous electronics for logic inverters, Advanced Materials 32 (2) (2020) 1906646, https://doi.org/10.1002/adma.201906646.

[100] K. Kang, S. Xie, L. Huang, Y. Han, P.Y. Huang, K.F. Mak, C.-J. Kim, D. Muller, J. Park, High-mobility three-atom-thick semiconducting films with wafer-scale homogeneity, Nature 520 (7549) (2015) 656–660, https://doi.org/10.1038/nature14417.

[101] H. Kim, D. Ovchinnikov, D. Deiana, D. Unuchek, A. Kis, Suppressing nucleation in metal–organic chemical vapor deposition of MoS_2 monolayers by alkali metal halides, Nano Letters 17 (8) (2017) 5056–5063, https://doi.org/10.1021/acs.nanolett.7b02311.

[102] Q. Smets, B. Groven, M. Caymax, I. Radu, G. Arutchelvan, J. Jussot, D. Verreck, I. Asselberghs, A.N. Mehta, A. Gaur, D. Lin, S.E. Kazzi, Ultra-scaled MOCVD MoS_2 MOSFETs with 42 nm contact pitch and 250 $\mu a/Mm$ drain current, in: 2019 IEEE International Electron Devices Meeting (IEDM), IEEE, San Francisco, CA, USA, 2019, pp. 23.2.1–23.2.4.

[103] J. Mun, H. Park, J. Park, D. Joung, S.-K. Lee, J. Leem, J.-M. Myoung, J. Park, S.-H. Jeong, W. Chegal, S. Nam, S.-W. Kang, High-mobility MoS_2 directly grown on polymer substrate with kinetics-controlled metal–organic chemical vapor deposition, ACS Applied Electronic Materials 1 (4) (2019) 608–616, https://doi.org/10.1021/acsaelm.9b00078.

[104] H. Yu, M. Liao, W. Zhao, G. Liu, X.J. Zhou, Z. Wei, X. Xu, K. Liu, Z. Hu, K. Deng, S. Zhou, J.-A. Shi, L. Gu, C. Shen, T. Zhang, L. Du, L. Xie, J. Zhu, W. Chen, R. Yang, D. Shi, G. Zhang, Wafer-scale growth and transfer of highly-oriented monolayer MoS_2 continuous films, ACS Nano 11 (12) (2017) 12001–12007, https://doi.org/10.1021/acsnano.7b03819.

[105] A. Koma, K. Ueno, K. Saiki, Heteroepitaxial growth by van der Waals interaction in one-dimensional, two-dimensional and three-dimensional materials, Journal of Crystal Growth 111 (1991) 1029–1032.

[106] A. Koma, Van der Waals epitaxy for highly lattice-mismatched systems, Journal of Crystal Growth 201 (1999) 236–241.

[107] H. Liu, L. Jiao, F. Yang, Y. Cai, X. Wu, W. Ho, C. Gao, J. Jia, N. Wang, H. Fan, W. Yao, M. Xie, Dense network of one-dimensional midgap metallic modes in monolayer $MoSe_2$ and their spatial undulations, Physical Review Letters 113 (2014) 066105.

[108] Y. Zhang, T.-R. Chang, B. Zhou, Y.-T. Cui, H. Yan, Z. Liu, F. Schmitt, J. Lee, R. Moore, Y. Chen, H. Lin, H.-T. Jeng, S.-K. Mo, Z. Hussain, A. Bansil, Z.-X. Shen, Direct observation of the transition from indirect to direct bandgap in atomically thin epitaxial $MoSe_2$, Nature Nanotechnology 9 (2014) 111–115.

[109] H. Liu, J. Chen, H. Yu, F. Yang, L. Jiao, G.-B. Liu, W. Ho, C. Gao, J. Jia, W. Yao, M. Xie, Observation of intervalley quantum interference in epitaxial monolayer tungsten diselenide, Nature Communications 6 (2015) 8180.

[110] H.J. Liu, L. Jiao, L. Xie, F. Yang, J.L. Chen, W.K. Ho, C.L. Gao, J.F. Jia, X.D. Cui, M.H. Xi, Molecular-beam epitaxy of monolayer and bilayer WSe_2: a scanning tunneling microscopy/spectroscopy study and deduction of exciton binding energy, 2D Materials 2 (2015) 034004.

[111] S. Vishwanath, X. Liu, S. Rouvimov, P.C. Mende, A. Azcatl, S. McDonnell, R.M. Wallace, R.M. Feenstra, J.K. Furdyna, D. Jena, Comprehensive structural and optical characterization of MBE grown $MoSe_2$ on graphite, CaF_2 and graphene, 2D Materials 2 (2015) 024007.

[112] H.C. Diaz, R. Chaghi, Y. Ma, M. Batzill, Molecular beam epitaxy of the van der Waals heterostructure $MoTe_2$ on MoS_2: phase, thermal, and chemical stability, 2D Materials 2 (2015) 44010.

[113] A. Roy, H.C.P. Movva, B. Satpati, K. Kim, R. Dey, A. Rai, T. Pramanik, S. Guchhait, E. Tutuc, S.K. Banerjee, Structural and electrical properties of $MoTe_2$ and $MoSe_2$ grown by molecular beam epitaxy, ACS Applied Materials & Interfaces 8 (2016) 7396–7402.

[114] K. Onomitsu, A. Krajewska, R.A.E. Neufeld, F. Maeda, K. Kumakura, H. Yamamoto, Epitaxial growth of monolayer $MoSe_2$ on GaAs, Applied Physics Express 9 (2016) 115501.

[115] M.T. Dau, C. Vergnaud, A. Marty, F. Rortais, C. Beigné, H. Boukari, E. Bellet-Amalric, V. Guigoz, O. Renault, C. Alvarez, H. Okuno, P. Pochet, M. Jamet, Millimeter-scale layered $MoSe_2$ grown on sapphire and evidence for negative magnetoresistance, Applied Physics Letters 110 (2017) 011909.

[116] S. Vishwanath, A. Sundar, X. Liu, A. Azcatl, E. Lochocki, A.R. Woll, S. Rouvimov, W.S. Hwang, N. Lu, X. Peng, H.-H. Lien, J. Weisenberger, S. McDonnell, M.J. Kim, M. Dobrowolska, J.K. Furdyna, K. Shen, R.M. Wallace, D. Jena, H.G. Xing, MBE growth of few-layer 2H-$MoTe_2$ on 3D substrates, Journal of Crystal Growth 482 (2018) 61–69.

[117] A.T. Barton, R. Yue, S. Anwar, H. Zhu, X. Peng, S. McDonnell, N. Lu, R. Addou, L. Colombo, M.J. Kim, R.M. Wallace, C.L. Hinkle, Transition metal dichalcogenide and hexagonal boron nitride heterostructures grown by molecular beam epitaxy, Microelectronic Engineering 147 (2015) 306–309.

[118] A.T. Barton, R. Yue, L.A. Walsh, G. Zhou, C. Cormier, C.M. Smyth, R. Addou, L. Colombo, R.M. Wallace, C.L. Hinkle, $WSe_{(2-x)}Te_x$ alloys grown by molecular beam epitaxy, 2D Materials 6 (2019) 045027.

[119] R. Yue, Y. Nie, L.A. Walsh, R. Addou, C. Liang, N. Lu, A.T. Barton, H. Zhu, Z. Che, D. Barrera, L. Cheng, P.-R. Cha, Y.J. Chabal, J.W.P. Hsu, J. Kim, M.J. Kim, L. Colombo, R.M. Wallace, K. Cho, C.L. Hinkle, Nucleation and growth of WSe_2: enabling large grain transition metal dichalcogenides, 2D Materials 4 (2017) 045019.

[120] M.-W. Chen, D. Ovchinnikov, S. Lazar, M. Pizzochero, M.B. Whitwick, A. Surrente, M. Baranowski, O.L. Sanchez, P. Gillet, P. Plochocka, O.V. Yazyev, A. Kis, Highly oriented atomically thin ambipolar $MoSe_2$ grown by molecular beam epitaxy, ACS Nano 11 (2017) 6355–6361.

[121] S.M. Poh, X. Zhao, S.J.R. Tan, D. Fu, W. Fei, L. Chu, D. Jiadong, W. Zhou, S.J. Pennycook, A.H.C. Neto, K.P. Loh, Molecular beam epitaxy of highly crystalline $MoSe_2$ on hexagonal boron nitride, ACS Nano 12 (2018) 7562–7570.

References 393

[122] J. Xia, X. Huang, L.-Z. Liu, M. Wang, L. Wang, B. Huang, D.-D. Zhu, J.-J. Li, C.-Z. Gu, X.-M. Meng, CVD synthesis of large-area, highly crystalline $MoSe_2$ atomic layers on diverse substrates and application to photodetectors, Nanoscale 6 (2014) 8949–8955.

[123] Y. Li, A. Chernikov, X. Zhang, A. Rigosi, H.M. Hill, A.M. van der Zande, D.A. Chenet, E.-M. Shih, J. Hone, T.F. Heinz, Measurement of the optical dielectric function of monolayer transition-metal dichalcogenides: MoS_2, $MoSe_2$, WS_2 and WSe_2, Physical Review B 90 (2014) 205422.

[124] X. Jing, E. Panholzer, X. Song, E. Grustan-Gutierrez, F. Hui, Y. Shi, G. Benstetter, Y. Illarionov, T. Grasser, M. Lanza, Fabrication of scalable and ultra low power photodetectors with high light/dark current ratios using polycrystalline monolayer MoS_2 sheets, Nano Energy 30 (2016) 494–502.

[125] D.A. Nguyen, H.M. Oh, N.T. Duong, S. Bang, S.J. Yoon, M.S. Jeong, Highly enhanced photoresponsivity of a monolayer WSe_2 photodetector with nitrogen-doped graphene quantum dots, ACS Applied Materials & Interfaces 10 (2018) 10322–10329.

[126] Y.H. Choi, G.H. Kwon, J.H. Jeong, K.S. Jeong, H. Kwon, Y. An, M. Kim, H. Kim, Y. Yi, S. Im, M.H. Cho, Trap-assisted high responsivity of a phototransistor using bi-layer $MoSe_2$ grown by molecular beam epitaxy, Applied Surface Science 494 (2019) 37–45.

[127] Q. He, P. Li, Z. Wu, B. Yuan, Z. Luo, W. Yang, J. Liu, G. Cao, W. Zhang, Y. Shen, Molecular beam epitaxy scalable growth of wafer-scale continuous semiconducting monolayer $MoTe_2$ on inert amorphous dielectrics, Advanced Materials 31 (2019) 1901578.

[128] D. Fu, X. Zhao, Y.-Y. Zhang, L. Li, H. Xu, A.-R. Jang, S.I. Yoon, P. Song, S.M. Poh, T. Ren, Z. Ding, W. Fu, T.J. Shin, H.S. Shin, S.T. Pantelides, W. Zhou, K.P. Loh, Molecular beam epitaxy of highly crystalline monolayer molybdenum disulfide on hexagonal boron nitride, Journal of the American Chemical Society 139 (2017) 9392–9400.

[129] M. Nakano, Y. Wang, Y. Kashiwabara, H. Matsuoka, Y. Iwasa, Layer-by-layer epitaxial growth of scalable WSe_2 on sapphire by molecular beam epitaxy, Nano Letters 17 (2017) 5595–5599.

[130] P. Paletti, R. Yue, C. Hinkle, S.K. Fullerton-Shirey, A. Seabaugh, Two-dimensional electric-double-layer esaki diode, npj 2D Materials and Applications 3 (2019) 19.

[131] J.H. Park, A. Rai, J. Hwang, C. Zhang, I. Kwak, S.F. Wolf, S. Vishwanath, X. Liu, M. Dobrowolska, J. Furdyna, H.G. Xing, K. Cho, S.K. Banerjee, A.C. Kummel, Band structure engineering of layered WSe_2 via one-step chemical functionalization, ACS Nano 13 (2019) 7545–7555.

[132] K.S. Novoselov, A. Mishchenko, A. Carvalho, A.H. Castro Neto, 2D materials and van der Waals heterostructures, Science 80 (353) (2016) aac9439.

[133] F. Pizzocchero, et al., The hot pick-up technique for batch assembly of van der Waals heterostructures, Nature Communications 7 (2016).

[134] R. Frisenda, et al., Recent progress in the assembly of nanodevices and van der Waals heterostructures by deterministic placement of 2D materials, Chemical Society Reviews 47 (2018) 53–68.

[135] S.J. Haigh, et al., Cross-sectional imaging of individual layers and buried interfaces of graphene-based heterostructures and superlattices, Nature Materials 11 (2012) 764–767.

[136] https://www.bnl.gov/newsroom/news.php?a=214343.

[137] A. Reina, et al., Transferring and identification of single- and few-layer graphene on arbitrary substrates, Journal of Physical Chemistry C 112 (2008) 17741–17744.

[138] X. Li, et al., Large-area synthesis of high-quality and uniform graphene films on copper foils, Science 80 (324) (2009) 1312–1314.

[139] K.S. Kim, et al., Large-scale pattern growth of graphene films for stretchable transparent electrodes, Nature 457 (2009) 706–710.

[140] S. Tongay, et al., Tuning interlayer coupling in large-area heterostructures with CVD-grown MoS_2 and WS_2 monolayers, Nano Letters 14 (2014) 3185–3190.

[141] A. Surrente, et al., Defect healing and charge transfer-mediated valley polarization in MoS_2/$MoSe_2$/MoS_2 trilayer van der Waals heterostructures, Nano Letters 17 (2017) 4130–4136.

[142] X. Hong, et al., Ultrafast charge transfer in atomically thin MoS_2/WS_2 heterostructures, Nature Nanotechnology. 9 (2014) 682–686.

[143] D. Pierucci, et al., Large area molybdenum disulphide- epitaxial graphene vertical van der Waals heterostructures, Scientific Reports 6 (2016) 26656.

[144] S.A. Svatek, et al., Graphene-InSe-graphene van der Waals heterostructures, Journal of Physics. Conference Series 647 (2015) 12001.

[145] H. Coy Diaz, R. Addou, M. Batzill, Interface properties of CVD grown graphene transferred onto $MoS_2(0001)$, Nanoscale 6 (2014) 1071–1078.

[146] X. Zhang, et al., Vertical heterostructures of layered metal chalcogenides by van der Waals epitaxy, Nano Letters 14 (2014) 3047–3054.

[147] S. Wang, X. Wang, J.H. Warner, All Chemical vapor deposition growth of MoS_2:h-BN vertical van der Waals heterostructures, ACS Nano 9 (2015) 5246–5254.

[148] F. Chen, Y. Wang, W. Su, S. Ding, L. Fu, Position-Selective Growth of 2D WS_2-based vertical heterostructures via a one-step CVD approach, Journal of Physical Chemistry C 123 (2019) 30519–30527.

[149] A. Koma, Van der Waals epitaxy—a new epitaxial growth method for a highly lattice-mismatched system, Thin Solid Films 216 (1992) 72–76.

[150] C.R. Dean, et al., Boron nitride substrates for high-quality graphene electronics, Nature Nanotechnology 5 (2010) 722–726.

[151] A.S. Mayorov, et al., Micrometer-scale ballistic transport in encapsulated graphene at room temperature, Nano Letters 11 (2011) 2396–2399.

[152] Y. Liu, et al., Toward barrier free contact to molybdenum disulfide using graphene electrodes, Nano Letters 15 (2015) 3030–3034.

[153] X. Cui, et al., Multi-terminal transport measurements of MoS_2 using a van der Waals heterostructure device platform, Nature Nanotechnology 10 (2015) 534–540.

[154] A. Avsar, et al., Air-stable transport in graphene-contacted, fully encapsulated ultrathin black phosphorus-based field-effect transistors, ACS Nano 9 (2015) 4138–4145.

[155] Z. Fei, et al., Edge conduction in monolayer WTe_2, Nature Physics 13 (2017) 677–682.

[156] B. Huang, et al., Electrical control of 2D magnetism in bilayer Crl3, Nature Nanotechnology 13 (2018) 544–548.

[157] Y. Liu, et al., Van der Waals heterostructures and devices, Nature Reviews Materials 1 (2016) 16042.

[158] L. Britnell, et al., Field-effect tunneling transistor based on vertical graphene heterostructures, Science 80 (335) (2012) 947–950.

[159] L. Britnell, et al., Resonant tunnelling and negative differential conductance in graphene transistors, Nature Communications 4 (2013) 1794.

[160] L.-N. Nguyen, et al., Resonant tunneling through discrete quantum states in stacked atomic-layered MoS_2, Nano Letters 14 (2014) 2381–2386.

[161] A. Carvalho, R.M. Ribeiro, A.H. Castro Neto, Band nesting and the optical response of two-dimensional semiconducting transition metal dichalcogenides, Physical Review B 88 (2013) 115205.

[162] D. Kozawa, et al., Photocarrier relaxation pathway in two-dimensional semiconducting transition metal dichalcogenides, Nature Communications 5 (2014) 4543.

[163] L. Britnell, et al., Strong light-matter interactions in heterostructures of atomically thin films, Science 80 (340) (2013) 1311–1314.

[164] W.J. Yu, et al., Highly efficient gate-tunable photocurrent generation in vertical heterostructures of layered materials, Nature Nanotechnology 8 (2013) 952–958.

[165] M.M. Furchi, A. Pospischil, F. Libisch, J. Burgdörfer, T. Mueller, Photovoltaic Effect in an electrically tunable van der Waals heterojunction, Nano Letters 14 (2014) 4785–4791.

[166] R. Cheng, et al., Electroluminescence and photocurrent generation from atomically sharp WSe_2/MoS_2 heterojunction p–n diodes, Nano Letters 14 (2014) 5590–5597.

[167] P. Lin, et al., Piezo-phototronic effect for enhanced flexible MoS_2/WSe_2 van der Waals photodiodes, Advanced Functional Materials 28 (2018) 1802849.

[168] Y. Deng, et al., Black phosphorus–monolayer MoS_2 van der Waals heterojunction p–n diode, ACS Nano 8 (2014) 8292–8299.

[169] J. Bullock, et al., Polarization-resolved black phosphorus/molybdenum disulfide mid-wave infrared photodiodes with high detectivity at room temperature, Nature Photonics 12 (2018) 601–607.

[170] F. Wang, et al., Tunable GaTe-MoS_2 van der Waals p–n junctions with novel optoelectronic performance, Nano Letters 15 (2015) 7558–7566.

[171] C.-H. Lee, et al., Atomically thin p–n junctions with van der Waals heterointerfaces, Nature Nanotechnology 9 (2014) 676–681.

[172] E. Suárez Morell, J.D. Correa, P. Vargas, M. Pacheco, Z. Barticevic, Flat bands in slightly twisted bilayer graphene: tight-binding calculations, Physical Review B 82 (2010) 121407.

[173] G. Li, et al., Observation of van hove singularities in twisted graphene layers, Nature Physics 6 (2010) 109–113.

[174] J. Xue, et al., Scanning tunnelling microscopy and spectroscopy of ultra-flat graphene on hexagonal boron nitride, Nature Materials 10 (2011) 282–285.

[175] R. Decker, et al., Local electronic properties of graphene on a BN substrate via scanning tunneling microscopy, Nano Letters 11 (2011) 2291–2295.

[176] M. Yankowitz, et al., Emergence of superlattice Dirac points in graphene on hexagonal boron nitride, Nature Physics 8 (2012) 382–386.

[177] R. Bistritzer, A.H. MacDonald, Moiré bands in twisted double-layer graphene, Proceedings of the National Academy of Sciences of the United States of America 108 (2011) 12233–12237.

[178] P. Moon, M. Koshino, Energy spectrum and quantum Hall effect in twisted bilayer graphene, Physical Review B 85 (2012) 195458.

[179] D.R. Hofstadter, Energy levels and wave functions of Bloch electrons in rational and irrational magnetic fields, Physical Review B 14 (1976) 2239–2249.

[180] L.A. Ponomarenko, et al., Cloning of Dirac fermions in graphene superlattices, Nature 497 (2013) 594–597.

[181] B. Hunt, T. Taniguchi, P. Moon, M. Koshino, R.C. Ashoori, Massive Dirac fermions and, Science 80 (340) (2013) 1427–1431.

[182] C.R. Dean, et al., Hofstadter's butterfly and the fractal quantum Hall effect in moiré superlattices, Nature 497 (2013) 598–602.

[183] A. Mishchenko, et al., Twist-controlled resonant tunnelling in graphene/boron nitride/graphene heterostructures, Nature Nanotechnology 9 (2014) 808–813.

[184] B. Fallahazad, et al., Gate-tunable resonant tunneling in double bilayer graphene heterostructures, Nano Letters 15 (2015) 428–433.

[185] K. Kim, et al., Van der Waals heterostructures with high accuracy rotational alignment, Nano Letters 16 (2016) 1989–1995.

[186] K. Kim, et al., Spin-conserving resonant tunneling in twist-controlled WSe_2-hBN-WSe_2 heterostructures, Nano Letters 18 (2018) 5967–5973.

[187] R. Ribeiro-Palau, et al., Twistable electronics with dynamically rotatable heterostructures, Science 80 (361) (2018) 690–693.

[188] K. Kim, et al., Tunable moiré bands and strong correlations in small-twist-angle bilayer graphene, Proceedings of the National Academy of Sciences of the United States of America 114 (2017) 3364–3369.

[189] Y. Cao, et al., Correlated insulator behaviour at half-filling in magic-angle graphene superlattices, Nature 556 (2018) 80–84.

[190] Y. Cao, et al., Unconventional superconductivity in magic-angle graphene superlattices, Nature 556 (2018) 43–50.

[191] K. Tran, et al., Evidence for moiré excitons in van der Waals heterostructures, Nature 567 (2019) 71–75.

[192] https://phys.org/news/2019-10-physics-magic-angle-graphene-switchable.html.

[193] Y. Xu, C. Cheng, S. Du, J. Yang, B. Yu, J. Luo, W. Yin, E. Li, S. Dong, P. Ye, et al., Contacts between two- and three-dimensional materials: ohmic, Schottky, and p–n heterojunctions, ACS Nano 10 (2016) 4895–4919.

[194] F. Giannazzo, G. Fisichella, A. Piazza, S. Di Franco, G. Greco, S. Agnello, F. Roccaforte, Impact of contact resistance on the electrical properties of MoS_2 transistors at practical operating temperatures, Beilstein Journal of Nanotechnology 8 (2017) 254–263.

[195] W. Liu, D. Sarkar, J. Kang, W. Cao, K. Banerjee, Impact of contact on the operation and performance of back-gated monolayer MoS_2 field-effect-transistors, ACS Nano 9 (2015) 7904–7912.

[196] A. Allain, J. Kang, K. Banerjee, A. Kis, Electrical contacts to two-dimensional semiconductors, Nature Materials 14 (2015) 1195–1205.

[197] Y. Liu, P. Stradins, S.-H. Wei, Van der Waals metal-semiconductor junction: weak Fermi level pinning enables effective tuning of Schottky barrier, Science Advances 2 (2016) e1600069.

[198] S. McDonnell, R. Addou, C. Buie, R.M. Wallace, C.L. Hinkle, Defect-dominated doping and contact resistance in MoS_2, ACS Nano 8 (2014) 2880–2888.

[199] R. Addou, L. Colombo, R.M. Wallace, Surface defects on natural MoS_2, ACS Applied Materials & Interfaces 7 (2015) 11921–11929.

[200] R. Addou, S. McDonnell, D. Barrera, Z. Guo, A. Azcatl, J. Wang, H. Zhu, C.L. Hinkle, M. Quevedo-Lopez, H.N. Alshareef, et al., Impurities and electronic property variations of natural MoS_2 crystal surfaces, ACS Nano 9 (2015) 9124–9133.

[201] P. Bampoulis, R. van Bremen, Q. Yao, B. Poelsema, H.J.W. Zandvliet, K. Sotthewes, Defect Dominated charge transport and Fermi level pinning in MoS_2/metal contacts, ACS Applied Materials & Interfaces 9 (2017) 19278–19286.

[202] S. KC, R.C. Longo, R. Addou, R.M. Wallace, K. Cho, Impact of intrinsic atomic defects on the electronic structure of MoS_2 monolayers, Nanotechnology 25 (2014) 375703.

[203] Y. Han, Z. Wu, S. Xu, X. Chen, L. Wang, Y. Wang, W. Xiong, T. Han, W. Ye, J. Lin, et al., Probing defect-induced midgap states in MoS_2 through graphene–MoS_2 heterostructures, Advanced Materials Interfaces 2 (2015) 1500064.

[204] C.-P. Lu, G. Li, J. Mao, Wang L.-M., E.Y. Andrei, Bandgap, mid-gap states, and gating effects in MoS_2, Nano Letters 14 (2014) 4628–4633.

[205] J. Kang, W. Liu, D. Sarkar, D. Jena, K. Banerjee, Computational study of metal contacts to monolayer transition-metal dichalcogenide semiconductors, Physical Review X 4 (2014) 031005.

[206] C. Gong, L. Colombo, R.M. Wallace, K. Cho, The unusual mechanism of partial Fermi level pinning at metal–MoS_2 interfaces, Nano Letters 14 (2014) 1714–1720.

[207] M. Farmanbar, G. Brocks, First-principles study of van der Waals interactions and lattice mismatch at MoS_2/metal interfaces, Physical Review B 93 (2016) 085304.

[208] Y. Guo, D. Liu, J. Robertson, 3D behavior of Schottky barriers of 2D transition-metal dichalcogenides, ACS Applied Materials & Interfaces 7 (2015) 25709–25715.

[209] W. Monch, On the physics of metal-semiconductor interfaces, Reports on Progress in Physics 53 (1990) 221.

[210] S.G. Louie, J.R. Chelikowsky, M.L. Cohen, Ionicity and the theory of Schottky barriers, Physical Review B 15 (1977) 2154–2162.

[211] J. Tersoff, Schottky barrier heights and the continuum of gap states, Physical Review Letters 52 (1984) 465–468.

[212] S. Das, H.-Y. Chen, A.V. Penumatcha, J.Appenzeller, High performance multilayer MoS_2 transistors with scandium contacts, Nano Letters 13 (2012) 100–105.

[213] N. Kaushik, A. Nipane, F. Basheer, S. Dubey, S. Grover, M.M. Deshmukh, S. Lodha, Schottky barrier heights for Au and pd contacts to MoS_2, Applied Physics Letters 105 (2014) 113505.

[214] C. Kim, I. Moon, D. Lee, M.S. Choi, F. Ahmed, S. Nam, Y. Cho, H.-J. Shin, S. Park, W.J.Yoo, Fermi level pinning at electrical metal contacts of monolayer molybdenum dichalcogenides, ACS Nano 11 (2017) 1588–1596.

[215] N. Kaushik, S. Grover, M.M. Deshmukh, S. Lodha, Metal contacts to MoS_2, in: 2D Inorganic Materials Beyond Graphene, World Scientific (Europe), London, UK, ISBN 978-1-78634-269-0, 2016, pp. 317–347.

[216] H. Liu, M. Si, Y. Deng, A.T. Neal, Y. Du, S. Najmaei, P.M. Ajayan, J. Lou, P.D.Ye, Switching mechanism in single-layer molybdenum disulfide transistors: an insight into current flow across Schottky barriers, ACS Nano 8 (2014) 1031–1038.

[217] S.M. Sze, K.K. Ng, Physics of Semiconductor Devices, John Wiley & Sons, Hoboken, NJ, USA, 2006, Online, ISBN: 9780470068328.

[218] F. Ahmed, M. Sup Choi, X. Liu, W. Jong Yoo, Carrier transport at the metal–MoS_2 interface, Nanoscale 7 (2015) 9222–9228.

[219] ITRS 2.0 Home Page, Available online http://www.itrs2.net/.

[220] D. Jena, K. Banerjee, G.H. Xing, 2D crystal semiconductors: intimate contacts, Nature Materials 13 (2014) 1076.

[221] Y. Zhao, K. Xu, F. Pan, C. Zhou, F. Zhou, Y. Chai, Doping contact and interface engineering of two-dimensional layered transition metal dichalcogenides transistors, Advanced Functional Materials 27 (2017) 1603484.

[222] S.-L. Li, K. Tsukagoshi, E. Orgiu, P. Samorì, Charge transport and mobility engineering in two-dimensional transition metal chalcogenide semiconductors, Chemical Society Reviews 45 (2016) 118–151.

[223] Z. Yu, Z.-Y. Ong, S. Li, J.-B. Xu, G. Zhang, Y.-W. Zhang, Y. Shi, X. Wang, Analyzing the carrier mobility in transition-metal dichalcogenide MoS_2 field-effect transistors, Advanced Functional Materials 27 (2017) 1604093.

[224] B. Radisavljevic, A. Kis, Mobility engineering and a metal–insulator transition in monolayer MoS_2, Nature Materials 12 (2013) 815–820.

[225] B.W.H. Baugher, H.O.H. Churchill, Y. Yang, P. Jarillo-Herrero, Intrinsic electronic transport properties of high-quality monolayer and bilayer MoS_2, Nano Letters 13 (2013) 4212–4216.

[226] D. Jena, M. Li, N. Ma, W.S. Hwang, D. Esseni, A. Seabaugh, H.G. Xing, Electron transport in 2D crystal semiconductors and their device applications, in: Proceedings of the 2014 Silicon Nanoelectronics Workshop (SNW), Honolulu, HI, USA, 8–9 June 2014, 2014, pp. 1–2.

Reference list continued.

[227] G. He, K. Ghosh, U. Singisetti, H. Ramamoorthy, R. Somphonsane, G. Bohra, M. Matsunaga, A. Higuchi, N. Aoki, S. Najmaei, et al., Conduction mechanisms in CVD-grown monolayer MoS_2 transistors: from variable-range hopping to velocity saturation, Nano Letters 15 (2015) 5052–5058.

[228] K. Khair, S. Ahmed, Dissipative transport in monolayer MoS_2: role of remote Coulomb scattering, in: Proceedings of the 2015 International Workshop on Computational Electronics (IWCE), West Lafayette, IN, USA, 2 September 2015, pp. 1–2.

[229] T. Mori, N. Ninomiya, T. Kubo, N. Uchida, E. Watanabe, D. Tsuya, S. Moriyama, M. Tanaka, A. Ando, Characterization of effective mobility and its degradation mechanism in MoS_2 MOSFETs, IEEE Transactions on Nanotechnology 15 (2016) 651–656.

[230] S. Ahmed, J. Yi, Two-dimensional transition metal dichalcogenides and their charge carrier mobilities in field-effect transistors, Nano-Micro Letters 9 (2017) 50.

[231] G. Mirabelli, F. Gity, S. Monaghan, P.K. Hurley, R. Duffy, Impact of impurities, interface traps and contacts on MoS_2 MOSFETs: modelling and experiments, in: Proceedings of the 2017 47th European Solid-State Device Research Conference (ESSDERC), Leuven, Belgium, 11–14 September, 2017, pp. 288–291.

[232] K. Kaasbjerg, K.S. Thygesen, K.W. Jacobsen, Phonon-limited mobility in n-type single-layer MoS_2 from first principles, Physical Review B 85 (2012) 115317.

[233] X. Li, J.T. Mullen, Z. Jin, K.M. Borysenko, M. Buongiorno Nardelli, K.W. Kim, Intrinsic electrical transport properties of monolayer silicene and MoS_2 from first principles, Physical Review B 87 (2013) 115418.

[234] Z. Jin, X. Li, J.T. Mullen, K.W. Kim, Intrinsic transport properties of electrons and holes in monolayer transition-metal dichalcogenides, Physical Review B 90 (2014) 045422.

[235] T. Gunst, T. Markussen, K. Stokbro, M. Brandbyge, First-principles method for electron-phonon coupling and electron mobility: applications to two-dimensional materials, Physical Review B 93 (2016) 035414.

[236] Y. Cai, G. Zhang, Y.-W. Zhang, Polarity-reversed robust carrier mobility in monolayer MoS_2 nanoribbons, Journal of the American Chemical Society 136 (2014) 6269–6275.

[237] N. Ma, D. Jena, Charge scattering and mobility in atomically thin semiconductors, Physical Review X 4 (2014) 011043.

[238] S. Ghatak, A.N. Pal, A. Ghosh, Nature of electronic states in atomically thin MoS_2 field-effect transistors, ACS Nano 5 (2011) 7707–7712.

[239] H.-P. Komsa, S. Kurasch, O. Lehtinen, U. Kaiser, A.V. Krasheninnikov, From point to extended defects in two-dimensional MoS_2: evolution of atomic structure under electron irradiation, Physical Review B 88 (2013) 035301.

[240] H. Qiu, T. Xu, Z. Wang, W. Ren, H. Nan, Z. Ni, Q. Chen, S. Yuan, F. Miao, F. Song, et al., Hopping transport through defect-induced localized states in molybdenum disulphide, Nature Communications 4 (2013) 2642.

[241] S. Ghatak, S. Mukherjee, M. Jain, D.D. Sarma, A. Ghosh, Microscopic origin of charged impurity scattering and flicker noise in MoS_2 field-effect transistors, arXiv:1403.3333, 2014.

[242] D. Liu, Y. Guo, L. Fang, J. Robertson, Sulfur vacancies in monolayer MoS_2 and its electrical contacts, Applied Physics Letters 103 (2013) 183113.

[243] Z. Yu, Y. Pan, Y. Shen, Z. Wang, Z.-Y. Ong, T. Xu, R. Xin, L. Pan, B. Wang, L. Sun, et al., Towards intrinsic charge transport in monolayer molybdenum disulfide by defect and interface engineering, Nature Communications 5 (2014) 5290.

[244] M. Amani, D.-H. Lien, D. Kiriya, J. Xiao, A. Azcatl, J. Noh, S.R. Madhvapathy, R. Addou, S. Kc, M. Dubey, et al., Near-unity photoluminescence quantum yield in MoS_2, Science 350 (2015) 1065–1068.

[245] F. Giannazzo, G. Fisichella, A. Piazza, S.D. Franco, G. Greco, S. Agnello, F. Roccaforte, Effect of temperature–bias annealing on the hysteresis and subthreshold behavior of multilayer MoS_2 transistors, Physica Status Solidi (RRL)—Rapid Research Letters 11 (2016) 797–801.

[246] P. Bolshakov, P. Zhao, A. Azcatl, P.K. Hurley, R.M. Wallace, C.D. Young, Electrical characterization of top-gated molybdenum disulfide field-effect-transistors with high-κ dielectrics, Microelectronic Engineering 178 (2017) 190–193.

[247] P. Bolshakov, P. Zhao, A. Azcatl, P.K. Hurley, R.M. Wallace, C.D. Young, Improvement in top-gate MoS_2 transistor performance due to high quality backside Al_2O_3 layer, Applied Physics Letters 111 (2017) 032110.

[248] C.D. Young, P. Zhao, P. Bolshakov-Barrett, A. Azcatl, P.K. Hurley, Y.Y. Gomeniuk, M. Schmidt, C.L. Hinkle, R.M. Wallace, Evaluation of few-layer MoS_2 transistors with a top gate and HfO_2 dielectric, ECS Transactions 75 (2016) 153–162.

[249] H. Nan, Z. Wu, J. Jiang, A. Zafar, Y. You, Z. Ni, Improving the electrical performance of MoS_2 by mild oxygen plasma treatment, Journal of Physics. D, Applied Physics 50 (2017) 154001.

[250] J. Guo, B. Yang, Z. Zheng, J. Jiang, Observation of abnormal mobility enhancement in multilayer MoS_2 transistor by synergy of ultraviolet illumination and ozone plasma treatment, Physica. E, Low-Dimensional Systems and Nanostructures 87 (2017) 150–154.

[251] Z. Cheng, J.A. Cardenas, F. McGuire, A.D. Franklin, Using ar ion beam exposure to improve contact resistance in MoS_2 FETs, in: Proceedings of the 2016 74th Annual Device Research Conference (DRC), Newark, dE, USA, 19–22 June 2016, pp. 1–2.

[252] Y.T. Ho, Y.C. Chu, C.A. Jong, H.Y. Chen, M.W. Lin, M. Zhang, P.Y. Chien, Y.Y. Tu, J. Woo, E.Y. Chang, Contact resistance reduction on layered MoS_2 by ar plasma pre-treatment, in: Proceedings of the 2016 IEEE Silicon Nanoelectronics Workshop (SNW), Honolulu, HI, USA, 12–13 June, 2016, pp. 52–53.

[253] S. Manzeli, D. Ovchinnikov, D. Pasquier, O.V. Yazyev, A. Kis, 2D transition metal dichalcogenides, Nature Reviews Materials 2 (2017) 17033.

[254] H.F. Liu, S.L. Wong, D.Z.C. VD Chi, Growth of MoS_2-based two-dimensional materials, Chemical Vapor Deposition 21 (2015) 241–259.

[255] S.L. Wong, H. Liu, D. Chi, Recent progress in chemical vapor deposition growth of two-dimensional transition metal dichalcogenides, Progress in Crystal Growth and Characterization of Materials 62 (2016) 9–28.

[256] H. Liu, Recent progress in atomic layer deposition of multifunctional oxides and two-dimensional transition metal dichalcogenides, Journal of Molecular and Engineering Materials 4 (2016) 1640010.

[257] L.A.Walsh, C.L. Hinkle, Van der Waals epitaxy: 2D materials and topological insulators, Applied Materials Today 9 (2017) 504–515.

[258] J.R. Brent, N. Savjani, P. O'Brien, Synthetic approaches to two-dimensional transition metal dichalcogenide nanosheets, Progress in Materials Science 89 (2017) 411–478.

[259] W. Zhou, X. Zou, S. Najmaei, Z. Liu, Y. Shi, J. Kong, J. Lou, P.M. Ajayan, B.I. Yakobson, J.-C. Idrobo, Intrinsic structural defects in monolayer molybdenum disulfide, Nano Letters 13 (2013) 2615–2622.

[260] S. Najmaei, J. Yuan, J. Zhang, P. Ajayan, J. Lou, Synthesis and defect investigation of two-dimensional molybdenum disulfide atomic layers, Accounts of Chemical Research 48 (2015) 31–40.

[261] A. Rai, A. Valsaraj, H.C.P. Movva, A. Roy, E. Tutuc, L.F. Register, S.K. Banerjee, Interfacial-oxygen-vacancy mediated doping of MoS_2 by high-κ dielectrics, in: Proceedings of the 2015 73rd Annual Device Research Conference (DRC), Columbus, OH, USA, 21–24 June, 2015, pp. 189–190.

[262] A. Rai, A. Valsaraj, H.C.P. Movva, A. Roy, R. Ghosh, S. Sonde, S. Kang, J. Chang, T. Trivedi, R. Dey, et al., Air stable doping and intrinsic mobility enhancement in monolayer molybdenum disulfide by amorphous titanium suboxide encapsulation, Nano Letters 15 (2015) 4329–4336.

[263] A. Valsaraj, J. Chang, A. Rai, L.F. Register, S.K. Banerjee, Theoretical and experimental investigation of vacancy-based doping of monolayer MoS_2 on oxide, 2D Materials 2 (2015) 045009.

[264] A. Alharbi, D. Shahrjerdi, Contact engineering of monolayer CVD MoS_2 transistors, in: Proceedings of the 2017 75th Annual Device Research Conference (DRC), South Bend, iN, USA, 25–28 June, 2017, pp. 1–2.

[265] C.J. McClellan, E. Yalon, K.K.H. Smithe, S.V. Suryavanshi, E. Pop, Effective n-type doping of monolayer MoS_2 by AlO_x, in: Proceedings of the 2017 75th Annual Device Research Conference (DRC), South Bend, IN, USA, 25–28 June, 2017, pp. 1–2.

[266] D. Jena, A. Konar, Enhancement of carrier mobility in semiconductor nanostructures by dielectric engineering, Physical Review Letters 98 (2007) 136805.

[267] Z. Yu, Z.Y. Ong, Y. Pan, T. Xu, Z. Wang, L. Sun, J. Wang, G. Zhang, Y.W. Zhang, Y. Shi, et al., Electron transport and device physics in monolayer transition-metal dichalcogenides, in: Proceedings of the 2016 IEEE International Nanoelectronics Conference (INEC), Chengdu, China, 9–11 May, 2016, pp. 1–2.

[268] Y. Sun, R. Wang, K. Liu, Substrate induced changes in atomically thin 2-dimensional semiconductors: fundamentals, engineering, and applications, Applied Physics Reviews 4 (2017) 011301.

[269] X. Li, X. Xiong, T. Li, S. Li, Z. Zhang, Y. Wu, Effect of dielectric interface on the performance of MoS_2 transistors, ACS Applied Materials & Interfaces (2017).

[270] Z. Yu, Z.-Y. Ong, Y. Pan, Y. Cui, R. Xin, Y. Shi, B. Wang, Y. Wu, T. Chen, Y.-W. Zhang, et al., Realization of room-temperature phonon-limited carrier transport in monolayer MoS_2 by dielectric and carrier screening, Advanced Materials 28 (2016) 547–552.

[271] B. Chamlagain, Q. Cui, S. Paudel, M.M.-C. Cheng, P.-Y. Chen, Z. Zhou, Thermally oxidized 2D TaS_2 as a high-κ gate dielectric for MoS_2 field-effect transistors, 2D Materials 4 (2017) 031002.

[272] H.-J. Kwon, J. Jang, C.P. Grigoropoulos, Laser direct writing process for making electrodes and high-κ sol–gel ZrO_2 for boosting performances of MoS_2 transistors, ACS Applied Materials & Interfaces 8 (2016) 9314–9318.

[273] L. Zeng, Z. Xin, S. Chen, G. Du, J. Kang, X. Liu, Remote phonon and impurity screening effect of substrate and gate dielectric on electron dynamics in single layer MoS_2, Applied Physics Letters 103 (2013) 113505.

[274] A. Konar, T. Fang, D. Jena, Effect of high-κ gate dielectrics on charge transport in graphene-based field effect transistors, Physical Review B 82 (2010) 115452.

[275] V. Perebeinos, P. Avouris, Inelastic scattering and current saturation in graphene, Physical Review B 81 (2010) 195442.

[276] NSM Archive, Physical properties of semiconductors, Available online http://www.ioffe.ru/SVA/NSM/Semicond/.

[277] X. Cui, G.-H. Lee, Y.D. Kim, G. Arefe, P.Y. Huang, C.-H. Lee, D.A. Chenet, X. Zhang, L. Wang, F. Ye, et al., Multi-terminal transport measurements of MoS_2 using a van der Waals heterostructure device platform, Nature Nanotechnology 10 (2015) 534–540.

[278] S. Xu, Z. Wu, H. Lu, Y. Han, G. Long, X. Chen, T. Han, W. Ye, Y. Wu, J. Lin, et al., Universal low-temperature ohmic contacts for quantum transport in transition metal dichalcogenides, 2D Materials 3 (2016) 021007.

[279] S. Bhattacharjee, K.L. Ganapathi, H. Chandrasekar, T. Paul, S. Mohan, A. Ghosh, S. Raghavan, N. Bhat, Nitride dielectric environments to suppress surface optical phonon dominated scattering in high-performance multilayer MoS_2 FETs, Advanced Electronic Materials 3 (2017) 1600358.

[280] A. Murakoshi, M. Iwase, H. Niiyama, M. Koike, K. Suguro, Ultralow contact resistivity for a metal/p-type silicon interface by high-concentration germanium and boron doping combined with low-temperature annealing, Japanese Journal of Applied Physics 52 (2013) 075802.

[281] J. Pelletier, A. Anders, Plasma-based ion implantation and deposition: a review of physics, technology, and applications, IEEE Transactions on Plasma Science 33 (2005) 1944–1959.

[282] J.S. Williams, Ion implantation of semiconductors, Materials Science & Engineering. A, Structural Materials: Properties, Microstructure and Processing 253 (1998) 8–15.

[283] S.W. Jin, J.C. Cha, H.S. Lee, S.H. Son, B.G. Kim, Y.S. Jung, Implant and anneal technologies for memory and CMOS devices, in: Proceedings of the 2016 21st International Conference on Ion Implantation Technology (IIT), Tainan, Taiwan, 26–30 September 2016, 2016, pp. 1–5.

[284] J.F. Ziegler, High energy ion implantation, Nuclear Instruments & Methods in Physics Research. Section B, Beam Interactions With Materials and Atoms 6 (1985) 270–282.

[285] S.J. Pearton, Ion implantation in iii–v semiconductor technology, International Journal of Modern Physics B 7 (1993) 4687–4761.

[286] M.I. Current, Ion implantation of advanced silicon devices: past, present and future, Materials Science in Semiconductor Processing 62 (2017) 13–22.

[287] J.F. Ziegler, Ion Implantation Science and Technology, Elsevier, Amsterdam, the Netherlands, 2012, ISBN 978-0-323-14401-8.

[288] H.J. Cho, H.S. Oh, K.J. Nam, Y.H. Kim, K.H. Yeo, W.D. Kim, Y.S. Chung, Y.S. Nam, S.M. Kim, W.H. Kwon, et al., Si FinFET based 10 nm technology with multi vt gate stack for low power and high performance applications, in: Proceedings of the 2016 IEEE Symposium on VLSI Technology, Honolulu, HI, USA, 14–16 June 2016, pp. 1–2.

[289] W. Hsu, T. Kim, H. Chou, A. Rai, S.K. Banerjee, Novel BF+ implantation for high performance ge pMOSFETs, IEEE Electron Device Letters 37 (2016) 954–957.

[290] W. Hsu, T. Kim, A. Benítez-Lara, H. Chou, A. Dolocan, A. Rai, M. Josefina Arellano-Jiménez, M. Palard, M. José-Yacamán, S.K. Banerjee, Diffusion and recrystallization of b implanted in crystalline and pre-amorphized ge in the presence of f, Journal of Applied Physics 120 (2016) 015701.

[291] W. Hsu, A. Rai, X. Wang, Y. Wang, T. Kim, S.K. Banerjee, Impact of Junction Depth and Abruptness on the Activation and the Leakage Current in Germanium n^+/p Junctionss, arXiv:1705.06733, 2017.

[292] N.W. Cheung, Plasma immersion ion implantation for ULSI processing, Nuclear Instruments & Methods in Physics Research. Section B, Beam Interactions With Materials and Atoms 55 (1991) 811–820.

[293] J.F. Gibbons, Ion implantation in semiconductors—part I: range distribution theory and experiments, Proceedings of the IEEE 56 (1968) 295–319.

[294] Jones, K.S., Perry, S., Murray, R., Hynes, K., Zhao, X, in: (Invited) An Overview of Doping Studies in MoS_2. Meet. Abstr., 2016, MA2016–01, 1298.

[295] A. Yoon, Z. Lee, Synthesis and properties of two dimensional doped transition metal dichalcogenides, Applied Microscopy 47 (2017) 19–28.

[296] M.R. Laskar, D.N. Nath, L. Ma, E.W.L. Ii, C.H. Lee, T. Kent, Z. Yang, R. Mishra, M.A. Roldan, J.-C. Idrobo, et al., P-type doping of MoS_2 thin films using nb, Applied Physics Letters 104 (2014) 092104.

[297] J. Suh, T.-E. Park, D.-Y. Lin, D. Fu, J. Park, H.J. Jung, Y. Chen, C. Ko, C. Jang, Y. Sun, et al., Doping against the native propensity of MoS_2: degenerate hole doping by cation substitution, Nano Letters 14 (2014) 6976–6982.

[298] G. Mirabelli, M. Schmidt, B. Sheehan, K. Cherkaoui, S. Monaghan, I. Povey, M. McCarthy, A.P. Bell, R. Nagle, F. Crupi, et al., Back-gated nb-doped MoS_2 junctionless field-effect-transistors, AIP Advances 6 (2016) 025323.

[299] C.D. English, G. Shine, V.E. Dorgan, K.C. Saraswat, E. Pop, Improved contacts to MoS_2 transistors by ultra-high vacuum metal deposition, Nano Letters 16 (2016) 3824–3830.

[300] J. Kang, W. Liu, K. Banerjee, High-performance MoS_2 transistors with low-resistance molybdenum contacts, Applied Physics Letters 104 (2014) 093106.

[301] J. Gao, Y.D. Kim, L. Liang, J.C. Idrobo, P. Chow, J. Tan, B. Li, L. Li, B.G. Sumpter, T.-M. Lu, et al., Transition-metal substitution doping in synthetic atomically thin semiconductors, Advanced Materials 28 (2016) 9735–9743.

[302] L. Yang, K. Majumdar, Y. Du, H. Liu, H. Wu, M. Hatzistergos, P.Y. Hung, R. Tieckelmann, W. Tsai, C. Hobbs, et al., High-performance MoS_2 field-effect transistors enabled by chloride doping: record low contact resistance (0.5 k$\Omega \cdot$ μm) and record high drain current (460 μa/ μm), in: Proceedings of the 2014 Symposium on VLSI Technology (VLSI-Technology), Honolulu, HI, USA, 9–12 June, 2014, pp. 1–2.

[303] L. Yang, K. Majumdar, H. Liu, Y. Du, H. Wu, M. Hatzistergos, P.Y. Hung, R. Tieckelmann, W. Tsai, C. Hobbs, et al., Chloride molecular doping technique on 2D materials: WS_2 and MoS_2, Nano Letters 14 (2014) 6275–6280.

[304] A.T. Neal, R. Pachter, S. Mou, P-type conduction in two-dimensional MoS_2 via oxygen incorporation, Applied Physics Letters 110 (2017) 193103.

[305] F. Giannazzo, G. Fisichella, G. Greco, S. Di Franco, I. Deretzis, A. La Magna, C. Bongiorno, C. Nicotra, C. Spinella, M. Scopelliti, et al., Ambipolar MoS_2 transistors by nanoscale tailoring of Schottky barrier using oxygen plasma functionalization, ACS Applied Materials & Interfaces 9 (2017) 23164–23174.

[306] A. Nipane, D. Karmakar, N. Kaushik, S. Karande, S. Lodha, Few-Layer MoS_2 p-type devices enabled by selective doping using low energy phosphorus implantation, ACS Nano 10 (2016) 2128–2137.

[307] A. Nipane, N. Kaushik, S. Karande, D. Karmakar, S. Lodha, P-type doping of MoS_2 with phosphorus using a plasma immersion ion implantation (PIII) process, in: Proceedings of the 2015 73rd Annual Device Research Conference (DRC), Columbus, OH, USA, 21–24 June 2015, 2015, pp. 191–192.

[308] K. Xu, Y. Zhao, Z. Lin, Y. Long, Y. Wang, M. Chan, Y. Chai, Doping of two-dimensional MoS_2 by high energy ion implantation, Semiconductor Science and Technology 32 (2017) 124002.

[309] H.-J. Chuang, B. Chamlagain, M. Koehler, M.M. Perera, J. Yan, D. Mandrus, D. Tománek, Z. Zhou, Low-resistance 2D/2D ohmic contacts: a universal approach to high-performance WSe_2, MoS_2, and $MoSe_2$ transistors, Nano Letters 16 (2016) 1896–1902.

[310] Y. Kim, A.R. Kim, J.H. Yang, K.E. Chang, J.-D. Kwon, S.Y. Choi, J. Park, K.E. Lee, D.-H. Kim, S.M. Choi, et al., Alloyed 2D metal–semiconductor heterojunctions: origin of interface states reduction and Schottky barrier lowering, Nano Letters 16 (2016) 5928–5933.

[311] K.F. Mak, C. Lee, J. Hone, J. Shan, T.F. Heinz, Atomically thin MoS_2: a new direct-gap semiconductor, Physical Review Letters 105 (13) (Sep. 2010).

[312] A. Splendiani, et al., Emerging photoluminescence in monolayer MoS_2, Nano Letters 10 (4) (Apr. 2010) 1271–1275.

[313] B. Radisavljevic, A. Radenovic, J. Brivio, V. Giacometti, A. Kis, Single-layer MoS_2 transistors, Nature Nanotechnology 6 (3) (Mar. 2011) 147–150.

[314] O. Lopez-Sanchez, D. Lembke, M. Kayci, A. Radenovic, A. Kis, Ultrasensitive photodetectors based on monolayer MoS_2, Nature Nanotechnology 8 (7) (Jul. 2013) 497–501.

[315] S. Bertolazzi, D. Krasnozhon, A. Kis, Nonvolatile memory cells based on MoS_2/graphene heterostructures, ACS Nano 7 (4) (Apr. 2013) 3246–3252.

[316] Y. Yoon, K. Ganapathi, S. Salahuddin, How good can monolayer MoS_2 transistors be?, Nano Letters 11 (9) (Sep. 2011) 3768–3773.

[317] H. Liu, P.D. Ye, MoS_2 dual-gate MOSFET with atomic-layer-deposited Al_2O_3 as top-gate dielectric, IEEE Electron Device Letters 33 (4) (Apr. 2012) 546–548.

[318] M. Fontana, et al., Electron-hole transport and photovoltaic effect in gated MoS_2 Schottky junctions, Sci. Rep. 3, srep01634, (Apr. 2013).

[319] D.J. Late, et al., Sensing behavior of atomically thin-layered MoS_2 transistors, ACS Nano 7 (6) (Jun. 2013) 4879–4891.

[320] J. Chang, L.F. Register, S.K. Banerjee, Atomistic full-band simulations of monolayer MoS_2 transistors, Applied Physics Letters 103 (22) (Nov. 2013) 223509.

[321] Q.H. Wang, K. Kalantar-Zadeh, A. Kis, J.N. Coleman, M.S. Strano, Electronics and optoelectronics of two-dimensional transition metal dichalcogenides, Nature Nanotechnology 7 (11) (Nov. 2012) 699–712.

[322] D. Jariwala, V.K. Sangwan, L.J. Lauhon, T.J. Marks, M.C. Hersam, Emerging device applications for semiconducting two-dimensional transition metal dichalcogenides, ACS Nano 8 (2) (Feb. 2014) 1102–1120.

[323] H. Liu, A.T. Neal, P.D. Ye, Channel length scaling of MoS_2 MOSFETs, ACS Nano 6 (10) (Oct. 2012) 8563–8569.

[324] D. Jena, A. Konar, Enhancement of carrier mobility in semiconductor nanostructures by dielectric engineering, Physical Review Letters 98 (13) (Mar. 2007) 136805.

[325] P.W. Peacock, K. Xiong, K. Tse, J. Robertson, Bonding and interface states of $Si:HfO_2$ and $Si:ZrO_2$ interfaces, Physical Review B 73 (7) (Feb. 2006) 075328.

[326] K. Dolui, I. Rungger, S. Sanvito, Origin of the n-type and p-type conductivity of MoS_2 monolayers on a SiO_2 substrate, Physical Review B 87 (16) (Apr. 2013) 165402.

[327] P.-H. Ho, et al., Self-encapsulated doping of n-type graphene transistors with extended air stability, ACS Nano 6 (7) (Jul. 2012) 6215–6221.

[328] A. Valsaraj, L.F. Register, S.K. Banerjee, J. Chang, Density-functional-theory-based study of monolayer MoS_2 on oxide, in: 2014 International Conference on Simulation of Semiconductor Processes and Devices (SISPAD), 2014, pp. 73–76.

[329] G. Kresse, J. Furthmüller, Efficient iterative schemes for ab initio total-energy calculations using a plane-wave basis set, Physical Review B 54 (16) (Oct. 1996) 11169–11186.

[330] G. Kresse, J. Furthmüller, Efficiency of ab-initio total energy calculations for metals and semiconductors using a plane-wave basis set, Computational Materials Science 6 (1) (Jul. 1996) 15–50.

[331] J.P. Perdew, Y. Wang, Accurate and simple analytic representation of the electron-gas correlation energy, Physical Review B 45 (23) (Jun. 1992) 13244–13249.

[332] J. Chang, Ballistic performance comparison of monolayer transition metal dichalcogenide MX2 (m = mo, w, x = s, se, te) metal-oxide-semiconductor field effect transistors, Journal of Applied Physics 115 (8) (Feb. 2014) 084506.

[333] J. Kang, W. Liu, D. Sarkar, D. Jena, K. Banerjee, Computational study of metal contacts to monolayer transition-metal dichalcogenide semiconductors, Physical Review X 4 (3) (Jul. 2014) 031005.

[334] T. Böker, et al., Band structure of MoS_2, $MoSe_2$, and $\alpha - MoTe_2$: angle-resolved photoelectron spectroscopy and ab initio calculations, Physical Review B 64 (23) (Nov. 2001) 235305.

[335] J.P. Perdew, K. Burke, M. Ernzerhof, Generalized gradient approximation made simple, Physical Review Letters 77 (18) (Oct. 1996) 3865–3868.

[336] A.V. Krukau, O.A. Vydrov, A.F. Izmaylov, G.E. Scuseria, Influence of the exchange screening parameter on the performance of screened hybrid functionals, Journal of Chemical Physics 125 (22) (Dec. 2006) 224106.

[337] S. McDonnell, et al., HfO_2 on MoS_2 by atomic layer deposition: adsorption mechanisms and thickness scalability, ACS Nano 7 (11) (Nov. 2013) 10354–10361.

[338] S. Grimme, J. Antony, S. Ehrlich, H. Krieg, A consistent and accurate ab initio parametrization of density functional dispersion correction (DFT-d) for the 94 elements h-pu, Journal of Chemical Physics 132 (15) (Apr. 2010) 154104.

[339] J. Kang, E.-C. Lee, K.J. Chang, First-principles study of the structural phase transformation of hafnia under pressure, Physical Review B 68 (5) (Aug. 2003) 054106.

[340] S.-D. Mo, W.Y. Ching, Electronic and optical properties of $\Theta - Al_2O_3$ and comparison to $\alpha - Al_2O_3$, Physical Review B 57 (24) (Jun. 1998) 15219–15228.

[341] A. Valsaraj, J. Chang, A. Rai, L.F. Register, S.K. Banerjee, Theoretical and experimental investigation of vacancy-based doping of monolayer MoS_2 on oxide, 2D Materials 2 (4) (2015) 045009.

CHAPTER

12

Concluding remarks

Rafik Addou and Luigi Colombo

University of Texas at Dallas, Richardson, TX, United States

In this book the authors have done an exhaustive review of 2D materials growth processes, defect identification and structure, defect simulations, defect creation processes and healing, presented data on the use of defects in 2D materials for catalysis, and the effect of a variety of defects on electronic device performance.

The review starts with Chapter 2 where the authors, *Hannu-Pekka Komsa and Arkady V. Krasheninnikov*, present an overview of the recent advances and current understanding of the physics of 2D materials and the role of their reduced dimensionality. Unlike in 3D crystals, grain boundaries are line defects in 2D, and edge dislocations are point defects. Also, interstitial atoms do not exist in the majority of 2DMs (e.g. graphene, h-BN, MoS_2), as it is energetically more favorable for the atom to take an adatom position than to be embedded two-dimensional network. Defects in 2DMs in general can give rise to a large out-of-plane bending of the material. The two-dimensional nature of these materials strongly affects the behavior of defects due the interaction with ambient species like oxygen, hydrocarbons molecules, hydrogen and OH species. The high surface to volume ratio also makes it easy to create defects through chemical treatments and irradiation with energetic particles such as ions and electrons. Passivation and healing of the same defects is also possible through chemical treatments and annealing in controlled chemical environments as is also discussed in Chapter 6. This is a much more efficient process than in 3D crystals, although control can be tricky due to monolayer nature of the materials since no material can be etched away to create a pristine surface as is normally done in 3D crystals. While defect control in 3D crystals is performed on a routine basis and is used principally to control the electronic transport properties, electronic transport control of 2D materials is not fully understood in this respect since the surfaces also play a significant role. It will take significantly more effort to understand doping in 2D materials. The unique nature of 2D materials makes direct observation of defects by STM or TEM much easier than in 3D crystals and with the aid of Raman spectroscopy, a better understanding can be gained although not trivial. Furthermore, the geometry of 2DMs requires making changes in the theoretical description of charged point and line defects, since the screening/electrostatics is strongly anisotropic and inhomogeneous, and the results may heavily depend on the shape of the simulation cell and electrostatic correction scheme used. The fact that these crystals have two surfaces, and no bulk makes them ideal for some ap-

Defects in Two-Dimensional Materials
https://doi.org/10.1016/B978-0-12-820292-0.00018-5

Copyright © 2022 Elsevier Inc. All rights reserved.

404   12. Concluding remarks

plications, for example catalysis, optoelectronics, and may show interesting memristive and magnetic behavior. The authors further add that there is still work to be done to understand the origins of the defect-related PL features observed in these materials, especially for materials where many PL lines are observed as in WSe_2 and h-BN which can arise, in addition to different defects, from strain variations or phonon-assisted emission processes. The situation is not helped by the fact that simulating these from first principles with predictive accuracy is still computationally very demanding. In addition, 2DMs are known to host also larger carrier complexes than exciton, such as trions and biexcitons. The interaction of these with defects is still largely unexplored territory, both computationally as well as experimentally.

In Chapter 3 *De Padova and co-authors* present an in-depth review on the preparation and characterization of elemental two-dimensional materials beyond graphene such as group-III, -IV, -V and -VI like borophene, silicene, germanene, phosphorene, stanene, arsenene, antimonene, bismuthene, selenene and tellurene and the effects of point defects on their 2D honeycomb structure. The authors expand the discussion to additional structural defects to those described in Chapter 2 found in 2D materials such as Stone-Wales, single vacancy and bi-vacancies, as well as grain boundaries (lines) and adatoms have been described based on theoretical and experimental published reports. Until now theoretical reports have focused their assessment on the probability of formation and thermodynamic stability of point defects in 2D materials structural lattice. The identification of such structural defects and understanding of the diffusion behavior and impact on electronic/magnetic properties is believed to be crucial for future applications of 2D materials. The motivation for trying to discover, prepare and study these new elemental 2D materials was spearheaded by observation of defect mediated non-trivial topological behavior in these materials. Topological insulators being of great interest for their potential use in several electronic applications. As also described in Chapter 2 and subsequent chapters, structural defects are almost inevitable, and whether they are native or intentional their existence can strongly affect the fundamental physical properties of 2D materials. As in the case of TMD, graphene and h-BN defects in elemental 2D materials beyond graphene can also be intentionally introduced by physical methods such as stress, irradiation, annealing or chemical treatments. The authors conclude that although two-dimensional materials are still quite young and many new physical phenomena have yet to be discovered, there are many challenges that have to be faced in the incoming years, where the impact of 2D materials and their defect engineering will play a major role in the improvement of the current nanotechnologies.

Stephen J. McDonnell and Petra Reinke in Chapter 4 continue describing defects in transition metal dichalcogenides and how their functional properties are dominated by defects. There have been numerous reports on defect engineering and how these can be used to control catalytic properties, conductivity, optical properties, and magnetic properties. However, we are still a long way from synthesizing high-purity low defect density films comparable to silicon or other 3D compound semiconductors. Both geological and synthetic material are reviewed and how they are impacted by the presence of impurities, vacancies, vacancies clusters, and grain boundaries. In the long term, achieving high purity material will significantly simplify property engineering of TMDs, but in the near term, the materials must be thoroughly characterized and researchers must continue to consider both random/uncontrolled and engineered defects when interpreting the functional properties of these materials. As already described by the authors of Chapter 4, the quality of 2D materials is still somewhat immature. In Chapter 5, *Vitaliy Babenko and Stephan Hofmann* describe processes in realizing electronic grade graphene and h-BN. Since the first attempts at exfoliating graphene from graphite [1–3] extraordinary efforts have been made to advance the growth of scalable and defect-free production of graphene and h-BN, built upon the vast knowledge accumulated over the years across a multitude of disciplines and applied research fields. From surface science studies and modeling at atomic scales to heterogeneous catalysis and crystal growth, from metallurgical processing to epitaxial single crystal film growth, from atomistic hetero-interfacial studies to industry-relevant processing to name a few. Availability of bulk layered crystals and their exfoliation will continue to help driving fundamental 2D materials discovery and to facilitate small-scale proof of concept devices. However, bottom-up growth will be critical for the transition to commercial applications and the challenge now is to reach the quality set by exfoliated materials or the quality demanded by high-value-added applications, while preserving scalability. Roll-to-roll growth processes of graphene by catalytic CVD have reached ultra-fast growth rates and with greatly improved performance compared to the early studies. Epitaxial thin film growth on metal catalysts has also resulted in significantly improved graphene and h-BN films at the wafer scale. Such single crystal 2D materials will be able to satisfy applications with stringent quality requirements and reach a stage where material quality is not the limiting factor as is the case for most 3D crystals used today in the electronics industry. Graphene and monolayer h-BN crystals grown by catalytic CVD are the building blocks for future heterostructures, such as twisted bi-layers and designer heterostructures, at the atomic layer limit and processes to achieve these films are being improved. In parallel crystalline multilayer exfoliated h-BN, which has played a critical role in improving the performance of almost every 2D-material based device to date, has been almost exclusively used for fundamental understanding of basic materials and device properties. However, the technology needed for the growth of controlled mono- or multilayer h-BN is still immature and either alternative materials must be found, or significantly more efforts must be dedicated to achieving the required materials quality. Recently, there have been reports on deposition of amorphous BN (α-BN) films with a mixed sp^2/sp^3 bonding character [4]. It would be important for the scientific to evaluate α-BN as a dielectric material to improve the transport properties of graphene and other 2D materials. The ultimate quality and performance requirement is typically tied to an application, with a wide range of applications that can tolerate some imperfections, such as 2D materials-based sensors, currently under evaluation for commercialization. Two-dimensional materials are natural candidates for monolithic 3D integration for advanced functionality, however, while 2D material films are reaching maturity, the next research challenge is not only scalable, clean 2D-2D layer interfacing, but also heterogeneous 2D-3D materials integration, i.e., moving from basic material control at atomic scale to heterogeneous complementary technology and bringing together many materials. Integration of 2D materials, especially in CMOS technology, will require focused research efforts and adaptation of the manufacturing approaches that are at the core of the semiconductor industry. Understanding the role of defects and their control is a crucial vector, with a topical example being the recently discovered single phonon emission from atomic defects in h-BN. Such atomic engineering requires high-quality 2D material to start with, which are now within reach of modern scalable growth technologies. Continued efforts in the synthesis of high quality 2D materials will enable the realization of new devices and perhaps improve current ones.

406   12. Concluding remarks

Chapter 6 by *Yu-Chuan Lin and co-authors on* the realization of electronic grade TMD materials describes the growth and growth mechanisms and defects of TMDs in great details. The growth mechanisms and controllability of thin-film deposition is essential to control defects since most are introduced during nucleation and growth in the non-equilibrium growth environment. The nature of each deposition technique, precursors, substrates, as well as kinetic and thermodynamic considerations for crystal growth can all influence the final defect type and density inside the 2D film. The authors provide a basic understanding on some of most popular thin-film deposition systems of 2D TMD materials and present representative results of TMD monolayer synthesis. They then follow to describe the many aspects of CVD for TMD such as precursor types and properties, growth temperatures, growth pressures, and substrates, which could be used as tuning knobs to control the crystallinity, impurity density, point defect density, grain boundary density, and resulting heterogeneity in the film. Several *ex-situ* methods for engineering and controlling defects in the as-grown film are reviewed and vertical vdW heterostructures are formed to control their semiconducting properties. This chapter further covers doping and alloying TMD not only to modify the electronic properties but also to suppress deep levels of chalcogen vacancies through isoelectronic doping. While it would be advantageous to eliminate defects in any material one can only hope to minimize them to meet the device performance requirements or equivalently electronic grade materials.

Following the growth of TMD materials presented in Chapter 6, *Andrew T.S. Wee and Yuli Huang* discuss materials engineering through defect passivation and healing in Chapter 7. The reduction of defects in 2D materials is essential to fully realize their intrinsic properties and demonstrate their potential for many applications. The authors provide an overview of the current research efforts devoted to defect healing and passivation, particularly in 2D TMDs. Towards this aim, the types and sources of the defects are identified, however they point out that more work is needed to develop methods for rapid characterization to provide guidance to crystal growers to achieve the highest quality materials. The common defects observed in 2D TMDs can be categorized into intrinsic defects arising from crystalline imperfections, extrinsic defects arising from exposure to the environment, and intentional doping. The formation of the intrinsic defects greatly depends on the growth environment, and self-healing of these defects can be achieved in-situ after growth, in ambient conditions, or by post-growth annealing in an appropriate environment. Growth of large single crystal 2D materials has been extremely challenging and significant progress has been achieved in the synthesis of wafer-scale graphene h-BN, and MoS_2 as well as clean transfer techniques. Large area single crystal growth though is still challenging but needed for many applications in order to achieve low and controlled low defect densities. Scale-up encapsulation approaches are also needed for surface protection to reduce extrinsic defects, which is facilitated by the synthesis of high-quality layered metal oxides as well as h-BN films with large grain size. Further progress in this field requires rapid development in material synthesis and processing techniques to reduce defect density on a large scale. Various strategies have been used to minimize the impact of defects on the material properties, including chemical treatment, electron beam irradiation, plasma treatment, and so on as also discussed in other chapters. Among them, chemical treatment and plasma treatment provide promising approaches for flexible modification of 2D materials over a large area and scale, allowing effective healing of vacancies, passivation of defective states, formation of surface protection layers, as well as doping of the materials. While a lot of progress has been made in identifying, controlling and healing defects extensive efforts are still needed to develop materials engineering techniques that enable precise control with high efficiency and stability. These studies are critical in the understanding and control of defect in any of the 2D materials.

David B. Geohegan and co-authors present a chapter on "Nonequilibrium synthesis and Processing Approaches to Tailor Heterogeneity in 2D Materials". In this chapter the authors illustrate how and why atomically thin 2D crystals can be promising platforms to explore the effects of non-equilibrium synthesis and processing techniques that affect, and can be used to control, heterogeneity in these materials. They discuss *in situ* process diagnostics of the non-equilibrium growth environment *and* evolving structure, atomistic characterization of the resulting structure to correlate with measured properties and discuss how computational simulations of the dynamics of synthesis provide unique structural information on 2D crystals. Today, the large number of 2D materials systems and their heterostructures, coupled with the various film growth approaches, is providing an incentive for artificial intelligence and machine learning to be used to explore materials science to optimize synthetic routes. While such methods can also be applied to 2D materials, understanding of single layer 2D materials assembly mechanisms and simple heterostructures require coordinated characterization and computational simulation. Today such dynamic, first-principles computational simulations of the simplest multicomponent 2D monolayers and bilayer heterostructures push the limits of atomic resolution electron microscopy analysis as well as computational methods. These current limitation in supercomputing power and atomistic characterization reinforces why atomically thin 2D materials are unique testbeds for explorations of non-equilibrium processing to induce and control heterogeneity. It is hoped that through the various examples presented in this chapter, the ability to 'see' and the use of crystal growth control knobs such as chemical potential, strain, and energetic 'building blocks' can be used to illustrate the novelty and opportunity of atomically thin 2D materials as a platform for heterogeneous quantum materials. The discussions in this chapter are also particularly important to help the crystal grower to control nucleation in order to control the grain size of the growing 2D materials.

In Chapter 9 by *Mahdi Ghorbani-Asl and co-authors* on "Two-dimensional materials under ion irradiation: from defect production to structure and property engineering" the authors go one step further in the study of defects and their intentional creation. They use energetic particles to produce defects and introduce impurities in 2D materials as is widely done in 3D crystals through ion implantation. As in the case of 3D crystals, the properties of 2D materials can be modified by energetic beams of ions. This chapter reviews the basic understanding of the effects of ion irradiation on the properties of 2D materials considering the differences between 2D and 3D materials. Unlike 3D crystals, where the ions can penetrate the film and control the properties at specific desired depths from the surface, in 2D materials the thickest monolayers are much less than 1 nm thick and only ultralow energy ions are needed to modify/implant into the 2D lattice. In addition, the presence of an underlying substrate can have a significant effect on the film's properties in comparison to a suspended film. However, whether the 2D materials are suspended or in direct contact with the substrate, species adsorbed on their surface can interact with the ions being irradiated and can modify the properties of the films. It is thus important to understand the state of the surfaces and learn how to clean them or control the adsorbed species. The understanding of surfaces prior to ion treatment will dictate the type of equipment used during and post treatment to minimize uncontrolled ambient effects, perhaps by using clustered tools and/or in-situ surface treatments prior to integration with other materials. Additional effects in 2D materials can be achieved by impurities introduced by a post-bombardment deposition of the desired chemical species. These species can interact with vacancies and other irradiation-induced defects thereby adding new functionalities such as magnetism or high catalytic activity to the irradiated 2D material. It is also possible to pattern the 2D materials by focused ion beams. This opens many opportunities for the beam-mediated engineering of devices and nano meshes for example DNA sequencing or molecule separation. The authors of this chapter further point out that the interpretation of experimental results is made difficult by the unavailability of high purity materials, the importance of high purity materials is discussed throughout the book. The authors of this chapter point out that there are challenges in the theoretical description of defect-creation mechanisms and interpretation of experimental results. The atomically thin nature of the films in comparison to 3D materials can provide a way to study ion-solid interactions more directly. At time most of the experiments have been performed on graphene with limited data on MoS_2 and WS_2. This suggests that there are many opportunities to study ion-solid interactions in TMD materials.

While the focus of many studies has concentrated on the identification and reduction of these defects for electronic applications, this next chapter on "Tailoring Defects for 2D Electrocatalysts" by *Ruitao Lv and co-authors* reviews the applicability of different defects, such as vacancies, dopants, grain boundaries, and heterojunctions to electrocatalytic activity of 2D materials for various electrochemical reactions, e.g., OER, HER, NRR, CO_2RR.

Many defects in 2D materials have been identified by characterization techniques, such as X-ray absorption fine spectroscopy, spherical aberration corrected transmission electron microscope, high-resolution scanning transmission electron microscope, and electron paramagnetic resonance. Nevertheless, a series of inevitable challenges are still impeding their development and use for practical applications. The defects tend to be reactive and easily poisoned or removed under certain conditions. Also, as catalysts may not have a wide pH tolerance, the catalytic active sites are rapidly corroded during the electrocatalytic process. Further, high oxidation potentials or reduction potentials significantly induce surface reconstruction of the catalysts, especially for OER and NRR, causing large deviation in composition from the pristine stoichiometry. In short, it is currently difficult to prepare films with controlled specific defects. Variations in coordination environments cause defects to exhibit different electronic properties, thus affecting the electrolytic performance. In some situations, different defects are present in these active materials and thus could produce positive synergistic effects, while others could be harmful for electrocatalysis due to their huge incompatibility.

The authors conclude that with a rational design of high-efficiency electrocatalysts basis can be used to fabricate defect-chemistry-based electrocatalysts. As materials growth processes and defect identification and characterization are improved, high-efficiency and cost-effective electrocatalysis could be achieved in the future.

The last chapter by *A. Rai et al.* provides an exhaustive review of on 2D defects and devices with comments on the outlook and perspectives of 2D materials-based devices. Two-dimensional semiconducting 2D TMDs offer many advantages over conventional 3D semiconductors (Si, Ge and III-Vs) such as inherent ultra-thin bodies enabling enhanced electrostatic gate control and carrier confinement, availability of a wide range of sizable bandgaps and diverse band alignments, and much lower surface "dangling bonds" unlike conventional 3D semiconductors enabling the formation of low defect interfaces (especially 2D/2D vdW interfaces). These attributes make the semiconducting 2D TMDs extremely promising for future "ultra-scaled" and "ultra-low-power" devices. Several device applications such as ultra-scaled FETs, digital logic, memory, analog/RF, conventional diodes, photodetectors, light emitting diodes (LEDs), lasers, photovoltaics, sensors, ultra-low-power tunneling-devices such as tunnel-FETs (TFETs), and piezotronics, among others, have been demonstrated using 2D MoS_2 highlighting its promise and versatility of 2D semiconductors.

This chapter provides an overview of two-dimensional (2D) materials and highlights some of the latest and promising advances in 2D material-based devices along with a discussion of some novel 2D defect characterization and 'defect-engineering' techniques. Beside defect identification and control, defect engineering as also extensively discussed in other chapters of this book is the ultimate objective of any materials process and it is no different for 2D materials. Thus, characterization techniques used for deciphering defects and defect-carrier dynamics in 2D materials and the relationship to device characteristics are of tantamount importance. The authors delve in the electrical characterization of 2D materials-based devices, both simple FETs as well as heterostructures. They also report on the fabrication of these materials *via* large-area growth techniques, such as chemical vapor deposition (CVD) and molecular beam epitaxy (MBE), as well as the fabrication of 2D van der Waals (vdW) heterostructures. Given the presence of a plethora of defects, they went on to employ concepts such as passivation, doping and alloying to study functionality enhancement of 2D material-based devices using the novel and promising 'defect-engineering' techniques. Lastly, an in-depth theoretical investigation of the effects of defects, such as vacancies in the interfacial dielectric layer, on 2D material properties is also provided. This last chapter also provides a guide to the types of electronic devices that can be fabricated and critical factors affecting device performance.

References

[1] X.K. Lu, M.F. Yu, H. Huang, R.S. Ruoff, Tailoring graphite with the goal of achieving single sheets, Nanotechnology 10 (3) (1999) 269–272.

[2] K.S. Novoselov, A.K. Geim, S.V. Morozov, D. Jiang, Y. Zhang, S.V. Dubonos, I.V. Grigorieva, A.A. Firsov, Electric field effect in atomically thin carbon films, Science 306 (5696) (2004) 666–669.

[3] K.S. Novoselov, D. Jiang, F. Schedin, T.J. Booth, V.V. Khotkevich, S.V. Morozov, A.K. Geim, Two-dimensional atomic crystals, Proceedings of the National Academy of Sciences of the United States of America 102 (30) (2005) 10451–10453.

[4] A. Antidormi, L. Colombo, S. Roche, Emerging properties of non-crystalline phases of graphene and boron nitride based materials, Nano Materials Science (2021).

Index

A
Ab initio approach, 21
Abnormal growth, 146
Accelerated growth, 241, 245
Acceptor, 375
Adatom, 22, 164
  defects, 44
All-surface, 119
Alloying, 102, 103, 129, 137, 179
  catalyst, 139
  TMDs, 103, 109
Ammonia borane (AB), 131
Annealing, 36, 63, 207, 271
  thermal, 50, 197, 198, 202, 207
  vacuum, 108
Antesignanus, 43
Antimonene, 64–67
  multilayer, 66
Antisite defects, 91
Approach
  ab initio, 21
  first-principles, 21, 23
Argon plasma treatment, 345
Arsenene, 64–67
  layer, 69
Atomic
  layers, 43, 60, 76, 265, 267
  scale, 47
  structure engineering, 280
Atomic force microscopy (AFM), 75, 90, 126
Atomic layer deposition (ALD), 120, 130, 160, 348, 359
Atomistic
  characterization, 4, 224, 250, 252
  simulations, 12, 32, 265
  structure, 223
Atoms
  boron, 20, 46, 276
  carbon, 43, 267, 275, 276, 323
  chalcogen, 17, 91, 169, 177, 201, 228, 247, 339, 344

  dopants, 247, 375, 376
  fluorine, 282
  hydrogen, 28, 263, 280
  impurities, 10, 11, 33
  interstitial, 2, 18, 35
  phosphorus, 60, 61
  recoil, 265, 269
  sulfur, 243, 268
  transition metal, 91, 160, 316, 339, 341
Auxiliary gas pressures, 133

B
Band-bending, 365
Band-to-band transitions, 30
Beam-interruption, 168
Benzyl viologen (BV), 207
Bilayer plumbene, 57
Bismuthene, 64, 70, 71
Black phosphorus (BP), 59
Born-Oppenheimer energy surface, 264
Boron atoms, 20, 46, 276
Borophene, 15, 20, 45, 47, 48
  epitaxial, 15
  synthesis, 46, 47
Brillouin zone (BZ), 381, 383
Building blocks, 241–243, 252, 407
Bulk, 9
  crystals, 2, 160, 221
    production, 121
  samples, 90

C
Can-opener effect, 271
Capacitance-voltage (C-V), 351
Carbon
  atoms, 43, 267, 275, 276, 323
Carbon dioxide reduction reaction (CO_2RR), 4, 304
Carrier diffusion length, 343
Carrier lifetime, 343
Catalyst
  alloying, 139

metal, 140, 147
pre-treatment, 136
substrate, 129, 133, 140
Catalytic CVD, 127, 130
Chalcogen, 90, 91, 94, 101, 163, 168, 173, 174, 177, 197
   atoms, 17, 91, 169, 177, 201, 228, 247, 339, 344
   vacancies, 36, 91, 92, 161, 168, 177, 183, 197, 198, 203, 228, 344–346, 356, 406
Chalcogenide
   vacancy, 198, 210
     defects, 210
Channel length (L_{CH}) scaling, 339
Charge density waves (CDW), 200
Charge transfer doping, 179
Charge transfer resistance, 306
Charged impurities (CI), 367, 370
Chemical content, 10
Chemical vapor deposition (CVD), 119, 124, 196, 259, 340, 344, 348, 409
Close-spaced sublimation (CSS), 147
Complementary metal oxide semiconductor (CMOS), 1
Complex defects, 93
Conduction band edge (CBE), 364, 365, 369
Conduction band minimum (CBM), 2, 201
Conductive atomic force microscopy (CAFM), 376
Configuration coordinate (CC), 34
Contaminants, 94
Controlling
   nucleation, 170
     density, 165
   vacancies, 222
Counting defects, 90
Covalent functionalization, 204
Crystals
   bulk, 2, 160, 221
   growth, 3, 119, 121, 134, 145, 149, 170, 183, 240, 241, 244, 405, 406
   monolayer, 234, 241
CVD, 123
   process control, 134

D

Deep defects, 29
Deep-level traps, 370, 371
Defect-bound excitons, 30
Defect-engineering techniques, 409

Defective
   graphene, 315
   phosphorene, 62
Defects, 7, 8, 91
   adatoms, 44
   antisite, 91
   complex, 93
     geometrically, 91
   counting, 90
   deep, 29
   engineering, 303, 304, 307, 311, 321, 330
   flower, 13
   formation energy, 18
   Frenkel, 18
   healing, 196, 403, 406
   in plumbene, 58
   in stanene, 55
   irradiation-induced, 10
   line, 15, 89, 104, 198
   point, 12, 32, 44, 89, 196, 344
     native, 44
   Schottky, 18
   shallow, 30
   vacancies, 59, 197, 227
Density functional theory (DFT), 47, 61, 91, 181, 196, 263, 305, 345, 364, 379
Density of states (DOS), 307
Device fabrication, 360
DFT calculations, 21, 228, 230, 232, 233, 308, 317, 323, 327, 329, 344, 379
Digital transfer growth, 244
Dimensionality, 9
Divacancy, 44, 49, 53, 54, 56, 71
Domain stitching, 145
Donor, 375
   shallow-level, 370, 371
Donor-acceptor pairs, 30
Dopants, 100, 101, 107, 179–181, 259, 309, 314, 315, 322
   atoms, 247, 375, 376
   electronic, 181
   isoelectronic, 180
   metal, 308, 309
   substitutional, 89, 96, 97, 101
   unintentional, 91, 96, 97
Doped 2D semiconductors, 182
Doping, 179
   charge transfer, 179
   concentration, 102, 167, 181, 182

effect, 369, 378
electronic, 182
fluorine, 328
heteroatoms, 311, 321
isoelectronic substitution, 181
substitutional, 102, 109, 179, 180, 375

E
Edges/grain boundaries, 345
Effect
can-opener, 271
doping, 369, 378
quantum Hall, 43
Electric double layer transistors (EDLT), 356
Electrocatalysis, 303, 304, 331, 332
Electrocatalytic CO_2RR, 325, 326
Electrocatalytic NRR, 317, 318, 320–322, 325
Electrochemical impedance spectroscopy (EIS), 306
Electron beam irradiation, 209
Electron double layer transistors (EDLT), 355
Electron energy loss spectroscopy (EELS), 11, 224, 247
Electronic
dopants, 181
doping, 182
grade, 183
films, 123
graphene, 3
ground state, 34, 264
Hamiltonian, 263
industry, 1, 2, 179
states, 4, 11, 19, 26, 27, 91, 105, 264
stopping, 260, 263, 265, 272, 278
power, 264, 272
Electronic-grade, 121, 160
Electrostatic doping, 101
Encapsulation, 211, 352
Engineering, 175
defect, 303, 304, 307, 311, 321, 330
substrate, 170
Epitaxial
borophene, 15
films, 147
graphene, 177, 179
growth, 61, 66, 70, 147, 167, 168
Equilibrium concentration, 19
Evolutionary selection, 148
Exchange current density, 306, 313

Excitation, 24
Excitons, 342

F
Faradic efficiency, 307
Fermi level pinning (FLP), 364
Field effect transistors (FET), 341, 348
Films
electronic grade, 123
epitaxial, 147
monolayer, 160
multidomain, 234
multilayer, 129
stanene, 55
stoichiometric, 168, 173
TMD, 160, 355, 356
First-principles approach, 21, 23
Flower defects, 13
Fluorine atoms, 282
Fluorine doping, 328
Focused ion beams (FIB), 280
Formation energies, 21
Free-to-bound transitions, 30
Frenkel defect, 18

G
Gallenene, 75
Generalized gradient approximation (GGA), 379
Geometrically complex defects, 91
Germanene, 49, 52–55
synthesis, 52
Gibbs free energy, 19
Grain boundary (GB), 195, 198, 238, 352, 366, 369
Graphene
crystal growth, 3, 133
defective, 315
epitaxial, 177, 179
growth, 127, 131, 133, 136–138, 148
crystal, 3, 133
rate, 134, 138
hexagonal, 140
layer, 43, 77, 105, 138, 271
monolayer, 119, 122, 123, 132, 137, 276, 288
nucleation, 136, 144
density, 136
polycrystalline, 123, 127, 148
Growth, 119
abnormal, 146
accelerated, 241, 245

crystals, 3, 119, 121, 134, 145, 149, 170, 183, 240, 241, 244, 405, 406
digital transfer, 244
epitaxial, 61, 66, 70, 147, 167, 168
graphene, 127, 131, 133, 136–138, 148
  pyrolytic, 126
kinetically stabilized, 129
kinetics, 225, 232, 246
mechanisms, 127, 129, 130, 170, 182, 406
MOCVD, 355
morphology, 106, 170, 232
nonequilibrium, 222, 224, 231
pressure, 175
pyrolytic, 124, 125
scalable, 124, 126, 160
single domain, 144
substrate, 120, 126, 140, 141, 161–163, 170, 171, 174, 183, 221, 351, 406
two-step, 170
GW method, 24

H

Hafnene, 77
Heliarc, 45
Helium ion microscopy (HIM), 270, 280
Heteroatoms doping, 311, 321
Heterostructures, 178, 179
  van der Waals (vdW), 183, 340, 359–363, 406, 409
  vertical, 170, 179, 360
Hexagonal
  boron nitride (h-BN), 1, 3, 165, 178, 195, 212, 259, 266, 361, 373
  graphene, 140
High-energy proton irradiation, 276
Highly charged ions (HCI), 278
Highly oriented pyrolytic graphite (HOPG), 121, 168
Hofstader butterfly, 363
Hollow site, 68
Huang-Rhys (HR) factor, 34
Hydrogen atom, 28, 263, 280
Hydrogen evolution reaction (HER), 4, 108, 109, 170, 209, 286, 304

I

Impurities, 94
  atoms, 10, 11, 33
  scavenging, 136

substitutional, 13, 94
Integration, 119
Intensified charge coupled device (ICCD), 244
Intercalants, 98
Interfacial charge transfer, 206
Interstitial atoms, 2, 18, 35
Irradiation tolerance, 284
Irradiation-induced defects, 10
Isoelectronic
  dopants, 180
  substitution, 181
    doping, 181

K

Kinetic energy (KE), 247
Kinetic Wulff construction (KWC), 231
Kinetically stabilized growth, 129

L

Large area production, 148
Layer
  arsenene, 69
  atomic, 43, 60, 76, 265, 267
  exfoliation, 121
  graphene, 43, 77, 105, 138, 271
  protective, 352
  TMD, 170, 175, 339
Light emitting diodes (LED), 340, 409
Line defects, 15, 89, 104, 198
Linear sweep voltammetry (LSV), 306
Local density approximation (LDA), 379
Low temperature (LT), 212

M

Mass flow controllers (MFC), 133, 161
Materials by design, 359
Mechanical exfoliation (ME), 61, 66, 122, 141, 195, 196
Metal
  catalyst, 140, 147
  dopants, 308, 309
Metal organic chemical vapor deposition (MOCVD), 90, 160, 162, 165, 355
Metal-insulator-semiconductor (MIS), 351
Method
  beam-interruption, 168
  GW, 24
  synthesis, 61
  zone refining, 146

Index   **415**

Mirror twin boundaries (MTB), 16, 199
Molecular beam epitaxy (MBE), 55, 66, 130, 160, 163, 166, 340, 355, 409
Molecular dynamic (MD), 238, 262
Monolayer
    crystals, 234, 241
    films, 160
    graphene, 119, 122, 123, 132, 137, 276, 288
Monovacancy, 56
Multidomain films, 234
Multilayer
    antimonene, 66
    films, 129

N

Native point defects, 44
Negative differential resistance (NDR), 370
Nitrogen reduction reaction (NRR), 4, 304, 317
Nonequilibrium growth, 222, 224, 231
Nonequilibrium synthesis, 224, 227, 234, 247, 407
NRR electrocatalysts, 319, 321
Nucleation density, 133–135, 137, 162, 170, 171, 173–175
    controlling, 165
Nudged elastic band (NEB), 262

O

Origin, 10
Overpotential, 306, 313
Oxidation, 136
Oxygen evolution reaction (OER), 4, 304, 311

P

Passivation, 196, 198, 214, 368, 379, 403, 406
Personal protective equipment (PPE), 165
Phonon contributions, 34
Phosphorene, 59, 61–64
    defective, 62
Phosphorus atoms, 60, 61
Physical vapor deposition (PVD), 146, 196, 221
Plasma immersion ion implantation (PIII), 377
Plasma treatment, 210, 345, 368
Plumbene, 57–59
    bilayer, 57
    overlayer, 58
    structure, 58, 59
    synthesis, 58
Plume, 165, 166

Point defect, 12, 32, 44, 89, 196, 344
    inventory, 89, 90
Polycrystalline graphene, 123, 127, 148
Precursor, 133
    chemistry, 172
    choice, 130
Process control, 89
    CVD, 134
Proportional-integral-derivative (PID), 163
Protective layer, 352
Pulling, 28
Pulsed laser deposition (PLD), 243
Pump-probe spectroscopy, 342
Pyrolytic graphene growth, 126
Pyrolytic graphite (PG), 121
Pyrolytic growth, 124, 125

Q

Quantum Hall effect (QHE), 43
Quantum information, 222
Quantum yield (QY), 202

R

Raman spectrum, 32
Raman-tensor weighted Γ-point DOS (RGDOS), 32
Reactor design, 161
Receiver, 244
Recoil atoms, 265, 269
Reflection high energy electron diffraction (RHEED), 163
Reversible hydrogen electrode (RHE), 307
Room temperature (RT), 55, 66, 212, 356

S

S vacancy self-healing (SVSH), 202
Saturated calomel electrode (SCE), 327
Scalable growth, 124, 126, 160
Scanning transmission electron microscopy (STEM), 10, 196, 223
Scanning tunneling microscopy (STM), 123, 167, 198, 223
Scanning tunneling microscopy/spectroscopy (STM/STS), 47
Scanning tunneling spectroscopy (STS), 11, 199, 359
Schottky barrier height (SBH), 364
Schottky barrier (SB), 364
Schottky barrier width (SBW), 365
Schottky defect, 18

416   Index

Second harmonic generation (SHG), 235, 237
Selenene, 45, 72, 74
Semiconductors
   complementary metal oxide, 1
   doped 2D, 182
Shallow defects, 30
Shallow-level donors, 370, 371
Shielding, 290
Silent partner, 120
Silicene, 49, 51, 55, 56, 62
   lattice, 51
   reconstruction, 51
   synthesis, 49
Single
   domain growth, 144
   photon emitters, 241
   vacancy, 44, 49, 53, 54, 56, 59, 62, 71
Single photon emitters (SPE), 241
Solid-source chemical vapor deposition
      (SS-CVD), 163
Spectroscopy
   electrochemical impedance, 306
   electron energy loss, 11, 224, 247
   pump-probe, 342
   scanning tunneling, 11, 199, 359
   ultrafast laser, 343
Stanene, 55, 56
   crystal structure, 55
   domain boundary, 56
   films, 55
Stoichiometric films, 168, 173
Stoichiometry control, 244
Stone-Wales (SW), 44, 49, 71
Substitutional
   dopants, 89, 96, 97, 101
   doping, 102, 109, 179, 180, 375
   impurities, 13, 94
   oxygen atoms, 198
Substrate
   catalyst, 129, 133, 140
   degradation, 134
   engineering, 170
   for graphene, 361
   growth, 120, 126, 140, 141, 161–163, 170, 171,
      174, 183, 221, 351, 406
   surface, 141, 162, 164, 169, 247, 278
   temperature, 46, 73, 75, 161, 168, 170, 244, 246
Sulfur
   atoms, 243, 268

vacancies, 29, 205, 226, 227, 321, 346, 354, 364,
     367, 376
   healing, 203
Surface optical (SO), 372
Swift heavy ions (SHI), 278
Synthesis
   borophene, 46, 47
   germanene, 52
   methods, 61
   nonequilibrium, 224, 227, 234, 247, 407
   plumbene, 58
   silicene, 49
   techniques, 161, 165, 172
   TMD monolayer, 165, 183, 406
   transition metal dichalcogenide, 159

T

Tafel slope, 306, 313
Tellurene, 45, 72–74
Thermal annealing, 50, 197, 198, 202, 207
Thermal spikes, 247
Time-dependent controls, 134
TMD, 1, 8, 89, 159, 170, 195, 222, 259, 307, 309, 339,
     341, 342, 378
   2D, 378
   alloying, 103, 109
   films, 160, 355, 356
   layers, 170, 175, 339
   monolayer synthesis, 165, 183, 406
Topoelectronic, 64
Topological insulators (TI), 70
Transfer, 119
Transition
   band-to-band, 30
   free-to-bound, 30
   metal, 180
      atoms, 91, 160, 316, 339, 341
      carbonyls, 165
      catalysts, 127, 145
      dichalcogenide synthesis, 159
      dichalcogenides (TMD), 1, 8, 89, 159, 161,
         163, 195, 222, 259, 339, 341, 378, 404
      nitrides (TMN), 320
      precursors, 174, 344
      sublattice, 90
Transmission electron microscopy (TEM), 90
Trion, 343
Turnover frequency, 306, 314
Twistronics, 340, 359

Index

417

Two-dimensional (2D) materials, 1, 119, 159, 195, 221, 303, 304, 310, 315, 317, 330, 409
Two-dimensional transition metal dichalcogenides (TMD), 307
Two-step growth, 170

U

Ultra-high vacuum (UHV), 47
Ultra-low-power, 340, 409
Ultra-scaled, 340, 409
Ultrafast laser spectroscopy (ULS), 343
Unintentional dopants, 91, 96, 97

V

Vacancies, 91, 92
 chalcogen, 36, 91, 92, 161, 168, 177, 183, 197, 198, 203, 228, 344–346, 356, 406
 chalcogenide, 198, 210
 concentration, 32, 36, 344, 387
 controlling, 222
 defects, 59, 197, 227
 density, 347, 386
 healing, 201, 204, 206
 single, 44, 49, 53, 54, 56, 59, 62, 71
 sulfur, 29, 205, 226, 227, 321, 346, 354, 364, 367, 376
  healing, 203
Vacuum annealing, 108
Valence band maximum, 21, 201
Valence band minimum, 2
Vapor-liquid-solid (VLS), 127
Vapor-solid (VS), 124
Vapor-solid-solid (VSS), 127
Vertical heterostructures, 170, 179, 360
Volcano plots, 127

Z

Zone refining method, 146